Publications from the International Commission for Optics

International Trends in Optics
By J. W. Goodmann (Ed.), Academic Press, Boston, 1991,
ISBN: 0-12-289690-4

Current Trends in Optics
By J. C. Dainty (Ed.), Lasers Opt. Eng. Ser., Academic Press, London, 1994,
ISBN: 0-12-200720-4

Trends In Optics
Research, Developments and Applications
By A. Consortini (Ed.), Lasers Opt. Eng. Ser., Academic Press, San Diego, 1996,
ISBN: 0-12-186030-2

International Trends in Optics and Photonics
By T. Asakura (Ed.), Springer Ser. Opt. Sci. Vol.74, Springer, Berlin, Heidelberg, 1999,
ISBN: 3-540-65897-1

Springer Series in
OPTICAL SCIENCES 74

founded by H.K.V. Lotsch

Springer-Verlag Berlin Heidelberg GmbH

Springer Series in
OPTICAL SCIENCES

The Springer Series in Optical Sciences, under the leadership of Editor-in-Chief *William T. Rhodes*, Georgia Institute of Technology, USA, and Georgia Tech Lorraine, France, provides an expanding selection of research monographs in all major areas of optics: lasers and quantum optics, ultrafast phenomena, optical spectroscopy techniques, optoelectronics, information optics, applied laser technology, industrial applications, and other topics of contemporary interest.

With this broad coverage of topics, the series is of use to all research scientists and engineers who need up-to-date reference books.

The editors encourage prospective authors to correspond with them in advance of submitting a manuscript. Submission of manuscripts should be made to the Editor-in-Chief or one of the Editors. See also http://www.springer.de/phys/books/optical_science/os.html

Editor-in-chief

William T. Rhodes

Georgia Tech Lorraine
2-3, rue Marconi
F-57070 Metz, France
Phone: +33 378 20 3922
Fax: +33 378 20 3940
e-mail: wrhodes@georgiatech-metz.fr
URL: http://www.georgiatech-metz.fr
http://users.ece.gatech.edu/~wrhodes

Editorial Board

Toshimitsu Asakura

Faculty of Engineering
Hokkei-Gakuen University
1-1, Minami-26, Nishi 11, Chuo-ku
Sapporo, Hokkaido 064-0926, Japan
e-mail: asakura@eli.hokkai-s-u.ac.jp
(*Special Editor for Optics in the Pacific Rim*)

Karl-Heinz Brenner

Chair of Optoelectronics
University of Mannheim
B6, 26
D-68131 Mannheim, Germany
Phone: +49 (621) 292 3004
Fax: +49 (621) 292 1605
e-mail: brenner@rumms.uni-mannheim.de
URL: http://www.ti.uni-mannheim.de/~oe

Theodor W. Hänsch

Max-Planck-Institut für Quantenoptik
Hans-Kopfermann-Strasse 1
D-85748 Garching, Germany
Phone: +49 (89) 2180 3211 or +49 (89) 32905 702
Fax: +49 (89) 32905 200
e-mail: t.w.haensch@physik.uni-muenchen.de
URL: http://www.mpq.mpg.de/~haensch

Ferenc Krausz

Institut für Angewandte Elektronik
und Quantenelektronik
Technische Universität Wien
Gusshausstr. 27/359
A-1040 Wien, Austria
Phone: +43 (1) 58801 35937
Fax: +43 (1) 58801 35997
e-mail: krausz@iaee.tuwien.ac.at
URL: http://www.tuwien.ac.at

Horst Weber

Optisches Institut
Technische Universität Berlin
Strasse des 17. Juni 135
D-10623 Berlin, Germany
Phone: +49 (30) 314 23585
Fax: +49 (30) 314 27850
e-mail: weber@physik.tu-berlin.de
URL: http://www.physik.tu-berlin.de/institute/
    OI/Weber/Webhome.htm

Toshimitsu Asakura (Ed.)

International Trends in Optics and Photonics

ICO IV

With 190 Figures

 Springer

Professor Toshimitsu Asakura
Faculty of Engineering
Hokkei-Gakuen University
1-1 Minami-26, Nishi 11, Chuo-ku,
Sapporo, Hokkaido 064-0926
Japan

ISSN 0342-4111

ISBN 978-3-662-14212-7

Library of Congress Cataloging-in-Publication Data applied for

Die Deutsche Bibliothek – CIP-Einheitsaufnahme
International trends in optics and photonics: ICO IV / Toshimitsu Asakura (ed.).
(Springer series
in optical sciences; Vol 74)
ISBN 978-3-662-14212-7   ISBN 978-3-540-48886-6 (eBook)
DOI 10.1007/978-3-540-48886-6

© Springer-Verlag Berlin Heidelberg 1999
Originally published by Springer-Verlag Berlin Heidelberg New York in 1999
Softcover reprint of the hardcover 1st edition 1999

Data conversion by Kurt Mattes, Heidelberg
Cover concept: eStudio Calamar Steinen
Cover production: *design & production* GmbH, Heidelberg

Computer to plate: Mercedes Druck GmbH, Berlin

SPIN: 10725733   57/3144/mf - 5 4 3 2 1 0 – Printed on acid-free paper

Preface

It is more than ten years since, in 1987, the bureau of the International Commission for Optics (ICO) decided to launch a series of books "International Trends in Optics", and it may be appropriate to reflect on this occasion upon the way in which this series of publications has developed and to ask whether it has fulfilled the aim that the ICO had in mind at its conception.

The books which are set to grow by one volume every three years are intended to provide an authoritative overview of research underway in optics throughout the world. The chapters are suitable for the specialists and non-specialists alike, providing general, readable overviews of many different aspects of optical science and engineering. They tend to be less formal than the standard technical reviews found in journals. In addition to examining their designated topics, the authors discuss unsolved problems and speculate on future directions in their fields.

The first three books[1] in the series already appeared in 1991, 1994 and 1996, respectively. The first book, "International Trends in Optics", was edited by J.W. Goodman of Stanford University, ICO president during the period of 1987–1990; the second, entitled "Current Trends in Optics", was edited by J.C. Dainty of Imperial College, London, ICO president during 1990–1993; the third, "Trends in Optics", was edited by A. Consortini of Universita degli Studi, Firenze, ICO president during 1993–1996. This volume is the fourth book of this series of the ICO. The development of the numerous activities in many branches of optics and photonics have been covered in the volumes of "International Trends in Optics". We are fortunate in obtaining contributions from leading scientists and engineers[2] and the books have surely contributed to increasing ICO visibility and to adding funds for ICO acivities. Judging from the reviews of the past volumes and the past ICO activities, this series is indeed fulfilling the aim originally set for it.

---

[1] The past three volumes in the series were published by Academic Press.

[2] Another purpose of starting this series is to provide greater visibility for the ICO and to raise additional funds to support ICO activities such as the traveling lecture program in developing countries. As a rule of the ICO for this purpose, the royalties, typically paid to the editor and the authors, are instead paid to the ICO.

With a continually increasing volume of research, workers in all branches of science and engineering are experiencing difficulties in keeping abreast of the numerous developments. As mentioned above, it is the aim of the ICO series to provide information in the form of review articles about current researches in optics and photonics. ICO IV consists of 25 chapters, written by outstanding scientists and engineers, covering a broad subset of the most recent advances in optics and photonics. The sequence of the chapters are somewhat arbitrary; they are split roughly between fields in fundamental optics, information optics, optical communication, optical materials and processing, optical technologies, optical metrology, biomedical optics, and others. These and other developments present new opportunities both for basic research and for technical developments. It is hoped that "International Trends in Optics and Photonics" will reflect these activities and will provide help and stimulus to workers not only in optics and photonics but also in related sciences and engineering.

I would like to take this opportunity to thank all the authors for their excellent contributions. Most of them are also the major contributors to the field itself, and I feel honoured to have their contributions. I would also like to thank Professor A. Consortini and Dr. P. Chavel for their advice and the management and editorial staff at Springer for publishing this book.

Sapporo, May 1999   *Toshimitsu Asakura*

Contents

Part II. Information Optics

**Holographic Optics for Beamsplitting
and Image Multiplication** 96
A. L. Mikaelian, A. N. Palagushkin, and S. A. Prokopenko

**Image Restoration, Enhancement and Target Location
with Local Adaptive Linear Filters** 111
L. Yaroslavsky

**Fuzzy Problem for Correlation Recognition
in Optical Digital Image Processing** 128
G. Cheng, G. Jin, M. Wu, and Y. Yan

Part V. Optical Technologies

Part VI. Optical Metrology
(Optical Systems)

Part VII. Biomedical Optics

Part VIII. Others

Contributors

T. Asakura
Faculty of Engineering, Hokkai-Gakuen University, Minami-26 Nishi-11, Chuo-ku, Sapporo, Hokkaido 064-0926, Japan

M. Bertolotti
INFM at Dipartimento di Energetica, Universita' degli Studi di Roma
"La Sapienza", Via Scarpa 16, 00161 Rome, Italy,
Phone: +39(06)49916541, Fax: +39(06)44240183,
E-mail: bertolotti@axrma.uniroma1.it

G. Bohn
Chair for Optics, Physics Institute, University of Erlangen-Nürnberg,
Staudtstr. 7, D-91058 Erlangen, Germany

V. Bužek
Institute of Physics, Slovak Academy of Sciences,
Dúbravská cesta 9, 842 28 Bratislava, Slovakia
Faculty of Mathematics and Physics, Comenius University,
Mlynská dolina, 842 15 Bratislava, Slovakia

A. Casini
CNR, Istituto di Ricerca sulle Onde Elettromagnetiche "Nello Carrara", Florence, Italy

G. Cheng
Optical-Electrical Engineering Center, Tsinghua University, 100084 Beijing, China

R. Dändliker
Institute of Microtechnology, University of Neuchâtel, Breguet 2, CH-2000 Neuchâtel, Switzerland

E. Desurvire
Corporate Research Centre, Alcatel-CIT, 91460 Marcoussis, France

G. Drobný
Institute of Physics, Slovak Academy of Sciences,
Dúbravská cesta 9, 842 28 Bratislava, Slovakia

P. Ettl
Chair for Optics, Physics Institute, University of Erlangen-Nürnberg,
Staudtstr. 7, D-91058 Erlangen, Germany

A. F. Fercher
Institute of Medical Physics, University of Vienna, A-1090 Wien, Austria

A. T. Friberg
Royal Institute of Technology, Department of Physics, Optics Section,
SE-100 44 Stockholm, Sweden

J. P. Goedgebuer
GTL-CNRS Telecom, Georgia Tech Lorraine, 2–3 rue Marconi, 57070 Metz, France
Laboratoire d'Optique P.M. Duffieux, UMR CNRS 6603, Université de Franche-
Comté, 25030 Besançon cedex, France

E. György
Institute of Atomic Physics, Bucharest, Romania

S. Haroche
Laboratoire Kastler Brossel, Département de Physique, École Normale Supérieure,
24 rue Lhomond, F-75005 Paris, France

G. Häusler
Chair for Optics, Physics Institute, University of Erlangen-Nürnberg,
Staudtstr. 7, D-91058 Erlangen, Germany

H. P. Herzig
Institute of Microtechnology, University of Neuchâtel, Neuchâtel, Switzerland

C. K. Hitzenberger
Institute of Medical Physics, University of Vienna, A-1090 Wien, Austria

R. Ito
Department of Applied Physics, The University of Tokyo, Tokyo 113-8656, Japan

G. Jin
Optical-Electrical Engineering Center, Tsinghua University, 100084 Beijing, China

P. Kipfer
Institute of Microtechnology, University of Neuchâtel, Neuchâtel, Switzerland

T. Kondo
Department of Materials Science, The University of Tokyo, Tokyo 113-8656, Japan

N. D. Kundikova
Joint Laboratory of Nonlinear Optics of Institute of Electrophysics of Urals Branch
of Russian Academy of Science and Southern Ural State University, 76 Lenin ave.,
Chelyabinsk, 454080, Russia

I. Laszlo
Technical University of Budapest, Physics Institute, 1111 Budapest, Hungary

O. Leclerc
Corporate Research Centre, Alcatel-CIT, 91460 Marcoussis, France

M. H. Lee
Department of Physics, Inha University, Nam-gu, Inchon 402-751, Korea,
Fax: +82-032-873-6992, E-mail: mhlee@inha.ac.kr

S. S. Lee
Department of Physics, Korea Advanced Institute of Science and Technology,
Yusong-gu, Taejon 305-701, Korea,
Fax: +82-042-869-2511, E-mail: yoonkim@sorak.kaist.ac.kr

L. Lin
Institute of Modern Optics, Optical Information Science Laboratory,
Nankai University, China

F. Lotti
CNR, Istituto di Ricerca sulle Onde Elettromagnetiche "Nello Carrara", Florence,
Italy

D. K. Lynch
The Aerospace Corporation, P.O. Box 92957, Los Angeles, CA 90009, USA

I. N. Mihailescu
Institute of Atomic Physics, Bucharest, Romania

A. L. Mikaelian
Institute of Optical Neural Technologies, Russian Academy of Sciences,
No 44/2 Vavilov Str., 117333 Moscow, Russia,
Phone: +7-(095)-135-5551, Fax: +7-(095)-135-7802, E-mail: iont@glas.apc.org

G.-G. Mu
Institute of Modern Optics, Optical Information Science Laboratory,
Nankai University, China

M. Nieto-Vesperinas
Instituto de Ciencia de Materiales de Madrid, Consejo Superior de Investigaciones
Cientificas, Cantoblanco, 28049 Madrid, Spain

A. N. Palagushkin
Institute of Optical Neural Technologies, Russian Academy of Sciences,
No 44/2 Vavilov Str., 117333 Moscow, Russia,
Phone: +7-(095)-135-5551, Fax: +7-(095)-135-7802, E-mail: iont@glas.apc.org

K.-P. Peiponen
Department of Physics, University of Joensuu, P.O. Box 111, SF-80101 Joensuu,
Finland

M. Picollo
CNR, Istituto di Ricerca sulle Onde Elettromagnetiche "Nello Carrara", Florence,
Italy

S. A. Prokopenko
Institute of Optical Neural Technologies, Russian Academy of Sciences,
No 44/2 Vavilov Str., 117333 Moscow, Russia,
Phone: +7-(095)-135-5551, Fax: +7-(095)-135-7802, E-mail: iont@glas.apc.org

J.-M. Raimond
Laboratoire Kastler Brossel, Département de Physique, École Normale Supérieure,
24 rue Lhomond, F-75005 Paris, France

Y. Salvadé
Institute of Microtechnology, University of Neuchâtel, Breguet 2, CH-2000 Neuchâtel,
Switzerland

M. Schenk
Chair for Optics, Physics Institute, University of Erlangen-Nürnberg,
Staudtstr. 7, D-91058 Erlangen, Germany

I. Shoji
Department of Applied Physics, The University of Tokyo, Tokyo 113-8656, Japan

C. Sibilia
INFM at Dipartimento di Energetica, Universita' degli Studi di Roma
"La Sapienza", Via Scarpa 16, 00161 Rome, Italy

R. Silvennoinen
Department of Physics, University of Joensuu, P.O. Box 111, SF-80101 Joensuu,
Finland

R. S. Sirohi
Applied Optics Laboratory, Physics Department, Indian Institute of Technology,
Madras, Chennai 600 036, India

B. H. Soffer
665 Bienveneda Avenue Pacific Palisades, CA 90272, USA,
Telephone (310) 454-0721, Facsimile (310) 459-4774, bhsoffer@alum.mit.edu

Y. R. Song
Department of Physics, Inha University, Nam-gu, Inchon 402-751, Korea,
Fax: +82-032-873-6992, E-mail: mhlee@inha.ac.kr

J. Uozumi
Research Institute for Electronic Science, Hokkaido University, Sapporo, Hokkaido
060-0812, Japan

Z.-Q. Wang
Institute of Modern Optics, Optical Information Science Laboratory,
Nankai University, China

A. M. Weiner
Purdue University, School of Electrical and Computer Engineering, West Lafayette,
IN 47907-1285, USA, E-mail: amw@ecn.purdue.edu,
http://ece.www.ecn.purdue.edu/faculty/weiner.html

H. Wiedemann
Abteilung für Quantenphysik, Universität Ulm, D-89069 Ulm, Germany

M. Wu
Optical-Electrical Engineering Center, Tsinghua University, 100084 Beijing, China

Y. Yan
Optical-Electrical Engineering Center, Tsinghua University, 100084 Beijing, China

L. Yaroslavsky
Department of Interdisciplinary Studies, Faculty of Engineering, Tel Aviv University, Tel Aviv 69978, Israel

C. Ya. Zel'dovich
Center for Research and Education in Optics and Lasers, University of Central Florida, P.O. Box 162700, Orlando, FL 32816-2700, USA

Part I

Fundamental Optics
(General, Physical and Quantum Optics)

Optical Twist

A. T. Friberg

Summary. Optical twist was introduced in 1993 as an unconventional phase that rotates and diffracts symmetric, partially coherent, Gaussian beams. Symmetrization of arbitrary astigmatic fields, such as TEM modes or radiation from diode-laser arrays, yields non-Gaussian beams endowed with a twist. The general properties and experimental realization of both Gaussian and non-Gaussian twisted beams are described with illustrative physical examples. As an application, the role of the twist in beam quality is elucidated. The relation of the twist to optical vortices and angular momentum is addressed. Some unsolved issues are pointed out.

1 Introduction

A new phenomenon, called optical twist, was recently discovered as a result of an additional phase impressed on axially symmetric, partially coherent Gaussian beams [1]. The twist phase, which has its origin in symmetry considerations, is physically quite unlike the ordinary phase curvature. The twist phase is rotationally symmetric but changes sign under reflections about any plane containing the beam axis. So this new phase does not possess axial or cylindrical symmetry, but rather it has an intrinsic chiral or handedness property (hence the name 'twist'). The twist phase is bounded in magnitude by a quantity proportional to the squared inverse of the beam's transverse coherence length. This implies that the twist cannot exist in an ideal coherent Gaussian laser (TEM$_{00}$ mode), which perhaps partly explains why it was not encountered until a few years ago.

The twist manifests itself, for example, in beam rotation and in increased diffractive spreading. In this review we discuss the nature and physical implications of the optical twist in symmetric and astigmatic systems. For general optical fields the twist phenomenon is defined using the variance matrix, which contains information about the spatial and directional (momentum) wave properties. Twisted Gaussian beams can be produced with the help of controlled acousto-optic interactions with elliptic Gaussian fields or by incoherent superpositions of displaced and skewed laser beams. Arbitrary astigmatic fields, such as those emitted by slab lasers or laser-diode arrays, can be transformed by special mode converters into circularly symmetric beams with a twist. The symmetrization is of importance in applications involving fiber coupling and materials processing. The twist phenomenon also has a central role in general beam characterization and in the assessment of beam quality, e.g., the so-called M^2 parameter [2].

In suitable conditions astigmatic mode converters or diffractive elements turn Hermite–Gaussian beams into Laguerre–Gaussian ones, and vice versa. The symmetric, ring-shaped Laguerre–Gaussian fields carry orbital angular momentum, and they can be used as optical spanners to rotate and manipulate small particles [3]. The relation of the optical twist to vortices and holes (screw dislocations), energy transfer, second-harmonic generation, and related applications will be briefly considered. We keep the discussion on a colloquial level, but give references to original research.

2   Gaussian Twisted Beams

A general description of partially coherent waves in optics is best undertaken using the cross-spectral density W, which characterizes second-order field correlations [4], and its symmetrized spatial Fourier transform $\mathcal{W}$, known as the Wigner distribution function [5]. In a plane in which $z = $ constant

$$\mathcal{W}(\boldsymbol{r}, \boldsymbol{p}) = (k/2\pi)^2 \iint W(\boldsymbol{r} - \boldsymbol{r}'/2, \boldsymbol{r} + \boldsymbol{r}'/2) \exp(\mathrm{i}k\boldsymbol{p} \cdot \boldsymbol{r}') \, \mathrm{d}^2 r', \qquad (1)$$

where $k = 2\pi/\lambda$ is the wave number and $\boldsymbol{r} = (x, y)$ and $\boldsymbol{p}$ are 2D vectors. Since W is Hermitian, the function $\mathcal{W}$ is real but not necessarily non-negative. However, integrations over $\boldsymbol{r}$ and $\boldsymbol{p}$ always yield the correct marginal distributions, and so it is useful to interpret $\mathcal{W}$ paraxially as a phase-space density of rays at point $\boldsymbol{r}$, in direction $\boldsymbol{p}$. Moreover, since (1) is invertible, both W and $\mathcal{W}$ contain identical information.

The twist phenomenon arises in a natural way when one considers rotationally symmetric cross-spectral densities $W(\boldsymbol{r}_1, \boldsymbol{r}_2)$. The bilinear rotational invariants are $\xi = r_1^2$, $\zeta = r_2^2$, $\gamma = \boldsymbol{r}_1 \cdot \boldsymbol{r}_2$, and $\tau = \boldsymbol{r}_1 \wedge \boldsymbol{r}_2 = x_1 y_2 - y_1 x_2$. Of these, all but τ are symmetric about any plane containing the z axis, whereas τ reverses its sign. These symmetries were already noted in classic aberration theory, but the τ dependence was normally ignored (resulting in axially or cylindrically symmetric systems) since it would lead to an inbuilt screw sense and "in the optical field, examples of systems with this property are likely to be rather artificial" [6]. The most general rotationally invariant Gaussian cross-spectral density takes the form [1]

$$W(\boldsymbol{r}_1, \boldsymbol{r}_2) = W_{\mathrm{GSM}}(\boldsymbol{r}_1, \boldsymbol{r}_2) \exp(-\mathrm{i}ku\boldsymbol{r}_1 \wedge \boldsymbol{r}_2), \qquad (2)$$

where W_{GSM} corresponds to an arbitrary Gaussian Schell-model beam [4] and $u = u(z)$ is the twist parameter. Besides an overall constant proportional to total power, the GSM beams are characterized by three physical parameters, namely $w(z)$ (beam width or spot size), $\sigma(z)$ (transverse coherence length), and $R(z)$ (radius of phase curvature). The twisted Gaussian beams then form a four-parameter family.

The new twist phase in (2), which is truly nonseparable, proves quite subtle, and it leads to striking mathematical and physical effects [1,7]. Unlike w, σ, and R, the twist parameter u is bounded in magnitude; the upper limit is inversely proportional to the beam's transverse coherence area. Indeed, as a consequence of the non-negative-definiteness property of the cross-spectral density [4], one can show that

$$k\sigma^2|u| \leq 1. \tag{3}$$

This result is related to the diffractive 'uncertainty principle' of conjugate variables r and p, and it implies that in the coherent limit as $k\sigma \to \infty$ there cannot be any twist in a symmetric Gaussian beam.

2.1   Propagation and Manifestations

Propagation of beam-like wave fields in free space and in first-order (paraxial) optical systems is efficiently handled using matrices. Two sets of matrices are required, namely those representing the fields and those characterizing the systems. A beam field of any state of coherence is described in second order by its 4×4 variance matrix [8]

$$V = \begin{bmatrix} \langle x^2 \rangle & \langle xy \rangle & \langle xp_x \rangle & \langle xp_y \rangle \\ \langle yx \rangle & \langle y^2 \rangle & \langle yp_x \rangle & \langle yp_y \rangle \\ \langle p_x x \rangle & \langle p_x y \rangle & \langle p_x^2 \rangle & \langle p_x p_y \rangle \\ \langle p_y x \rangle & \langle p_y y \rangle & \langle p_y p_x \rangle & \langle p_y^2 \rangle \end{bmatrix} = \begin{bmatrix} W & M \\ M^T & U \end{bmatrix}, \tag{4}$$

where the angle brackets denote averages calculated using the Wigner distribution function, and W, M, and U are 2×2 submatrices. The variance matrix is real, symmetric, and positive-definite, containing in general 10 independent elements. Clearly, W and U deal with the beam's irradiance (intensity) and far-field variations, respectively, while the elements of M are associated with the phase curvature and twist [9,10] (Table 1).

In the absence of any tilt and decentering the first-order optical systems are characterized by 4×4 matrices, known as the ABCD or Kogelnik's ray matrices [11]. They can be written in the form

$$S = \begin{bmatrix} A & B \\ C & D \end{bmatrix}, \tag{5}$$

where A, B, C, and D are 2×2 submatrices. The system matrices S form a symplectic group [1,8,12]. Making use of the usual ray-transfer property, one can readily see that the variance matrices transform in optical systems as

$$V_{\text{out}} = S V_{\text{in}} S^{\text{T}}, \tag{6}$$

with the subscripts 'in' and 'out' refering to the input and output planes, respectively.

Table 1. Physical meaning of the elements of variance matrix V

Element[a]	Description
$\langle x^2 \rangle$, $\langle y^2 \rangle$	Beam width in two orthogonal directions
$\langle \theta_x^2 \rangle$, $\langle \theta_y^2 \rangle$	Far-field divergency in two orthogonal directions
$\langle xy \rangle$, $\langle \theta_x \theta_y \rangle$	Orientation of irradiance ellipticity in near and far fields
$\langle x\theta_x \rangle$, $\langle y\theta_y \rangle$	Inverse of the radius of curvature
$\langle x\theta_y \rangle$, $\langle y\theta_x \rangle$	Twist

[a] We have changed the notation to $p_x = \theta_x$ and $p_y = \theta_y$ to emphasize the consistency with the usual angular far-field averages.

Depending on the number of independent elements in V the beams can be divided into categories with characteristic physical properties [8]: stigmatic (ST), orthogonal or simple astigmatic (SA), and general astigmatic (GA). The twisted Gaussian beams belong to the broad class of GA beams, even though their transverse irradiance profiles are rotationally symmetric. The passage of Gaussian light endowed with twist in first-order optical systems has been considered in general terms [1,13], but here we concentrate on free-space propagation and on systems that generate twisted fields.

The Wigner distribution function $\mathcal{W}$ and variance matrix V corresponding to a twisted Gaussian beam are readily calculated [1]. Taking the system matrix S to be that of free-space propagation over a distance z, one finds that the beam width and radius of curvature are [14]

$$w(z) = w(0) \left[1 + (z/z_R)^2 \right]^{1/2} , \tag{7a}$$
$$R(z) = z \left[1 + (z_R/z)^2 \right] , \tag{7b}$$

where

$$z_R = \frac{\pi w^2(0)}{\lambda} \beta \left[1 + \eta^2 \left(\frac{1-\beta^2}{2\beta} \right)^2 \right]^{-1/2} \tag{8}$$

denotes the associated Rayleigh range and $\beta = \{1 + [w(z)/\sigma(z)]^2\}^{-1/2}$ and $\eta = k\sigma^2(z)u(z)$ are two propagation-invariant quantities. The coherence parameter β is normalized between 0 and 1, whereas the scaled twist parameter η is a signed quantity, bounded between -1 and 1 in accordance with (3). The sign of η determines the direction of the handedness.

The dominant manifestations of twist are beam rotation and increased divergency. Though the rotation itself cannot be directly seen, the twist phase influences the propagation laws of the GSM parameters. In other words, observations of $w(z)$, $R(z)$, and β may lead to results that are inconsistent with ordinary GSM beams and thus imply the presence of the twist ($\eta \neq 0$). The beam rotation, which in free space (measured from the waist) assumes the value $\pi/4$ by Rayleigh distance z_R and $\pi/2$ by infinity, can be made visible by using a differential magnifier [1,14]. Such an element, which stretches and squashes the field in orthogonal directions while leaving the twist unaltered, produces ellipses that rotate much like the familiar coherent, simple astigmatic (SA) beams. Within a gradient-index fiber the twisted Gaussian beams rotate by a full 2π revolution in each self-imaging distance [1].

The far-field diffraction angle $\theta_{\text{diff}} = \lim_{z\to\infty} w(z)/z = w(0)/z_R$ is readily found from (7a) and (8), and one can identify in it three contributions that correspond to a coherent Gaussian laser $[\lambda/\pi w(0)]$ and the effects of partial spatial coherence (β^{-1}) and optical twist (the square-root term). The latter terms imply that the twist can be viewed as introducing an effective, reduced degree of coherence $\beta_{\text{eff}} \le \beta$, which increases diffractive spreading. With $\eta = 0$ the formulas give the divergence angle of an ordinary GSM beam and with $\beta = 1$ that of a coherent laser. In focusing, the twist increases the size of the smallest achievable spot while rotating the beam and moving the best focus away from the geometrical focal plane. Hence the twist increases the focal shift, which is consistent with a decreased effective Fresnel number. The field rotation, increased divergency, and behavior of the focal spot of a twisted, partially coherent Gaussian beam have been demonstrated experimentally [14,15].

The Wigner distribution function $\mathcal{W}$ is closely related to the generalized radiance that describes the energy flow in fluctuating wave fields. Owing to this connection and the role of the twist in beam rotation and reduction of effective coherence, it is natural that a general radiometric description and analysis of the radiation efficiency of twisted Gaussian fields also take proper account of this phenomenon [16,17].

2.2   Decompositions and Realization

Insight into the twist phenomenon and its practical realization can be gained by decomposing the cross-spectral density W into a superposition of spatially coherent but mutually uncorrelated fields. Several decomposition methods of this type exist for twisted Gaussian beams. Perhaps the most formal one is the so-called coherent-mode decomposition, which is a direct consequence of the cross-spectral density being mathematically a Hermitian, nonnegative-definite, Hilbert–Schmidt kernel [4]. The coherent-mode and eigenvalue structure of twisted Gaussian fields brings into evidence an analogy to statistical quantum-mechanical operator techniques [18,19]. The coherent modes turn out to be the familiar Laguerre–Gaussian functions (those of ordinary GSM beams are Hermite–Gaussian functions), which have recently also been studied in the context of orbital angular momentum. However, they do not appear yet to have been exploited in twist synthesis.

Another decomposition technique makes use of a six-element astigmatic optical system that consists of two identical lens blocks rotated by $\phi = \pi/4$

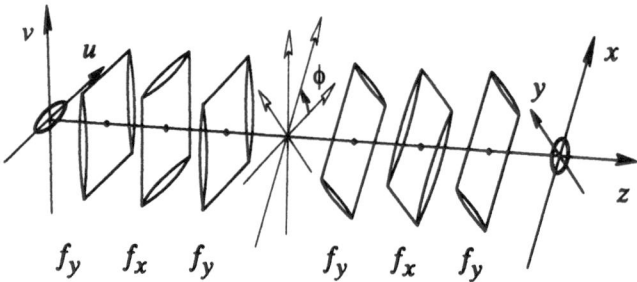

Fig. 1. Astigmatic system that converts an elliptic GSM beam into a symmetric Gaussian twisted beam. Focal lengths satisfy $f_x = 2f_y$ and separations between all lenses and planes shown are f_y. (After [14])

with respect to each other, as shown in Fig. 1. Each block performs imaging in one transverse coordinate and a Fourier transform in the other. In suitable conditions the system converts an elliptic GSM field in the input (u, v) plane into a rotationally symmetric twisted Gaussian beam in the output (x, y) plane [14]. The required, partially coherent input beam is most conveniently synthesized from an ordinary elliptic laser by means of the real-time acousto-optic coherence-control method [20]. In this technique an optimized binary-phase grating in an acousto-optic cell produces a weighted superposition of uncorrelated (due to a small Doppler shift) elliptic Gaussian beams propagating in different directions around the first Bragg order.

It is required that the Rayleigh ranges z_R of the input beam be equal in both directions and that $z_{Rx} = z_{Ry} = f_x = 2f_y$. An anisotropic GSM field allows a decomposition in terms of elliptic Gaussian waves, tilted in the u and v directions. The passage of these waves through the astigmatic lenses is governed by the extended Fresnel diffraction formula, which is conveniently written in block form (5) corresponding to the appropriate matrix S [11,14]. Though a general situation can be handled similarly, we focus on the case of a maximum twist ($|\eta| = 1$), which is obtained when the input beam is fully coherent in one coordinate, say v. Its decomposition takes the form

$$W(u_1, u_2, v_1, v_2) = \int P(\theta) U^*(u_1, v_1; \theta) U(u_2, v_2; \theta) \, d\theta, \tag{9}$$

where

$$U(u, v; \theta) \propto \exp(-u^2/w_u^2) \exp(-v^2/w_v^2) \exp(ik\theta u), \tag{10a}$$

$$P(\theta) \propto k\sigma \exp[-(k\sigma)^2 \theta^2/2], \tag{10b}$$

i.e., U is an inclined elliptic Gaussian beam of spot sizes w_u and w_v, and P is an angular weighting function. The acousto-optic device forms a discrete sampling of this decomposition. In the exit (x, y) plane the beam consists of a weighted superposition of elliptic waves that are displaced along one (y)

axis and have a propagation component along an orthogonal (x) axis. This tilt grows with increasing displacement. In addition, there is an astigmatic phase term. On propagation the ellipses diffract, rotate, and move continually further away from the yz plane; these effects combine to produce the rotation and increased divergence of a twisted Gaussian beam.

Some experimental results are illustrated in Figs. 2 and 3. The measured beam widths, along the x and y directions, on propagation from the waist-plane of a twisted Gaussian beam of parameters $w(0) = 0.62\,\mathrm{mm}$, $\beta = 0.29$, and $|\eta| = 1$ are plotted in Fig. 2, together with the calculated divergency [14]. For comparison, the divergencies of an analogous ordinary GSM beam

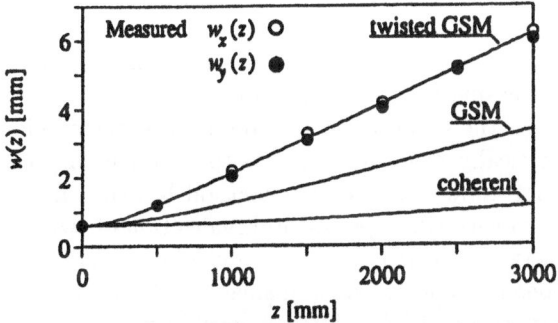

Fig. 2. Experimental results comparing the measured beam widths to calculated divergencies of a coherent Gaussian laser beam, a conventional GSM beam, and a twisted Gaussian beam. The theoretical curves are based on equal irradiance and coherence widths at the waist. (See [14])

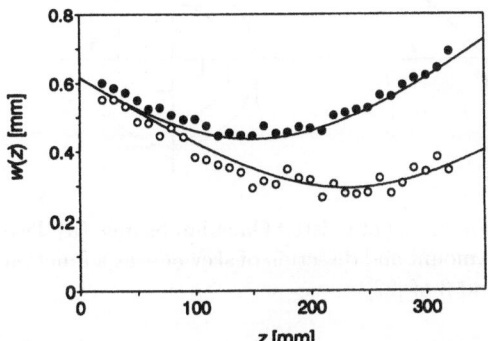

Fig. 3. Measured beam profiles of an ordinary GSM beam (*open circles*) and a twisted Gaussian beam (*solid circles*) within the focal region. The twisted-beam Rayleigh range equals the focal length, $z_\mathrm{R} = f = 300\,\mathrm{mm}$, showing focal shifts. The lines are the corresponding theoretical beam profiles. (From [15])

($\eta = 0$) and of a coherent laser beam ($\beta = 1$) are also given. In Fig. 3 the profile of the twisted Gaussian beam focused with a lens of focal length $f = z_\mathrm{R} = 300\,\mathrm{mm}$ is shown [15]. It is seen that the focal spot is larger and closer to the lens than that of a similar conventional GSM beam. Using only a small number of sample beams in the decomposition one can demonstrate that the twisted Gaussian beam is rotated by $\pi/4$ in the plane of the best focus (here $z = f/2$) and by $\pi/2$ in the plane $z = f$.

An alternative, physically appealing, decomposition model has been put forward for the twisted Gaussian beams [21,22]. In this method the twisted beam is formed as an independent superposition of ordinary (symmetric) Gaussian beams that are laterally displaced from the z axis. Moreover, these beams are tilted (skewed out of meridional plane) by an amount proportional to the displacement, as illustrated in Fig. 4. A Gaussian weighting function is used. The rate of the superposition-beam skewness is directly related to the twist parameter u in (2), and the model automatically satisfies the upper bound (3) of the twist. Explicit calculations show that the beam-parameter product $M^2 = 1/\beta_\mathrm{eff}$ is an increasing function of $|u|$ [14,22].

We emphasize that although the two practical decomposition techniques discussed above are mathematically exact and appear to have similarities, their precise relationship is not known. The former method has the advantage of being directly related to versatile experimental realization of twisted fields by acousto-optic interactions, whereas the latter model can readily be extended to include elliptic beams, more complex displacement and inclination schemes, etc. Such superpositions lead to further novel beams.

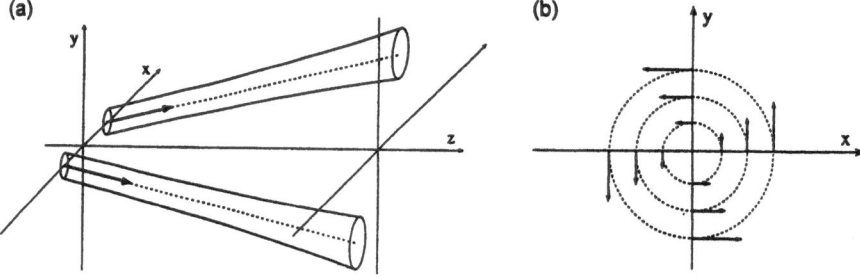

Fig. 4. Illustration of the superposition model of twisted Gaussian beams. (a) Two typical skewed Gaussian beams. (b) Amount and direction of skewness as a function of the beam displacement. (Figs. 2 and 3 of [22])

3   General Twisted Fields

For an unambiguous characterization of an arbitrary beam field one needs to know, in principle, all the 10 independent parameters of its variance matrix V. Usually this is impractical and even the optical standards presently under development aim at the measurement of some typical parameters only.

In many novel applications it is desirable that the radiation fields be transformed into some particular transverse shapes. These include rotationally symmetric beams for materials processing and fiber coupling, nondiffractive Bessel beams for metrology, and doughnut-shaped fields for particle trapping and guiding. Such beam transformations reduce the number of elements in the variance matrix but they also explicitly introduce optical twist. Below we consider the symmetrization of an arbitrary astigmatic field, coherent or otherwise, and discuss the effects of the twist on beam quality.

3.1   Symmetrization

It is known that any variance matrix V can be made diagonal by a suitable optical system S [8]. In general, the Rayleigh ranges z_{Rx} and z_{Ry} (ratios of beam widths to far-field divergencies) will be different in the orthogonal directions, but these can be equalized by using an astigmatic telescope. Let the input beam enter an optical system (S) that consists of one half of the system shown in Fig. 1 (i.e., three cylindrical elements with $z_R = f_x = 2f_y$), combined with a $\pi/4$ rotation (R). The variance matrix V_{out} of the exiting beam then is [23,24] (see Table 1)

$$V_{\text{out}} = (SR)V_{\text{in}}(SR)^{\text{T}} = \begin{bmatrix} \langle x^2 \rangle_{\text{out}} & 0 & 0 & t \\ 0 & \langle y^2 \rangle_{\text{out}} & -t & 0 \\ 0 & -t & \langle \theta_x^2 \rangle_{\text{out}} & 0 \\ t & 0 & 0 & \langle \theta_y^2 \rangle_{\text{out}} \end{bmatrix}, \tag{11}$$

where

$$\langle x^2 \rangle_{\text{out}} = \langle y^2 \rangle_{\text{out}} = [\langle x^2 \rangle_{\text{in}} + \langle y^2 \rangle_{\text{in}}]/2 \,, \tag{12a}$$

$$\langle \theta_x^2 \rangle_{\text{out}} = \langle \theta_y^2 \rangle_{\text{out}} = [\langle \theta_x^2 \rangle_{\text{in}} + \langle \theta_y^2 \rangle_{\text{in}}]/2 \,, \tag{12b}$$

$$t = [\langle x^2 \rangle_{\text{in}} - \langle y^2 \rangle_{\text{in}}]/2z_R = z_R \,[\langle \theta_x^2 \rangle_{\text{in}} - \langle \theta_y^2 \rangle_{\text{in}}]/2 \,, \tag{12c}$$

and t corresponds to the twist (sign depends on the sense of R).

The output beam is rotationally symmetric with respect to the second moments of its waist and divergence. In terms of the variance quantities the field propagates in free space and through symmetric optical elements like a stigmatic beam. It differs from a real stigmatic beam by the presence of the twist, which is seen from (12c) to be proportional to the input-field asymmetry. The twist becomes apparent only when astigmatic elements or interference are used.

The results of (11)–(12c) apply to coherent and partially coherent fields alike. Beam symmetrization has been demonstrated by TEM$_{n0}$ modes and high-power linear diode-laser arrays [24]. We emphasize that the ensuing fields show azimuthal structure in their local irradiance patterns, but the second-order moments are circularly symmetric.

3.2 Beam Quality

Traditionally laser beam quality is measured by a propagation factor $M^2 = (M_x^2 M_y^2)^{1/2} = 2k(\det V)^{1/4} \leq 1$ [8], which for a fundamental Gaussian mode takes on the minimum value 1. The determinant of the variance matrix is constant in ABCD optics, and so the M^2 factor is not modified by paraxial systems. This theoretical method was used to develop a useful experimental approach for beam characterization [2], and that work forms the basis of the new ISO standard.

However, often the beam propagation factor is introduced as the product of the waist size and the far-zone divergence, normalized to the fundamental-mode value, i.e., $M^{2*} = 2k (\langle x^2 \rangle \langle \theta_x^2 \rangle \cdot \langle y^2 \rangle \langle \theta_y^2 \rangle)^{1/4} = (M_x^{2*} M_y^{2*})^{1/2}$, which for a diagonal variance matrix obviously is the same as M^2. Making use of (11)–(12c) we readily find that

$$M_{\text{out}}^{2*} = [M_{x\,\text{in}}^{2*} + M_{y\,\text{in}}^{2*}]/2 > M_{\text{in}}^{2*} = M_{\text{in}}^2. \tag{13}$$

Hence the beam propagation factor defined as (waist) $\times$ (divergence) is not conserved [23,24]. The beam parameter product is increased and the beam quality of the transformed (twisted) output beam is the average of the orthogonal beam parameter products of the astigmatic input beam. This result illustrates the importance of the twist effect in beam characterization and shows that, in general, it is not sufficient to determine the beam propagation factor from the waist and far-field divergence measurements.

3.3 Mode Conversion

The astigmatic multi-element systems discussed above are closely related to mode converters consisting of two cylindrical lenses and introduced for the transformation of Hermite–Gaussian beams into Laguerre–Gaussian beams [25,26]. Some conditions related to the input-field and lens properties must be fulfilled, otherwise the output is a Hermite–Gaussian beam with complex arguments. The Gouy-shifts along the two orthogonal directions of the cylindrical lenses differ by $\pi/2$, leading to a factor i^n in the mode expansions and resulting in the Laguerre–Gaussian beams [27].

In general, using these converters a Hermite–Gaussian mode with spatial dependence $H_n(u)H_m(v)$ may be transformed into a well-defined Laguerre–Gaussian mode $L_m^{n-m}(r^2)$, for $n \geq m$, where m and $(n-m)$ characterize the radial and azimuthal dependencies, respectively [27]. As a simple example we may consider the following pair [23,24]

$$E_{\text{in}}(u,v) = C_n H_n(\sqrt{2}u/w_0)\exp[-(u^2+v^2)/w_0^2]\,, \tag{14a}$$

$$E_{\text{out}}(r,\phi) = C_n'(\sqrt{2}r/w_0)^n \exp[-(r/w_0)^2]\exp(-in\phi)\,, \tag{14b}$$

where C_n and C_n' are constants. The input is TEM_{n0} mode, while the output is a crater-like, rotationally symmetric, Laguerre–Gaussian field. The radius of the central area (dark beam) obeys $r_{\text{max}} = \sqrt{n/2}\,w_0$. The interesting point is that the exiting beam is endowed with a twist [see (11)]

$$t = n/2k, \tag{15}$$

which is proportional to mode number n. By explicit calculations one also finds for the output field, in accordance with (13), that $M_{x\,\text{out}}^{2*} = M_{y\,\text{out}}^{2*} = n+1$ and so $M_{\text{out}}^{2*} = n+1 \geq \sqrt{2(n+1/2)} = M_{\text{in}}^{2*}$. The ring-like structures of the Laguerre–Gaussian beam correspond to (paraxial) eigenmodes of free-space propagation. They become larger but are not otherwise modified, when the field propagates and rotates.

A linear Hermite–Gaussian field can be produced by a mode generator, which is a fiber-coupled, diode-pumped Nd:YAG laser with a short cavity. The mode number is controlled by changing the position of the pump fiber. In Fig. 5 we illustrate experimental results on some typical input and output modes. Figure 6 shows a measurement of the output beam quality. The experimental results confirm the theoretical predictions. The influence of system parameters, as well as shifted and multiple inputs, have also been studied. Using these models the transformed radiation of a diode-laser array may be represented as an incoherent superposition of Laguerre–Gaussian modes of elliptical symmetry [24].

Fig. 5. Some examples of mode conversions. *Left*: Hermite–Gaussian beams TEM_{n0}. *Right*: Transformed beams of circular symmetry with twist. (From [23])

4   Vortices and Angular Momentum

As is well-known, a circularly polarized beam transports angular momentum. Each photon of such a beam has an angular momentum of $\pm\hbar$, the sign de-

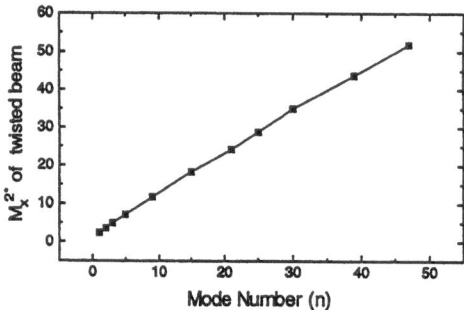

Fig. 6. Beam propagation factor M_x^{2*} measured for the rotationally symmetric twisted beam as a function of the input mode number n. (From [23])

pending on the sense of polarization, while linearly polarized light has no spin angular momentum. However, a linearly polarized beam containing a phase singularity at its (dark) center also carries angular momentum that is associated with its helical phase structure. This is orbital angular momentum. The angular momentum introduces a torque that rotates nano-particles [27,28]. Novel beams of this type constitute optical tweezers and spanners, which are used to trap, manipulate, move, and rotate small specimens.

Examining (14b) in the limit as $r \to 0$ one observes that the field E_{out} contains an optical vortex (screw dislocation) [29]. The vortex is a phase singularity of order (charge) n. In general, Laguerre–Gaussian fields $L_m^{n-m}(r^2)$ are eigenstates of angular momentum about the axis (L_z). Such a beam can be identified to carry $(n-m)\hbar$ of orbital angular momentum per photon, regardless of the polarization state. One can show that for coherent, linearly polarized beams L_z is proportional to the difference between the off-diagonal twist elements in variance matrix V [24]. However, in partially coherent beams the relation between the twist and angular momentum is not so clear. Also, the nondiffracting Bessel beams, $E \propto J_m(k\alpha r) \exp(-im\phi)$, are nonparaxial wave fields containing an optical vortex of order m ($m > 0$). However, they do not correspond to a well-defined angular momentum state. The Poynting vector of a linearly polarized Laguerre–Gaussian beam, as well as the corresponding flux vector of a Bessel beam in scalar treatment, form spiraling curves characteristic of the helical phase structure. The nature of twisted partially coherent fields consisting of helicoidal modes has recently been examined [30]. Besides using the astigmatic mode converters, beams containing optical vortices of any order can efficiently be produced with the help of computer-generated holograms [31,32].

Beams containing optical vortices also play an important role in nonlinear optics, such as in second-harmonic generation [33,34] and optical parametric oscillation (OPO) [35]. When a Hermite–Gaussian beam $H_n H_0$ is transformed into a Laguerre–Gaussian mode L_0^n and then frequency-doubled in a nonlinear crystal, the resulting second-harmonic beam is L_0^{2n}. This is consistent with the intuitive picture that two photons, each of angular momentum $n\hbar$, combine to form a single photon of twice the energy and angular momentum $2n\hbar$. The spot size is reduced by a factor of $\sqrt{2}$, while the Rayleigh range is preserved. Since the mode converter works in both directions, the beam may again be transformed into a Hermite–Gaussian field $H_{2n}H_0$. The experiments, performed in LBO by temperature tuning (type I phase matching) and in KTP by angle tuning (type II phase matching), demonstrate the association and conservation of orbital angular momentum in beams containing a phase singularity [34]. Similar results have recently been reported with high-order Laguerre–Gaussian modes [36].

5   Conclusions

Optical twist, which may be associated equally well with Gaussian and non-Gaussian beams, appears in many areas of physics and laser technology. The main examples are partially coherent Gaussian fields in general beam characterization and coherent or partially coherent symmetric fields for beam handling and power applications. Although some proposals have been put forward, it seems that a clear, unambiguous definition of the optical twist, applicable to any beam, is still somewhat lacking. Moreover, we are awaiting specific applications in which the general twist property would be explicitly exploited to an advantage.

Another area where additional work is required is the relation between the twist and optical vortices and angular momentum. This applies particularly to cases that are not eigenstates of the angular momentum such as beams of arbitrary irradiance profiles and partially coherent and partially polarized fields.

References

1. Simon, R., Mukunda, N. (1993) Twisted Gaussian Schell-model Beams. J. Opt. Soc. Am. A **10**, 95–109
2. Siegman, A.E. (1990) New Developments in Laser Resonators. In: Holmes, D.A., (Ed.), Optical Resonators, Proc. SPIE **1224**, 2–14
3. Simpson, N.B., Dholakia, K., Allen, L., Padgett, M.J. (1977) Mechanical Equivalence of Spin and Orbital Angular Momentum of Light: an Optical Spanner. Opt. Lett. **22**, 52–54
4. Mandel, L., Wolf, E. (1995) Optical Coherence and Quantum Optics. Cambridge University Press, Cambridge, UK
5. Bastiaans, M.J. (1993) Wigner Distribution Function Applied to Partially Coherent Light. In: Mejías, P.M., Weber, H., Martínez-Herrero, M., González-Ureña, A., (Eds.), Workshop on Laser Beam Characterization, SEDO, Madrid, pp. 65–87

6. Buchdahl, H.A. (1967) Optical Aberration Coefficients, XIII. Theory of Reversible Semi-symmetric Systems. J. Opt. Soc. Am. **57**, 517–522

7. Simon, R., Mukunda, N. (1998) Twist Phase in Gaussian-beam Optics. J. Opt. Soc. Am. A **15**, 2373–2382

8. Nemes, G., Siegman, A.E. (1994) J. Opt. Soc. Am. A **11**, 2257–2264

9. Belanger, P.A. (1991) Beam Propagation and ABCD Ray Matrices. Opt. Lett. **16**, 196–198

10. Weber, H. (1992) Propagation of Higher-order Intensity Moments in Quadratic-index Media. Opt. Quantum Electron. **24**, 1027–1049

11. Siegman, A.E. (1986) Lasers. University Science Books, Mill Valley, CA

12. Kauderer, M. (1990) Fourier-optics Approach to the Symplectic Group. J. Opt. Soc. Am. A **7**, 231–239

13. Bastiaans, M.J. (1998) On the Propagation of the Twist of Gaussian Light in First-order Optical Systems. In: Nijhawan, O.P., Gupta, A.K., Musla, A.K., Singh, K., (Eds.), Optics and Optoelectronics: Theory, Devices and Applications, Vol. 1, Narosa Publishing House, New Delhi, India, pp. 121–125

14. Friberg, A.T., Tervonen, E., Turunen, J. (1994) Interpretation and Experimental Demonstration of Twisted Gaussian Schell-model Beams. J. Opt. Soc. Am. A **11**, 1818–1826

15. Friberg, A.T., Tervonen, E., Turunen, J. (1994) Focusing of Twisted Gaussian Schell-model Beams. Opt. Commun. **106**, 127–132

16. Simon, R., Friberg, A.T., Wolf, E. (1996) Transfer of Radiance by Twisted Gaussian Schell-model Beams in Paraxial Systems. Pure Appl. Opt. **5**, 331–343

17. Friberg, A.T. (1996) Radiation Efficiency and Radiometry of Twisted Gaussian Fields. In: Chang, J.-S., Lee, J.-H., Lee, S.-Y., Nam, C.H., (Eds.), Optics for Science and New Technology, Proc. SPIE **2278**, 311–312

18. Simon, R., Sundar, K., Mukunda, N. (1993) Twisted Gaussian Schell-model Beams. I. Symmetry Structure and Normal-mode Spectrum. J. Opt. Soc. Am. A **10**, 2008–2016

19. Sundar, K., Simon, R., Mukunda, N. (1993) Twisted Gaussian Schell-model Beams. II. Spectrum Analysis and Propagation Characteristics. J. Opt. Soc. Am. A **10**, 2017–2023

20. Turunen, J., Tervonen, E., Friberg, A.T. (1990) Acousto-optic Control and Modulation of Optical Coherence by Electronically Synthesized Holographic Gratings. J. Appl. Phys. **67**, 49–59

21. Gori, F., Bagini, V., Santarsiero, M., Frezza, F., Schettini, G., Spagnolo, G.S. (1994) Coherent and Partially Coherent Twisting Beams. Opt. Rev. **1**, 143–145

22. Ambrosini, D., Bagini, V., Gori, F., Santarsiero, M. (1994) Twisted Gaussian Schell-model Beams: A Superposition Model. J. Mod. Opt. **41**, 1391–1399

23. Eppich, B., Friberg, A.T., Gao, C., Weber, H. (1996) Twist of Coherent Fields and Beam Quality. In: Morin, M., Giesen, A., (Eds.), Third International Workshop on Laser Beam and Optics Characterization, Proc. SPIE **2870**, 260–267

24. Friberg, A.T., Gao, C., Eppich, B., Weber, H. (1997) Generation of Partially Coherent Fields with Twist. In: Shladov, I., Rotman, S.R., (Eds.), Tenth Meeting on Optical Engineering in Israel, Proc. SPIE **3110**, 317–328

25. Tamm, C., Weiss, C.O. (1990) Bistability and Optical Switching of Spatial Patterns in a Laser. J. Opt. Soc. Am. B **7**, 1034–1038

26. Beijersbergen, M.W., Allen, L., van der Veen, H.E.L.O., Woerdman, J.P. (1993) Astigmatic Laser Mode Converters and Transfer of Orbital Angular Momentum. Opt. Commun. **96**, 123–132

27. Allen, L., Beijersbergen, M.W., Spreeuw, R.J.C., Woerdman, J.P. (1992) Orbital Angular Momentum of Light and the Transmission of Laguerre–Gaussian Laser Modes. Phys. Rev. A **45**, 8185–8189

28. He, H., Friese, H.E.J., Heckenberg, N.R., Rubinstein-Dunlop, H. (1995) Direct Observation of Transfer of Angular Momentum to Absorptive Particles from a Laser Beam with a Phase Singularity. Phys. Rev. Lett. **75**, 826–829

29. Indebetouw, G. (1993) Optical Vortices and Their Propagation. J. Mod. Opt. **40**, 73–87

30. Gori, F., Santarsiero, M., Borghi, R., Vicalvi, S. (1998) Partially Coherent Sources with Helicoidal Modes. J. Mod. Opt. **45**, 539–554

31. He, H., Heckenberg, N.R., Rubinstein-Dunlop, H. (1995) Optical Particle Trapping with Higher-order Doughnut Beams Produced Using High Efficiency Computer-generated Holograms. J. Mod. Opt. **42**, 217–223

32. Vasara, A., Turunen, J., Friberg, A.T. (1989) Realization of General Non-diffracting Beams with Computer-generated Holograms. J. Opt. Soc. Am. A **6**, 1748–1754

33. Wulle, T., Herminghaus, S. (1993) Nonlinear Optics of Bessel Beams. Phys. Rev. Lett. **70**, 1401–1404

34. Dholakia, K., Simpson, N.B., Padgett, M.J., Allen, L. (1996) Second-harmonic Generation and the Orbital Angular Momentum of Light. Phys. Rev. A **54**, R3742–R3745

35. Gadonas, R., Marcinkevičius, A., Piskarskas, A., Smilgevičius, V., Stabinis, A. (1998) Travelling Wave Optical Parametric Generator Pumped by a Conical Beam. Opt. Commun. **146**, 253–256

36. Courtial, J., Dholakia, K., Allen, L., Padgett, M.J. (1997) Second-harmonic Generation and the Conservation of Orbital Angular Momentum with High-order Laguerre–Gaussian Modes. Phys. Rev. A **56**, 4193–4196

Principles and Fundamentals of Near Field Optics

M. Nieto-Vesperinas

Summary. The foundations of near field optical microscopy are discussed based on the concept of resolution from scattered wavefields. The different set-ups and problems associated with them, as well as their perspectives, are discussed.

1 Introduction

The smallest object detail that optical microscopes can yield is limited to a linear size of about one half of the illumination wavelength; namely, to 0.2–0.5 µm for visible light. This *resolution limit* has been well known since the works by Abbe [1] and Lord Rayleigh [2] on the diffraction theory of image formation. After the quality of optical microscopes achived by the end of last century, which attained resolving powers of about 1 µm, and the development and success of electron microscopy during this century, little progress beyond this limit of resolution was obtained. It was not until the 1980s that the electron scanning tunnelling microscope (STM) was invented [3], and the related near field optical microscope was proposed [4,5]. Since then, very active research has been undertaken on these subjects [6–9].

Increasing resolution beyond half a wavelength requires a widening of the range of plane wave components detected in the wavefront scattered by the object, thus retrieving those that do not propagate into the radiation zone. This demands detection at subwavelength distances from the surface of the object, that is, in the *near field*. However, this proximity may involve a strong coupling between the detector and the emitting body. Hence, unlike conventional far zone or image plane measurements, the detection process may not be passive; namely, the signal 'seen' by the recording system may not be the same as that in the absence of the detector. Addressing object details smaller than half a wavelength also implies entering into the *mesoscopic* scale in which the interaction of light with the structure cannot be described by the usual physical optics or the Kirchhoff approximation of diffraction theory [9]. The optical image (that is, the wavefront closest to the surface of the object) built by the optical system does not generally reproduce the object transmittance, and cannot be directly related to the object itself. Thus, the operation of increasing the Rayleigh limit of resolution carries a large cost regarding the understanding of the image formation process and what the detector actually sees.

This chapter presents the fundamentals of these problems and the current progress towards solving them.

2 Angular Spectrum Representation and the Limit of Resolution

The spatial resolution that can be ascertained about the structure of an object with a given illuminating wave stems from the angular information content of the associated wavefield. To see this, let us consider an object whose boundary surface contains points $r = (x, y, z)$ such that $z < 0$. The field U, either radiated or scattered by this object into the free propagation half space $z > 0$, can be represented by an angular spectrum $A(K)$ of plane waves [10,11]:

$$U(r) = \int_{-\infty}^{\infty} A(K) \exp(\mathrm{i}K \cdot R + \mathrm{i}qz) \, \mathrm{d}K. \tag{1}$$

In (1) $R = (x, y)$ and $k = (K, q)$, k being the wavevector of each plane wave component of amplitude A. Since $k^2 = K^2 + q^2 = (2\pi/\lambda)^2$, where λ represents the wavelength, then

$$q = \sqrt{k^2 - K^2}, \quad K^2 \leq k^2, \tag{2}$$

$$q = \mathrm{i}\sqrt{K^2 - k^2}, \quad K^2 > k^2. \tag{3}$$

Waves with q given by (3) are evanescent in $z > 0$ since they decay exponentially as z increases.

Equation (1) expresses that the boundary value of U at any plane of propagation in $z > 0$ unequivocally determines U throughout the whole half space of free propagation $z > 0$. Let $f(R) = U(R, z = 0)$; on introducing this boundary value of U into (1) and inverting the corresponding Fourier integral, one obtains $A(K)$ as

$$A(K) = \frac{1}{4\pi^2} \int_{-\infty}^{\infty} f(R) \exp(-\mathrm{i}K \cdot R) \, \mathrm{d}K. \tag{4}$$

In turn through (1), $A(K)$ determines U at any point of propagation in $z > 0$. On introducing (4) into (1) one gets

$$U(r) = \frac{1}{2\pi} \int_{-\infty}^{\infty} f(R') \frac{\partial}{\partial z'} G(r, r') \, \mathrm{d}R'. \tag{5}$$

Equation (5) has been obtained by using the fact that the spherical wave $G(r, r') = \exp[\mathrm{i}k(|r - r'|)]/|r - r'|$ in $z > 0$ assumes the plane wave representation [12]:

$$G(r, r') = \frac{\mathrm{i}}{2\pi} \int_{-\infty}^{\infty} \frac{1}{q} \exp[\mathrm{i}K \cdot (R - R') + \mathrm{i}q(z - z')] \, \mathrm{d}K. \tag{6}$$

Readers familiar with diffraction theory will notice that (5) is identical to the usual Rayleigh–Sommerfeld representation for diffracted wavefields [10,11].

To appreciate the significance of (1)–(3) for retrieving wavefront and object data from measurements, let us assume that after some experiment, $U(\boldsymbol{R}, z_0)$ is determined at some plane $z = z_0 > 0$. On introducing this value of U into (1) and inverting the corresponding Fourier integral, one gets

$$A(\boldsymbol{K}) \exp(iqz_0) = \frac{1}{4\pi^2} \int_{-\infty}^{\infty} U(\boldsymbol{R}, z_0) \exp(-i\boldsymbol{K} \cdot \boldsymbol{R}) \qquad (7)$$

In comparison with the wavefield at $z = 0$, (see (4)), the field at $z_0 > 0$ in (7) yields the angular spectrum $A(\boldsymbol{K})$ multiplied by the low-pass filter, $\exp(iqz_0)$, which according to (3) exponentially decays as K increases in the evanescent wave region $K > k$, thus being practically zero at any distance $z_0 \geq \lambda$. At these distances, the impulse response associated with $\exp(iqz_0)$ (that is, its Fourier transform) has a width equal to λ. Hence, $\lambda/2$ constitutes the minimum distance at which two points of either the object or the wavefront at $z = 0$ can be resolved. This is Rayleigh's limit of resolution. As z_0 decreases below λ, however, the exponential decay of $\exp(iqz_0)$ is slower and one can retrieve values of $A(\boldsymbol{K})$ in the range of evanescent waves, $K > k$. For example, for $z_0 = 0.01\lambda$ the filter $\exp(iqz_0)$ has a significant value background at $K = 10k$, so that one can get a resolution of $\lambda/20$. In general, to distinguish two points separated by a fraction λ/N, $(N > 2)$, one has to detect U at distances $z_0 \leq \lambda/(5N)$. The fundamental question therefore is: does any field, scattered or radiated by an object, always have waves in the evanescent region? The answer is that, effectively, these fields always possess both propagating and evanescent components [13]. This establishes the ground for their detection and hence for the operation of near field optics (NFO).

3   Near Field Microscopy Configurations

The concept of increasing resolution by a fraction of λ as explained above was known well before the invention of near field optical microscopes. As a matter of fact, the first proposal to scan an object with a subwavelength sized source close to its surface was made by Synge in 1928 [14]. Also, superresolution by data inversion using (1)–(7) for the holographic determination of acoustic waves at a distance of $z_0 = 10\,\text{cm}$ and a wavelength of $\lambda = 156\,\text{cm}$ ($z_0 = 0.06\lambda$), was demonstrated by Williams et al. [15]. However, it was not until the late 1980s and early 1990s that both enough intensity of the signal and control of the detector as it scans close to the object surface, were achieved.

A tip laterally scans the zone near the object surface at subwavelength distances. A tapered optical fiber, either bare or metal-coated, is used for this purpose. The tip can act as a nanosource, as a nanodetector, or both. Three main designs of NFO microscopy exist according to this role of the tip. In the

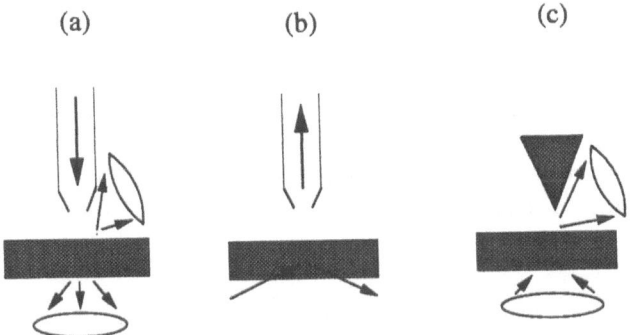

Fig. 1. Different configurations of near field optical microscopy. (**a**) Illumination mode (SNOM) (**b**) Collection mode (PSTM) (**c**) Apertureless microscope

illumination mode (see Fig. 1a) the light comes out from the aperture of the fiber, illuminating the object with a spot of subwavelength dimensions. The scattered light from this nanometric object area or volume is then collected through a microscope system either by reflection or transmission, depending on the transparency properties of the sample. A variety of *scanning near field optical microscopy* techniques (SNOM) are included in this category [4,5,16–19].

In the *collection mode* (Fig. 1b) the tapered fiber tip is used as a detector. The detected light originates from a light beam which illuminates the object from the opposite side and is transmitted through the object. The *photon scanning tunnelling microscope* (PSTM) belongs to this kind of technique [20–22]. Due to the property of reciprocity of scattering which interchanges the roles of source and detector [11], SNOM and PSTM are equivalent configurations yielding similar images [23]. A third procedure used to scan the near field lies in the domain of the so-called *apertureless techniques* (see Fig. 1c) in which a metallic tip is used. This is either a metal-coated apex of a tapered fiber [24] or a metallic nanoparticle, levitated above the sample surface by optical trapping [25]. This tip scatters the near field projecting it into the far zone where it is detected [26,27]. In order to filter the signal scattered by the tip from the background light directly scattered by the object surface, a vertical vibration with a lock-in detection is added to the scanning movement of the tip.

4 The Optical Signal at the Tip

The existence of wave scattering with structures smaller than the wavelength, and hence the failure of Fourier optics, causes the spatial configuration of the near field to be, in general, quite different than that of the optical parameters of the object. Therefore, the interpretation of the NFO image is difficult. This problem still increases due to the perturbation introduced into this near field by the proximity of the tip. Thus, multiple scattering methods are necessary in order to accurately predict the modulus and phase of the detected signal [28–31].

Figures 2a–c show a two-dimensional simulation of the reflected field intensity distribution detected by a tip as it scans along the line $z = d$ parallel to the mean plane of a surface-relief grating of profile $z = D(x) = h\cos(2\pi x/b)$

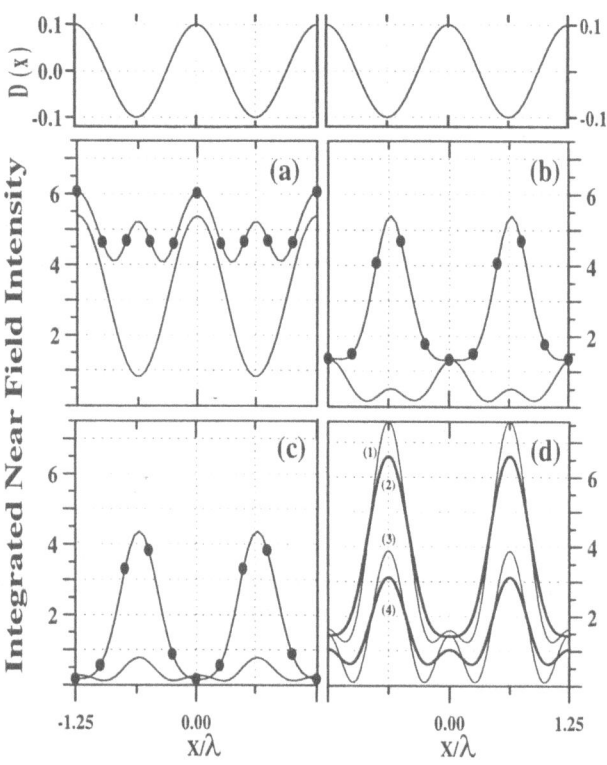

Fig. 2. Variation of the integrated intensity with the cylinder position moved along $z = d$ for several radii a of the cylinder and distances d of the cylinder-plane. The cylinder has a dielectric permitivity $\epsilon_1 = 2.12$. Parameters of the grating are: $b = 1.25\lambda$, $h = 0.1\lambda$, $\epsilon_2 = (-17.2, 0.498)$, $\theta_0 = 0°$ (p-polarization). (**a**) $a = 0.5\lambda$, $d = 0.8\lambda$, $z_0 = 0.2\lambda$. (**b**) $a = 0.2\lambda$, $d = 0.5\lambda$, $z_0 = 0.2\lambda$. (**c**) $a = 0.2\lambda$, $d = 1\lambda$, $z_0 = 0.7\lambda$. Each figure shows the intensity inside the cylinder and integrated over its diameter (*solid line with circles*) and the scattered intensity outside the cylinder and integrated on a segment of length $2a$ on a line below it at a distance of $z = z_0$ (*solid line*). (**d**) Intensity distribution on the plane $z = 0.5\lambda$ due to the grating alone. (1) Total, i.e. incident plus scattered, intensity. (2) Total intensity integrated on the interval $2a$. (3) Scattered intensity. (4) Scattered intensity integrated on the interval $2a$. The grating profile is shown on the top [31]

which separates two half spaces: one being vacuum and the other filled with silver of permittivity $\epsilon = -17.2 + i0.498$ at $\lambda = 652.6\,\text{nm}$. The tip is constituted by a dielectric cylinder, with axis along OY and $\epsilon_1 = 2.12$. This cylinder thus represents a two-domensional version of either a detecting nanoparticle or an optical fiber apex. The system is illuminated from the vacuum side by a Gaussian beam, TM polarized, at normal incidence. These figures contain the intensity inside the cylinder, integrated over the diameter $2a$ parallel to $z = 0$, as well as the scattered intensity outside the cylinder, integrated over the interval $2a$ on a line $z = z_0$ just below the cylinder. For a tip with $a = 0.5\lambda$ we show these intensities at $d = 0.8\lambda$ and $z = 0.2\lambda$ (Fig. 2a), whereas for $a = 0.2\lambda$ these intensities are shown at $d = 0.5\lambda$ and $z = 0.2\lambda$ (Fig. 2b) and at $d = \lambda$ and $z = 0.7\lambda$ (Fig. 2c). The intensity distribution oscillates with both d and z, and its contrast can be inverted, depending on the scanning distance to the grating. This is due to the standing wave along the OZ-axis that exists between the tip and the grating.

The question of how the integrated intensities inside the cylinder are related to the unperturbed near field intensities (incident plus scattered) that exist in absence of the cylinder is illustrated next. Figure 2d shows a calculation of four intensities in the plane $z = 0.5$ in absence of the cylinder: (1) the total, i.e. incident plus scattered, intensity, (2) the total intensity integrated over the interval $2a$ ($a = 0.2\lambda$) along $z = 0.5\lambda$, (3) the scattered intensity, and (4) the scattered intensity integrated over the same interval $2a$, ($a = 0.2\lambda$). Note that, apart from a change in contrast, curve (2) is similar to the integrated intensity obtained inside the cylinder in Fig. 2b. This can be interpreted as a consequence of the fact that this cylinder does not introduce an appreciable perturbation on the field reflected by the grating alone. The smoothing effect of the integration is also seen. Hence, tips of size $a \leq 0.2\lambda$ and a dielectric constant similar to that of glass can be considered as *passive*. This means that the field inside them (or projected into the far zone by scattering of the near field on them) is very similar to the field in the absence of the tip; however, the field distribution surrounding the tip is perturbed by their presence.

5   Inverse Scattering and Coherence

From the discussion of the previous section it is clear that the interpretation of the object image from the near field measurements requires data inversion. Inverse scattering procedures have been established for retrieving the topography of smooth surfaces, either from knowledge of both the amplitude and phase of the scattered near field by direct profile iterative reconstruction from the Rayleigh scattering equations [27], or by inverse diffraction of NFO holographic intensities [32].

A direct topographic retrieval by adding the NFO scattered intensities, each of which is produced by illumination at a different angle of incidence, was put forward by Garcia et al. [33]. This procedure possesses a strong connection to the STM operation [3] and amounts to producing an effective incoherent field at the surface, where it is scattered. As a matter of fact, this result matches other works [34,35] showing the better performance of incoherent light at the surface area for direct observation of the object topography from the NFO optical image.

When subsurface structure $\epsilon(r)$ is searched below the profile $D(R)$, the scattered field $U^{(s)}$ due to the bulk volume of the object is expressed as:

$$U^{(s)}(r) = k^2 \int_V [\epsilon^2(r') - 1] \frac{\exp(ik|r - r'|)}{|r - r'|} U(r') \, dr', \tag{8}$$

where V represents the bulk volume and U is the field inside the object. It is well known [11] that the three-dimensional function $\epsilon(r)$ cannot be recovered from the essentially two-dimensional structure of $U^{(s)}(r)$. This question is the basis of diffraction tomography [36], which requires object illumination under different k-vectors, usually achieved by changing the angle of incidence. Thus, one obtainins a set of scattered wavefunctions that yields a three-dimensional extension of the angular spectrum of the scattered field, and this allows the recovery of $\epsilon(r)$. In the case of weak volume scattering, wave propagation is straight through the scattering volume V and the integral of (8) reduces to the eikonal approximation: $D_{eq}(R) = \int_0^{D(R)} [\epsilon^2(r') - 1] \, dz'$, which yields $\epsilon(r)$ integrated along the transmission z-direction. This result states that the scattered near field is the same as that produced by an equivalent surface profile $D_{eq}(R)$ [37]. The retrieval of $\epsilon(r)$ from a set of its projections along different lines of straight-through propagation, other than the z-direction, by changing the angle of incidence, and hence the operation of tomographic methods, appears as difficult to implement in NFO.

6   Applications, Artifacts, and Conclusions

Three operation modes are used in NFO: constant height, constant distance, and constant intensity. In the first, the tip is scanned in a plane parallel to the object surface; in the second, a non-optical control mechanism, such as the tunneling current of an STM [17], is used to maintain a constant tip-object surface distance; and in the third mode, a feedback is employed to keep the optical signal. The constant intensity and constant height signal produce equivalent images [38], whereas the widely used constant distance mode gives rise to artifacts due to the non-optical signal arising from the path followed by the tip [39,40]. These false image features can be easily understood simply by considering the signal obtained from the flat surface of a glass object illuminated from the dielectric side under total internal reflection while the tip follows a controlled path that varies its height with respect to the object interface. Evidently, the recorded intensity increases as the height decreases. The spurious signal, then, comes from the vertical movement of the tip and has nothing to do with the flat topography of the object. It is likely that at least half of all published SNOM images may be influenced or dominated by this artifact [41].

NFO has been successfully developed for imaging and optical storage [6–8], observation of spectroscopic features of solid state and biological systems, as well as single molecule dynamics [9]; however, much work is still necessary to compare experiments and theoretical models in order to fully understand the production and control of the images obtained by this technique.

Acknowledgments

I thank R. Carminati, N. Garcia, J.J. Greffet, and A. Madrazo for many shared discussions and work over the years. This work has been supported by the European Union.

References

1. Abbe, E. (1873) Archiv f. Mikroskop. Anat. **9**, 413
2. Lord Rayleigh (1879) Phil. Mag. **8**, 477
3. Binning, G., Roherer, H. (1981) Helv. Phys. Acta **55**, 762
4. Pohl, D.W., Denk, W., Lanz, M. (1984) Appl. Phys. Lett. **44**, 651
5. Betzig, E., Lewis, A., Harootian, A., Isaacson, M., Kratschmer, E. (1986) Biophys. J. **49**, 269
6. Pohl, D., Courjon, D. (Eds.) (1992) Near Field Optics. Kluwer Academic Publishers, Dordrecht
7. Marti, O., Moeller, R. (Eds.) (1995) Photons and Local Probes. Kluwer Academic Publishers, Dordrecht
8. Nieto-Vesperinas, M., Garcia, M. (Eds.) (1996) Optics at the Nanometer Scale: Imaging and Storing with Photonic Near Fields. Kluwer Academic Publishers, Dordrecht
9. Paesler, M.A., Moyer, P.J. (1996) Near Field Optics. Wiley-Interscience, New York
10. Goodman, J.W. (1968) Introduction to Fourier Optics. Mc Graw-Hill, New York
11. Nieto-Vesperinas, M. (1991) Scattering and Diffraction in Physical Optics. Wiley-Interscience, New York
12. Baños, A. (1966) Dipole Radiation in the Presence of a Conducting Half Space. Pergamon Press, Oxford, Chapter 2
13. Wolf, E., Nieto-Vesperinas, M. (1985) J. Opt. Soc. Am. A **2**, 886
14. Synge, E. (1928) Philos. Mag. **6**, 356
15. Williams, E.G., Maynard, J.D. (1980a) Phys. Rev. Lett. **45**, 554; (1980b) J. Acoust. Soc. Am. **68**, 340
16. Lewis, A., Isaacson, M., Harootunian, A., Murray, A. (1984) Ultramicroscopy **13**, 227
17. During, U., Pohl, D.W., Rohner, F. (1986) J. Appl. Phys. **59**, 3318
18. Fischer, U., Durig, U., Pohl, D.W. (1988) Appl. Phys. Lett. **52**, 249

19. Vaez-Iravani, M., Toledo-Crow, R., Chen, Y. (1993) J. Vac. Sci. Technol. A **11**, 742
20. Reddick, R., Warmack, R., Ferrell, T.L. (1989) Phys. Rev. B **39**, 767
21. Courjon, D., Sarayeddine, K., Spajer, M. (1989) Opt. Comm. **71**, 23
22. Dawson, P., de Fornel, F., Goudonnet, J.P. (1989) Phys. Rev. Lett. **72**, 2927
23. Greffet, J.J., Carminati, R. (1997) Prog. Surf. Sci. **56**, 133
24. Inouye, Y., Kawata, S. (1994) Opt. Lett. **19**, 159; Madrazo, A., Nieto-Vesperinas, M. (1997) J. Opt. Soc. Am. A **14**, 2768
25. Sugiura, T., Okada, T., Inouye, Y., Nakamura, O., Kawata, S. (1997) Opt. Lett. **22**, 1663
26. Zenhausern, F., O'Boyle, M., Wickramsinghe, H. (1994) Appl. Phys. Lett. **65**, 1623
27. Garcia, N., Nieto-Vesperinas, M. (1995) Appl. Phys. Lett.**66**, 339
28. Girard, C., Dereux, A. (1996) Rep. Prog. Phys. **59**, 657
29. Garcia, N., Nieto-Vesperinas, M. (1993) Opt. Lett. **18**, 2090
30. Carminati, R., Greffet, J.J. (1995a) J. Opt. Soc. Am. A **12**, 2716; (1995b) Ultramicroscopy **61**, 11
31. Madrazo, A., Nieto-Vesperinas, M. (1996) J. Opt. Soc. Am. A **13**, 785
32. Greffet, J.J., Sentenac, A., Carminati, R. (1995) Opt. Comm. **116**, 20
33. Garcia, N., Nieto-Vesperinas, M. (1995) Opt. Lett. **20**, 949
34. Pohl, D.W. (1991) Scanning Near Field Optical Microscopy. In: Sheppard, C., Mulvey, T. (Eds.) Advances in Optical and Electron Microscopy. Academic Press, London
35. de Fornel, F., Salomon, L., Adam, P., Goudonnet, J.P., Guerin, P. (1994) Opt. Lett. **19**, 14
36. Wolf, E. (1996) Principles and Development of Diffraction Tomography. In: Consortini A. (Ed.) Trends in Optics Vol. 3, Academic Press, San Diego
37. Carminati, R., Greffet, J.J. (1995) J. Opt. Soc. Am. A **12**, 2716
38. Carminati, R., Greffet, J.J. (1996) Opt. Lett. **21**, 1208
39. Hecht, B, Bielefeldt, H., Inouye, Y., Pohl, D.W., Novotny, L. (1997) J. Appl. Phys. **81**, 2492
40. Carminati, R., Madrazo, A., Nieto-Vesperinas, M., Greffet, J.J. (1997) J. Appl. Phys. **82**, 501
41. Pohl, D.W., Hecht, B., Heinzelmann, H. (1998) In: Garcia, N., Nieto-Vesperinas, M., Roherer, H. (Eds.) Nanoscale Science and Technology. Kluwer Academic Publishers, Dordrecht

Spin-Orbit Interaction of a Photon: Theory and Experimentation on the Mutual Influence of Polarization and Propagation

N. D. Kundikova and B. Ya. Zel'dovich

Summary. The results of the theoretical and experimental investigation on the optical effects connected with the mutual influence of polarization and propagation are presented.

1 Introduction

It is generally assumed that light polarization and the process by which it propagates are mutually independent under narrow light beam propagation through a homogeneous medium. However this is not true in the case of an inhomogeneous medium. The simplest example of the influence of light propagation on its polarization is a light beam propagating through a lens. As can be seen from Fig. 1, the vector of polarization E_1 of the light falling on the lens does not coincide with the vector of polarization E_2 of the light beam after it has passed through the lens. Change in the azimuth of linear polarized light is a consequence of the transverse nature of the light waves.

It was shown theoretically by F. Bortoloti [1], S. Rytov [2], V. Vladimirsky [3] that the influence of the trajectory on the light polarization consists in the rotation of the polarization plane under light propagation along a non-planar trajectory. If S_{in} and S_{ex}, vectors tangential to this trajectory at the input and exit points, respectively, are parallel, then the angle of rotation (in radians) is numerically equal to the solid angle Ω (in steradians) subtended

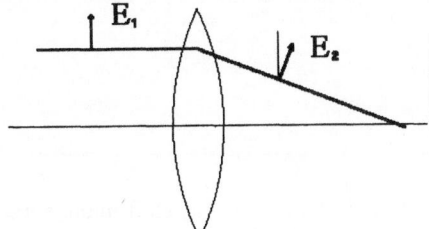

Fig. 1. The change in direction of the vector of polarization E of a light ray after passing through a lens

by the trajectory of the vector S at the unit sphere [3–5]. The change in the azimuth of linear polarized light has been observed experimentally under light propagation through a helical (coiled into a spiral) single mode fiber [6] and has been interpreted on the basis of the adiabatic theorem (geometric phase) of M. Berry [4]. This effect has also been observed in a nonplanar Mach-Zender interferometer [7].

The influence of light polarization on trajectory was demonstrated for the first time under total internal reflection. It was shown that a reflected ray suffers a longitudinal shift (Fig. 2) [8,9]. The size of the shift is on the order of the light wavelength and depends on the azimuth of linear polarized light. Experimental observation of this shift was made by Goos and Hanchen [10,11].

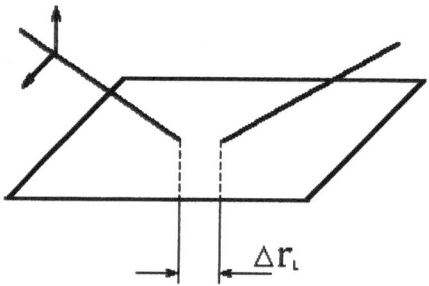

Fig. 2. Longitudinal ray shift under total internal reflection

Light penetration into an optical medium with a low refractive index was demonstrated in [12,13]. The geometrical interpretation of the longitudinal shift was made in [14].

It was shown that in addition to a longitudinal shift a transverse shift should be also observed and the direction (sign) of that shift depends on the sign of the circular polarized light (Fig. 3) [15,16]. The value of the transverse shift L_t is related to the longitudinal shift L_l as follows:

$$L_t = \alpha \cdot L_l , \tag{1}$$

where $\alpha < 1$. The experimental observation of this effect was made in [17–19].

It should be stressed that the longitudinal and transverse shifts of a ray under total internal reflection were not considered to be a result of the influ-

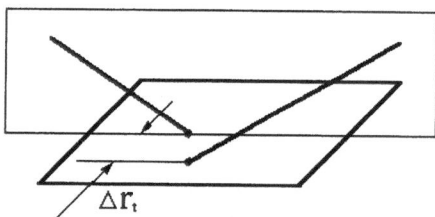

Fig. 3. Transverse ray shift under total internal reflection

ence of the state of polarization on propagation and were not connected with the influence of propagation on polarization.

The influence of the trajectory on polarization and the influence of the polarization on trajectory were considered as mutually inverse effects for the first time in 1990 [20]. This consideration led to the prediction of a new effect: rotation of the speckle pattern in a multimode optical fiber due to the change of circular polarization from left-handed to right-handed. In terms of quantum mechanics this effect can be regarded as an interaction between the orbital momentum of photon and its spin (polarization) [20]. The effect is known as the optical Magnus effect and was observed for the first time in [21]. Its name is connected with the following analogy. By looking at a rotating ping-pong ball falling down through the air, we can see that it is deflected from the vertical into a path of rotation due to the Magnus effect. If we assume that a circularly polarized photon propagating in a waveguide is like our rotating ball, we can expect it to deviate from the initial trajectory.

The prediction and experimental observation of the optical Magnus effect lead to the observation of new optical effects connected with the mutual influence of light polarization and propagation. This paper is devoted to these new optical effects.

2 Optical Magnus Effect

2.1 Light Propagation Through a Multimode Fiber with a Step-like Refractive Index Profile

Let us consider the propagation of light through an axially symmetrical fiber with a step-like refractive index profile $n(r)$:

$$
\begin{aligned}
n(r) &= n_{\text{co}}, & r/\rho &< 1, \\
n(r) &= n_{\text{cl}}, & r/\rho &> 1.
\end{aligned}
\tag{2}
$$

Here $(x, y) = r$ are the transverse coordinates, $r = |r|$, and ρ is the radius of the core; n_{co} and n_{cl} are the refractive indices of the core at its axis and of the cladding respectively.

The polarization is not coupled to the propagation of the beam in the zeroth paraxial approximation, corresponding to the scalar parabolic equation. More accurately, the spatial structure of the field and its polarization are coupled due to the inhomogeneity of the refractive index. The polarization correction to the propagation constant β of mode $e(r) \exp(i\beta z)$ in a first approximation has the form (see [22], their Equation 32.24)

$$
\delta\beta = -\frac{(2\Delta)^{1/2}\rho}{2V} \frac{\int (\nabla_\perp e_\perp) \left[e_\perp^* \nabla_\perp \ln n^2(r) \right] d^2 r}{\int e_\perp^* e_\perp d^2 r}.
\tag{3}
$$

Here $e(r) = e_\perp(r) + e_z(r)e_z$, $\nabla_\perp = \partial/\partial r = e_x(\partial/\partial x) + e_y(\partial/\partial y)$, $\Delta = (n_{\text{co}}^2 - n_{\text{cl}}^2)/2n_{\text{co}}^2 \approx (n_{\text{co}} - n_{\text{cl}})/n_{\text{co}} \ll 1$, and $V = \rho k n_{\text{co}}(2\Delta)^{1/2}$ is a dimensionless parameter ($V \gg 1$ for a multimode fiber, $k = 2\pi/\lambda$, and λ is the wavelength of light in the air).

The calculation of the corrections $\delta\beta$ needs the proper modes of zeroth (uncoupled) approximation. Axial symmetry of the fiber allows (see [22]) these nodes to be taken in the following form:

$$e_{m,N}^{\pm}(r,\varphi) = \left(1/\sqrt{2}\right)(e_x \pm ie_y)\exp(im\varphi)F_{|m|,N}(r) \,, \tag{4}$$

corresponding to right-handed and left-handed circular polarizations and the values of the orbital momentum $m = 0, \pm 2, \pm 3, \pm 4, \ldots$ Here $x = r\cos\varphi$, $y = r\sin\varphi$, $F_{|m|,N}(r)$ is the radial function, and $N = 0, 1, \ldots$ is the radial quantum number. It has been shown [22] that only modes $m \neq \pm 1$ constitute the proper basis for the diagonalization of grad(n) perturbation.

In our approximation the circular polarization of the wave, right- or left-handed, is conserved during the propagation. The only modes which may violate that property are those with $m = +1$ or $m = -1$. We shall neglect their contribution in the subsequent consideration. It should be emphasized that the consideration of modes with $m = +1$ or $m = -1$ leads to the prediction of new effects, which will be considered in Sect. 4.

Consider now the important specific case in which we illuminate our fiber with right- and left-handed circular polarization in turn, but with strictly the same modal distribution $C_{m,N}$ at the input. It is interesting to compare the field and/or intensity distributions for $(e_x + ie_y)$ and $(e_x - ie_y)$ in some cross-section z. Our qualitative analogy with the mechanical Magnus effect allows us to suppose that these distributions will be similar, but somewhat rotated relative to each other around the fiber axis.

Initial theoretical hypothesis was made for the case of a fiber with an unconfined parabolic refractive index profile. It was shown [20] that the patterns $|E^{\pm}(r,\varphi,z)|^2$ are the same, with mutual angular shift

$$\varphi_+ - \varphi_- = 2\frac{\Delta}{2\pi\rho^2 n}\lambda z \,. \tag{5}$$

For the case of the step-index fiber, the analytical expressions cannot be obtained. The modes and the corresponding corrections (3) have been calculated [23] by a computer for the following particular fiber parameters: $n = 1.5$, $\Delta n = n_{co} - n_{cl} = 0.006$, $\lambda = 0.63\,\mu\text{m}$ (He-Ne laser), $2\rho = 200\,\mu\text{m}$. It is important to note that for a general index profile the dependence of $\beta_{mN}^+ - \beta_{mN}^-$ on m and N is rather complicated. Therefore the intensity patterns $|E^+(r,\varphi,z)|^2$ and $|E^-(r,\varphi,z)|^2$ will not be identical, even for input modes with the same amplitudes. However, we can also expect that at some moderate propagation length these patterns will be somewhat similar, with a relative angular shift of the same order of magnitude as in (5).

For a step-index fiber, the radial functions in (4) are

$$F_{|m|,N}(r) = J_{|m|}(Ur), \qquad r/\rho < 1 \tag{6}$$

$$F_{|m|,N}(r) = K_{|m|}(Wr), \qquad r/\rho > 1. \tag{7}$$

Here, $J_{|m|}$ and $K_{|m|}$ are the Bessel and MacDonald functions, respectively, and U and W ($V^2 = W^2 + U^2$) are determined from the equation:

$$U \frac{J_{|m|+1}(U)}{J_{|m|}(U)} = W \frac{K_{|m|+1}(W)}{K_{|m|}(W)} . \tag{8}$$

Polarization corrections from (3) are

$$\delta\beta^+_{|m|,N} = \delta\beta^-_{-|m|,N} = \frac{(2\Delta)^{3/2}}{2\rho} \frac{WU^2}{V^3} \frac{K_{|m|}(Wr)}{K_{|m+1|}(Wr)} , \tag{9}$$

$$\delta\beta^-_{|m|,N} = \delta\beta^+_{-|m|,N} = \frac{(2\Delta)^{3/2}}{2\rho} \frac{WU^2}{V^3} \frac{K_{|m|}(Wr)}{K_{|m-1|}(Wr)} . \tag{10}$$

These equations were used for the numerical modeling of the physical experiment.

2.2 Computer Experiment

The angular distribution of speckle patterns $|E^\pm(r, \varphi, z)|^2$ at a given radius for left and right circular polarizations was calculated. It has been shown that from a change in the circular polarization the whole picture suffers an angular shift (rotation); the main features remain, but there is a slight distortion. The angular shift is about $1.5°$ for the abovementioned fiber parameters and a fiber length of 96 cm.

To extract the pure rotation from the whole change of the speckle pattern the correlation functions

$$K_{ij}(r, \psi, z) = \int I_i(r, \varphi, z) I_j(r, \varphi + \psi, z) d\varphi \quad \text{and}$$

$$I_i(r, \varphi, z) = |E^i(r, \varphi, z)|^2 \quad (i = +, -; \quad j = +, -) \tag{11}$$

were calculated. Hence, averaging over the statistical ensemble was substituted by averaging over the angle φ in the interval $0 < \varphi < 2\varphi_0$, with $\varphi_0 \gg \pi/m_{max}$ and $2\varphi_0 = 360°$. As was supposed, the correlation function had a sharp maximum at the angle $\psi_0 \neq 0$, which was proportional to the length and at $z = 96$ cm was equal to $1.5°$ independent of radius and specific realization.

2.3 Experiment

The experiment was carried out with a step-index fiber with the following parameters: $2\rho = 200\,\mu m$; the difference between refractive indices of quartz core and polymeric cladding, $\Delta n = 0.006$, was measured via the critical angle of propagating rays.

The experimental set-up is shown in Fig. 4. The beam of the He-Ne laser (1) with linear polarization and wavelength, $\lambda = 0.63\,\mu m$ propagated through

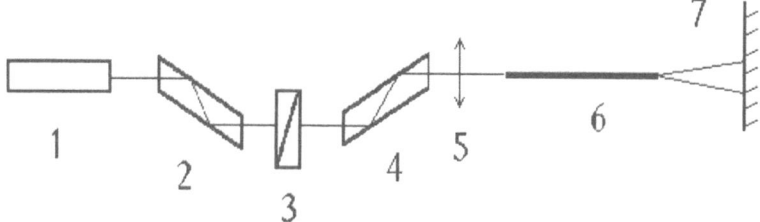

Fig. 4. Experimental setup. 1 – the He-Ne laser with linear polarization and wavelength, $\lambda = 0.63\,\mu m$; 2,4 – Fresnel rhombi; 3 – the polarizer; 5 – the lens; 6 – the fiber under investigation; 7 – the screen with the polar coordinate net

the Fresnel rhomb (2). The plane of linear polarization was oriented in such a way that the outgoing radiation was circularly polarized. The polarizer (3) allowed us to cut desirable linear polarization out of the circular polarization. Next, the light propagated through the second Fresnel rhomb (4). The light was focused at the entrance of the fiber (6) by the lens (5). Rotation of the polarizer (3) through 90° allowed us to switch from left-handed circular polarization of the propagating beam to right-handed and back. The speckle pattern of the light transmitted through the fiber was observed at the screen (7) with the polar coordinate net.

By changing circular polarization of propagating light from left-handed to right-handed we could see the picture moving clockwise in accordance with the theoretical prediction. During that movement some details of the speckle pattern changed, but the main features were conserved under rotation. The speckle patterns may be seen in Fig. 5. The arrow points to the bright spot that was rotated slightly, changing its shape (as calculated in the computer experiment).

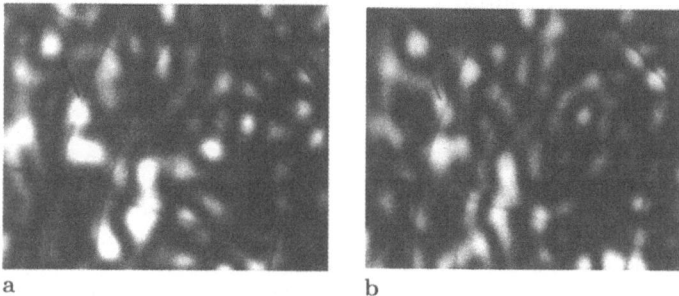

a   b

Fig. 5. A fragment of the real speckle patterns $|E^i(r, \varphi, z)|^2$ of (a) right-hand ($i =$ "+") and (b) left-hand ($i =$ "−") circularly polarized light at the screen. The arrow points to the bright spot that serves as an example of rotation under a circular-polarization sign change

For more accurate measurement of the "rotation angle" an experimental analog of the correlation function technique [23] was used. For the fiber under investigation the rotation angle was 1.4°. From a number of replicate measurements, the angle measurement showed the same result with an accuracy of ±0.5°.

The agreement of the theoretically computed rotation $(+1.5 \pm 0.5)°$ with the experimentally measured $(+1.4 \pm 0.5)°$ proved to be incredibly strong. Even the analytical result for a fiber with a parabolic refractive index profile gives a quite reasonable estimation $(+3.3°)$ for the effect.

2.4 Inhomogeneity of Optical Magnus Effect

The experimental and numerical results considered above show that the angle of rotation of the speckle pattern is the same within the experimental error and within the accuracy of calculations for all points of the speckle pattern. More detailed theoretical study of the influence of polarization on the trajectory of light [24] has shown that the peripheral parts are shifted by a larger angle compared with the central part. The peripheral parts of the speckle pattern correspond to the limiting angles of entry into the fiber and the central part represents the rays traveling close to the fiber axis. The angle of rotation of the speckle pattern, φ, depends on the angle between the fiber axis and the direction of propagation of the incident beam, α [24]:

$$\varphi = \frac{\lambda z}{\pi n_{co}^3 \rho^2} \alpha^2 . \tag{12}$$

In order to prove this statement experimentally, the large angle of speckle pattern rotation is needed. According to (12) it possible to increase the angle of speckle pattern rotation by decreasing the radius of the fiber core, ρ.

A multimode fiber with a step-like refractive index profile of length, $z = 7.5\,\mathrm{cm}$, a core diameter, $2\rho = 100\,\mu\mathrm{m}$, and a difference between the refractive indices of the core and cladding, $\Delta n = 0.016$ was used.

An experiment was carried out which investigated the dependence of the angle of rotation of the speckle pattern, φ, of the transmitted light on the angle of entry, α, of a narrow light beam into the fiber.

The reversal of the sign of circular polarization was performed by a rotation of the $\lambda/4$ plate through 90°. This $\lambda/4$ plate was an adjustable composite polarization system [25]. The image of the speckle pattern was recorded by means of CCD matrix.

Figure 6 shows negative images of a ring speckle pattern recorded with the aid of the CCD matrix when the angle of incidence of the left- and right-hand circularly polarized light on the end face of the fiber was $\sim 12°$.

By means of this technique we have measured the angle of the speckle pattern rotation for angles of incidence up to 12°. The measurements have demonstrated that the angle of rotation of the speckle pattern depends on

Fig. 6. The observed speckle patterns (the observer looking in the direction from which the light is coming). (**a**) Left-handed polarized light. (**b**) Right-handed polarized light

the radius of this pattern or, equivalently, on the angle of entry of light into the fiber. This dependence is in good agreement with that predicted by (12).

3   Topological Optical Activity in a Rectilinear Optical Fiber

The effects connected with the influence of a ray's trajectory on the polarization and connected with the influence of polarization on the propagation process are results of the spin-orbit interaction of a photon and can be described in the frame of the unified Hamiltonian, $\mathcal{H}$ [26]. Moreover, it is possible to observe these effects under light propagation through a rectilinear optical fiber [27].

Let us consider a skew ray in a rectilinear multimode fiber. In this case the ray trajectory is nonplanar and following [2–5] we can expect a rotation of the polarization plane after light transmission through the fiber. The depolarization of linearly polarized light transmitted through a fiber with a parabolic profile was explained on the basis of this consideration in [28]. Following [27], we examine the rotation of the polarization plane under light propagation through a multimode rectilinear fiber.

Let us describe the skew ray by two parameters, b and α: b is the distance between the center of the fiber core and the point of ray entrance into the fiber; α is the angle between the direction of ray propagation in the fiber and the axis of the fiber ($\alpha \approx n\beta$; β is the angle of ray incidence in air). According to [24], the angle of rotation of the polarization plane is

$$\psi = \frac{\alpha^3}{2\rho^2} bL \ . \tag{13}$$

The value of α is assumed to be small; ρ is the radius of the fiber core, and L is the length of the fiber. Equation (13) was derived on the basis of the wave theory in [24]. It is also possible to get the same result using the Rytov-Vladimirsky-Berry theory [27].

The following experimental arrangement was used to study the rotation of the polarization plane [27]. Linearly polarized light of an He-Ne laser with a wavelength, $\lambda = 0.63\,\mu$m was transformed into circularly polarized light by means of a Fresnel rhomb and was focused on the fiber input by a lens. A polarizer allowed us to make linearly polarized light with any azimuth. The length of the quartz fiber under investigation was $L = 7.5$ cm, the diameter of the fiber core was $2\rho = 100\,\mu$m, the difference Δn between the refractive indices of the core $n_{co} = n$ and the cladding n_{cl} was $\Delta n = 0.016$. The value of Δn determines the maximum angle of incidence β_{max} when the meridional ray is confined to the fiber:

$$\beta_{max} = \sqrt{2n\Delta n} \approx 12.7^\circ . \tag{14}$$

A microscope was used to control the illumination of the fiber input. The scattered light was studied under the microscope by means of a metal mirror. It allowed the illumination of the fiber input to be controlled and did not disturb the beam polarization.

The angle of the input beam divergence, θ_0, in the air was measured by the criterion e^{-1} of the intensity at maximum: $\theta_0 = \theta(\text{HWe}^{-1}\text{M}) \approx 0.5^\circ$, i.e. $\theta_0 \ll \beta_{max}$. The value of the radius of the illuminated spot at the fiber input end, $a_0 = a(\text{HWe}^{-1}\text{M})$, was calculated from θ_0, so that $a_0 = (k\theta_0)^{-1} \approx 1.5\,\mu$m and $a_0 \ll \rho$. Here $k = 2\pi\lambda$.

The notion of a ray is taken from geometric optics. It cannot be applied to the light propagation through the fiber under investigation [22]. As a result the mode approach becomes better. The interference of the different fiber modes produces a speckle pattern at a screen after the transmission of the light through the fiber. In our case that pattern looks like a narrow ring. The radius of the ring allowed us to calculate the angle β of the beam incidence at the fiber input.

The azimuth of the output linearly polarized light, ψ_{out}, did not coincide with the azimuth of the input plane polarized light, ψ_{in}. We determined the angle ψ_{out} by the rotation of the analyzer until the intensity of the light at the screen achieved it's minimum value. The angle of the plane polarization rotation, ψ, was determined as $\psi = \psi_{in} - \psi_{out}$. We measured the values of ψ_{out} for different values of ψ_{in} under each value of β. As was expected, the value of ψ did not depend on ψ_{in}. We controlled the illumination of the fiber input and then checked the direction of the plane polarization rotation. This direction coincided with the one predicted from theory. When we illuminated first the top part and then the bottom part, the plane of polarization changed it's direction of rotation. It should be mentioned that in our experiment linear polarization was conserved only partially (70–90%); for larger angles of incidence, the depolarization was larger.

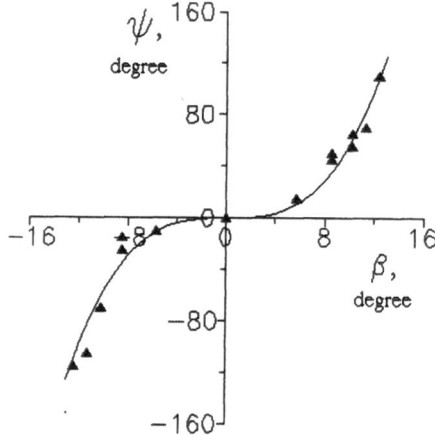

Fig. 7. The angle of the plane polarization rotation, ψ, versus of the angle of incidence in air, β

The experimental results are presented in Fig. 7. The solid line is a result of (13) under the conditions $b = 0.8\rho$ and $\beta \approx n\alpha$. The core of the fiber was illuminated near the cladding, so it is possible to say that Fig. 7 shows good agreement of the theory and the experiment.

In this experiment, the plane polarization rotation under the propagation of skew ray through a rectilinear multimode fiber with a step-like refractive index profile has been demonstrated. Therefore, it is possible to observe the influence of the light's trajectory on its polarization and vice versa under light propagation through a rectilinear fiber using the same experimental set up.

4   The Optical Effects Connected with Meridional Rays

As it was stressed in Sect. 2, the modes with $m = \pm 1$ were neglected in the consideration of the optical Magnus effect. These modes, "hedgehog" and "steering-wheel" (Fig. 8, [29]), correspond to the meridional rays and play an important role in the effects presented here [29–31].

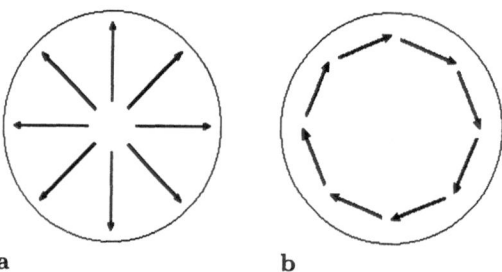

a   b

Fig. 8. Electrical fields of the (a) "hedgehog" and (b) "steering-wheel" modes

4.1   "Magnetic" Rotation of the Speckle Pattern of Light Transmitted Through an Optical Fiber

It is well known that a magnetic field H applied to a transparent medium influences the state of polarization of the light propagated in this medium. The transverse shift of a beam is very small in the typical case of a moderate magnetic field. It has been shown [29] that a longitudinal magnetic field can lead to some kind of speckle pattern rotation of the light transmitted through an optical fiber with a step-like refractive index profile. According to [29], the angle φ of speckle pattern rotation is $\varphi \approx VHL$, where V is the Verdet constant, H is module of magnetic field H and l is the length of a fiber. Here we present the results of the experimental investigation of this effect [30].

The beam of an He-Ne laser with a wavelength, $\lambda = 0.63\,\mu m$ was transmitted through a few-mode optical fiber with a step-like refractive index profile. The diameter of the quartz fiber core was $2\rho = 9.5\,\mu m$ and the difference between the refractive indices of the core and the cladding was $\triangle n = 0.004$. The fiber was placed at the axis of a coil with a magnetic field. The average strength of the magnetic field over the length of the coil was $\sim 500\,G$. To intensify the effect, the 17 m long fiber was passed through the coil seven times, so that the length, l, of the fiber in the magnetic field was $l \approx 1.4\,m$. In order to exclude fake effects, the behavior of the speckle pattern of the light transmitted through optical fiber was observed during change in the magnetic field direction. The entrance to the fiber was illuminated with linearly polarized light. The analyzer allowed linearly polarized light at the fiber exit to be selected.

It was found that the speckle pattern rotated and the direction of the rotation corresponded to the sign of Faraday rotation (Fig. 9). The magnitude of the speckle pattern rotation was about $15°$. The magnitude of Faraday rotation over this distance should be $\sim 40°$. The value of this angle corresponded in order of magnitude to the quantitative estimation.

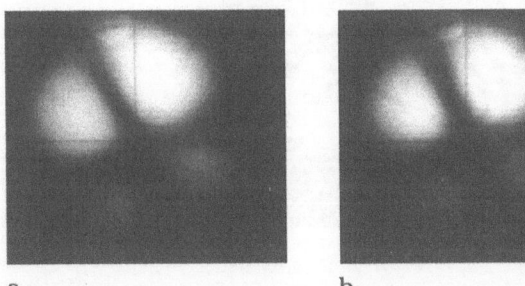

a   b

Fig. 9. Speckle pattern of the light transmitted through the fiber. The direction of the magnetic field and the direction of the light propagation (**a**) coincide and (**b**) do not coincide

4.2   Generation of Light Waves with a Single Screw Dislocation in the Wavefront

The speckle pattern produced by the scattering of coherent laser light from a random medium or by transmission through a multimode optical fiber consists of points with phase singularities. The surface of constant phase near these points has a left or right screw form. The amplitude, $E_0 = 0$ and the phase, φ, is multivalued at these points. These phase singularities are referred to as screw dislocations [32]. It has been shown theoretically and experimentally that in speckle patterns there is, on average, one optical screw dislocation accompanying each speckle spot [33,34]. It should be emphasized that the number of right screw dislocations is equal to the number of left screw dislocations. Here we present the experimental method used to generate light waves with a single screw dislocation [31].

The method is based on the transmission of light through a multimode optical fiber. Theoretical analysis of the circularly polarized light propagation through a multimode fiber with a step-like refractive index profile shows that transmitted light contains a circularly polarized wavefield with opposite sign and with a single screw dislocation. This wavefield can be selected by means of a "circular analyzer". The sign of dislocation depends on the light circularity sign at the fiber input. The "hedgehog" and "steering-wheel" modes play an important role in the generation of the light waves with a single dislocation [31].

The experiment was carried out with an optical fiber with a step-like refractive index profile. The diameter of the optical fiber core was $2\rho = 9\,\mu\text{m}$. The difference between the refractive indices of the fiber core and cladding was $\triangle n = n_{\text{co}} - n_{\text{cl}} = 0.004$.

In order to obtain circularly polarized light of high quality and to select circularly polarized light of the definite sign with high accuracy special devices were used. The main elements of our "circular polarizer" and "circular

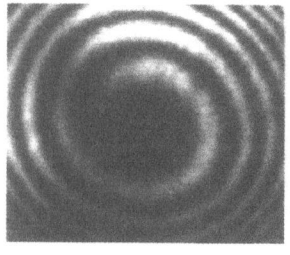 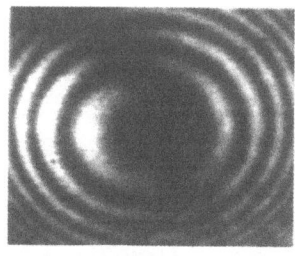

a   b

Fig. 10. Photographs of the interference pattern on the screen. (**a**) Right-handed circularly polarized light incident at fiber input and left-handed light extracted from fiber output, and (**b**) left-handed circularly polarized light incident at fiber input and right-handed light extracted from fiber output

analyzer" were the adjustable composed $\lambda/4$-plates [25]. The interferometer of Mach-Zehnder was used to register the light waves with a single dislocation.

We observed the interference pattern in the form of a spiral line (Fig. 10). This spiral line unwinded either (a) clockwise or (b) counterclockwise. The direction of unwinding agrees with the theoretical prediction.

References

1. Bortolotti F. (1926) Rend R Acc Naz Linc 4: 552
2. Rytov S.M. (1938) Dokl Akad Nayk 18: 2
3. Vladimirsky V.V. (1941) Dokl Akad Nayk 21: 222
4. Berry M.V. (1984) Proc Roy Soc A 392: 45
5. Chiao R.Y., Wu Y.-S. (1986) Phys Rev Lett 57: 933
6. Tomita A., Chiao R.Y. (1986) Phys Rev Lett 57: 936
7. Chiao R.Y., Antaramian A., Ganga K.M., Jiao H., Wilkinson S.R., Nathel H. (1988) Phys Rev Lett 60: 1214
8. Picht J. (1929) Ann Physik 3: 433
9. Picht J. (1929) Physik Z 30: 905
10. Goos F., Hanchen H. (1947) Ann Physik 1: 333
11. Goos F., Hanchen H. (1949) Ann Physik 5: 251
12. Acloque P., Guillemet C. (1960) Compt Rebd 250: 4328
13. Osterberg H., Smith L.W. (1964) J Opt Soc Am 54: 1078
14. Risset C.A., Vigoureux J.M. (1992) Optics Comm 91: 157
15. Fedorov F.I. (1955) Dokl Akad Nayk 105: 465
16. Kristofel N. (1956) Proc Tartu Univ 42: 94
17. Imbert C. (1970) Phys Lett 31A: 337
18. Imbert C. (1972) Phys Rev D 5: 787
19. Costa de Beauregard O., Imbert C. (1972) Phys Rev Lett 28: 1211
20. Zel'dovich B.Ya., Liberman V.S. (1990) Sov J Quantum Electron 20: 427
21. Zel'dovich B.Ya., Dooghin A.V., Kundikova N.D., Liberman V.S. (1991) JETP Lett 53: 186
22. Snyder A.W., Love J.D. (1983) Optical Waveguide Theory. Chapman and Hall, London New York
23. Dooghin A.V., Kundikova N.D., Liberman V.S., Zel'dovich B.Ya. (1992) Phys Rev A 45: 8204
24. Liberman V.S., Zel'dovich B.Ya. (1993) Pure Appl Optics 2: 367
25. Goltser I.V., Darscht M.Ya., Kundikova N.D., Zeldovich B.Ya. (1993) Optics Comm 97: 291
26. Liberman V.S., Zel'dovich B.Ya. (1992) Phys Rev A 46: 5199
27. Zel'dovich B.Ya., Kundikova N.D. (1995) Sov J Quantum Electron 22: 184
28. Esayan A.A., Zel'dovich B.Ya. (1988) Sov J Quantum Electron 18: 149
29. Baranova N.B., Zel'dovich B.Ya. (1994) JETP Lett 59: 681
30. Darsht M.Ya., Zhirgalova I.V., Zel'dovich B.Ya., Kundikova N.D. (1994) JETP Lett 59: 763
31. Darsht M.Ya., Zel'dovich B.Ya., Kataevskaya I.V., Kundikova N.D. (1995) JETP Lett 107: 817
32. Nye J.F. (1981) Proc R Soc Lond A 378: 219
33. Baranova N.B., Zel'dovich B.Ya. (1981) JETP Lett 80: 1780
34. Baranova N.B., Zel'dovich B.Ya., Mamaev A.V., Pilipezkii N.F., Shkunov V.V. (1982) JETP Lett 83: 1702

Atoms and Cavities: The Birth
of a Schrödinger Cat of the Radiation Field

J.-M. Raimond and S. Haroche

Summary. Rydberg atoms in superconducting cavities make it possible to test fundamental features of quantum mechanics. The non-resonant interaction of a single atom with a mesoscopic coherent field results in a quantum superposition of two fields with different classical phases. This mesoscopic quantum superposition is similar to the famous "Schrödinger cat", suspended between life and death. By probing the field state with a second atom, we monitor the relaxation of the field quantum superposition towards a statistical mixture. The "decoherence" time decreases rapidly with the "distance" between the two field components, being instantaneous for any macroscopic object. This experiment gives first direct evidence of the process which, at the heart of the quantum measurement, bans quantum superpositions from the macroscopic world.

1 Introduction

One of the most striking features of quantum mechanics is the existence of state superpositions. Any linear combination of two quantum states is yet another possible quantum state. The superposition principle is at the heart of any quantum interference experiment. Many experimental techniques allow us to prepare and to detect superposition states for atoms, photons, spins, etc. It is of course more difficult to have an intuitive understanding of these states. A superposition state describes, for instance, a particle located *at the same time* at two different locations, a spin pointing in one direction *and* in another, and an atom in two different energy levels. All these situations are highly counterintuitive. The quantum superpositions are radically different from mere statistical mixtures, describing, for instance, a spin pointing in one direction *or* in another. In the latter case, only a lack of information prevents us from giving the spin direction. It has nevertheless a well-defined one. In the former case, on the contrary, the spin simultaneously exhibits properties pertaining to the two states. It is possible to design quantum interference experiments giving different results for a statistical mixture and a quantum superposition.

Quantum superpositions are even more shocking when one tries to extrapolate them to the macroscopic scale. In principle, any macroscopic object could be in a superposition of different states. Schrödinger vividly illustrated this difficulty in a famous paper [1]. He imagined a cat, trapped in a box with a radioactive atom. A contraption kills the cat when the atom decays. After some time, the atom is in a quantum superposition of the initial and "decayed" states. The cat should thus accordingly be in a quantum superposition of its "alive" and "dead" states. Such quantum superpositions have never been observed. We expect the cat instead to be in a statistical mixture of the two states, *either* dead *or* alive, but not dead *and* alive.

The conspicuous lack of quantum superpositions at the macroscopic level explains why they look so odd to us. More importantly, it has also important consequences for our understanding of the quantum measurement process [2]. A measurement "copies" the state of a quantum system onto the one of a macroscopic degree of freedom of the measuring apparatus (the position of a needle, for instance). The unitary evolution of the "system+meter" state should result in an entangled state, involving a quantum superposition of different macroscopic states of the meter, when the system is in a quantum superposition of the eigenstates of the measurement. The "Schrödinger cat" illustrates this situation, the cat being a macroscopic indicator of the atomic state. These quantum superpositions are never reached. According to the von Neumann quantum measurement description [3], the meter state should be considered (before the needle's position is read out) as a statistical mixture of the possible outcomes.

The role of the environment was soon recognized in this problem. The meter is a complex system, strongly coupled to a vast environment. Detailed models of simple meters show that the interaction of the meter with the environment's thermodynamic bath tends to cancel very rapidly the quantum coherences between different states [4,5]. The time scale of this "decoherence" process becomes shorter and shorter when a parameter characterizing the "distance" between the states in the superposition increases. For microscopic systems, as expected, the lifetime of the superpositions is the damping time of the system itself. Conversely, for macroscopic states, the lifetime of the quantum superposition is much shorter than the characteristic damping time, so short in fact that no observation of these coherences will ever be possible. The basis on which the decoherence occurs ("preferred basis") is determined by the dynamics of the meter relaxation. It is the damping dynamics that determines the "authorized" quantum states of the meter, and, finally, the observable actually measured. In spite of the lack of a completely general decoherence theory, the universal features found in the tractable models tend to indicate that the process is quite general.

An experimental investigation of the decoherence requires a mesoscopic pointer, large enough for the decoherence time to be shorter than the "ordinary" relaxation time, small enough for it to be measurable. It also obviously requires very well-isolated and well-controlled systems and meters. Recent developments in quantum optics make it possible now to work with individual quantum systems in a controlled environment. Trapped ions, for instance, made it possible recently to study entanglement or to realize a mesoscopic variant of the "Schrödinger cat", with an ion in a quantum superposition of two quite different vibration states [6].

Another interesting playground for these fundamental quantum mechanical tests is cavity quantum electrodynamics with circular Rydberg atoms and superconducting cavities. With such a system, we have been able to investigate recently, for the first time, the dynamics of the decoherence for a mesoscopic "meter" [7]. The "needle" of this meter is the amplitude of a quantum field stored in the cavity, whose size can be varied from the microscopic to the mesoscopic domains. This experiment is described in detail in this chapter. We recall first some features of the Rydberg atoms and describe the experimental set-up. We then show how a quantum meter of the atomic state can be implemented with the non-resonant atom–field interaction. We describe next the experiment on decoherence and give finally a simple model explaining the decoherence features in simple terms related to the important concept of complementarity.

2   Experimental Techniques

Circular Rydberg atoms [9] are almost perfect tools for fundamental experiments on matter–field coupling. These are very excited levels (of rubidium in our experiments) with a large principal quantum number (of the order of 50). At the same time, the orbital and magnetic quantum numbers take the maximum possible values. In a classical picture, this means that the orbit of the electron around the ionic core is the Bohr circle. This has many consequences. First, the dipole matrix element on a millimeter-wave transition between neighboring circular Rydberg states is extremely large. The transition between the circular states with principal quantum numbers 51 and 50 (e and g in the following), at 51.099 GHz, has a dipole matrix element of 1250 atomic units. To sum it up, these atoms are large antennae, very strongly coupled to the millimeter-wave radiation, a frequency domain where very high-quality factor superconducting cavities can be used. In spite of this very strong coupling to radiation, the lifetime of the circular states is quite long, of the order of 30 ms for e or g. This is due to the circular nature of the orbit, which minimizes the average acceleration of the electron and, hence, the radiative losses.

In a small directing electric field lifting the degeneracy of the hydrogenic Stark manifold, the $e \to g$ transition appears as an excellent approximation of a two-level system, since fine and hyperfine structures are completely negligible. This small electric field (about 0.3 V/cm) is also required for the stability of these extremely anisotropic levels. In the absence of a physical quantization axis, they are rapidly admixed with other high angular momentum levels in stray electric or magnetic fields.

Finally, these states can be detected in a sensitive and selective way by the field ionization method. A moderate (150 V/cm) electric field is enough to extract the outer electron, which can be easily accelerated and detected. Since the ionizing field depends rapidly on the principal quantum number, it is possible to design separate detectors, clicking either for an atom in e or for an atom in g. The overall quantum efficiency of these detectors is of the order of 40%.

Obviously, the preparation of these levels, with 50 or 49 units of angular momentum, from the 5S ground state, is a quite complex process. We use a variant [8] of the method originally designed by Hulet and Kleppner [9]. It involves, first, a stepwise optical excitation of the 52F "ordinary" Rydberg level, followed by a series of radiofrequency adiabatic rapid passages on the transitions between Stark levels, towards the circular state of the 52 manifold. A final microwave transition to the 51 or 50 circular states in a high electric field lifting all the degeneracies is used to get a pure circular state population in e or g. The "circularization" process is time-resolved and the atom preparation time is known within a 2 μs interval. The velocity of the atoms being selected with a 1 m/s accuracy by a combination of velocity-selective optical pumping and time-of-flight techniques, the position of the atom is known at any time during its flight through the apparatus with a ±1 mm accuracy. This precise control of the timing is essential in order to apply selected transformations to the atom and to control the atom/cavity interaction time. In most experiments, a single atom should be used. The preparation scheme is thus set to produce much less than one atom on the average (0.1 to 0.2), with Poisson statistics. When an atom is detected, the probability that a second was present at the same time is therefore negligible. Of course, this low mean atom number is obtained at the expense of an increase in the acquisition time.

The general scheme of the experiment [7] is depicted in Fig. 1. The atoms effuse out of the oven O. They are velocity-selected in zone V. In box B, a circular atom is prepared. Since it is extremely sensitive to blackbody millimeter-wave radiation, the whole circular atom path is in a cryogenic environment (^{3}He cryostat at 0.6 K). On their path, the atoms cross the superconducting cavity C. It is a Fabry–Perot cavity, compatible with the small static electric field stabilizing the circular atoms. It sustains two orthogonally polarized, nearly degenerate TEM$_{900}$ Gaussian modes, with a 6 mm waist. With careful preparation of the mirrors, the cavity quality factor Q may reach rather high values. In this experiment, the cavity field energy lifetime, T_r, was 160 μs, much longer than the typical atom–cavity interaction time for an atom at thermal velocity. The cavity C may be filled with a small coherent field by the classical source S. Harmonic generation of millimeter waves from an X-band source allows us to control precisely the injected field intensity at the single photon level. After the interaction with the cavity, the atoms are detected, according to their state, by the two field-ionization counters D$_e$ and D$_g$.

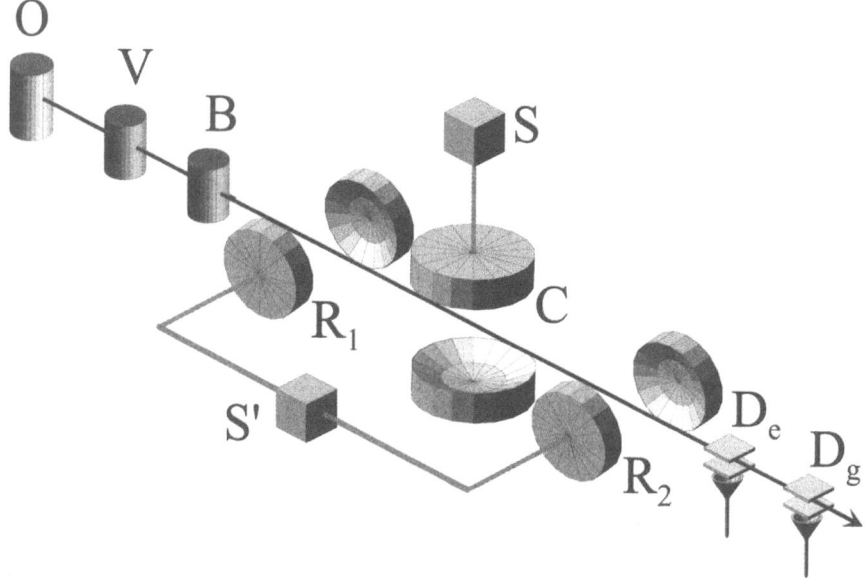

Fig. 1. Scheme of the experimental set-up

Before the atoms enter the cavity, the levels e and g can be mixed by a pulse of classical microwave radiation resonant on the $e \to g$ transition, applied in a low-Q cavity R_1. With proper adjustment of the amplitude and phase of this microwave field (source S'), any state of the two-level system can be prepared. Levels e and g can be mixed again after the interaction with C in a second classical field zone R_2. This transformation can be used to analyze an atomic coherence. Note also that the combination of zones R_1 and R_2, fed by the same source, is a Ramsey separated oscillatory fields interferometer [10].

3 Non-resonant Interaction: A Quantum Meter

The non-resonant atom–field interaction may be used to turn the cavity field into a quantum meter of the atomic state. When the atom and the cavity are not at resonance, they do not exchange energy. However, the position of their energy levels is modified by the coupling. For the atom, this modification results from a combination of the light shifts (at the single photon level) with the Lamb shift, present even if the cavity is empty [11]. Reciprocally, the cavity mode frequency is also modified by the atom. This is a simple index of refraction effect. The atom is a piece of transparent dielectric material modifying the cavity effective length. Due to the very strong coupling of Rydberg atoms to the field, this index of refraction effect is surprisingly large. A single atom in the cm^3 mode volume produces an index around

1 ± 10^{-7}. This is a quite large effect compared to the cavity width (10^{-8} in relative value), about 10^{16} times more than for an "ordinary" atom in the same volume.

When the atom flies across the cavity, the transient frequency shift experienced by the field results in a classical phase shift. The phase shift for an atom in level e is $\Phi = \Omega^2 t_i/\delta$, where Ω, half the single-photon Rabi frequency for an atom at the cavity center, measures the atom–field coupling. δ is the atom–cavity detuning and t_i the effective interaction time taking into account the field mode geometry. An atom in level g has the opposite effect and changes the field phase by $-\Phi$. In this experiment, $\Omega/2\pi = 25\,\mathrm{kHz}$ (measured directly by the quantum Rabi oscillations of an atom in the empty cavity [12]). With an effective interaction time $t_i = 19\,\mu s$ for a $300\,\mathrm{m/s}$ atom, Φ reaches up to 0.7 rd for $\delta/2\pi = 100\,\mathrm{kHz}$. For such detuning values, quite larger than the atom–field coupling or the cavity width, the atom–cavity energy exchange is negligible, as checked experimentally.

A small coherent field in the cavity is determined by its classical amplitude α, represented by an arrow in the Fresnel plane [13]. Quantum fluctuations add to this classical amplitude. This may be pictorially represented by an "uncertainty disk" on top of the arrow in the Fresnel plane (Fig. 2a). In proper units, the radius of the uncertainty disk is one and the length of the arrow squared is the average photon number $n = |\alpha|^2$. After the interaction with an atom in level e, the arrow is rotated by Φ, whereas it is rotated by $-\Phi$ for an atom in g. The field amplitude is thus analogous to a needle, pointing in the phase space towards a direction correlated to the atomic state. Provided the indication of the needle is unambiguous (i.e. provided Φ is greater than the quantum uncertainty on the field phase), the cavity field is a quantum meter of the atomic state.

If the atom is sent in a quantum superposition of e and g (prepared in zone R_1), the final state is an entangled atom/cavity one involving a superposition of two coherent amplitudes with different classical phases [14].

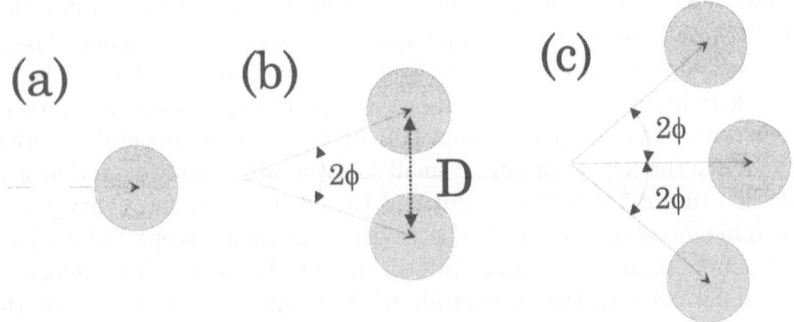

Fig. 2. Phase components in the cavity. (**a**) Initial coherent field. (**b**) Field components after interaction with the first atom. (**c**) Final phase components

This is a mesoscopic quantum superposition state, quite analogous to the Schrödinger cat. This superposition is bound to experience decoherence and to turn rapidly into a statistical mixture (one phase or the other, instead of a superposition of the two phases). Since the relaxation of the cavity field is well-known, the decoherence dynamics can be calculated explicitly [15]. In the limit of large fields, its time scale is simply $2T_r/D^2$, where D is the distance between the two components in the Fresnel plane (D^2 is a photon number). For a microscopic single-photon field, the superposition is expected to live as long as the field, whereas it is much shorter lived than the photon number for a field involving a few photons.

4   An Experiment on Complementarity

We first make sure that the cavity field, after the interaction, contains information about the atomic state. A single atom is sent in the apparatus in level e. It is prepared, in R_1, in a superposition of e and g with equal weights. At the same time, a small coherent field is injected in C. When crossing C, the atom gets entangled with the field phase. A direct detection of the atomic state at the exit of C would not lead to any interesting situation. The detected atomic state being also the one in the cavity, the phase shift would be a classical quantity, either Φ or $-\Phi$. In order to get an interesting quantum situation, a superposition of phase shifts, the atomic detection should not provide any information on the atomic state inside the cavity. This is easily realized by again mixing e and g with equal weights in zone R_2. In this case, R_1 and R_2 are used as a Ramsey interferometer around the cavity.

When C is empty, an atom detected in g may have followed two quantum paths. The transition from e to g occurs either in R_1 or in R_2. Since the classical field in these zones does not keep track of the single photon emitted by the atom, these two paths, corresponding to the same final state, are indistinguishable. The final probability thus reflects a quantum interference between them and oscillates with the relative phase of the two interfering amplitudes. When the frequency ν of source S′ is scanned, the probability $P_g^{(1)}(\nu)$ to detect the atom in g oscillates between zero and one. These oscillations, depicted in Fig. 3a, are the well-known Ramsey fringes. Their period (3.2 kHz here) is the reciprocal of the time of flight between the two zones. The contrast is reduced to about 60% by various experimental imperfections.

When the cavity contains a small coherent field, the two interfering atomic paths cause different phase shifts of the cavity field, since they correspond to different atomic levels in the cavity. The cavity keeps "which path" information about the atomic state in the interferometer. This information is incompatible with the observation of the fringes. One thus expects that the fringe contrast is reduced to zero when the field phase shift, Φ, is greater than the phase quantum fluctuations. In intermediate situations, the fringes should be observed, albeit with a reduced contrast.

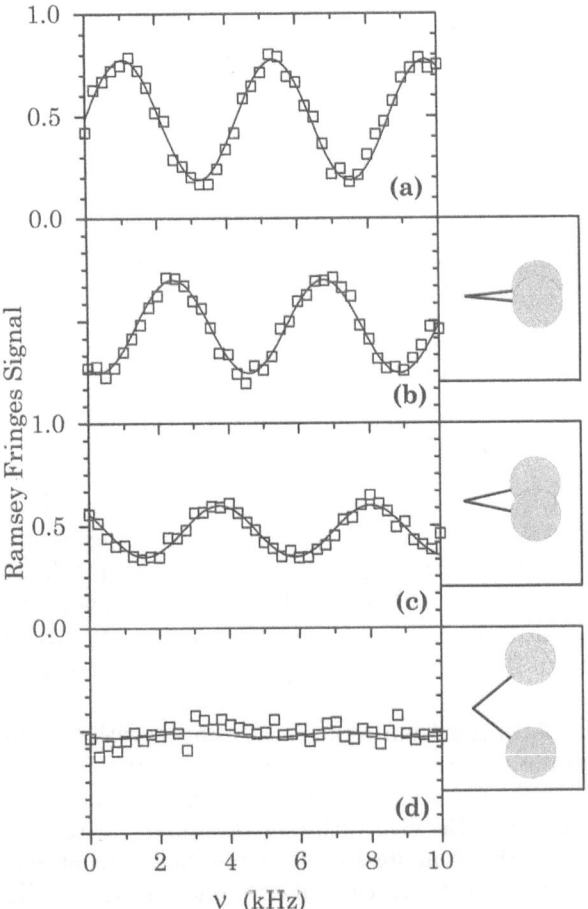

Fig. 3. $P_g^{(1)}(\nu)$ signal exhibiting Ramsey fringes: **(a)** C empty, $\delta/2\pi = 712$ kHz; **(b)** to **(d)** C stores a coherent field with $|\alpha| = \sqrt{9.5} = 3.1$, $\delta/2\pi = 712$, 347 and 104 kHz respectively. Points are experimental and curves are sinusoidal fits. Inserts show the phase space representation of the field components left in C

The phase shift Φ can be varied by adjusting the atom/cavity detuning δ. Figures 3b–d present the observed fringes for decreasing detunings and increasing phase component separations (the field components after the interaction are pictorially represented in the inserts). As expected, the fringes' contrast reduces when the amount of "which path" information increases. For a large enough separation, the fringes are no longer observed. The contrast of the fringes is in excellent agreement with theoretical predictions. When the fringes are still visible, they are phase-shifted with respect to the empty cavity case. This is due to the light shifts experienced by the atomic levels in the cavity field and provides a very precise calibration of the average photon number (9.5 in this case).

This experiment illustrates in a very simple way the complementarity between the quantum interferences and the "which path" information [16]. It also shows that the cavity field performs a quantum measurement of the atomic state. It does not show, however, that the resulting cavity state is a quantum superposition of the two states associated to the two phases. The same effects would be observed for a classical needle or for a field in a statistical mixture of the components. The nature of the cavity state can only be revealed in a quantum interference experiment involving the field itself.

5   Dynamics of Decoherence

In order to probe the cavity field state, a second atom is sent in the apparatus after a tunable delay τ. It undergoes the same transformations as the first atom in R_1 and R_2. It thus enters C in a superposition of two indices of refraction. Each of the phase components left by the first atom again undergoes a quantum superposition of two phase shifts. The final field includes therefore four coherent components (see Fig. 2c). Two of these components, with phases $\pm 2\Phi$, correspond to the cases where the two atoms have crossed the cavity in the same states ((e, e) or (g, g)). Unless $\Phi \simeq \pi/2$, these phase components do not overlap and play no role in the following.

The other two components of this "four-legged" cat coincide with the initial phase. They correspond to the cases where the atoms cross the cavity in different states ((e, g) or (g, e)). The second atom exactly undoes the phase shift produced by the first, leaving the field with its initial phase. These two quantum paths correspond to the same final cavity state. Since the atomic states determining these paths are not directly related to the detection states (the levels are mixed again in R_2), the two quantum paths are indistinguishable. We have here again an interplay between quantum interferences and "which path" information. The final detection probabilities of the two atoms reflect the quantum interferences associated to these paths. Obviously, if the cavity state involves only a statistical mixture (either one phase or the other), all the phase shifts are classical effects and the interferences should disappear.

A correlation signal η directly revealing this quantum interference is deduced from the four possible detection probabilities: $\eta = \left[P_{ee}^{(2)}/(P_{ee}^{(2)} + P_{eg}^{(2)}) \right] - \left[P_{ge}^{(2)}/(P_{ge}^{(2)} + P_{gg}^{(2)}) \right]$. For a pure quantum superposition, η is expected to be 0.5, independently of the Ramsey frequency ν, whereas it is zero for a statistical mixture. Figure 4 presents η as a function of the delay τ for a 3.3 average photon number, and two phase configurations (depicted in the inserts). The solid lines depict the results of a simple calculation presented in the next section.

At short times, one observes a non-zero η value. The first atom actually leaves in the cavity a quantum superposition of two coherent fields with different classical phases. This is an utterly non-classical field state, very similar to the famous "Schrödinger cat". The maximum value of η is 0.18 instead of 0.5. This is due to the limited contrast of our Ramsey interferometer (60%).

At longer times, η evolves rapidly towards zero (note that, being a conditional probability difference, η may become negative). This evolution is much faster than the cavity energy lifetime (the time unit in Fig. 4). Moreover, the decay time is shorter when the phase difference is increased, while keeping the photon number constant. This shows that the evolution to the statistical mixture is a non-trivial relaxation effect, whose characteristic time depends upon the initial cavity state. This is one of the main features of decoherence, clearly evidenced in this experiment. The same kind of behavior can be observed with a constant phase difference but different intensities. The larger the photon number, the faster the decoherence.

This experiment thus illustrates the basic features of an ideal quantum measurement with a mesoscopic meter. Immediately after the system–meter interaction, the meter state involves a quantum superposition of the possible outcomes. Very rapidly, then, this quantum superposition turns into a mere statistical mixture. This evolution is faster and faster when the separation between the meter's states is increased. For any macroscopic difference, as in any actual meter, the extrapolated decoherence is instantaneous, validating the standard description of quantum measurement.

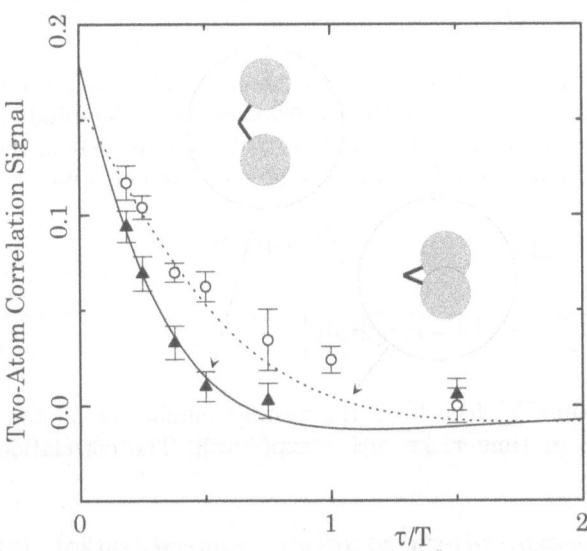

Fig. 4. Two-atom correlation signal η versus τ/T_r for $\delta/2\pi = 170$ kHz (circles) and $\delta/2\pi = 70$ kHz (triangles). Dashed and solid lines are theoretical. Inserts: representations of the field components left by the first atom

6   A Simple Model

The decoherence signal presented above can be calculated easily in the standard relaxation theory framework. Characteristic functions allow one to compute the field density matrix at any time and to deduce the second atom's behavior. These calculations are presented in [15,17] and are in excellent agreement with the experimental data. Much more physical insight into the decoherence mechanism can be gained by a simpler description emphasizing the links between decoherence and complementarity [18].

When there is no relaxation, it is quite easy to show that η is half the real part of the scalar products of the two overlapping field components of Fig. 2c: $\eta = (1/2)\Re\langle\alpha|\alpha\rangle = 1/2$ (this result holds when the other phase components are well-separated and when the field is large enough – the model presented here can be straightforwardly extended to remove these conditions). A standard description of the field damping in C uses a linear coupling of C to a large number of cavities (or field modes) $C_1, C_2, \ldots, C_i, \ldots$, with arbitrary frequencies and arbitrary coupling constants. This situation models the standard situation of an harmonic oscillator relaxing in a large bath of zero temperature oscillators. An initially coherent field $|\alpha\rangle$ in C remains coherent, with an amplitude $\alpha(t) = \alpha\exp(-\gamma t/2)$ $(\gamma = 1/T_r)$, whereas the "reservoir" cavities contain small coherent fields whose amplitudes $\beta_i(t)$ are linear functions of $\alpha(t)$.

When C initially contains a phase cat involving the coherent components $|\alpha\exp(i\phi)\rangle$ and $|\alpha\exp(-i\phi)\rangle$, the field in C_i evolves into a superposition of two phase components $|\beta_i(t)\exp(i\phi)\rangle$ and $|\beta_i(t)\exp(-i\phi)\rangle$, due to the linearity of the coupling. Each of the "reservoir" cavities contains a tiny "Schrödinger cat". It is quite easy to show that the correlation signal obtained for a second atom crossing C at time τ involves the scalar product of the overlapping phase components in C and of the field states left in the "reservoir" cavities, which are not affected by the second atom crossing C:

$$\eta(\tau) = \frac{1}{2}\Re\langle\alpha(\tau)|\alpha(\tau)\rangle \prod_i \langle\beta_i(\tau)\exp(i\phi)|\beta_i(\tau)\exp(-i\phi)\rangle$$

$$= \frac{1}{2}\Re\exp\left(-\sum_i |\beta_i(\tau)|^2[1 - \exp(2i\phi)]\right) . \tag{1}$$

Due to energy conservation, $\sum_i |\beta_i(\tau)|^2$ is the average number of photons having escaped from C_0 at time τ, i.e. $n[1 - \exp(-\gamma\tau)]$. The correlation signal is thus finally

$$\eta(\tau) = \frac{1}{2}\exp\{-2n[1 - \exp(-\gamma t)]\sin^2\phi\}\cos\{n[1 - \exp(-\gamma t)]\sin 2\phi\} . \tag{2}$$

Note that this simple result coincides, in its validity limits, with the result of the complete calculation [17], adapted from [15]. This expression is used for the theoretical fits in Fig. 4 with a normalization to a maximum value of 0.18 to account for experimental imperfections (see [7]). The agreement with the experimental data is quite good. Note that the decay is far from being exponential, the signal becoming negative for large time, due to the cosine term. However, for large photon numbers and short enough times, the correlation signal decays almost exponentially, with a time constant $2T_r/D^2$, with $D = 2n\sin\Phi$ being the distance in the Fresnel plane between the two phase components.

This model interprets decoherence in terms of complementarity. The many cavities coupled to C acquire, during the relaxation, tiny versions of the cat. They thus get partial information about the field phase in C. The environment is in fact continuously performing an unread measurement of the field in C. The amount of information which has leaked in the environment is measured by the product of scalar products in equation (1). As soon as the two states of the environment linked to the two phase components in C are orthogonal, the environment has "determined" the phase in C. Since this information cannot be retrieved, the state in C becomes a mere statistical mixture of the two phases. The decoherence time scale can be understood easily, when the phase difference between the components is large enough. The environment states will become orthogonal when the quantum fluctuations of the phases of the "leak field components" become smaller than the phase difference Φ. When Φ is not too small, this occurs as soon as about one photon has escaped into the environment (a one-photon field has a less than π phase uncertainty). The escape time of the first photon out of n is clearly T_r/n. We thus find easily, with this simple argument, the scaling law of the decoherence time.

7   Conclusion and Perspectives

Rydberg atoms and superconducting cavities allow us to implement the simplest model of the matter–field interaction: a single atom coherently coupled to a single field mode. The experiment is close enough to its ideal theoretical description to give a direct insight of the most intimate features of quantum mechanics. In particular, this system allows the realization of an ideal mesoscopic quantum measurement device. The study of the rapid decoherence of the quantum superposition of the meter's states, resulting from the measurement of a microscopic system in a quantum superposition state, sheds some light on the appearance of the classical behavior from the quantum world. This experiment clearly exhibits the extreme sensitivity of macroscopic quantum superpositions to dissipation. The unavoidable coupling of these superpositions to a thermal reservoir results in an irreversible leakage of information from the mesoscopic system into the outside world. Since this information is inaccessible, the state left for the mesoscopic system is a mere statistical mixture of states, determined by the dynamics of the relaxation process.

Only simple models of decoherence can be completely analyzed, one of which is experimentally implemented here. Similar processes could explain in all cases a very important step in quantum measurement: the reduction to a statistical mixture of the meter outcomes. Decoherence theory gives, however, no clue to the basic probabilistic nature of quantum mechanics. It does not tell us why God is playing dice; it only explains why He is ultimately playing with classical dice.

The Rydberg atoms and cavity experiments are also well suited to test other features of the quantum world. Resonant atom–field interaction may, for instance, be used to entangle one atom and the cavity mode energy, or two separate atoms. Einstein–Podolsky–Rosen pairs [19] of entangled atoms have been realized in this way [20]. They open the way to new tests of the quantum non-locality with massive particles. These experiments could also be generalized, with an improved experimental set-up, to multiparticle entangled states, like the GHZ triplet [21,22]. These states would display quantum weirdness in a more vivid way than the "ordinary" EPR pairs.

An experiment involving two superconducting cavities would allow the realization of even more striking quantum states. A single atom crossing the two cavity fields could entangle them and prepare a non-local mesoscopic quantum state, a non-local "Schrödinger cat", or a mesoscopic EPR pair [23]. The study of the decoherence processes for these states would be extremely interesting, opening the way to completely new tests of our understanding of quantum mechanics.

References

1. Schrödinger, E. (1935) Naturwissenschaften, **23**, 807, 823, 844; reprinted in english in [2]
2. Wheeler, J.A., Zurek, W.H. (1983) Quantum Theory of Measurement, Princeton University Press, Princeton, NJ
3. von Neumann, J. (1932) Mathematische Grundlagen der Quantenmechanik, Springer, Berlin; reprinted in English in [1]
4. Zurek, W.H. (1991) Phys. Today, **44**, 36
5. Zurek, W.H. (1981) Phys. Rev. D, **24**, 1516 and **26**, 1862 (1982); Caldeira, A.O., Leggett, A.J. (1983) Physica A, **121**, 587; Joos, E., Zeh, H.D. (1985) Z. Phys. B, **59**, 223; Omnès, R. (1994) The Interpretation of Quantum Mechanics, Princeton University Press, Princeton, NJ
6. Monroe, C. et al. (1996) Science, **272**, 1131
7. Brune, M. et al. (1996) Phys. Rev. Lett., **77**, 4887
8. Nussenzveig, P. et al. (1993) Phys. Rev. A, **48**, 3991
9. Hulet, R.G., Kleppner, D. (1983) Phys. Rev. Lett., **51**, 1430
10. Ramsey, N.F. (1985) Molecular Beams, Oxford University Press, New York
11. Brune, M. et al. (1994) Phys. Rev. Lett., **72**, 3339
12. Brune, M. et al. (1996) Phys. Rev. Lett., **76**, 1800
13. Glauber, R.J. (1963) Phys. Rev., **131**, 2766
14. Buzek, V., Knight, P.L. (1995) Progress in Optics, vol. XXXIV, E. Wolf ed., North-Holland, Amsterdam, p. 1 and refs. therein

15. Davidovich, L. et al. (1996) Phys. Rev. A, **53**, 1295
16. Scully, M.O. et al. (1991) Nature, **351**, 111; Haroche, S. et al. (1992) Appl. Phys. B, **54**, 355; Pfau, T. et al. (1994) Phys. Rev. Lett., **73**, 1223; Chapman, M.S. et al. (1995) Phys. Rev. Lett., **75**, 3783
17. Maître, X. et al. (1997) J. Modern Optics, **44**, 2023
18. Raimond, J.M. et al. (1997) Phys. Rev. Lett., **79**, 1964
19. Einstein, A., Podolsky, B., Rosen, N. (1935) Phys. Rev., **47**, 777
20. Hagley, E. et al. (1997) Phys. Rev. Lett., **79**, 1
21. Greenberger, D.M., Horne, M.A., Zeilinger, A. (1990) Am. J. Phys., **58**, 1131
22. Haroche, S. (1995) *Fundamental problems in quantum theory*, D. Greenberger, Zeilinger, A. (Eds.), Ann. NY Acad. Sci., **755**, 73
23. Davidovich, L. et al. (1993) Phys. Rev. Lett., **71**, 2360

Quantum Tomography of Wigner Functions from Incomplete Data

V. Bužek, G. Drobný, and H. Wiedemann

Summary. In this chapter we will study how the states of light can be reconstructed from the knowledge of a restricted set of mean values of system observables. We will show how the maximum entropy (MaxEnt) principle can be applied for a reconstruction of quantum states of light fields. The chapter is organized as follows. In Sect. 2 we briefly describe the main ideas of the MaxEnt principle. In Sect. 3 we set a scene for a description of the reconstruction of quantum states of light fields. In this section we briefly discuss the phase-space formalism which can be used for a description of quantum states of light. In Sect. 4 we introduce various observation levels suitable for a description of light fields and we present the reconstruction of Wigner functions of these fields prepared in nonclassical states. Finally, in Sect. 5 we present the results of the numerical reconstruction of quantum states of light from incomplete tomographic data. We compare two reconstruction schemes: reconstruction via the MaxEnt principle and reconstruction via direct sampling (i.e. the tomography reconstruction via pattern functions – see below in Sect. 3).

1 Introduction

The concept of a quantum state represents one of the most fundamental pillars of the paradigm of quantum theory [1,2]. Contrary to its mathematical elegance and convenience in calculations, the physical interpretation of a quantum state is not so transparent. The problem is that the quantum state (described either by a state vector, or by a density operator of a phase-space probability density distribution) does not have a well-defined objective status, i.e. a state vector is not an *objective* property of a particle. According to Peres (see [1], p. 374): "There is no physical evidence whatsoever that every physical system has at every instant a well defined state ... In strict interpretation of quantum theory these mathematical symbols [i.e. state vectors] represent *statistical information* enabling us to compute the probabilities of occurrence of specific events." Once this point of view is adopted then it becomes clear that any "measurement" or a reconstruction of a density operator (or its mathematical equivalent) can be understood exclusively as an expression of our knowledge about the quantum-mechanical state based on a certain set of measured data. To be more specific, any quantum-mechanical reconstruction scheme is nothing more than an a posteriori estimation of the density operator of a quantum-mechanical (microscopic) system based on data obtained with the help of macroscopic measurement apparatus [2]. The quality of the reconstruction depends on the "quality" of the measured data and the efficiency of the reconstruction procedure, with the help of which the data analysis is performed. In particular, we can specify three different situations. Firstly, when all system observables are precisely measured; in this case the complete reconstruction of an initially unknown state can be performed (we will call this the reconstruction on the complete observation level). Secondly, when just part of the system observables is precisely measured then one cannot perform a complete reconstruction of the measured state. Nevertheless, the reconstructed density operator still uniquely determines mean values of the measured observables (we will denote this scheme as the reconstruction on incomplete observation levels). Finally, when measurement does not provide us with sufficient information to specify the exact mean values (or probability distributions) but only the frequencies of appearances of eigenstates of the measured observables, then one can perform an estimation (e.g. reconstruction based on quantum Bayesian inference) which is the "best" with respect to the given measured data and the a priori knowledge about the state of the measured system.

2 MaxEnt Principle and Observation Levels

The state of a quantum system can always be described by a statistical density operator $\hat{\rho}$. Depending on the system preparation, the density operator represents either a pure quantum state (complete system preparation) or a statistical mixture of pure states (incomplete preparation). The degree of deviation of a statistical mixture from the pure state can be best described by the *uncertainty measure* $\eta[\hat{\rho}]$ (see [3])

$$\eta[\hat{\rho}] = -\text{Tr}\left(\hat{\rho}\ln\hat{\rho}\right).\tag{1}$$

The uncertainty measure $\eta[\hat{\rho}]$ possesses the following properties:

1. In the eigenrepresentation of the density operator $\hat{\rho}$, i.e. $\hat{\rho}|r_m\rangle = r_m|r_m\rangle$, we can write

$$\eta[\hat{\rho}] = -\sum_m r_m \ln r_m \geq 0,\tag{2}$$

   where r_m are eigenvalues and $|r_m\rangle$ the eigenstates of $\hat{\rho}$.
2. For uncertainty measure $\eta[\hat{\rho}]$ the following inequality holds:

$$0 \leq \eta[\hat{\rho}] \leq \ln N,\tag{3}$$

where N denotes the dimension of the state space of the system.

When instead of the density operator $\hat{\rho}$, expectation values G_ν of a set $\mathcal{O}$ of operators $\hat{G}_\nu$ ($\nu = 1, \ldots, n$) are given (measured), then the uncertainty measure can be determined as well. The set of linearly independent operators is referred to as the *observation level* $\mathcal{O}$ [4,5]. The operators $\hat{G}_\nu$ which belong to a given observation level do not commute necessarily. A large number of density operators which fulfill the conditions

$$\mathrm{Tr}\,\hat{\rho}_{\{\hat{G}\}} = 1; \qquad \mathrm{Tr}\,(\hat{\rho}_{\{\hat{G}\}}\hat{G}_\nu) = G_\nu, \quad \nu = 1, 2, \ldots, n; \tag{4}$$

can be found for a given set of expectation values $G_\nu = \langle \hat{G}_\nu \rangle$, i.e. the conditions (4) specify a set $\mathcal{C}$ of density operators which has to be considered. Each of these density operators $\hat{\rho}_{\{\hat{G}\}}$ can possess a different value of the uncertainty measure $\eta[\hat{\rho}_{\{\hat{G}\}}]$. If we wish to use only the expectation values G_ν of the chosen observation level for determining the density operator, we must select a particular density operator $\hat{\rho}_{\{\hat{G}\}} = \hat{\sigma}_{\{\hat{G}\}}$ in an unbiased manner. According to the Jaynes principle of the maximum entropy [6,4] this density operator $\hat{\sigma}_{\{\hat{G}\}}$ must be the one which has the largest uncertainty measure $\eta_{\max} \equiv \eta[\hat{\sigma}_{\{\hat{G}\}}]$ and simultaneously fulfills constraints (4). Consequently, the following fundamental inequality holds

$$\eta[\hat{\sigma}_{\{\hat{G}\}}] = -\mathrm{Tr}\,(\hat{\sigma}_{\{\hat{G}\}} \ln \hat{\sigma}_{\{\hat{G}\}}) \geq \eta[\hat{\rho}_{\{\hat{G}\}}] = -\mathrm{Tr}\,(\hat{\rho}_{\{\hat{G}\}} \ln \hat{\rho}_{\{\hat{G}\}}) \tag{5}$$

for all possible $\hat{\rho}_{\{\hat{G}\}}$ which fulfill (4). The variation determining the maximum of $\eta[\hat{\sigma}_{\{\hat{G}\}}]$ under the conditions (4) leads to a generalized canonical density operator [6]

$$\begin{aligned}
\hat{\sigma}_{\{\hat{G}\}} &= \tfrac{1}{Z_{\{\hat{G}\}}} \exp\left(-\sum_\nu \lambda_\nu \hat{G}_\nu\right); \\
Z_{\{\hat{G}\}}(\lambda_1, \ldots, \lambda_n) &= \mathrm{Tr}\left[\exp\left(-\sum_\nu \lambda_\nu \hat{G}_\nu\right)\right],
\end{aligned} \tag{6}$$

where λ_n are the Lagrange multipliers and $Z_{\{\hat{G}\}}(\lambda_1, \ldots, \lambda_n)$ is the generalized partition function.

The maximum uncertainty measure regarding an observation level $\mathcal{O}_{\{\hat{G}\}}$ will be referred to as the entropy $S_{\{\hat{G}\}}$

$$S_{\{\hat{G}\}} \equiv \eta_{\max} = -\mathrm{Tr}\,(\hat{\sigma}_{\{\hat{G}\}} \ln \hat{\sigma}_{\{\hat{G}\}}). \tag{7}$$

This means that different entropies are related to different observation levels. By inserting $\sigma_{\{\hat{G}\}}$ (cf. (6)) into (7), we obtain the following expression for the entropy:

$$S_{\{\hat{G}\}} = \ln Z_{\{\hat{G}\}} + \sum_\nu \lambda_\nu G_\nu. \tag{8}$$

2.1 Extension and Reduction of the Observation Level

If an observation level $\mathcal{O}_{\{\hat{G}\}} \equiv \hat{G}_1, \ldots, \hat{G}_n$ is extended by including further operators $\hat{M}_1, \ldots, \hat{M}_l$, then additional expectation values $M_1 = \langle \hat{M}_1 \rangle, \ldots,$ $M_l = \langle \hat{M}_l \rangle$ can only increase the amount of available information about the state of the system. This procedure is called the *extension* of the observation level (from $\mathcal{O}_{\{\hat{G}\}}$ to $\mathcal{O}_{\{\hat{G},\hat{M}\}}$) and is associated with a decrease of the entropy. More precisely, the entropy $S_{\{\hat{G},\hat{M}\}}$ of the extended observation level $\mathcal{O}_{\{\hat{G},\hat{M}\}}$ can be only smaller than or equal to the entropy $S_{\{\hat{G}\}}$ of the original observation level $\mathcal{O}_{\{\hat{G}\}}$,

$$S_{\{\hat{G},\hat{M}\}} \leq S_{\{\hat{G}\}} . \tag{9}$$

We can also consider a *reduction of the observation level* if we decrease the number of independent observables which are measured, e.g. $\mathcal{O}_{\{\hat{G},\hat{M}\}} \rightarrow \mathcal{O}_{\{\hat{G}\}}$. This reduction is accompanied by an increase of the entropy due to the decrease of the information available about the system.

3 States of Light: Phase-Space Description

Utilizing a close analogy between the operator for the electric component $\hat{E}(r,t)$ of a monochromatic light field and the quantum-mechanical harmonic oscillator we will consider a dynamical system which is described by a pair of canonically conjugated Hermitian observables $\hat{q}$ and $\hat{p}$, such that $[\hat{q}, \hat{p}] = \mathrm{i}\hbar$. Eigenvalues of these operators range continuously from $-\infty$ to $+\infty$. The annihilation and creation operators $\hat{a}$ and $\hat{a}^\dagger$ can be expressed as a complex linear combination of $\hat{q}$ and $\hat{p}$:

$$\hat{a} = \frac{1}{\sqrt{2\hbar}} \left(\lambda\hat{q} + \mathrm{i}\lambda^{-1}\hat{p} \right); \qquad \hat{a}^\dagger = \frac{1}{\sqrt{2\hbar}} \left(\lambda\hat{q} - \mathrm{i}\lambda^{-1}\hat{p} \right), \tag{10}$$

where λ is an arbitrary real parameter. The operators $\hat{a}$ and $\hat{a}^\dagger$ obey the Weyl–Heisenberg commutation relation $[\hat{a}, \hat{a}^\dagger] = 1$, and therefore possess the same algebraic properties as the operator associated with the complex amplitude of a harmonic oscillator (in this case $\lambda = \sqrt{m\omega}$, where m and ω are the mass and the frequency of the quantum-mechanical oscillator, respectively) or the photon annihilation and creation operators of a single mode of the quantum electromagnetic field. In this case $\lambda = \sqrt{\epsilon_0 \omega}$ (ϵ_0 is the dielectric constant and ω is the frequency of the field mode) and the operator for the electric field reads (we do not take into account polarization of the field)

$$\hat{E}(r,t) = \sqrt{2}\,\mathcal{E}_0 \left(\hat{a}\mathrm{e}^{-\mathrm{i}\omega t} + \hat{a}^\dagger \mathrm{e}^{\mathrm{i}\omega t} \right) u(r), \tag{11}$$

where $u(r)$ describes the spatial field distribution and is the same in both classical and quantum theories. The constant $\mathcal{E}_0 = (\hbar\omega/2\epsilon_0 V)^{1/2}$ is equal to the "electric field per photon" in the cavity of volume V.

A particularly useful set of states is the overcomplete set of coherent states $|\alpha\rangle$ which are the eigenstates of the annihilation operator $\hat{a}$, i.e. $\hat{a}|\alpha\rangle = \alpha|\alpha\rangle$. These coherent states can be generated from the vacuum state $|0\rangle$ (defined as $\hat{a}|0\rangle = 0$) by the action of the unitary displacement operator $\hat{D}(\alpha)$ [7]

$$\hat{D}(\alpha) \equiv \exp(\alpha\hat{a}^\dagger - \alpha^*\hat{a})\,; \qquad |\alpha\rangle = \hat{D}(\alpha)|0\rangle\,. \tag{12}$$

The parametric space of eigenvalues, i.e. the *phase space* for our dynamical system, is the *infinite* plane of eigenvalues (q, p) of the Hermitian operators $\hat{q}$ and $\hat{p}$.

The phase-space description of the quantum-mechanical oscillator prepared in the state described by the density operator $\hat{\rho}$ (in what follows we will consider mainly pure states such that $\hat{\rho} = |\Psi\rangle\langle\Psi|$) is based on the definition of the Wigner function [8] $W_{\hat{\rho}}(\xi)$. Here the subscript $\hat{\rho}$ in the expression $W_{\hat{\rho}}(\xi)$ explicitly indicates the state which is described by the given Wigner function.

The Wigner function is related to the characteristic function $C_{\hat{\rho}}^{(W)}(\eta)$ of the Weyl-ordered moments of the annihilation and creation operators of the harmonic oscillator as follows [9]:

$$W_{\hat{\rho}}(\xi) = \frac{1}{\pi} \int C_{\hat{\rho}}^{(W)}(\eta)\, \exp(\xi\eta^* - \xi^*\eta)\, d^2\eta\,. \tag{13}$$

where $C_{\hat{\rho}}^{(W)}(\eta) \equiv \mathrm{Tr}\,[\hat{\rho}\hat{D}(\eta)]$, and $\hat{D}(\eta)$ is the displacement operator given by (12).

Alternatively the Wigner function can be defined as a particular Fourier transform of the density operator expressed in the basis of the eigenvectors $|q\rangle$ of the position operator $\hat{q}$:

$$W_{\hat{\rho}}(q, p) \equiv \int_{-\infty}^{\infty} d\zeta \langle q - \zeta/2|\hat{\rho}|q + \zeta/2\rangle e^{ip\zeta/\hbar}\,. \tag{14}$$

Both definitions (13) and (14) of the Wigner function are identical (see Hillery et al. [8]), providing the parameters ξ and ξ^* are related to the coordinates q and p of the phase space as $\xi = (\lambda q + i\lambda^{-1}p)/\sqrt{2\hbar}$ and $\xi^* = (\lambda q - i\lambda^{-1}p)/\sqrt{2\hbar}$. The Wigner function can be interpreted as the quasiprobability (see below) density distribution through which a probability can be expressed to find a quantum-mechanical system (harmonic oscillator) around the "point" (q, p) of the phase space. With the help of the Wigner function $W_{\hat{\rho}}(q, p)$ the position and momentum probability distributions $w_{\hat{\rho}}(q)$ and $w_{\hat{\rho}}(p)$ can be expressed from $W_{\hat{\rho}}(q, p)$ via marginal integration over the conjugated variable (in what follows we assume $\lambda = 1$)

$$w_{\hat{\rho}}(q) \equiv \frac{1}{\sqrt{2\pi\hbar}} \int dp\, W_{\hat{\rho}}(q, p) = \sqrt{2\pi\hbar}\, \langle q|\hat{\rho}|q\rangle\,, \tag{15}$$

where $|q\rangle$ is the eigenstate of the position operator $\hat{q}$. The marginal probability distribution $W_{\hat{\rho}}(q)$ is normalized to unity, i.e. $\int dq\, w_{\hat{\rho}}(q) = \sqrt{2\pi\hbar}$

3.1   Quantum Homodyne Tomography

The relation (15) for the probability distribution $w_{\hat{\rho}}(q)$ of the position operator $\hat{q}$ can be generalized to the case of the distribution of the rotated quadrature operator $\hat{x}_{\theta}$. This operator is defined as

$$\hat{x}_{\theta} = \sqrt{\frac{\hbar}{2}} \left(\hat{a} e^{-i\theta} + \hat{a}^{\dagger} e^{i\theta} \right), \tag{16}$$

and the corresponding conjugated operator $\hat{x}_{\theta+\pi/2}$, such that $[\hat{x}_{\theta}, \hat{x}_{\theta+\pi/2}] = i\hbar$, reads

$$\hat{x}_{\theta+\pi/2} = \frac{\sqrt{\hbar}}{i\sqrt{2}} \left(\hat{a} e^{-i\theta} - \hat{a}^{\dagger} e^{i\theta} \right). \tag{17}$$

The position and the momentum operators are related to the operator $\hat{x}_{\theta}$ as $\hat{q} = \hat{x}_0$ and $\hat{x}_{\pi/2} = \hat{p}$. The rotation (i.e. the linear homogeneous canonical transformation) given by (16) and (17) can be performed by the unitary operator $\hat{U}(\theta)$:

$$\hat{U}(\theta) = \exp\left(-i\theta \hat{a}^{\dagger} \hat{a} \right), \tag{18}$$

so that

$$\hat{x}_{\theta} = \hat{U}^{\dagger}(\theta) \hat{x}_0 \hat{U}(\theta); \qquad \hat{x}_{\theta+\pi/2} = \hat{U}^{\dagger}(\theta) \hat{x}_{\pi/2} \hat{U}(\theta). \tag{19}$$

It has been shown by Ekert and Knight [10] that Wigner functions are transformed under the action of the linear canonical transformation (19) as

$$W_{\hat{\rho}}(q, p) \rightarrow W_{\hat{\rho}}(x_{\theta} \cos\theta - x_{\theta+\pi/2} \sin\theta; x_{\theta} \sin\theta + x_{\theta+\pi/2} \cos\theta), \tag{20}$$

which means that the probability distribution $w_{\hat{\rho}}(x_{\theta}, \theta) = \sqrt{2\pi\hbar} \langle x_{\theta} | \hat{\rho} | x_{\theta} \rangle$ can be evaluated as

$$w_{\hat{\rho}}(x_{\theta}, \theta) = \frac{1}{\sqrt{2\pi\hbar}} \int_{-\infty}^{\infty} dx_{\theta+\pi/2} \tag{21}$$
$$\times W_{\hat{\rho}}(x_{\theta} \cos\theta - x_{\theta+\pi/2} \sin\theta; x_{\theta} \sin\theta + x_{\theta+\pi/2} \cos\theta).$$

As shown by Vogel and Risken [11] the knowledge of $w_{\hat{\rho}}(x_{\theta}, \theta)$ for all values of θ (such that $0 < \theta \leq \pi$) is equivalent to the knowledge of the Wigner function itself. This Wigner function can be obtained from the set of distributions $w_{\hat{\rho}}(x_{\theta}, \theta)$ via the inverse Radon transformation:

$$W_{\hat{\rho}}(q, p) = \frac{1}{(2\pi\hbar)^{3/2}} \int_{-\infty}^{\infty} dx_{\theta} \int_{-\infty}^{\infty} d\xi \, |\xi|$$
$$\times \int_0^{\pi} d\theta \, w_{\hat{\rho}}(x_{\theta}, \theta) \exp\left[\frac{i}{\hbar} \xi (x_{\theta} - q\cos\theta - p\sin\theta) \right]. \tag{22}$$

We will show later in this chapter that the optical homodyne tomography is implicitly based on a measurement of all (in principle, infinite number) independent moments (cumulants) of the system operators. If the state under consideration is characterized by an infinite number of nonzero cumulants then the homodyne tomography can fail because it does not provide us with a consistent truncation scheme (see below). As we will show later, the Max-Ent principle may help us to reconstruct the Wigner function reliably from incomplete tomographic data.

Quantum Tomography via Pattern Functions. In a sequence of papers D'Ariano et al. [12], Leonhardt et al. [13] and Richter [14] have shown that Wigner functions can be very efficiently reconstructed from tomographic data with the help of the so-called pattern functions. This reconstruction procedure is more effective than the usual Radon transformation [15]. To be specific, D'Ariano et al. [12] have shown that the density matrix ρ_{mn} in the Fock basis can be reconstructed directly from the tomographic data, i.e. from the quadrature-amplitude "histograms" (probabilities), $w(x_\theta, \theta)$, via the so-called *direct sampling method* when

$$\rho_{mn} = \int_0^\pi \int_{-\infty}^\infty w(x_\theta, \theta) F_{mn}(x_\theta, \theta) \, dx_\theta \, d\theta \,, \tag{23}$$

where $F_{mn}(x_\theta, \theta)$ is a set of specific *sampling* functions (see below). Once the density matrix elements are reconstructed with the help of (23) then the Wigner function of the corresponding state can be directly obtained using the relation

$$W_{\hat\rho}(q, p) = \sum_{m,n} \rho_{mn} W_{|m\rangle\langle n|}(q, p) \,, \tag{24}$$

where $W_{|m\rangle\langle n|}(q, p)$ is the Wigner function of the operator $|m\rangle\langle n|$.

A serious problem with the direct sampling method as proposed by D'Ariano et al. [12] is that the sampling functions $F_{mn}(x_\theta, \theta)$ are difficult to compute. Later D'Ariano, Leonhardt and Paul [13,16] simplified the expression for the sampling function and found that it can be expressed as

$$F_{mn}(x_\theta, \theta) = f_{mn}(x_\theta) \exp[i(m - n)\theta] \,, \tag{25}$$

where the so-called *pattern* function "picks up" the pattern in the quadrature histograms (probability distributions) $w_{mn}(x_\theta, \theta)$ which just match the corresponding density-matrix elements. Recently Leonhardt et al. [15] have shown that the pattern function $f_{mn}(x_\theta)$ can be expressed as derivatives

$$f_{mn}(x) = \frac{\partial}{\partial x} g_{mn}(x) \,, \tag{26}$$

of functions $g_{mn}(x)$ which are given by the Hilbert transformation

$$g_{mn}(x) = \frac{\mathcal{P}}{\pi} \int_{-\infty}^\infty \frac{\psi_m(\zeta)\psi_n(\zeta)}{x - \zeta} \, d\zeta \,, \tag{27}$$

where $\mathcal{P}$ stands for the principal value of the integral and $\psi_n(x)$ are the real energy eigenfunctions of the harmonic oscillator, i.e. the normalizable solutions of the Schrödinger equation

$$\left(-\frac{\hbar^2}{2}\frac{d^2}{dx^2} + \frac{x^2}{2}\right)\psi_n(x) = \hbar(n+1/2)\psi_n(x), \tag{28}$$

(we assume $m = \omega = 1$). Further details of possible applications and a discussion devoted for numerical procedures of the reconstruction of density operators via the direct sampling method can be found in [15].

3.2   States of Light to Be Considered

In this chapter we will consider a specific nonclassical state of a single-mode light field – a superposition of two coherent states, the so-called even coherent state [17]. In order to understand quantum-statistical properties of this state we first briefly describe the Wigner function of a coherent state.

Coherent State. The coherent state $|\alpha\rangle$ is an eigenstate of the annihilation operator $\hat{a}$, i.e. $|\alpha\rangle$ is not an eigenstate of an observable [7]. The Wigner function (13) of the coherent state in the complex ξ-phase space reads

$$W_{|\alpha\rangle}(\xi) = 2\exp(-2|\xi - \alpha|^2); \qquad \alpha = \alpha_x + i\alpha_y, \tag{29}$$

or, alternatively, in the (q, p) phase space we have

$$W_{|\alpha\rangle}(q,p) = \frac{1}{\sigma_q\sigma_p}\exp\left[-\frac{1}{2\hbar}\frac{(q-\bar{q})^2}{\sigma_q^2} - \frac{1}{2\hbar}\frac{(p-\bar{p})^2}{\sigma_p^2}\right], \tag{30}$$

where $\bar{q} = \sqrt{2\hbar}\alpha_x/\lambda$, $\bar{p} = \sqrt{2\hbar}\alpha_y\lambda$, and $\sigma_q^2 = 1/2\lambda^2$ and $\sigma_p^2 = \lambda^2/2$.

Even Coherent State. In nonlinear optical processes superpositions of coherent states can be produced [17]. In particular, the superposition of coherent states with opposite phase can be generated

$$|\alpha_e\rangle = \mathcal{N}_e^{1/2}(|\alpha\rangle + |-\alpha\rangle); \qquad \mathcal{N}_e^{-1} = 2[1 + \exp(-2|\alpha|^2)], \tag{31}$$

which is called the even coherent state [17]. This state has been analyzed as a prototype of superposition states of light which exhibit various non-classical effects (for the review see [17]). In particular, quantum interference between component states leads to oscillations in the photon number distributions. Another consequence of this interference is a reduction (squeezing) of quadrature fluctuations. Nonclassical effects associated with superposition states can be explained in terms of quantum interference between the "points" (coherent states) in phase space. The character of quantum interference is very sensitive with respect to the relative phase between coherent components of superposition states. To illustrate this effect we give the expression for the Wigner function of the even coherent state (in what follows we assume α to be real):

$$W_{|\alpha_e\rangle}(q,p) = N_e[W_{|\alpha\rangle}(q,p) + W_{|-\alpha\rangle}(q,p) + W_{\text{int}}(q,p)], \tag{32}$$

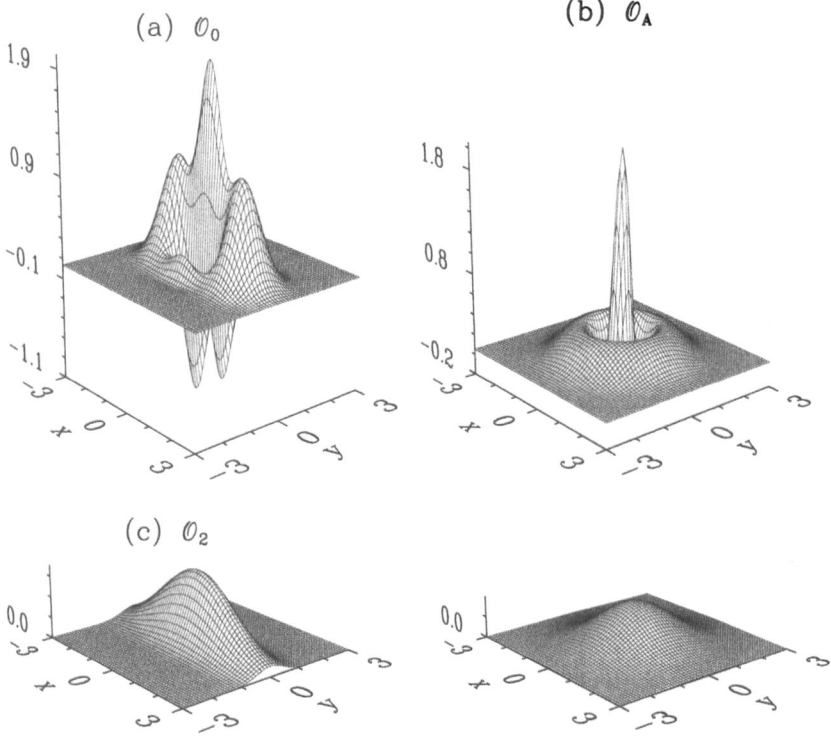

Fig. 1. The reconstructed Wigner functions of the even coherent state $|\alpha_e\rangle$ with $\bar{n} = 2$. We consider the observation levels as indicated in the figure. This non-Gaussian Wigner function can be completely reconstructed only on the complete observation level when all moments of systems operators are measured

where the Wigner functions $W_{|\pm\alpha\rangle}(q,p)$ of coherent states $|\pm\alpha\rangle$ are given by (30). The interference part of the Wigner function (32) is given by the relation

$$W_{\text{int}}(q,p) = \frac{2}{\sigma_q \sigma_p} \exp\left(-\frac{q^2}{2\hbar\sigma_q^2} - \frac{p^2}{2\hbar\sigma_p^2}\right) \cos\left(\frac{\bar{q}p}{\hbar\sigma_q\sigma_p}\right), \tag{33}$$

where $\bar{q} = \sqrt{2\hbar}\alpha$ and the variances $\sigma_q^2 = \sigma_p^2 = 1/2$. We plot the Wigner function of the even coherent state in Fig. 1a. From the figure it is clearly seen that the interference term (33) results in oscillations of the Wigner function around the origin of the phase space.

4   Observation Levels for Single-mode Field

In this chapter we will consider two different classes of observation levels. Namely, we will consider the phase-sensitive and phase-insensitive observation levels. These two classes do differ by the fact that phase-sensitive observation levels are related to such operators which provide some information about off-diagonal matrix elements of the density operator in the Fock basis (i.e. these observation levels reveal some information about the phase of states under consideration). On the contrary, phase-insensitive observation levels are based exclusively on a measurement of diagonal matrix elements in the Fock basis. Before we proceed to a detailed description of the phase-sensitive and phase-insensitive observation levels we introduce two exceptional observation levels, the complete and thermal observation levels.

4.1   Two Extreme Observation Levels

Complete Observation Level $\mathcal{O}_0 \equiv \{(\hat{a}^\dagger)^k \hat{a}^l; \ \forall k, l\}$. The set of operators $|n\rangle\langle m|$ (for all n and m) is referred to as the *complete* observation level. Expectation values of the operators $|n\rangle\langle m|$ are the matrix elements of the density operator in the Fock basis

$$\langle m|\hat{\rho}|n\rangle = \text{Tr}[\hat{\rho}|n\rangle\langle m|]; \qquad \forall n, m, \tag{34}$$

and therefore the generalized canonical density operator is identical to the statistical density operator

$$\hat{\sigma}_0 = \frac{1}{Z_0} \exp\left[-\sum_{m,n=0}^{\infty} \lambda_{m,n} |n\rangle\langle m| \right] = \hat{\rho}. \tag{35}$$

The complete observation level $\mathcal{O}_0$ can also be given by a set of operators $\{(\hat{a}^\dagger)^k \hat{a}^l; \ \forall k, l\}$ or $\{\hat{q}^k \hat{p}^l; \ \forall k, l\}$. The Wigner function on the complete information level is equal to the Wigner function of the state itself, i.e. $W_{\hat{\rho}}^{(0)}(\xi) = W_{\hat{\rho}}(\xi)$. We plot the Wigner function of the even coherent state on the complete observation level in Fig. 1a.

Thermal Observation Level $\mathcal{O}_{\text{th}} \equiv \{\hat{a}^\dagger \hat{a}\}$. The total reduction of the complete observation level $\mathcal{O}_0$ results in a thermal observation level $\mathcal{O}_{\text{th}}$ characterized by just one observable, the photon number operator $\hat{n}$, i.e. quantum-mechanical states of light on this observation level are characterized only by their mean photon number $\bar{n} \equiv \langle \hat{n} \rangle$. The generalized canonical density operator of this observation level is the well-known density operator of the harmonic oscillator in thermal equilibrium

$$\hat{\sigma}_{\text{th}} = \frac{1}{Z_{\text{th}}} \exp(-\lambda_{\text{th}} \hat{n}). \tag{36}$$

To find an explicit expression for the Lagrange multiplier λ_{th} we have to solve the equation $\text{Tr}[\sigma_{\text{th}}\hat{n}] = \bar{n}$, from which we find that $\lambda_{\text{th}} = \ln[(\bar{n} + 1)/\bar{n}]$, so we obtain the generalized canonical density operator $\hat{\sigma}_{\text{th}}$ in the Fock basis,

$$\hat{\sigma}_{\text{th}} = \sum_{n=0}^{\infty} \frac{\bar{n}^n}{(\bar{n} + 1)^{n+1}} |n\rangle\langle n|. \tag{37}$$

For the entropy of the thermal observation level we find a familiar expression:

$$S_{\text{th}} = (\bar{n} + 1) \ln(\bar{n} + 1) - \bar{n} \ln \bar{n}. \tag{38}$$

On the thermal observation level *all* states with the same mean photon number $\bar{n} > 0$ are indistinguishable. This is the reason why Wigner functions of different states on the thermal information level are identical. The Wigner function of the state $|\Psi\rangle$ on the thermal observation level is given by the relation

$$W_{\hat{\rho}}^{(\text{th})}(\xi) = \frac{2}{1 + 2\bar{n}} \exp\left(-\frac{2|\xi|^2}{1 + 2\bar{n}}\right). \tag{39}$$

This Wigner function is plotted in Fig. 1d. By extending the thermal observation level we can obtain more "realistic" Wigner functions which in the limit of the complete observation level are equal to the Wigner function of the measured state itself, i.e. they are not biased by the lack of information (measured data) about the state.

4.2   Phase-sensitive Observation Levels

We can extent the thermal observation level if in addition to the observable $\hat{n}$ we also consider the measurement of mean values of the higher moments of the operators $\hat{a}$ and $\hat{a}^\dagger$ (i.e. we consider a measurement of the moments of the observables $\hat{q}$ and $\hat{p}$).

Observation Level $\mathcal{O}_2 \equiv \{\hat{a}^\dagger \hat{a}, (\hat{a}^\dagger)^2, \hat{a}^2, \hat{a}^\dagger, \hat{a}\}$. One possible extension of the thermal observation level $\mathcal{O}_{\text{th}}$ can be performed with the help of observables $\hat{q}^2$ and $\hat{p}^2$, i.e. when not only the mean photon number $\bar{n}$ and mean values of $\hat{q}$ and $\hat{p}$ are known, but also the variances $\langle(\Delta\hat{q})^2\rangle$ and $\langle(\Delta\hat{p})^2\rangle$ are measured. On the observation level $\mathcal{O}_2$ we can express the reconstructed Wigner function $W_{|\alpha_e\rangle}^{(2)}(\xi)$ of the even coherent state as

$$W_{|\alpha_e\rangle}^{(2)}(\xi) = \frac{1}{[(N + 1/2)^2 - M^2]^{1/2}}$$
$$\times \exp\left\{-\frac{\xi_x^2}{[(N + 1/2) + M]} - \frac{\xi_y^2}{[(N + 1/2) - M]}\right\}, \tag{40}$$

where $\xi = \xi_x + i\xi_y$, and the parameters N and M read

$$N = \alpha^2 \tanh \alpha^2; \qquad M = \alpha^2. \tag{41}$$

We plot the Wigner function $W_{|\alpha_e\rangle}^{(2)}(\xi)$ in Fig. 1c. This Wigner function is slightly "squeezed" in the ξ_y-direction and stretched in the ξ_x-direction. Nevertheless, the reconstructed Wigner function is different from the Wigner function of the squeezed vacuum state which is a minimum uncertainty state.

4.3   Phase-insensitive Observation Levels

The choice of the observation level is very important in order to optimize the strategy for the measurement and the reconstruction of the Wigner function of a given quantum-mechanical state of light. For instance, if we would like to reconstruct the Wigner function of the Fock state $|n\rangle$ at the observation level $\mathcal{O}_k \equiv \{\hat{a}^\dagger\hat{a}, (\hat{a}^\dagger)^m\hat{a}^n; m+n \leq k$ and $m \neq n\}$ we find that irrespective of the number (k) of "measured" moments $\langle(\hat{a}^\dagger)^m\hat{a}^n\rangle$ (for $m \neq n$) the reconstructed Wigner function is always equal to the thermal Wigner function (39). So it can happen that in a very tedious experiment negligible information is obtained. On the other hand, if a measurement of diagonal elements of the density operator in the Fock basis is performed, relevant information can be obtained much easier.

Observation Level $\mathcal{O}_A \equiv \{\hat{P}_n = |n\rangle\langle n|; \ \forall n\}$. As an example we will consider the most general phase-insensitive observation level which corresponds to the case when *all* diagonal elements $P_n = \langle n|\hat{\rho}|n\rangle$ of the density operator $\hat{\rho}$ describing the state under consideration are measured. The observation level $\mathcal{O}_A$ can be obtained via a reduction of the complete observation level $\mathcal{O}_0$ and it corresponds to the measurement of the photon number distribution P_n such that $\sum_n P_n = 1$.

The generalized canonical density operator on this observation level reads

$$\hat{\sigma}_A = \sum_{n=0}^{\infty} P_n|n\rangle\langle n|; \qquad \sum_{n=0}^{\infty} P_n = 1. \tag{42}$$

The photon number distribution of the even coherent state is given by the relation (we assume α to be real)

$$P_{2n} = \frac{1}{\cosh\alpha^2}\frac{\alpha^{4n}}{(2n)!}; \qquad P_{2n+1} = 0. \tag{43}$$

We can express $W^{(A)}_{|\alpha_e\rangle}(\xi)$ as the phase-averaged Wigner function of the even coherent state $W_{|\alpha_e\rangle}(\xi)$ given by (32). After some algebra we find that $W^{(A)}_{|\alpha_e\rangle}(\xi)$ can be written in the closed form

$$W^{(A)}_{|\alpha_e\rangle}(\xi) = \frac{e^{-2|\xi|^2}}{\cosh\alpha^2}\left[e^{-\alpha^2}J_0(4i\alpha|\xi|) + e^{\alpha^2}J_0(4\alpha|\xi|)\right]. \tag{44}$$

We plot the Wigner function $W^{(A)}_{|\alpha_e\rangle}(\xi)$ in Fig. 1b. From this figure the dominant contribution of the Fock state $|2\rangle$ is transparent (in the present case we have $P_0 \simeq 2\exp(-2)$, $P_2 = 2P_0$, and $P_4 = 2P_0/3$, while all other probabilities P_n are much smaller) which results in a negative Wigner function.

The phase-insensitive observation level $\mathcal{O}_A$ can be further reduced if only a finite number of operators $\hat{P}_n$ (where $n \in \mathcal{M}$) are considered. In this case, in general, we have $\sum_{n\in\mathcal{M}} P_n < 1$ and therefore it is essential that apart from mean values P_n, the mean photon number $\bar{n}$ is also known from the measurement [5].

5   Optical Homodyne Tomography and MaxEnt Principle

From the point of view of the formalism presented above it follows that from the probability density distribution $w_{\hat{\rho}}(x_\theta)$ (see (15)) which corresponds to a measurement of *all* moments $\langle \hat{x}_\theta^n \rangle$, the generalized canonical density operators $\hat{\sigma}_{x_\theta}$ (see also (6))

$$\hat{\sigma}_{x_\theta} = \frac{1}{Z_{x_\theta}} \exp\left[-\int_{-\infty}^{\infty} \mathrm{d}x_\theta \, |x_\theta\rangle\langle x_\theta| \lambda(x_\theta) \right] \tag{45}$$

can be constructed. The Lagrange multipliers $\lambda(x_\theta)$ are given by an infinite set of equations

$$w_{\hat{\rho}}(x_\theta) = \sqrt{2\pi\hbar} \langle x_\theta | \hat{\sigma}_{x_\theta} | x_\theta \rangle ; \qquad \forall x_\theta \in (-\infty, \infty). \tag{46}$$

If probability distributions $w_{\hat{\rho}}(x_\theta)$ are known for all values of $\theta \in [0, \pi]$ then the density operator on the complete observation level can be obtained in the form

$$\hat{\rho} = \frac{1}{Z_0} \exp\left[-\int_0^\pi \mathrm{d}\theta \int_{-\infty}^{\infty} \mathrm{d}x_\theta \, |x_\theta\rangle\langle x_\theta| \lambda(x_\theta) \right], \tag{47}$$

and the corresponding Wigner function can be reconstructed. The optical homodyne tomography can be understood as a method of how to find a relation between measured distributions $w_{\hat{\rho}}(x_\theta)$ and the Lagrange multipliers $\lambda(x_\theta)$ for all values of x_θ and θ. As we have shown earlier in this section, the Gaussian and the generalized Gaussian states can be completely reconstructed on reduced observation levels based on a measurement of just a finite number of moments of system observables, and therefore the optical homodyne tomography is essentially not needed as a method for the reconstruction of Wigner functions in these cases. On the other hand, the non-Gaussian states can in principle be reconstructed, but in practice the reconstruction of their Wigner functions is associated with a measurement of an infinite number of independent moments of system observables, which is not realistic. In the experiments by Raymer et al. [18] only a finite number of values of θ have been considered, i.e. these types of experiments are associated with an observation level for which the corresponding generalized canonical density operator reads

$$\hat{\sigma} = \frac{1}{Z} \exp\left(\lambda_0 \hat{n} + \sum_{l=1}^{N_x} \sum_{m=1}^{N_\theta} \lambda_{l,m} |x_{\theta_m}^{(l)}\rangle\langle x_{\theta_m}^{(l)}| \right). \tag{48}$$

5.1 Implementation and Numerical Examples

We will demonstrate our reconstruction scheme and compare it with the known tomography scheme[1] for an example of the superposition of two coherent states of the form

$$|\Psi\rangle = \mathcal{N}^{1/2} \left(|\alpha_1\rangle + |\alpha_2\rangle \right). \tag{49}$$

All calculations were carried out in Fock representation, where the projection operators

$$\hat{O}_{lm} = |x_{\theta_m}^{(l)}\rangle\langle x_{\theta_m}^{(l)}| \tag{50}$$

read

$$\left(\hat{O}_{lm}\right)_{n_1, n_2} = \psi_{n_1}^*(x_l)\psi_{n_2}(x_l) \exp[i\theta_m(n_1 - n_2)], \tag{51}$$

where θ_m is the quadrature phase and x_l is the eigenvalue of the operator. In the numerical examples we chose $\alpha_1 = 1.25$ and $\alpha_2 = 1.25\,i$.

Our numerical approach forces us to truncate the Hilbert space at a finite value $n_{\max}$ and we must insure that an increase of this cut-off does not change our results significantly. On the other hand, the number N_θ of different angles θ and the number N_x and separation Δx of different x are given by the experiment. The error of any reconstruction scheme goes to zero when all x for all angles θ are covered, i.e. when our knowledge about the state is complete. However, for incomplete knowledge the different reconstruction schemes give different results and in this sense we want to compare the schemes.

In Fig. 2 we present Wigner functions of the state (49) and its reconstruction. The plots in the upper part show the Wigner functions as surface plots, whereas the lower part shows the same functions as gray scale plots. The uniform gray background corresponds to the value zero whereas darker areas indicate positive values of the Wigner function. In (a) we show the state $|\Psi\rangle$ as defined by (49) and in (b) the reconstruction $\hat{\sigma}$ as obtained via

---

[1] The direct sampling method as described in Sect. 3 can be straightforwardly applied also in the case when the quadrature components $\hat{x}_\theta$ are measured at N_θ discrete phases θ_m. As shown by Leonhardt and Munroe [19], if it is a priori known that $\rho_{mn} = 0$ for $|m - n| \geq N_\theta$ then the density-matrix elements ρ_{mn} for $|m - n| < N_\theta$ can be precisely reconstructed from the measured distributions $w(x_\theta, \theta_m)$ at N_θ phases θ_m. On the other hand, if the parameter x is discretized (which corresponds to a measurement of N_x projectors $|x_{\theta_m}^{(l)}\rangle\langle x_{\theta_m}^{(l)}|$ in the direction θ_m), then the direct-sampling reconstruction can be applied as well, but may lead to "pathological" density operators which are not positively defined. Alternatively, the least-squares inversion method (see for instance [20]) can be efficiently applied. The advantage of this method is that it is a linear method, which means that the density matrix can be reconstructed in real time together with an estimate of the statistical error. We note that this method may also lead to nonpositive density operators.

(a)   (b)   (c)

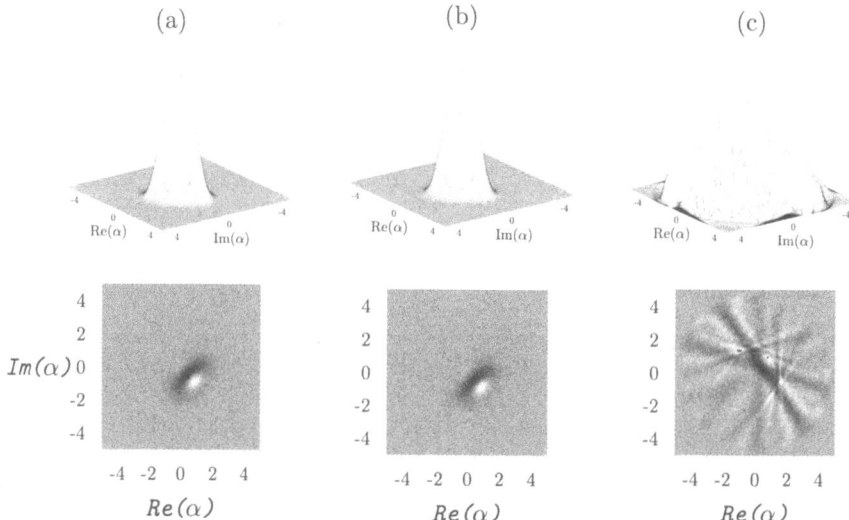

Fig. 2. The Wigner function of a coherent superposition of two coherent states given by (49) is presented in (**a**), its reconstruction via the MaxEnt principle (**b**) and its reconstruction via projection onto pattern functions in (**c**). The upper part shows the Wigner functions as surface plots, whereas the lower part shows the same functions as a gray scale plot, where dark areas correspond to higher values and bright areas to lower values. We see that the reconstruction via the MaxEnt principle is much more reliable than a straightforward application of direct sampling via pattern functions

the MaxEnt principle. For the reconstruction we used only four different angles and 13 points on each axis. Despite this extremely small number the graphical representation of the state $\hat{\sigma}$ reveals no difference to the original state $\hat{\rho} = |\Psi\rangle\langle\Psi|$. For completeness we include (**c**), the state as obtained via projection onto pattern functions as described in [15]. Contrary to (**a**) and (**b**) we also obtain white areas which correspond to a negative value of the Wigner function.

From this plot it is obvious that the reconstruction via the MaxEnt principle matches the original state much better. For a quantitative comparison we calculated

$$\Delta = \sum_{n_1,n_2} [(\rho_1)_{n_1,n_2} - (\tilde{\rho}_1)_{n_1,n_2}]^2 \tag{52}$$

as a measure for the error of the reconstruction. We varied the number N_θ of different angles and N_x of different values on each angle; their separation was chosen in such a way that they uniformly cover the integral $[-2, 2]$, where – as can be seen in Fig. 2 – almost the whole state is located. The numerical cut-off for the Hilbert space was $n_{\max} = 30$.

Whereas the errors Δ (Table 1) for usual quantum tomography are of the order of one (thus, on average, of the order of 10^{-3} per matrix element)

Table 1. Deviation Δ of the reconstructed state from the coherent superposition of two coherent states (49) for reconstruction via the MaxEnt principle and for reconstruction via projection onto pattern functions. From the data presented we clearly see the advantage of the reconstruction via the MaxEnt principle

Number of observables	Maximum entropy principle	Direct sampling
$N_\theta = 3$; $N_x = 11$	$1.74 \cdot 10^{-3}$	6.90
$N_\theta = 4$; $N_x = 11$	$7.8 \ \cdot 10^{-3}$	4.33
$N_\theta = 4$; $N_x = 13$	$1.3 \ \cdot 10^{-3}$	1.00
$N_\theta = 5$; $N_x = 15$	$1.5 \ \cdot 10^{-3}$	0.377

the inclusion of the MaxEnt principle reduces the errors by several orders of magnitude. We want to add that the large errors of usual quantum tomography of course decrease significantly when the amount of measurement data is increased, i.e. increasing N_θ and N_x. On the other hand, the reconstruction via projection onto pattern functions is in general not positive definite, which reflects some fundamental problems associated with this reconstruction scheme.

6   Conclusions

In this chapter we have shown how the MaxEnt principle can be utilized for a reliable reconstruction of quantum states of light fields. We have also presented a numerical application of the reconstruction scheme via the MaxEnt principle for a reconstruction of Wigner functions of quantum-mechanical states of light from incomplete tomographic data. We have shown that when the tomographic data are incomplete, then the reconstruction via the MaxEnt principle is much more reliable than the standard inversion Radon transformation scheme or the pattern function scheme. For a more detailed account of the problem of reconstruction of Wigner functions from incomplete data see the recent review [21].

Acknowledgements

We thank Gerhard Adam, Rado Derka, Peter Knight, Artur Ekert, Serge Massar, Ulf Leonhardt, Zdeněk Hradil, Tomáš Opatrný, Konrad Banaszek, and Jason Twamley for useful discussions. This work was supported in part by the Slovak Academy of Sciences and by the Royal Society.

References

1. Peres, A. (1993) Quantum Theory: Concepts and Methods. Kluwer, Dordrecht
2. Ballentine, L.E. (1990) Quantum Mechanics. Prentice Hall, Englewood Cliffs, NJ
3. Wehrl, A. (1978) Rev Mod Phys 50: 221
4. Fick, E., Sauermann, G. (1990) The Quantum Statistics of Dynamic Processes. Springer-Verlag, Berlin
5. Bužek, V., Adam, G., Drobný, G. (1996) Ann Phys (NY) 245: 37
   Bužek, V., Adam, G., Drobný, G. (1996) Phys Rev A 54: 801
6. Jaynes, E.T. (1957) Phys Rev 108: 171; (1957) 108: 620; (1963) Am J Phys 31: 66; see also
   Jaynes, E.T. (1963) Information theory and statistical mechanics. In: Ford, K.W. (Ed.) 1962 Brandeis Lectures, Vol. 3. Benjamin, p 181
7. Glauber, R.J. (1963) Phys Rev Lett 10: 84
   Sudarshan, E.C.G. (1963) Phys Rev Lett 10: 277
8. Wigner, E.P. (1932) Phys Rev 40: 749; see also
   Hillery, M., O'Connell, R.F., Scully, M.O., Wigner, E.P. (1984) Phys Rep 106: 121
9. Cahill, K.E., Glauber, R.J. (1969) Phys Rev 177: 1857; (1969) 177: 1882
10. Ekert, A.K., Knight, P.L. (1991) Phys Rev A 43: 3934
11. Vogel, K., Risken, H. (1989) Phys Rev A 40: 2847; see also
    Bertrand, J., Bertrand, P. (1987) Found Phys 17: 397
    Freyberger, M., Vogel, K., Schleich, W.P. (1993) Phys Lett A 176: 41
    Vogel, W., Welsch, D.-G. (1995) Acta Phys Slov 45: 313
    Leonhardt, U., Paul, H. (1994) Phys Lett A 193: 117
    Leonhardt, U., Paul, H. (1994) J Mod Opt 41: 1427
    Kühn, H., Welsch, D.-G., Vogel, W. (1994) J Mod Opt 41: 1607
    Leonhardt, U. (1993) Phys Rev A 48: 3265
    Leonhardt, U., Paul, H. (1993) Phys Rev A 48: 4598
    Paul, H., Leonhardt, U., D'Ariano, G.M. (1995) Acta Phys Slov 45: 261
12. D'Ariano, G.M., Machiavelo, C., Paris, M.G.A. (1994) Phys Rev A 50: 4298
13. Leonhardt, U., Paul, H., D'Ariano, G.M. (1995) Phys Rev A 52: 4899
14. Richter, T. (1996) Phys Lett A 211: 327
    Richter, T. (1996) Phys Rev A 53: 1197
15. Leonhardt, U., Munroe, M., Kiss, T., Richter, T., Raymer, M.G. (1996) Opt Commun 127: 144
16. D'Ariano, G.M., Leonhardt, U., Paul, H. (1995) Phys Rev A 52: R1801
17. Bužek, V., Vidiella-Barranco, A., Knight, P.L. (1992) Phys Rev A 45: 6570
    Bužek, V., Knight, P.L. (1995) Quantum interference, superposition states of light and nonclassical effects. In: Wolf, E. (Ed.) Progress in Optics, Vol. 34. North Holland, Amsterdam, p 1
18. Smithey, D.T., Beck, M., Raymer, M.G., Faridani, A. (1993) Phys Rev Lett 70: 1244
    Beck, M., Smithey, D.T., Raymer, M.G. (1993) Phys Rev A 48: R890
    Raymer, M.G., Beck, M., Mc Alister, D.F. (1994) Phys Rev Lett 72: 1137
    Munroe, M., Boggavarapu, D., Anderson, M.E., Raymer, M.G. (1995) Phys Rev A 52: R924

19. Leonhardt, U., Munroe, M. (1996) Phys Rev A 54: 3682; see also
    Wünsche, A. (1996) Phys Rev A 54: 5291
20. Opatrný T., Welsch, D.-G. (1997) Phys Rev A 55: 1462
    Tan, S.M. (1997) J Mod Opt 44: 2233 and references therein.
21. Bužek, V., Drobný, G., Derka, R., Adam, G., Wiedeman, H. (1999) Chaos,
    Solitons and Fractals 10: 981

Part II

Information Optics

Some New Aspects of the Resolution in Gaussian Pupil Optics

S. S. Lee, M. H. Lee, and Y. R. Song

Summary. An optical image is formed by the diffraction of an optical wave at the pupil. Rayleigh's pupil gives Rayleigh's limit of resolution, ε_R, an important resolution criterion. A Gaussian pupil may give a resolution of less than ε_R. The latest computer-aided optical working technology provides the required aspherical pupil with unprecedented speed and accuracy. In this paper, some new aspects of Gaussian pupil imagery are discussed in connection with its resolution, which is of the order of $\lambda/10$; this resolution is beyond Rayleigh's criterion, which is of the order of λ. Also, a criterion for the Gaussian function constant, σ, is discussed.

1 Introduction

The recent computer-aided optical working technology has made great advancement and an aspherical optical surface can now be generated with high precision at a considerably high speed [1]. Among the conic surfaces, the spherical surface is certainly most important, and the diffraction image and resolution of the optical system are usually discussed in the context of a pupil with a constant complex amplitude throughout the aperture, which is known as Rayleigh's pupil [2,3]. However, with the advent of the above optical working technology, it may be very worthy to consider again the imagery of an aspherical pupil, typically the Gaussian pupil, which has a Gaussian amplitude in the pupil function [4]. The Gaussian-form amplitude at the pupil produces another form of diffraction amplitude distribution at the image plane [5–8] (see Fig. 1) which is different from that of Rayleigh's pupil, so that the resolution is different from Rayleigh's criterion.

As the current theories of resolution and the optical transfer function are nearly all based on Rayleigh's pupil, Gaussian pupil imagery requires some new approaches; however, they are parallel to those already developed for Rayleigh's pupil. The lithography step-scan camera system with a Gaussian pupil will become very useful for the higher resolution required for higher density chip fabrication and for larger wafer processing.

In this paper we offer a one-dimensional approach, paying most attention to resolving the performance [9,10] of an aberration-free system. It can, how-

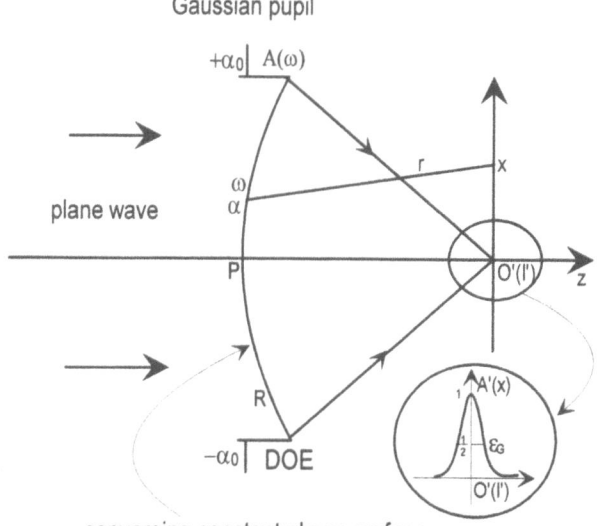

Gaussian pupil

converging constant phase surface

Fig. 1. Diffraction at a Gaussian pupil. Pupil function, $A = ae^{i\phi}$, where $a = \exp(-\frac{\omega^2}{4\sigma^2})$ and $\phi = 0$. α: pupil coordinate; $\omega = \frac{2\pi}{\lambda}\frac{\alpha}{l'} = \frac{2\pi}{\lambda}\frac{\alpha_0}{l'}\frac{\alpha}{\alpha_0} = \omega_0\frac{\alpha}{\alpha_0}$; R: reference sphere; $r^2 = (x-\alpha)^2 + l'^2$; PO': optical axis; O': image point with the geometric coordinate l' along the optical axis; DOE (diffraction optical element) is an amplitude modulation plate (AMP) and a lens

ever, be extended to rectangular and circular pupils with aberration [11–13] through some mathematical elaboration.

2   Diffraction at the Gaussian Pupil and Transforms

In Fig. 1, a one-dimensional Gaussian imaging system (case of a cylindrical lens) is shown. If the amplitude pupil function $A(\omega)$ is $ae^{-i\phi}$, where $a = \exp(-\omega^2/4\sigma^2)$ and $\phi = 0$, then the complex amplitude distribution function $A'(x)$ at the image plane is given by Huygens' principle or Green's theorem as follows:

$$A'(x) = \int \exp\left(-\frac{\omega^2}{4\sigma^2}\right)\frac{1}{r}\exp(ikr)\,d\alpha \tag{1}$$

where $k = \frac{2\pi}{\lambda}$, $\omega = k\frac{\alpha}{l'}$ and

$$r = \{l'^2 + (x-\alpha)^2\}^{1/2} \simeq l' + \frac{(x-\alpha)^2}{2l'}.$$

r, the denominator of (1), may be approximated to be l', then we have

$$A'(x) = \int \exp\left(-\frac{\omega^2}{4\sigma^2}\right)\frac{e^{ikl'}}{l'}\exp\left\{\frac{ik}{2}\frac{(x-\alpha)^2}{l'}\right\}\,d\alpha$$

$$= c(l') \int \exp\left(-\frac{\omega^2}{4\sigma^2}\right) \exp\left\{d(l')(x-\alpha)^2\right\} d\alpha, \tag{2}$$

where $c(l') = \frac{e^{ikl'}}{l'}$, $d(l') = \frac{ik}{2l'}$ and $i = \sqrt{-1}$.

Equation (2) is the Fresnel diffraction integral of the real amplitude of the pupil function $\exp(-\frac{\omega^2}{4\sigma^2})$. As O' (in Fig. 1) is the diffraction image point, $A'(x)$ extends over a small region, i.e. x is small. Also, $\frac{\alpha_0}{l'}$ ($\alpha \le \alpha_0$) is of the order of 10^{-1} in any practical system, so that we can make a paraxial approximation of (2) and obtain the following equation:

$$A'(x) = c(l') \int \exp\left(-\frac{\omega^2}{4\sigma^2}\right) \exp\left(-ik\frac{x\alpha}{l'}\right) d\alpha$$

$$= c(l') \int \exp\left(-\frac{\omega^2}{4\sigma^2}\right) \exp(-i\omega x) d\omega, \tag{3}$$

where $\omega = k\frac{\alpha}{l'}$.

Equation (3) can be normalized and may be rewritten as

$$A'(x) = \int_{-\omega_0}^{+\omega_0} \exp\left(-\frac{\omega^2}{4\sigma^2}\right) \exp(-i\omega x) d\omega. \tag{4}$$

This equation shows that $A'(x)$ may be obtained by taking the Fourier transform of the pupil amplitude, $\exp(-\frac{\omega^2}{4\sigma^2})$, of the pupil function in the finite aperture $-\alpha_0$ to $+\alpha_0$ or $-\omega_0$ to $+\omega_0$.

In the case of Rayleigh's pupil, the value a of the pupil function is unity and $\phi = 0$, so that the diffraction amplitude distribution is

$$A'(x) = 2\omega_0 \frac{\sin \omega_0 x}{\omega_0 x}, \tag{5}$$

and

$$\omega_0 \varepsilon_R = \pi \tag{6}$$

gives the Rayleigh's resolution criterion, ε_R, which is

$$\varepsilon_R = \frac{\lambda}{2(NA)},$$

where $NA = \frac{\alpha_0}{l'}$, which is the numerical aperture in air, and when $NA = 0.5$

$$\varepsilon_R = \lambda. \tag{7}$$

Rayleigh's resolution criterion is an important milestone; it allows the resolution of an optical imaging system to be quantified. As λ becomes shorter and NA becomes larger, ε_R becomes smaller and the resolving power ε_R^{-1} becomes larger.

3 Paraxial Gaussian Diffraction Amplitude and Resolution

Equation (4) may be rewritten as

$$
A'(x) = 2\sigma \exp(-\sigma^2 x^2) \int_{-\frac{\omega_0}{2\sigma}+i\sigma x_0}^{\frac{\omega_0}{2\sigma}+i\sigma x_0} \exp\left\{-\left(\frac{\omega}{2\sigma}+i\sigma x\right)^2\right\} \mathrm{d}\left(\frac{\omega}{2\sigma}+i\sigma x\right)
$$

$$
= 2\sigma \exp(-\sigma^2 x^2) \int_{Z_1}^{Z_2} \exp(-Z^2) \,\mathrm{d}Z
$$

$$
= 2\sigma \exp(-\sigma^2 x^2) \cdot J(x_0), \tag{8}
$$

where

$$
J(x_0) = \int_{Z_1}^{Z_2} \exp(-Z^2) \,\mathrm{d}Z. \tag{9}
$$

The Z's in (8) and (9) are

$$
Z = \frac{\omega}{2\sigma} + i\sigma x,
$$

$$
Z_1 = -\frac{\omega_0}{2\sigma} + i\sigma x_0 \quad \text{and} \tag{10}
$$

$$
Z_2 = \frac{\omega_0}{2\sigma} + i\sigma x_0.
$$

In order to obtain the analytical expression of $A'(x)$, we carry out the integration of (9) in the complex plane ($Z = \frac{\omega}{2\sigma} + i\sigma x$, Fig. 2).

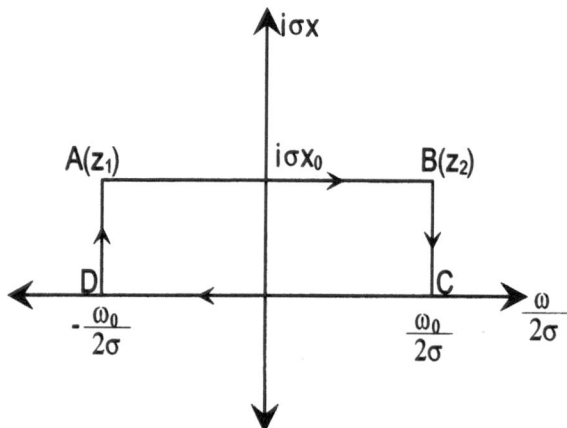

Fig. 2. Line integration along ABCD in the complex plane, $Z = \frac{\omega}{2\sigma} + i\sigma x$

In Fig. 2

$$\oint_{ABCD} \exp(-Z^2)\,dZ = 0,$$

namely,

$$\int_{AB} + \int_{BC} + \int_{CD} + \int_{DA} = 0,$$

so that $\int_{AB}$ of (9) is given by

$$J(x_0) = \int_{AB} = \int_{DC} + \left(\int_{CB} - \int_{DA} \right), \tag{11}$$

where

$$\int_{DC} = \int_{-\frac{\omega_0}{2\sigma}}^{\frac{\omega_0}{2\sigma}} \exp\left\{ -\left(\frac{\omega}{2\sigma}\right)^2 \right\} d\left(\frac{\omega}{2\sigma}\right) \equiv P(\sigma, \omega_0) \tag{12}$$

is the integration along real axis, and those along the imaginary axis are

$$\int_{CB} - \int_{DA} = \int_{\frac{\omega_0}{2\sigma}}^{\frac{\omega_0}{2\sigma}+i\sigma x_0} \exp\left\{ -\left(\frac{\omega_0}{2\sigma} + i\sigma x\right)^2 \right\} d(i\sigma x)$$

$$- \int_{-\frac{\omega_0}{2\sigma}}^{-\frac{\omega_0}{2\sigma}+i\sigma x_0} \exp\left\{ -\left(-\frac{\omega_0}{2\sigma} + i\sigma x\right)^2 \right\} d(i\sigma x)$$

$$= i\sigma \exp\left\{ -\left(\frac{\omega_0}{2\sigma}\right)^2 \right\} \int_0^{x_0} \exp(\sigma^2 x^2)\{ \exp(-i\omega_0 x) - \exp(i\omega_0 x)\}dx$$

$$= 2\sigma \exp\left\{ -\left(\frac{\omega_0}{2\sigma}\right)^2 \right\} \int_0^{x_0} \exp(\sigma^2 x^2) \sin(\omega_0 x)\,dx \equiv Q(\sigma, \omega_0, x_0). \tag{13}$$

In (13), $x \le x_0$ and x_0 is arbitrary and may be of the order of the half width at half maximum (HWHM) of the Gaussian image, $\exp(-\sigma^2 x^2)$. Substituting (12) and (13) into (11), we get

$$J(x_0) = P(\sigma, \omega_0) + Q(\sigma, \omega_0, x_0) \tag{14}$$

and then from (8), $A'(x)$ of (4) at the Gaussian image point, O', in Fig. 1 is given as follows:

$$A'(x) = 2\sigma \exp(-\sigma^2 x^2)\{P(\sigma, \omega_0) + Q(\sigma, \omega_0, x_0)\} \tag{15}$$

and, upon normalizing, we obtain

$$A'(x) = \exp(-\sigma^2 x^2), \tag{16}$$

which has the following HWHM (resolution) of the Gaussian pupil, ε_G, and resolving power:

$$\text{Resolution} = \varepsilon_G = \frac{\sqrt{\ln 2}}{\sigma} \quad (\text{cm}),$$

$$\text{Resolving power} = \varepsilon_G^{-1} = \frac{\sigma}{\sqrt{\ln 2}} \quad (\text{cm}^{-1}). \tag{17}$$

4   Numerical Estimates and a Criterion for σ

A recent highly developed lithography optical system has a resolution of the order of λ, which is of the order of 10^{-5} cm $= 0.1\,\mu$m. This resolution is close to the Rayleigh's criterion, ε_R. In this paper, by using Gaussian pupil optics, we want to show that a resolution ε_G of the order of $\varepsilon_R/10 = 0.01\,\mu$m is readily obtainable.

Since we assume $\varepsilon_G = 0.01\,\mu$m and $NA = 0.5$,

$$\text{Resolving power} \approx 10^6 \quad (\text{cm}^{-1}),$$
$$\lambda \approx 10^{-5} \quad (\text{cm}),$$
$$\omega_0 = \frac{2\pi}{\lambda}\frac{\alpha_0}{l'} \approx 10^4 \quad (\text{cm}^{-1}), \tag{18}$$
$$\sigma = \frac{\sqrt{\ln 2}}{\varepsilon_G} \approx 10^5 \quad (\text{cm}^{-1}).$$

These numerical data give the pupil function

$$\exp(-\frac{\omega^2}{4\sigma^2}) \approx \exp\left\{-\frac{10^8}{10^{10}}\left(\frac{\alpha}{\alpha_0}\right)^2\right\} = \exp\left\{-\frac{1}{10^2}\left(\frac{\alpha}{\alpha_0}\right)^2\right\}, \tag{19}$$

where

$$-1 \le \frac{\alpha}{\alpha_0} \le +1,$$

which can be generated without too much difficulty by using an amplitude modulation plate as discussed in the following section.

There is, however, a criterion for the magnitude of σ. This criterion is obtained by considering the threshold power of the image required for recording. As resolution ε_G improves, σ becomes larger. This larger σ makes the pupil function broaden, and it approaches Rayleigh's pupil. The total power of a pupil $2\alpha_0$ wide is approximately

$$2\alpha_0\beta \tag{20}$$

for intensity (per cm) β. This power is, from (5), nearly equal to the total power of the image, which is

$$(4\omega_0^2\beta)x_h, \tag{21}$$

where $4\omega_0^2\beta$ is the image intensity (per cm) and x_h is approximately the full width at the half maximum (FWHM) of the central lobe in the Rayleigh's diffraction pattern.

From (20) and (21),

$$x_h = \frac{\alpha_0}{2\omega_0^2}. \tag{22}$$

In the image, for intensity (per cm) i, the total power is $2\varepsilon_G \cdot i$, which is larger than $x_h \cdot i_{th}$ (i_{th} being the threshold intensity depending on the characteristics of the recorder), namely,

$$2\varepsilon_G \cdot i \geq x_h \cdot i_{th} = \frac{\alpha_0}{2\omega_0^2} i_{th} \tag{23}$$

and the HWHM of the image intensity, Δx, is $\Delta x = \frac{1}{\sqrt{2}} \varepsilon_G = \frac{1}{\sigma}\sqrt{\frac{\ln 2}{2}}$. From these relations we get the following inequality:

$$\sigma \leq 2\sqrt{\ln 2}\, \frac{2\omega_0^2}{\alpha_0} \frac{i}{i_{th}} < 2\sqrt{\ln 2}\, \frac{2\omega_0^2}{\alpha_0}, \quad \text{as} \quad i \approx i_{th} \quad \text{for minimal } i. \tag{24}$$

Assuming $\alpha_0 \approx 10^0$ cm, $\lambda \approx 10^{-5}$ cm, and $\omega_0 \approx 10^4$ cm^{-1} (see (18)), we get

$$\sigma \leq 10^8. \tag{25}$$

When $\sigma \approx 10^8$, $\Delta x \approx 10^{-8}$ cm. This result indicates a resolving power beyond the order of 10^8 is not possible in Gaussian pupil optics in the region of $\lambda \approx 10^{-5}$ cm.

5 Amplitude Modulation Plate (AMP) for the Gaussian Pupil

In Fig. 3a, the point source on the optical axis is collimated by using the lens L_c, and the amplitude of the plane wave is modulated by using an amplitude modulation plate (AMP), which is plane parallel. The AMP generates the Gaussian wave which comes to the pupil for image formation. As shown in Fig. 3b, the AMP is made from two types of optical glass, G_1 (refractive index n_1) which is convex and G_2 (complex refraction index n_2 in$_2'$, i $= \sqrt{-1}$) which is concave. They are sandwiched and the equation of the boundary between G_1 and G_2, $f(\alpha)$ is to be derived.

Ray P in Fig. 3b is incident at the AMP at a height α in G_1, which has no absorption; however, in G_2, absorption takes place according to Beer's law. As P traverses through the AMP, of thickness d, the intensity transmitted should be equal to $\{\exp(-\frac{\omega^2}{4\sigma^2})\}^2$. Namely,

$$\exp\left\{-\kappa[d - f(\alpha)]\right\} = \exp\left(-\frac{\omega^2}{2\sigma^2}\right), \tag{26}$$

where κ is the extinction coefficient of G_2 and equal to $\frac{4\pi}{\lambda} n_2'$. Thus,

$$-\kappa\{d - f(\alpha)\} = -\frac{\omega^2}{2\sigma^2}$$

and, as $\omega = \frac{2\pi}{\lambda}\frac{\alpha}{l'}$,

$$f(\alpha) = d - \frac{2\pi^2 \alpha^2}{\kappa \sigma^2 l'^2 \lambda^2}, \tag{27}$$

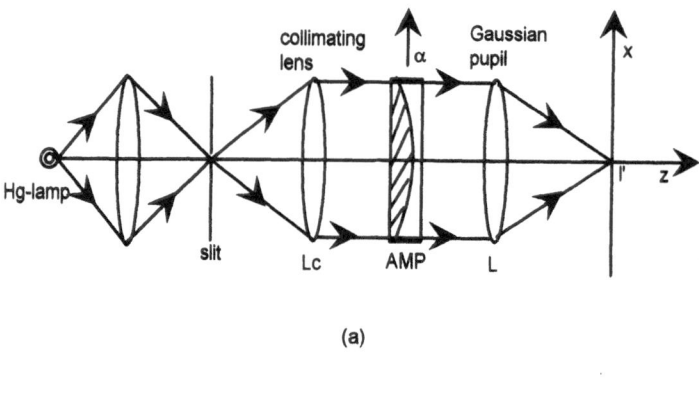

(a)

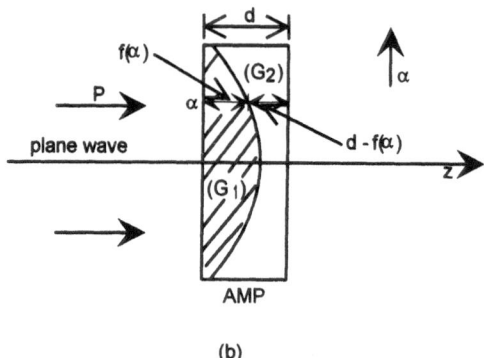

(b)

Fig. 3. (a) Optical imaging system using an amplitude modulation plate (AMP). (b) Fabrication of an AMP

which is parabolic. This parabolic surface can be generated precisely using the recent computer-aided optical working technology.

Another important matter for the AMP is that a plane wave incident at it should transmit through the plate as a plane wave without a distortion in the constant phase plane. The optical path length, $\phi(\alpha)$, for a ray transmitted through the AMP at height α is

$$n_1\phi(\alpha) + n_2\{d - \phi(\alpha)\}. \tag{28}$$

When $n_1 = n_2$, the phase change, $\phi(\alpha) = \frac{2\pi}{\lambda}n_2 d$, which is constant. This situation indicates that the real refractive indices of G_1 and G_2 should be same and G_2 alone should have an absorbing property. G_2 may be prepared with the same constituent materials as G_1, but with a minute amount of highly absorbing material added. Alternative, we may first make a choice of G_2, with the refractive index $n_2 - in_2'$ ($n_2' \neq 0$), and then find a transparent glass with the refractive index n_2.

6   Optical Transfer Function (OTF)

The optical transfer function, $\Psi_G(\Omega)$, is the normalized Fourier transform of the paraxial diffraction intensity distribution, $I(x) = |A'(x)|^2$. In the case of the Gaussian pupil, $I(x)$, from (16), is

$$I(x) = |\exp(-\sigma^2 x^2)|^2 = \exp(-2\sigma^2 x^2). \tag{29}$$

The Fourier transform of $I(x)$ is

$$F_G(\Omega) = \int_{-\infty}^{+\infty} \exp(-2\sigma^2 x^2)\exp(-i\Omega x)\,\mathrm{d}x, \tag{30}$$

where Ω is the spatial frequency ($[2\pi \cdot \mathrm{cm}^{-1}]$). Equation (30) is equal to

$$F_G(\Omega) = \frac{1}{\sqrt{2}\sigma} \exp\left(-\frac{\Omega^2}{8\sigma^2}\right) \int_{-\infty}^{+\infty} \exp\left\{-\left(\sqrt{2}\sigma x + \frac{i\Omega}{2\sqrt{2}\sigma}\right)^2\right\}\mathrm{d}(\sqrt{2}\,\sigma x)$$

and, upon normalization, the OTF, $\Psi_G(\Omega)$, is

$$\Psi_G(\Omega) = \exp\left(-\frac{\Omega^2}{8\sigma^2}\right) \tag{31}$$

and, as can be seen, there is no Ω which gives $\Psi_G(\Omega) = 0$. When the OTF is e^{-1}, we get

$$\Omega_G = \sqrt{8}\sigma \approx 3\sigma,$$

and, as is discussed in Sect. 4, we take σ to be of the order of 10^5 (cm^{-1}); therefore $\Omega_G \approx 3 \times 10^5$ (2π lines/cm). A regular holographic plate has an Ω of the order of 10^4, so the resolution of the Gaussian pupil is higher by an order of magnitude. The OTF in the case of Rayleigh's pupil is $\Psi_R = 1 - \frac{\Omega_R}{2\omega_0}$ (one dimension), and when it is e^{-1} we get

$$\Omega_R = 2\omega_0\left(1 - \frac{1}{e}\right) \approx 10^4. \tag{32}$$

Thus, it can be seen that $\Psi_G > \Psi_R$.

7   Conclusions

We have shown that Gaussian pupil optics can give upgraded image resolution characteristics beyond Rayleigh's criterion. Currently, the application of various high resolving power methods, such as the use of ferro-electric thin film and spectral hole burning is being discussed as a way to give higher density data storages [14]. However, in this paper it is shown that, by using Gaussian pupil optics, we may be able to improve, without much difficulty, the resolution by an order of magnitude to 10^{-6} cm.

Aspherical surface generation becomes easier and faster as the computer-aided optical working technology is further developed at The University of Rochester, NY, USA [1]. This technology gives the impression that it should become very important and useful in the coming century.

The higher resolution optics using paraxial diffraction optics with a Gaussian pupil function discussed in this paper can be useful in a one-dimensional scanning system – object scans in the $+x$ direction and recording material or the wafer does in the $-x$ direction. The slit width in the object plane is constrained as the plane-parallel amplitude modulation plate (AMP) should accept the plane wave normal to the surface of the AMP. In the case of a two-dimensional Gaussian pupil, scanning along the y-axis is necessary (y-scan). In the system, the AMP to generate $\exp(-\frac{y^2}{4\sigma^2})$ and a cylindrical lens are to be added. Also, instead of a slit as for the one-dimensional case, a rectangular aperture must be employed.

In realizing the overall optical system, a good number of further detailed considerations are required. The time program for the scanning process, tolerance, a pulsed or CW light source and, finally, the elaborate electro-mechanical moving mechanism are important subjects for further consideration.

We think even more applications of aspherical surfaces and aspherical optical elements will be recognised in optics and laser instrumentation. Time-old aspherical optics requires renewed attention for the step-scan optical lithography system [15]. We are looking for glass suitable for G_1 (refractive index $= n_2$) and G_2 (refractive index $= n_2 - in_2'$) in the preparation of the AMP. These different types of glass should be useful for any amplitude modulation necessary in apodization optics; however, their preparation requires some optical glass manufacturing research.

Acknowledgement

The authors thanks Dr. Hae Bin Chong, senior scientist in charge of the High Resolution Lithography Optical System Development Laboratory in the Electronic and Telecommunication Research Institute (ETRI, in Taejon, Korea) for his kind support on this work.

References

1. (1998) Machine Manufactures Aspheres in Minutes. OE Reports, SPIE 4
2. Welford, W.T. (1986) Aberrations of Optical System. Adam Hilger Ltd, Bristol Boston
3. Born, M., Wolf, E. (1964) Principles of Optics. Pergamon Press, New York
4. Mandel, L., Wolf, E. (1995) Optical Coherence and Quantum Optics. Cambridge Univ. Press, Cambridge
5. Chung, C.S., Hopkins, H.H. (1989) Influence of nonuniform amplitude on the optical transfer function. Appl Optics 28(6): 1244

6. Sheppard, C.J.R., Saghafi, S. (1998) Gaussian beams beyond the paraxial application. Proc OII'98 Topical Meeting ICO, p 42

7. Zhang, J., Huo, J., Liao, R. (1998) Control of wavefront propagation with phase-only diffractive optical elements. Proc. OII'98 Topical Meeting ICO, p 69

8. Andreic, Z. (1994) Superresolution performance of an absorbing glass positive lens element. Appl Optics 36(19): 4354

9. den Dekker, A.J., van den Bos, A. (1997) Resolution: a survey. J Opt Soc Am A 14(3): 547

10. Sementilli, P.J., Hunt, B.R., Nadar, M.S. (1993) Analysis of the limit to super-resolution in incoherent imaging. J Opt Soc Am A 10(11): 2265

11. Mills, J.P., Thompson, B.J. (1986) Effect of aberrations and apodization on the performance of coherent optical systems. II. Imaging. J Opt Soc Am A 3(5): 704

12. Tschunko, H.F.A. (1983) Imaging performance of annular apertures. 4: Apodization and point spread functions. Appl Optics 22(1): 133

13. Ruschin, S. (1992) Transverse profile shaping of laser beams by means of aspherical mirror resonators. SPIE 1834: 169

14. van Houten, H., Schleipen, J. (1998) Optical data storage. Phys World 11(10): 33

15. Hardin, R.W. (1997) Lasers in Lithography. Photonics Spectra 95

Multichannel Photography with Digital Fourier Optics

G.G. Mu, L. Lin, and Z.-Q. Wang

Summary. In this chapter we describe the technique of multichannel photography with N-superimposed grating and then present new progress in decoding with digital Fourier optics.

1 Introduction

In the world of today, the key branch of the economy is changing from industry to information sciences, and information technology, including electronic information and optical information, is of great importance. Wavelength division multiplexing (WDM) has been widely applied to optical communication to enhance the information transmission speeds and capacity. Some examples are the curved diffraction-grating WDM [1], the Mach–Zehnder WDM [2], and the arrayed-waveguide WDM [3]. Wavelength division multiplexing has also been applied to optical data storage, such as holographic recording to enhance storage capacity. Some examples are orthogonal wavelength multiplexing [4] and hybrid sparse-wavelength angle-multiplexing [5]. One recent development in the wavelength division multiplexing technique is the use of superimposed gratings on a plenary waveguide [6]. In this model, gratings are designed so that each of them diffracts one particular wavelength to one particular direction for a common input. The concept of superimposed grating wavelength division multiplexing can be applied to optical data recording or storage for a multispectrum target. For this purpose the superimposed grating should be designed to be able to record one particular spectrum information in one channel and diffract one particular wavelength to one particular spatial direction. In fact, it has been realized in color photography [8, 9], where a specifically designed and fabricated superimposed grating, tricolor-grating encoder, has been used to record a color object on a black-and-white film. For the retrieval of the color information, the developed film is input to a white-light processor, where different color information is diffracted to different spatial positions in the Fourier plane. By color filtering at the Fourier plane, the original color image can be retrieved at the output plane of the processor.

2 Superimposed Grating for Multichannel Photography

The key device in multichannel photography with wavelength division multiplexing is the superimposed grating. Figure 1 shows a three-wavelength superimposed grating, the tricolor grating, which is a composition of three gratings with different wavelength transmittance, $\lambda_1, \lambda_2, \lambda_3$, oriented in three different directions, x_1, x_2, x_3. The intensity transmittance of the superimposed grating can be expressed as

$$t_g(x,y) = \left(\frac{1}{2} + \frac{1}{2}\text{sgn}(\cos p_0 x_1)\right)_{\lambda_1} + \left(\frac{1}{2} + \frac{1}{2}\text{sgn}(\cos p_0 x_2)\right)_{\lambda_2} \\ + \left(\frac{1}{2} + \frac{1}{2}\text{sgn}(\cos p_0 x_3)\right)_{\lambda_3}, \tag{1}$$

where p_0 is the angular spatial frequency of the gratings and

$$\text{sgn}(\cos p_0 x) = \begin{cases} 1 & \cos p_0 x \geq 0 \\ -1 & \cos p_0 x < 0 \end{cases}. \tag{2}$$

We emphasize that p_0 could have a different value for the gratings with different orientations. For simplicity, we assume that it has the same value. Similarly, the intensity transmittance of an N-wavelength superimposed grating can be expressed as

$$t_g(x,y) = \sum_{n=1}^{N} \left[\frac{1}{2} + \frac{1}{2}\text{sgn}(\cos p_0 x_n)\right]_{\lambda_n}. \tag{3}$$

The superimposed grating could operate in the wavelength range of microwave, infrared and ultraviolet. For visible wavelength division multiplexing, the superimposed grating can be fabricated by photography. For instance, an N-wavelength superimposed grating can be fabricated by N sequential exposures to a fresh color slide close-contacted with a Ronchi ruling. With

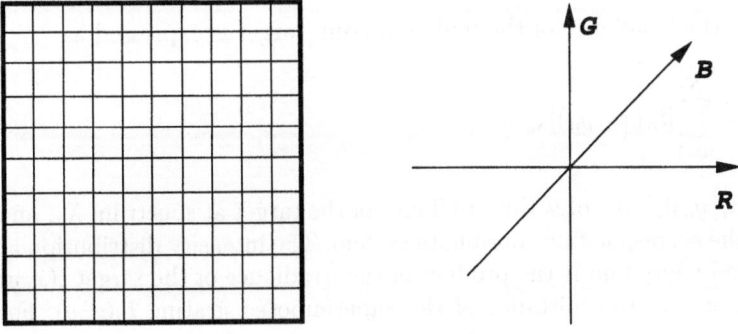

Fig. 1. Three-wavelength superimposed grating, the tricolor grating

each exposure, the Ronchi ruling should be at a certain orientation, and a correspondent color filter should be used in the illumination system. After development, the N-wavelength superimposed grating will be obtained with the intensity transmittance given by (3).

3   Multichannel Photography with Superimposed Grating

3.1   With an Ordinary Camera

For WDM recording of a multispectrum target, the superimposed grating is placed at the image plane of an ordinary camera, close-contacted with panchromatic black-and-white film. The camera is then used to take a photography of the target with a single exposure. Figure 2 shows the recording scheme.

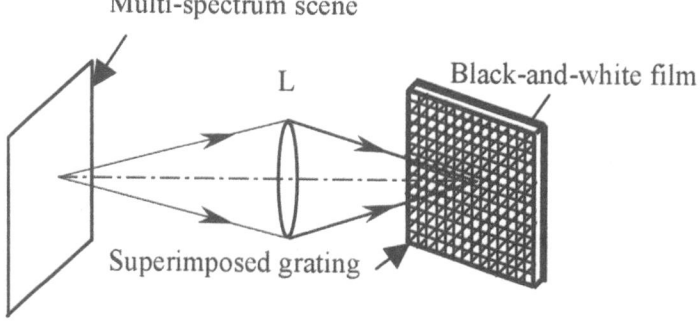

Multi-spectrum scene

L

Black-and-white film

Superimposed grating

Fig. 2. Configuration of the multichannel photography with superimposed grating

Suppose the irradiance of the multispectrum target is expressed as

$$t(x,y) = \sum_{n=1}^{N} [t_n(x_n, y_n)]_{\lambda_n} ,$$

(4)

where $[t_n(x_n,y_n)]_{\lambda_n}$ denotes the irradiance of the target at spectrum λ_n, and (x_n, y_n) is the corresponding coordinate system. The intensity distribution at the black-and-white film is the product of the irradiance of the target $t(x,y)$ with the intensity transmittance of the superimposed grating $t_g(x,y)$. For particular spectrum information of the target at λ_n only the correspondent grating $[1/2 + 1/2\,\mathrm{sgn}(\cos p_0 x_n)]_{\lambda_n}$ plays a role, and others do not allow any passage. Thus, after exposure and development, the intensity transmittance of the recorded transparency can be expressed as

$$t_{\text{neg}}(x, y) = \left[\sum_{n=1}^{N} [t_n(x_n, y_n)]_{\lambda_n} \left(\frac{1}{2} + \frac{1}{2} \text{sgn}(\cos p_0 x_n) \right)_{\lambda_n} \right]^{-\gamma_{n1}}, \qquad (5)$$

where γ_{n1} is the gamma value of the film. To obtain a positive transparency, a contact printing process with another fresh film is needed. The intensity transmittance of the positive transparency is given by

$$t_{\text{pos}}(x, y) = \left[\sum_{n=1}^{N} [t_n(x_n, y_n)]_{\lambda_n} \left(\frac{1}{2} + \frac{1}{2} \text{sgn}(\cos p_0 x_n) \right)_{\lambda_n} \right]^{\gamma_{n1}\gamma_{n2}}, \qquad (6)$$

where γ_{n2} is the gamma value of the second film. If the films used in the negative recording and positive printing are chosen such that the product of γ_{n1} with γ_{n2} is equal to 2, the intensity transmittance of the positive transparency will be expressed as

$$t_{\text{pos}}(x, y) = \left[\sum_{n=1}^{N} [t_n(x_n, y_n)]_{\lambda_n} \left(\frac{1}{2} + \frac{1}{2} \text{sgn}(\cos p_0 x_n) \right)_{\lambda_n} \right]^{2}. \qquad (7)$$

It can be seen that one particular spectrum information of the target is spatially modulated by one grating with a particular azimuth orientation. Multichannel photography of the multispectrum object is realized by use of the superimposed grating.

3.2   With a CCD Camera

The recording can also be realized with a specially fabricated CCD camera. In this case, a superimposed grating-like filter is put in front of the receiver of the camera. Then the camera is used to take a photograph. The electronic signal coming from the CCD camera can be expressed as

$$t(x, y) = \sum_{n=1}^{N} [t_n(x_n, y_n)]_{\lambda_n} \left(\frac{1}{2} + \frac{1}{2} \text{sgn}(\cos p_0 x_n) \right)_{\lambda_n}, \qquad (8)$$

which is as same as (7) except for the square power in (7).

4   Retrieval of the Multispectrum Image with Digital Decoding

The retrieval of the multispectrum image can be realized by either optical or digital techniques. With the optical method, a white-light optical processor is needed; the procedure has been described in detail in [8]. In this chapter we concentrate on the digital decoding technique.

For the digital decoding, the recorded black-and-white transparency is first digitized into a discrete digital image, based on sampling theory, whose amplitude transmittance is obtained by the square root operation

$$f(k, l) = \sqrt{t_{\text{pos}}(k, l)} , \tag{9}$$

where $k = 0, 1, \ldots, K - 1$; $l = 0, 1, \ldots, L - 1$. Substituting (7) into (9), we have

$$f(k, l) = \sum_{n=1}^{N} [t_n(k_n, l_n)]_{\lambda_n} \left(\frac{1}{2} + \frac{1}{2} \text{sgn}(\cos p_0 k_n) \right)_{\lambda_n} . \tag{10}$$

Then the digital Fourier transform has been taken. Two-dimensional discrete Fourier transform of the function $f(k, l)$ can be expressed as

$$F(u, v) = \frac{1}{KL} \sum_{k=0}^{K-1} \sum_{l=0}^{L-1} f(k, l) \exp\left(-i \left(\frac{2\pi}{K} \right) ku \right) \exp\left(-i \left(\frac{2\pi}{L} \right) lv \right) , \tag{11}$$

where u and v denote the discrete coordinates in the Fourier domain, K and L are the row and column of the transform window, respectively, $u = 0, 1, \ldots, K - 1$, and $v = 0, 1, \ldots, L - 1$.

It can be seen from (10) and (11) that the Fourier spectra $F(u, v)$ are N sets of spatial spectra, each of which is a convolution of the spatial spectra of an oriented grating with that of the corresponding spectrum image. For the retrieval of the original image a multichannel wavelength filtering should be adopted. Suppose the row and the column of the transform window are the same, i.e. $K = L$. The coordinates of the first-order spatial spectra of the grating corresponding to the wavelength λ_n are given by

$$u_n = \frac{p_0 K \sin(\theta_n)}{4\pi s} ,$$
$$v_n = \frac{p_0 K \sin(\theta_n)}{4\pi s} , \tag{12}$$

with

$$\theta_n = \frac{\pi n}{N} , \tag{13}$$

where s is the sampling frequency of the numeralization and p_0 is the angular spatial frequency of the grating. This relationship is shown in Fig. 3. The multichannel wavelength filtering takes place at the circled areas with the center coordinates of (u_n, v_n) and the radius given by

$$r_n = \begin{cases} \dfrac{u_n}{2} & n \leq 2 \\ u_n \sin \dfrac{\theta_n}{2} & n > 2 \end{cases} . \tag{14}$$

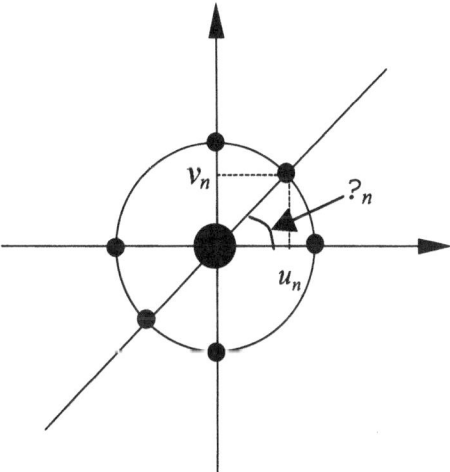

Fig. 3. Diagram of the Fourier spectra of the recording image

Suppose the radius of the circle is greater than the cut-off frequency of the target. The filtered spatial spectra are then expressed as

$$F'(u,v) = \sum_{n=1}^{N} [F_n(u - u_n, v - v_n)]_{\lambda_n} , \qquad (15)$$

where $F_n(u,v)$ is the Fourier transform of $t_n(k_n, l_n)$ in (10). The retrieval of the multispectrum image is finally obtained by an inverse Fourier transform of (15) and is given by

$$E(i,j) = \sum_{n=1}^{N} t_n(i_n, j_n)_{\lambda_n} \exp\left[i\frac{p_0}{2s}(i_n \sin\theta_n + j_n \cos\theta_n)\right] . \qquad (16)$$

By a square-power operation, the intensity distribution displayed in the output is given by

$$I(i,j) = \sum_{n=1}^{N} |t_n(i_n, j_n)|^2_{\lambda_n} , \qquad (17)$$

which is the square power of the irradiance of the original multispectrum target [see (4)].

In some applications of multichannel photography, separated spectrum images with gray levels are required. In this case only the multichannel inverse Fourier transforms are needed without the wavelength filtering. Equation (17) is then rewritten as

$$I_n(i,j) = |t_n(i_n, j_n)|^2_{\lambda_n} , \quad n = 1, 2, \ldots, N , \qquad (18)$$

where $I_n(i,j)$ denotes the λ_n spectrum image.

5   Local Decoding

It is clear that with the ordinary digital decoding, a Fourier transform and N inverse Fourier transforms are required. As the pixel number α increases, the time taken for the Fourier transform increases with a function of a square power of α. Furthermore, error extraction at a point in the Fourier spectrum domain will influence the whole retrieval image. For these reasons, we propose a new digital decoding technique called local decoding.

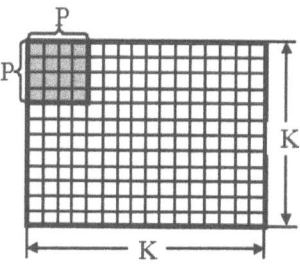

Fig. 4. Transform window for the local decoding

We start from (11). Suppose a smaller transform window with a square area $P \times P$ is adopted (shown in Fig. 4). The spatial spectrum values at the first orders are given by

$$F(u_n, v_n) = \frac{1}{P^2} \sum_{k=0}^{P-1} \sum_{l=0}^{P-1} f(k, l) \exp\left(-i\left(\frac{2\pi}{P}\right)u_n k\right) \exp\left(-i\left(\frac{2\pi}{P}\right)v_n l\right), \quad (19)$$

where u_n and v_n are given by (12) except for changes in the transform window size. It is clear that $F(u_n, v_n)$ is the average amplitude of the transform window, corresponding to the modulation frequency of the grating. The smaller the transform window is, the nearer the original spectra of the target are. However, the size of the transform window is limited by the spatial frequency of the superimposed grating:

$$\frac{1}{P} \leq \frac{p_0}{4\pi s}, \quad (20)$$

where $1/P$ is the multispectra resolution. As the spatial frequency of the grating increases, the multispectra resolution grows. It can be seen that (19) represents a multispectra pixel. By processing the whole image, a multispectra bitmap is constructed. If the transform window P reduces to $4\pi s/p_0$, the constructed multispectra bitmap can be regarded as the original multispectrum image, whose intensity distribution is given by

$$I(u, v) = \sum_{n=1}^{N} |F(u_n, v_n)|^2 = \sum_{n=1}^{N} |t_n(x_n, y_n)|^2_{\lambda_n}. \quad (21)$$

It can be seen with the local digital decoding that the large-size Fourier transform and the N inverse Fourier transforms are avoided. Instead, a number of small size Fourier transforms are adopted. Therefore, the time needed is reduced considerably.

6   Experimental Results

For a demonstration of multichannel photography with local digital decoding, a color scene is recorded on a black-and-white film by a "phoenix 205" camera equipped with a 20 lines/mm tricolor grating. After development, the negative transparency is obtained. Then the positive transparency with the intensity transmittance of (7) is obtained by close-contact printing (shown in Fig. 5a). This positive transparency is digitized at a resolution of 50 lines/mm by a Polaroid sprintscan 35 film scanner, and the discrete image data are fed into a Pentium II 266 MHz computer for decoding. The size of the discrete image is 2400×3500 pixels. In the decoding, the transform window P is chosen to be 5, according to (20), to get the finest color resolution. The time consumption is less than 5 seconds, which is compared with 30 minutes with the ordinary digital decoding based on a two-step Fourier transform. The retrieved images are shown in Fig. 5b–d, which are the gray-level images corresponding to the red, the green and the blue spectrum information of the multispectrum target, respectively.

7   Conclusions

We have outlined the technique of multichannel photography by means of wavelength division multiplexing. The key device is the superimposed grating. In the visible wavelength range, the superimposed grating can be fabricated by N sequential exposures to a fresh color slide close-contacted with a Ranchi ruling. The proposed photography is characterized as follows:

(1) It is photography with wavelength-space parallel recording and decoding.
(2) Multichannel photography is a single optical recording system with an encoding sensor, rather than a multirecording system. Therefore it is simple and convenient.
(3) It records a multispectrum target $\sum_{n=1}^{N}[t_n(x_n, y_n)]_{\lambda_n}$ in intensity $I(x,y)$. This enhances the capacity of the information recording considerably.

We have demonstrated the decoding technique with digital Fourier optics. It possesses the advantages of being commercially available, suitable for preprocessing and post-processing of the image and convenient in practice.

We have presented the experimental results of photography of a color scene with the proposed technique. The retrieved image is quite satisfactory, and the time consumption of the decoding is acceptable. Therefore it is

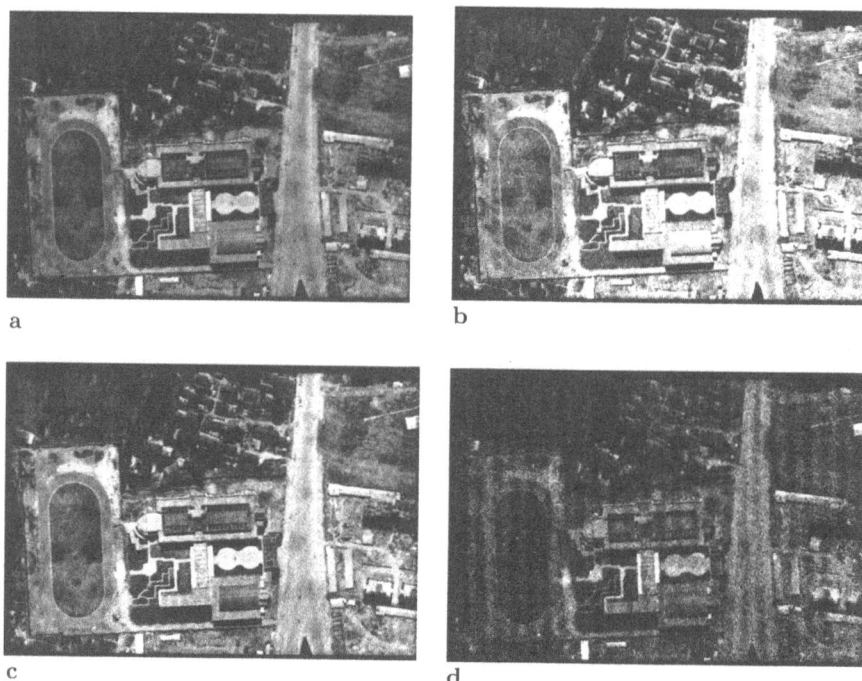

Fig. 5. Experimental results: (a) the encoded black-and-white transparency; (b) the retrieved image corresponding to the red spectrum information of the target; (c) the retrieved image corresponding to the green spectrum information of the target; (d) the retrieved image corresponding to the blue spectrum information of the target

quite competitive to the optical decoding. In our demonstration, wavelength division multiplexing is limited to the visible wavelength range. It can be extended to other wavelength ranges, such as infrared, if a suitable superimposed grating is fabricated. In the future it will find its applications more important in the extended wavelength range.

References

1. Fallahi, M., McGreer, K.A., Delage, A., Templeton, I.M., Chatenoud, F., Barber, R. (1993) Grating Demultiplexing Integrated with MSM Detector Array in InGaAs/AeGaAs/GaAs for WDM, IEEE Photonics Technol. Lett. 5: 794
2. Verveek, B.H. et al. (1988) Integrated Four-Channel Mach–Zehnder Multi/demultiplexer Fabricated with Phosphorus-doped SiO_2 Waveguides on Si, J. Lightwave Technol. 6: 1115
3. Takahashi, H., Oda, K., Toba, H., Inoue, Y. (1995) Transmission Characteristics of Arrayed Waveguides N*N Wavelength Multiplexer, J. Lightwave Technol. 13: 447

4. Rakuljic, G.A., Leyva, V., Yariv, A. (1992) Optical Data Storage by Using Orthogonal Wavelength – Multiplexed Volume Holograms, Opt. Lett. 17: 1471
5. Campbeu, S., Yi, X., Yeh, P. (1994) Hybrid Sparse-Wavelength Angle-Multiplexed Optical Data Storage System, Opt. Lett. 19: 2161
6. Minier, V., Kevorkian, A., Xu, J.M. (1992) Diffraction Characteristics of Superimposed Holographic Grating in Planar Optical Waveguides, IEEE Photonics Technol. Lett. 4: 1115
7. Othonos, A., Bismuth, J., Sweeny, M., Kevorkian, A., Xu, J.M. (1998) Superimposed Grating Wavelength Division Multiplexing in Ge-doped SiO_2/Si Planar Waveguides, Opt. Eng. 37: 717
8. Mu, G.G., et al. (1996) Trends in Optics, Chapter 29, Academic Press, London
9. Mu, G.G., et al. (1992) Electro-Optical Displays, Part 1, Marcel Dekker, New York

Holographic Optics for Beamsplitting and Image Multiplication

A. L. Mikaelian, A. N. Palagushkin, and S. A. Prokopenko

Summary. This paper describes how we see the prospects for just a small part of holographic optics. In our opinion, CGHs have the best practical potential for the near future. For this reason, we concentrate on the methods in the design and manufacture of CGHs. In designing and making phase CGHs, we show that the best realization of their capabilities requires a fresh approach to such important problems as the correct choice of the number of phase relief gradations, accuracy evaluations and monitoring of the fabrication process. The results of our research are given.

1 Introduction

Today, the fast progress of holographic optics design methodology and manufacturing techniques is evident. Diffractive optical elements (DOEs) and computer-generated holograms (CGHs) have become an integral part of many advanced optical systems. Holographic optical elements (HOEs) are finding ever-widening application in such fields as pattern recognition, optical processing, communications, medicine, data storage, laser treatment of materials, optical switching and neural networks. Holographic methods are effectively used for making spectroscopic prisms, Fresnel lenses, laser resonators and microinterferometers and for correcting aberrations.

The demand for larger arrays of light sources can be met with the help of holographic laser beamsplitters and monolithic arrays of diffraction gratings and flat lenses. The advantage of flat diffraction elements with a surface phase relief is the possibility of combining a number of functions in a single element (e.g. beamsplitter, lenslet, diffraction grating and aberration corrector) and simple methods of copying (e.g. using polymer materials).

HOEs can be used not only with coherent and partially coherent (e.g. multimode laser radiation) but also with incoherent light. The latter case involves greater dispersion effects and lower coherent noise (speckle patterns). The present state of the art of HOE development is reported in general surveys [1–3, 5] and in the proceedings of the symposium OIST'97 [4] held in Moscow in 1997 (conference on computer and holographic optics).

2   Review of the Literature

The availability of powerful computers makes it possible to use extensive computational algorithms for designing and optimizing CGHs. A lot of research teams worldwide work in this field. The methods of CGH computations are considered in [5–8]. Software for the iterative calculations and analysis of CGHs is described in [8–10]. The usual result of the computations is a particular phase distribution in the CGH plane which is realized as an appropriate surface microrelief using photo- or electron lithography and ion or plasmachemical etching methods [11].

The application of optical lithography for making multilevel visible-range phase elements does not give the necessary accuracy [12,13]. Though very precise, masks written by an electron beam cannot solve the problem because of insufficiently exact alignment. The repeated use of e-beam lithography for forming each phase level gives the necessary accuracy; however, this way is rather expensive and time consuming [11,14]. Currently, the methods of direct-write e-beam [15] and laser [16–18] lithography are employed for recording halftone phase reliefs in resist (light or electron sensitive materials). The development of the latent relief takes only one exposure cycle. The most promising modern method of making multilevel phase elements seems to be the direct electron-beam writing of the multilevel relief in electron-sensitive resist [19]. The method is relatively expensive; however, once made, the element can simply be copied using polymer materials.

In addition to conventional CGHs with feature dimensions $d \gg \lambda$ elements with $d < \lambda$ (which are not diffractive elements in terms of classical optics) are also developed. Examples of such subwavelength elements are reflection and antireflection coatings and elements that can change the state of polarization. There is promising research into polarization-sensitive birefringent CGHs (BCGHs) (where phase reliefs are formed on the surface of a birefringent material) [20,21] and optoelectronic devices (VLSI) for high-speed communication systems. Optical volume hologram-based elements that are sensitive to light polarization and wavelength are used in optical communication lines as switches (free-space optical multistage interconnection network) [21,28]. A set of CGH gratings are used for exciting the zero-order mode in a planar optical waveguide [29,30].

Various combinations of CGHs with other optical elements are also a subject of today's investigations. For example, [31] describes the combination of high-efficiency radiation-resistant polarization diffraction gratings with multilayer dielectric mirrors used for wavefront correction of pulsed high-power solid-state lasers. Reference [32] offers to weakly couple a thin-film planar optical waveguide to a CGH. The resonance effects that result in that system allow the control of transmittance by slightly changing the angle of coupling.

A new line of investigation connected with the development of dynamic CGHs has recently emerged. A novel optical device – MACH (multiple active computer-generated hologram) – is proposed in [33]. The device combines a phase CGH that can produce several output diffraction patterns with a liquid-crystal phase-shifting medium that enables the selection of a specific pattern and can be used for spatial filtering, 3-D image formation, and interconnection. However, the design has a disadvantage typical of all LC devices – slow response (~ 1.5 ms). Acousto-optical light deflectors (AOD) [34] and programmable electrooptical phase diffraction modulators [35,36] are used in developing fast devices with switching speeds of ~ 0.1–1 µs.

As an alternative, we proposed a 2-D acousto-optical modulator that uses two perpendicular acoustic waves generated in a transparent medium and a variant design combining the deflector with a CGH (see below).

We note that research in the field of synthesized holographic optics has become more active over the past few years. The CGH capabilities and areas of application have expanded significantly. The integration of CGHs with other optical elements allowed absolutely new characteristics to be developed. The manufacturing of CGHs is beginning to depart from the conventional methods of microelectronics, concentrating on direct-write techniques and polymer copying of phase reliefs.

At present, efforts are being made to develop methods to control diffraction patterns in real time. Of these methods, dynamic electro- and acousto-optical CGHs look to be the most promising.

3   Methods of Phase Hologram Synthesis

While the methods of hologram synthesis are constantly growing in number, the ever- growing power of modern computers allows the computation and analysis of more and more complex diffractive elements.

Depending on the relation between the CGH feature, d, and the wavelength, λ, various methods of CGH synthesis and analysis are used:

- $d \gg \lambda$. The realm of paraxial Fourier optics. The scalar approximation of the diffraction theory is used. The output field is proportional to the complex transmittance function of HOEs. The transfer function is described by Fresnel and Fraunhofer integrals.
- $d \leq 10\lambda$. Methods of geometric optics are used together with Fresnel coefficients.
- $d \approx \lambda$. It is necessary to employ strict electromagnetic theory and find solutions to Maxwell's vector equations. Just a few problems have exact solutions. The number of computations is very large.
- $d \ll \lambda$. It is also necessary to use strict electromagnetic theory; however, there are approximation methods.

In most cases, the hologram synthesis can be performed with the paraxial approximation of Fourier optics ($d \gg \lambda$) and it requires consideration of such parameters as (1) the phase of working orders; (2) the efficiency (which is to be as high as possible); and (3) the amplitude and phase of noise (higher diffraction orders, the field beyond the area of useful signal).

Any complex transmittance function can be used at the beginning of computations which are then to be optimized using a suitable technique (e.g. the gradient algorithm, annealing simulation). The inverse method is more efficient: the input field (and the corresponding transfer function $t(x, y)$) is computed using a predefined output pattern and then optimized. The iterative inverse method is used most often. The input field (and the transfer function $t(x, y)$) is computed using a predefined output pattern. The limitations that are largely defined by the material and the manufacturing method used (e.g. the uniform transmittance of the element, graded phase lag) are posed on the transfer function. The modified transfer function is then taken to find the output diffraction pattern. If the result is not satisfactory (e.g. low efficiency, great intensity variations), the required output amplitude distribution is taken to recompute the input field; during this process, the variable parameters (e.g. phase distribution) are kept fixed. The computation cycles are repeated until the result is good or the process stops. The limitations on the transfer function, $t(x, y)$, can be introduced gradually in the course of the iteration process. The fast Fourier transform algorithm is used to speed up the computations of Fresnel and Fraunhofer integrals. If necessary, the domain of signal legitimacy can be expanded to suppress noise (unintended diffraction orders) around the working diffraction orders.

Iterative methods using the phase retrieval algorithm [37] are now most effective. The limitation of the method – the discrete structure representation – approximately matches the ability of the current methods of CGH fabrication, lithography and direct electron-beam writing. The error diffusion algorithm allows lower corruption of the output distribution by modifying the transfer function. Annealing simulation, a method of nonlinear parametric optimization, is also widely used.

Direct computation of the binary structure is rather slow because each single period of practically any CGH (except for those that give the simplest output patterns) contains too many pixels. The phase encoding method is sometimes used in CGH computations.

The combined method [38] looks most promising. It includes:

- Conventional iterative algorithm (including gradual modification of the CGH transfer function). A pseudorandom noise signal whose amplitude gradually falls with iterations, can be introduced to attain the deepest minimum of the desired phase function.
- Gradient optimization of the output phase function followed by recomputation of the input field and modification of the transfer function.
- Binary search for the greatest sensitivity of the transmittance function to phase changes (e.g. at the boundaries of equiphase areas) by varying the phase of two or three pixels in a particular small area.

- Optimization of the CGH phase relief topology (smoothing the boundaries between equiphase areas) in order to increase the efficiency.
- Gradation of the continuous phase function can be accompanied by gradient optimization of the quantization threshold levels.

This method and the direct-write technique make it possible to manufacture two- to four-level CGHs with an efficiency of over 70% and beam intensity variations of less than 3%. The quantization of the phase function into more levels (over 8) is not reasonable because it makes the design and manufacturing more complicated with only minor improvements in the performance parameters.

4   CGH Design and Manufacture

The methods used in the computation and manufacture of holographic elements are closely related because many manufacturing errors can be predicted and corrected at the design stage. The result of computations is usually a large body of topological data needed for manufacturing a specific element and allowing for particular technological equipment.

4.1   Phase Function Gradation and Efficiency Evaluation

The computer synthesis methods give the continuous or gradated phase functions of a CGH. The continuous function allows better performance of the element but is rather hard to implement. Most manufacturing technologies require the gradated phase profile (of two to eight phase levels). Naturally, the gradation deteriorates the output diffraction pattern, increases the beam intensity variations and decreases efficiency. Efficiency is considered to mean the ratio of the total intensity of working-order beams to the intensity of the light incident on the CGH. We have worked out a simple method of evaluating the efficiency of gradated-phase elements.

The method is based on the representation of the stepped phase function as a superposition of the original continuous function P and phase correction function ΔP. The latter varies from $-Q/2$ to $+Q/2$ (where $Q = 2\pi/n$ is the phase difference between quantization levels, n is the number of levels). On this assumption, the effect of phase quantization on the CGH efficiency is similar to the introduction of a certain phase diffuser. It is important to note that the effect of the diffuser is dependent on the number of quantization levels rather than the phase function, no matter how complicated it may be. Pseudorandom phases of the diffuser range from $-Q/2 + C$ to $+Q/2 + C$ (where C is a constant). The scattering of light on the diffuser decreases the intensity of the zero-order beam and can be computed for each particular number of phase levels. The efficiency of a CGH with a stepped phase function, E_q, may be written as

$$E_q(n) = \eta(n) * E_c,$$

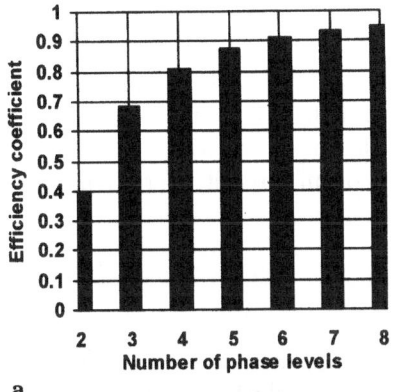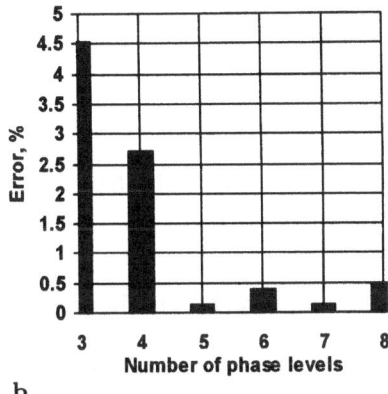

a   b

Fig. 1. (a) The relation between $\eta(n)$ and the number of phase levels for a CGH with 64×64 pixels in the period. (b) The relative error of evaluation

where E_c is the efficiency of the HOE with continuous phase function, $\eta(n)$ is the portion of the diffuser's energy in the zero-order beam.

The coefficient $\eta(n)$ is dependent on the number of phase levels and independent of the phase function itself. Figure 1a shows how $\eta(n)$ depends on the number of phase levels, n, and Fig. 1b the relative error of evaluation.

It should be noted that the method is only applicable to CGHs with originally continuous phase functions; that is, it evaluates how gradation affects the efficiency of the continuous-relief element. The result of many computation methods is a gradated phase function (see Sect. 3 and [38]). These methods make it possible to avoid the negative effects of phase gradation and manufacture even binary phase elements with an efficiency of as high as 79% (see Table 1 below).

4.2 CGH Manufacture. Specificity and Accuracy

The manufacture of phase reliefs of holographic elements must be very precise. The ion and plasmachemical etching techniques and methods of control that are used in microelectronics usually need significant improvement to provide such a high degree of accuracy.

Reference [39] gives the estimates of necessary accuracy by modeling the work of a beamsplitter with equal-intensity output beams. The criterion of the estimates was the output beam intensity variations, which should not be greater than 5%. The maximal admissible variations of equiphase pixel group boundaries and relief depth were found for different numbers of phase gradation levels. The requirements for manufacturing accuracy were shown to grow with the number of phase levels. For example, if a CGH has the index of refraction $n = 1.5$ and operates at $0.63\,\mu m$, the displacements of the phase areas should not exceed 0.1% of the phase function period. For a 200-μm period, this corresponds to a 0.1-μm accuracy.

The requirements for the phase relief depth are more rigid: the variations here must not exceed 1%. For a typical value of $\sim 0.63\,\mu m$, the relief depth should be made to $\pm 0.006\,\mu m$. In practice, certain smoothing of the relief takes place in manufacturing, which results in slight defocusing and scattering of laser light. Our estimation says that if the mean variations of the relief depth do not exceed $\pm 0.01\,\mu m$, the output beam intensity variations keep within the 5% limit.

4.3   Optical Methods of Fabrication

The use of conventional optical lithography can provide, at most, 1-μm line-width resolution and 0.2-μm positioning accuracy, which is suitable only for manufacturing infrared phase elements or elements with very small angles of diffraction ($< 0.1°$).

The direct laser writing of a phase relief on resist or the making of halftone masks [16–18] gives somewhat better results. The use of halftone masks solves the problem of automatic alignment of phase level boundaries because the recording and exposition require one cycle each.

4.4   Use of E-beam and Direct-write Lithography

E-beam lithography allows the drawing of masks with 0.1-μm resolution and 0.01-μm line-position accuracy. The use of such masks provides the accuracy necessary for CGH manufacturing; yet, the problems of fine alignment and low-resolution optical lithography remain. They can be solved by using halftone e-beam generated masks and deep UV for exposition.

Successive cycles of e-beam lithography and etching is another method of making phase reliefs which meet the accuracy requirements. However, this expensive method is too slow and can only be justifiably used for manufacturing highly accurate samples and production masters.

We use the technique of direct e-beam writing in which a computer synthesised phase structure is recorded into the resist layer deposited on a glass substrate [19].

The maximal depth of the phase relief here is determined by the resist layer thickness. To supress the reflection of light at the resist–glass interface, the index of refraction of the substrate is chosen to be close to that of resist. The etching treatment with specially adjusted exposure doses and etching modes is used to obtain the desired depth of phase levels. The departure of working-order beam intensities from design values serves as an accuracy criterion. The method makes use of all of the advantages of e-beam lithography – high resolution and precise positioning. Multiple-level phase reliefs are produced in a single exposure cycle, which eliminates the mask-alignment problem. The whole process is not so time-consuming because it needs fewer manufacturing steps.

4.5 Plasmachemical Etching of Reliefs in Glass.
Methods of Copying Phase Reliefs

CGHs have diverse applications, so the choice of elements depends on the specific operation conditions. CGHs that are made in electron or optical resist layers deposited on glass substrates can be used mostly in airtight optical devices because of their susceptibility to moisture. Rugged operation conditions require CGHs to be made of pure glass, quartz or transparent optical crystals without using other susceptible materials. The use of birefringent crystals is also welcome in manufacturing polarization-sensitive elements (BCGHs).

The use of wet chemical etching is impossible because of low resolution, insufficient boundary sharpness, and poor reproducibility. Only plasma etching is applicable in CGH manufacturing. In microelectronics, plasma etching is used for treating thin films of various materials deposited on conductor or semiconductor substrates. In CGH fabrication, these are mostly insulator materials that have to be etched. Moreover, microelectronic devices do not always require high-precision manufacturing (e.g. controlled excessive etching and specific process termination are admissible). The fabrication of CGHs places more stringent requirements upon accuracy and, therefore, the equipment and techniques used.

In our opinion, high-frequency reactive ion etching in conjunction with careful etch- depth control can provide the accuracy necessary for CGH manufacturing. Phase levels can be etched one at a time or all in one cycle; the latter technique looks more promising and involves the transfer of a phase relief from the ion-resistant resist layer into the substrate.

Holographic elements that are manufactured by the plasma processing of optical materials have best performance characteristics. However, the sophisticated fabrication process and stringent technical requirements for substrate materials make the cost of CGHs rather high.

The replication of surface reliefs provides a simple way to cheapen the large-scale production of CGHs. The copying technique used in production of laser CDs is best suited to this purpose. The original relief formed in the resist layer is coated with a thin film of conducting material by using, for example, vacuum evaporation. Then, the metal film is thickened to 0.1–0.5 mm by chemical deposition (nickel is usually used for this purpose) and separated from the original element. Before being coated with metal, the resist surface is sometimes coated with antiadhesive material to facilitate the separation. The metal replica, which is an exact copy of the original relief, can now be used as a production master. It is natural that the relief depth must correspond to the refractive index of future elements. One original element is usually used for making a few masters, or the first master is employed to make other masters. Epoxy resins or photopolymer materials are often used for making auxilliary replicas. The first copies are arranged in a single array to provide injection molding of CGHs in optical plastic. This replication methods allows cheap production of CGHs.

4.6 CGH Performance Characteristics

To exemplify the CGH performance, we present the results of the synthesis of laser beamsplitters with different numbers of phase gradation levels. The beamsplitters are designed for $\lambda = 0.633\,\mu\text{m}$ and $\lambda = 0.532\,\mu\text{m}$ [38]. We used direct e-beam writing to make the phase relief.

Table 1 shows the performance parameters of beamsplitters with 2 to 8 phase levels. They produce diffraction patterns with 4 to 135 equal-intensity beams. Two beamsplitters generate words "IONT" and "MOSCOW" as output patterns. The beamsplitter "EAGLE" gives an eagle-shaped emblem. The diffraction pattern of the beamsplitter "SPEC" is an array of beams of different predefined intensities.

Table 1. Phase multilevel beamsplitters

Output pattern	Number of phase levels	Number of working orders	Efficiency (%)	
			Theory	Experiment
3×11 beams	2	33	79.0	73.3
3×4 beams	2	12	74.8	70.3
"IONT"	3	43	69.5	
"MOSCOW"	3	59	67.8	
"SPEC"	4	14	70.3	
"IONT"	4	43	71.8	69.1
"EAGLE"	4	135	76.1	
2×2 beams	4	4	81.1	
3×11 beams	8	33	86.3	70.5
2×2 beams	8	4	88.2	

Figures 2 and 3 illustrate the work of 3×11 and 3×4 beamsplitters. The laser wavelength is 633 nm, the dimensions of the phase function period is $230.4 \times 230.4\,\mu\text{m}$, the aperture of the beamsplitters is 4.61×4.61 mm, which corresponds to 20×20 periods of the phase function. The angle between the neighboring beams is $9.2'$ (3×11 beamsplitter) and $18.4'$ (3×4 beamsplitter). The phase relief depth is 528 nm, the refractive index of resist is 1.6. The efficiency of the beamsplitters are 73.3% and 70.3% respectively. Figure 3d illustrates the use of the 3×4 beamsplitter as an image multiplier.

4.7 Dynamic CGHs

CGHs allow specific stationary patterns in the far diffraction zone. The aim of today's research is to produce dynamic phase elements allowing real-time control of the diffraction pattern.

We have made efforts to employ acoustic waves in the creation of dynamic beamsplitters. Sound waves are known to cause the index of refraction

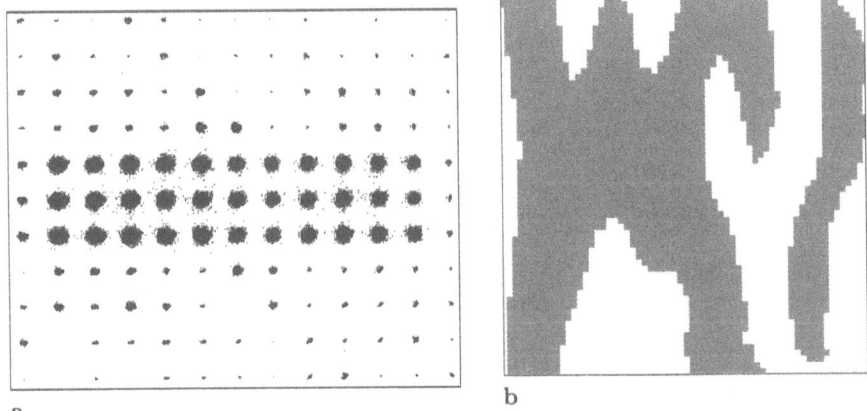

a   b

Fig. 2. (a) The output diffraction pattern of a two-level 3 × 11 beamsplitter. (b) Photograph of one period of the phase structure

of a material to vary sinusoidally. A plane light wave propagating through a material with varied refractive index acquires different phases across the wavefront as if it had encountered a phase diffraction grating on its way. Since two perpendicular low-energy sound waves do not interact, the variations of refractive index add up to produce a superposition of phase diffraction gratings.

We considered the simple case of when two supersonic waves and a light beam propagate in mutually perpendicular directions (Fig. 4). The instant and averaged (over the sound wave period) diffraction patterns were computed. Both traveling and standing acoustic waves were used in the experiment. The amplitude, frequency and phase of sound waves were variables in the computations.

Figure 5a shows the computed diffraction pattern for the case of the acoustic waves of equal amplitudes $A = 0.570\pi$ (0.285λ) (amplitude is given in units representing the maximal phase shift the light wave acquires in passing through the medium in which the given sound wave propagates), and phase difference, $\Delta\phi = 0.5\pi$. The pattern is averaged over the period of the acoustic wave and contains 5 beams of equal intensity (intensity variations are not greater than 1.4%). The computed efficiency of such an acoustic beamsplitter is 69.8%. Decreasing the amplitude of both acoustic waves to $A = 0.45\pi$ (0.225λ) results in the intensity of four outer beams falling to 40% of the intensity of the central beam (Fig. 5b). Figure 5c shows the experimental diffraction pattern corresponding to the computed pattern shown in Fig. 5b.

The experimental acoustic beamsplitter (see Fig. 4) employed LiNbO$_3$ as the active medium and provided Raman–Natt diffraction, which does not allow high efficiency. It is more promising to work with Bragg diffraction, which can enable not only 100% efficiency but also the control of each diffraction order.

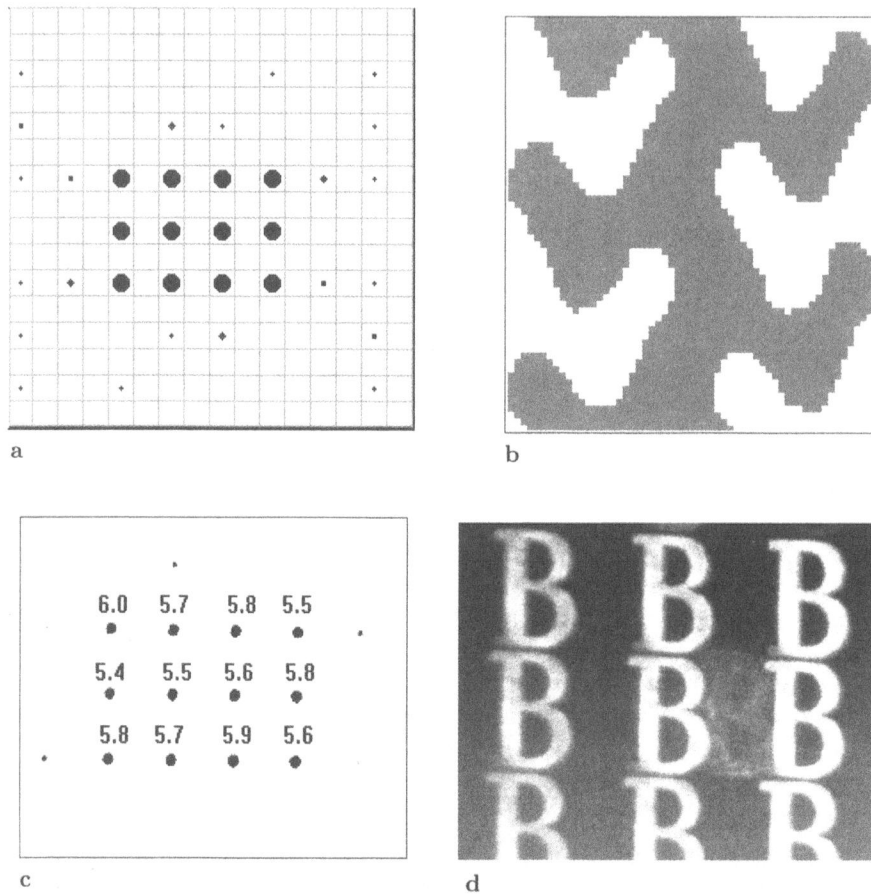

Fig. 3. Two-level 3 × 4 beamsplitter. The (**a**) calculated and (**c**) experimental diffraction pattern. (**b**) Photo of the phase structure period. (**d**) The work of the beamsplitter as an image multiplier

Interesting results can be obtained by putting an ordinary CGH immediately or some distance behind the acoustic element. The computations showed that the CGH output pattern in the far diffraction field can be controlled by varying the amplitude and phase of the acoustic waves. When the spacing of the CGH and the sound wavelength are roughly the same, the resulting diffraction pattern becomes highly complicated, and the choice of a particular pattern from all possible solutions rather difficult.

We experimented with the simplest design when a 3 × 4 two-level beamsplitter was located immediately at the front or back side of the crystal (see Fig. 4). Figure 6 shows the experimental diffraction patterns detected under the following conditions: (a) no acoustic waves are generated; (b) the acoustic wave propagates along the x-axis; (c) the acoustic wave propagates along the y-axis; (d) the computed pattern corresponding to case (c).

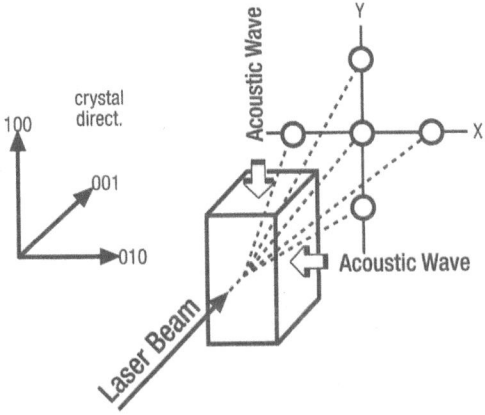

Fig. 4. Two-dimensional acoustic modulator

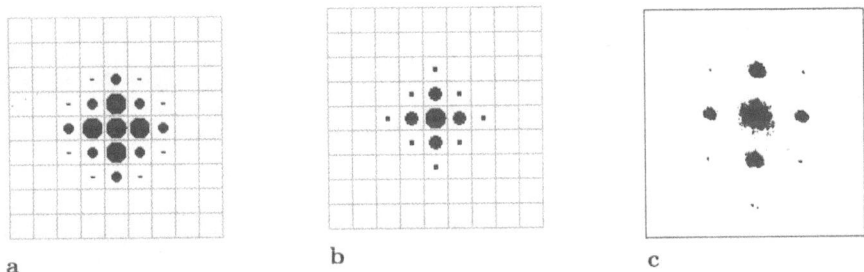

a   b   c

Fig. 5. (a,b) Theoretical and (c) experimental diffraction patterns of the acoustic beamsplitter

It is seen from Fig. 6 that each of the original beams (produced by the stationary 3 × 4 beamsplitter) is doubled in the coordinate axis parallel to the direction of propagation of the sound wave. It is clear that the location and intensity of beams can be controlled by varying the sound amplitude and wavelength.

Devices such as this may be used in optical switching, for example. An interesting idea is to record a holographic element directly into the $LiNbO_3$ crystal where the acoustic waves are generated. Another promising design is the integration of a CGH with a device using surface acoustic waves.

5 Conclusion

The paper shows that computer-generated diffractive optics has begun to come to the forefront and replace conventional optics in many fields of application. Advanced computational and manufacturing methods have consider-

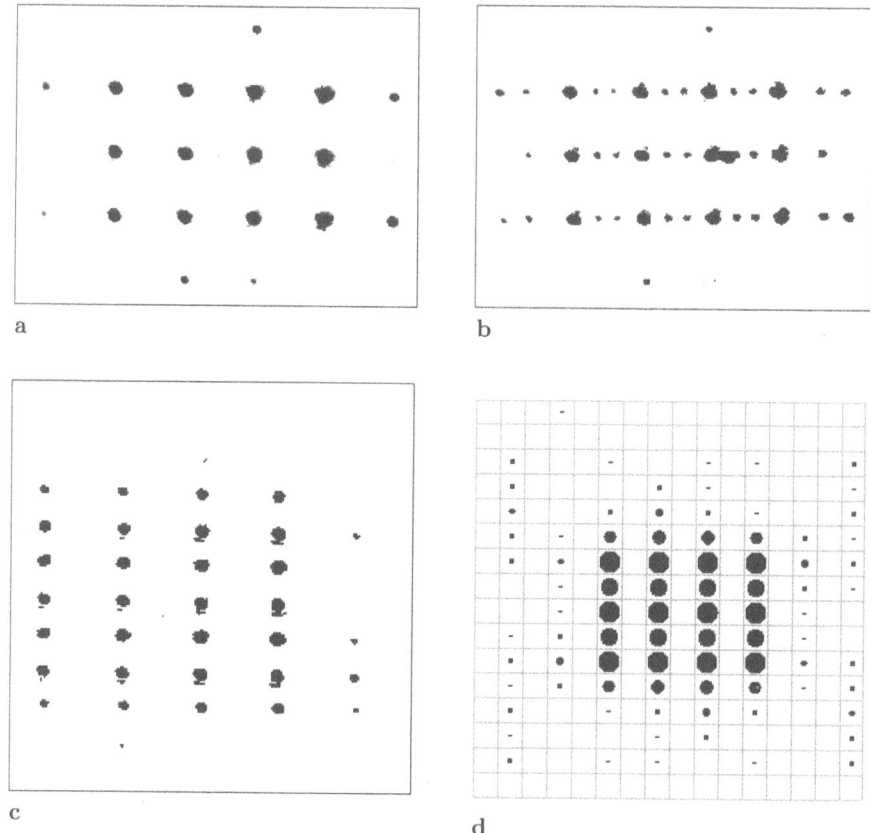

Fig. 6. (a–c) Experimental and (d) computed diffraction patterns of the combined dynamic beamsplitter

ably widened the capabilities of holographic optics and have allowed it the accomplishment of what was earlier impossible.

It is quite predictable that further progress in the theory of diffractive optics will allow the computation of subwavelength structures – highly promising devices for integrated optics. Dynamic CGHs are likely to find use in optical processing and communications. The technology of manufacturing diffractive optics will develop as a separate line of engineering, solving its own specific problems.

References

1. Lessard, R.A., Manivannan, G. (1997) Photochromism used in Holographic Recording: A review. Proc. SPIE 3347, pp. 11–19
2. Counture, J.A., Lessard, R.A. (1997) Organic dye doped colloids solid films used as real-time materials for optical engineering applications. Proc. SPIE 3347, pp. 36–51

3. Barachevsky, V. (1997) Light-sensitive media for high capacity optical memory. Proc. SPIE 3347, pp. 2–9

4. Mikaelian, A.L. (Ed.) (1997) Optical Information Science & Technology '97, Computer and Holographic Optics and Image Processing. 27–30 August 1997, Moscow, Russia. Proc. SPIE 3348

5. Doskolovich, L.L., Perlo, P., Petrova, O.I., Repetto, P., Soifer, V.A. (1997) Calculation of quantized DOEs on the basis of a continuous series approach. Proc. SPIE 3348, pp. 37–47

6. Kotlyar, V.V., Khonina, S.N. (1997) A method for design of DOE for the generation of countor images. Proc. SPIE 3348, pp. 48–55

7. Soifer, V.A., Pavelyev, V.S., Duparre, M., Ludge, B. (1997) Iterative calculation and technological realization of DOE focusing the laser beam into non-axial radially symmetrical domains. Proc. SPIE 3348, pp. 69–75

8. Prokopenko, S.A., Palagushkin, A.N., Mikaelian, A.L., Kiseliov, B.S., Shkitin, V.A. (1997) Computation of 2D High-Efficiency Multilevel Phase Beamsplitters. Proc. SPIE 3348, pp. 22–29

9. Volotovskiy, S.G., Kazanskiy, N.L., and Pavelyev, V.S. (1997) Software on iterative calculation and investigations DOE. Computer Optics 17: 48–53

10. Khonina, S.N., Kotlyar, V.V., Lushpin, V.V., Soifer, V.A. (1997) A Method for Design of Composite DOEs for the Generation of Letter Images. Opt Mem Neural Networks 6(3): 213–220

11. Golub, M.A., Duparee, M., Kley, E.B., Kowarschik, R., Ludge, B., Rockstoh, W., Fuchs, H.J. (1996) New diffractive beam shapper generated with aid of e-beam lithography. Opt Eng 35(5): 1400–1406

12. Miller, J.M., Taghizaden, M.R., Turunen, J., Ross, N. (1993) Multilevel-grating array generators: fabrication error analysis and experiments. Appl Optics 32(14): 2519–2525

13. Wong, V.V., Swancon, G.J. (1993) Design and fabrication of a Gaussian fan-out optical interconnect. Appl Optics 32(14): 2502–2510

14. Lalanne, P., Chavel, P. (Eds.) (1993) Perspectives for Parallel Optical Interconnects. Esprit, Basic Research Series, Springer-Verlag, Heidelberg, pp. 85–95

15. Daschner, W., Long, P., Stein, R., Wu, C., Lee, S.H. (1997) Cost-effective mass fabrication of multilevel diffractive optical elements by use of single optical exposure with a grey-scale mask on high-energy beam-sensitive glass. Appl Optics 36(20): 4675–4680

16. Smuk, A.Y., Lawandy, N.M. (1997) Direct laser writing of diffractive optics in glass. Optics Lett 22(13): 1030–1032

17. Chrkashin, V.V., Kharissov, A.A., Korolkov, V.P., Koronkevich, V.P., A.G. Poleshchuk (1997) Accurancy Potential of Circular Laser Writing of DOEs. Proc SPIE 3348, pp. 58–68

18. Perlo, P., Ripetto, M., Sinezi, S., Uspleniev, G.V. (1997) Use circular Laser record system for manufacture grey-scale mask for DOE on the basis of DLW glass plate. Comp Optics 17: 85–90

19. Palagushkin, A.N., Politov, M.V., Prokopenko, S.A., Mikaelian, A.L., Arlamenkov, A.N., Kiseliov, B.S., Shkitin, V.A. (1997) Fabrication of multilevel HOEs using direct e-beam writing. Proc. SPIE 3348, pp. 76–82

20. Krishnamoorthy, A.V., Xu, F., Ford, J.E., Fainman, Y. (1997) Polarization-controlled multistage switch based on polarization-selective computer-generated holograms. Appl Optics 36(5): 997–1010

21. Fainman, Y., Xu, F., Tyan, R., Sun, P.C., Ford, J.E., Krishnamoorthy, A., Scherer, A. (1997) Multifunctional Diffractive Optics for Optoelectronic System Packaging. Proc. SPIE 3348, pp. 152–162

22. Habraken, S., Michaux, O., Renotte, Y., Lion, Y. (1995) Polarizing holographic beam splitter on photoresist. Optics Lett 20(22): 2348–2350

23. Lima, R.A., Soares, L.L., Cescato, L., Gobbi, A.L. (1997) Reflecting polarizing beam splitter. Optics Lett 22(4): 203–205

24. Wyrowski, F., Kley, E.B., Schnabel, B., Turunen, J., Honkanen, M., Kuittinen, M. (1997) Subwavelength-Structured Interfaces: Fascination and Challenge. Opt Mem Neural Networks 6(4): 231–238

25. Miranda dos Santos, J.A., Bernardo, L.M. (1997) Quasi sub-wavelength one-dimensional 3x1 array generator for 0.6328 µm. Proc. SPIE 3348, pp. 30–36

26. Honkanen, M., Kettunen, V., Kuittinen, M., Laakkonen, P., Lautanen, J., Turunen, J., Vahimaa, P. (1997) Subwavelength-structured polarization-selective surfaces. Proc. SPIE 3348, pp. 104–107

27. Haritonov, S.I. (1997) Diffraction on quasi-periodical one-dimensional structure. Comp Optics 17: 10–15

28. Huang, Y.T., Su, D.C., Deng, J.S., Chang, J.T. (1997) Holographic polarization-selective and wavelength-selective elements in optic network applications. Proc. SPIE 3348, pp. 2–12

29. Miller, M., Beaucoudrey, N., Chavel, P., Turunen, J., Cambril, E. (1997) Design and fabrication of binary slanted surfase-relief gratings for a planar potical interconnection. Appl Optics 36(23): 5717–5727

30. Kamenev, N.N., Nalivaiko, V.I. (1997) Waveguide-hologram-based controllable unidirectional beamsplitter. Proc. SPIE 3348, pp. 140–142

31. Perry, M.D., Boyd, R.D., Britten, J.A., Decker, D., Shore, B.W. (1995) High-efficiency multilayer dielectric diffraction gratings. Optics Lett 20(8): 940–942

32. Brauer, R., Bryngdahl, O. (1997) From evanescent waves to specified diffraction orders. Optics Lett 22(11): 754–756

33. Slinger, C., Brett, P., Hui, V., Monnington, G., Pain, D., Saga, I. (1997) Electrically controllable multiple, active, computer-generated hologram. Optics Lett 22(14): 1113–1115

34. Riza, N.A. (1997) Acusto-optic device-based high-speed high-isolation photonic switching for signal processing. Optics Lett 22(13): 1003–1005

35. Thomas, J.A., Fainman, Y. (1995) Programmable diffractive optical element using a multichannel lanthanum-modified lead zirconate titanate phase modulator. Optics Lett 20(13): 1510–1512

36. Bogomolnyi, V. (1997) To calculation controlled with electrical field elements of optical system on base of metal-dielectric-metal (MDM) structures. Proc. SPIE 3348, pp. 212–217

37. Gerchberg, R.W., Saxton, S.O. (1972) A Practical Algorithm for the Determination of Phase from Image and Diffraction Plane Pictures. Optik 35(2): 237–246

38. Mikaelian, A.L., Prokopenko, S.A., Palagushkin, A.N. (1997) Multilevel Phase Beamsplitters: Design and Fabrication. Opt Mem Neural Networks 6(3): 221–229

39. Prokopenko, S.A., Palagushkin, A.N., Mikaelian, A.N., Kiseliov, B.S., Shkitin, V.A, (1997) Computer Simulation of 2D Multiple-levels Phase Beam-splitters Fabrication Errors. Proc. SPIE 3348, pp. 83–93

Image Restoration, Enhancement and Target Location with Local Adaptive Linear Filters

L. Yaroslavsky

Summary. Local adaptive filters for image restoration (denoising and deblurring), enhancement and target location are described. The filters work in the domain of an orthogonal transform (DFT, DCT or other transforms) in a moving window and nonlinearly modify the transform coefficients to obtain an estimate of the central pixel of the window. A framework for the filter design for multi-component images is presented and experimental results in denoising, enhancement and deblurring monochrome and color images, and in target location are provided. Also discussed is the filter implementation for a window size of 3 × 3 pixels in the form of five-channel convolution with masks implementing the local mean and four directional Laplacians: horizontal, vertical, and two (±45°) diagonal ones.

1 Introduction

Local adaptive linear filters for image restoration and target location implement the idea of space (time)–frequency signal representation that dates back to 1940s [1], though they only became practical (at least in image processing) in the 1980s [2] when computer technology provided the appropriate speed and memory size. The filters work in a moving window in the domain of an orthogonal transform and, in each position of the window, nonlinearly modify the signal transform coefficients to obtain an estimate of the central pixel of the window.

Image processing in the transform domain rather than in the signal domain suggests certain advantages in terms of the convenience of incorporating a prior knowledge of images into the design of processing algorithms and in terms of computational expense. The transfer from the signal domain to the transform domain is especially promising if it is applied locally rather than globally.

In this chapter, we overview the basic ideas for filter design and implementation, justify using the discrete cosine transform (DCT), for this purpose and present experimental results in denoising and deblurring monochrome and color images and in target location. We then discuss the use of other orthogonal transforms such as the Walsh–Hadamard and Haar transforms and show also that, for the window size of 3 × 3 pixels, filtering in the DCT transform domain is equivalent to five-channel convolution with masks implementing the local mean and four directional Laplacians: horizontal, vertical, and ±45° diagonal ones.

2   Multi-component Local Adaptive Filters

2.1   Optimal Local Adaptive Filters

For the design of an optimal filter one should formulate an optimization criterion. The design of local adaptive filters is based upon local criteria that evaluate image processing quality for each particular image sample as an average, over a certain neighborhood of the pixel, of a certain loss function that measures the deviation of the estimation of each pixel in the neighborhood from its true value [2,4–7]. In what follows we assume that the images to be processed are multi-component and represented by components such as the R, G, and B components of color images.

Let $\{a_{n_1,n_2}^{(c)}\}$ and $\{\hat{a}_{n_1,n_2}^{(c)}\}$ be samples of ideal image components and their estimations, respectively, where $\{n_1, n_2\}$ are pixel coordinate indices, C is the number of image components, and (k_1, k_2) are running coordinates in the image plane. Let also $\mathbf{LOC}\big(a_{k_1,k_2}^{(c)}; n_1, n_2\big)$ be a locality function that defines the pixel's a_{k_1,k_2} neighborhood and is nonzero only for those pixels with coordinates $\{n_1, n_2\}$ that belong to the neighborhood, and let $\mathbf{LOSS}\big(a_{n_1,n_2}^{(c)}, \hat{a}_{n_1,n_2}^{(c)}\big)$ be a loss function that describes the residual degradation of image samples after the processing. Then, for multi-component images, the local criterion can be expressed in the following form:

$$\mathbf{AVLOSS}(k_1, k_2) = \mathbf{AV}_{\text{imsys}}\bigg(\mathbf{AV}_{\text{obj}}\bigg(\sum_{c=1}^{C} \sum_{\{n_1,n_2\}} \mathbf{LOC}\big(a_{k_1,k_2}^{(c)}; n_1, n_2\big)$$
$$\times \mathbf{LOSS}\big(a_{n_1,n_2}^{(c)}, \hat{a}_{n_1,n_2}^{(c)}\big)\bigg)\bigg) \tag{1}$$

where $\mathbf{AV}_{\text{imsys}}(.)$ denotes the operation of averaging over realizations of the imaging system sensor's noise, and $\mathbf{AV}_{\text{obj}}(.)$ denotes averaging over unknown parameters of the image (such as the imaged object's position, orientation, size) that are not known before the processing. The design of optimal local adaptive filters is aimed at finding a mapping of the set of observed image samples $\{b_{n_1,n_2}^{(c)}\}$ to the set of estimations $\{\hat{a}_{n_1,n_2}^{(c)}\}$ to provide the minimum of the criterion for each pixel of the image. For multi-component images, if the neighborhood is defined as a spatial one,

$$\mathbf{LOC}\big(n_1, n_2, c; a_{k_1.k_2}^{(c)}\big) = \begin{cases} 1, \text{ if } |k_1 - n_1| \le M_1, \ |k_2 - n_2| \le M_2, \ \forall c \\ 0, \text{ otherwise,} \end{cases} \tag{2}$$

and the loss function is a quadratic one,

$$\mathbf{LOSS}\big(\hat{a}_{n_1,n_2}^{(c)}, a_{n_1,n_2}^{(c)}\big) = \big|\hat{a}_{n_1,n_2}^{(c)} - a_{n_1,n_2}^{(c)}\big|^2, \tag{3}$$

then the local criterion requires minimization of the functional

$$\mathbf{AVLOSS}(k_1, k_2)$$
$$= \mathbf{AV}_{\mathrm{imsys}} \mathbf{AV}_{\mathrm{obj}} \left\{ \sum_{c=1}^{C} \sum_{\substack{|k_1-n_1| \le M_1 \\ |k_2-n_2| \le M_2}} \left| \hat{a}_{n_1,n_2}^{(c)} - a_{n_1,n_2}^{(c)} \right|^2 \right\} \tag{4}$$

over all mappings of the observed signal for its estimation. To make the design constructive, one should parameterize the observation-to-estimation mapping. As an option, one may assume that the estimation of each pixel is obtained as a weighted sum of the observed pixels in its neighborhood. In this way we arrive at local adaptive linear filters.

For local adaptive filters, filter coefficients have to be determined in each particular position of the running window from the observed image samples within the window on the base of a priori information regarding images under processing. While formulation of this information in the image domain is apparently problematic, it is much simpler in the domain of certain orthogonal transforms, such as the DFT and DCT, because power spectra of image fragments in such transforms exhibit much more regular behavior than that of the pixels themselves. For instance, it is well known that image DFT and DCT spectra usually decay more or less rapidly at high frequencies.

Therefore, in what follows, we will assume filtering in a transform domain and design filters that operate, in each position (k_1, k_2) of the window, in the following three ways:

1. Computation of spectral coefficients $\{ \beta_{r_1,r_2,\sigma}^{(k_1,k_2)} \} = \mathbf{T} \mathbf{b}^{(k_1,k_2)}$ of the observed image fragment $\mathbf{b}^{(k_1,k_2)}$ within the window over the chosen orthogonal transform $\mathbf{T}$.
2. Multiplication of the obtained spectral coefficients by the filter coefficients $\{ \eta_{r_1,r_2,\sigma}^{(k_1,k_2)} \}$:
$$\hat{a}_{r_1,r_2,\sigma}^{(k_1,k_2)} = \eta_{r_1,r_2,\sigma}^{(k_1,k_2)} \beta_{r_1,r_2,\sigma}^{(k_1,k_2)} . \tag{5}$$
3. Inverse transformation $\mathbf{T}^{-1}$ of the output signal spectral coefficients $\{ \hat{a}_{r_1,r_2,\sigma}^{(k_1,k_2)} \}$ to obtain an estimated value of the central pixel of the window.

Here, subscripts (r_1, r_2, σ) are the corresponding indices in the transform domain.

With this approach, the synthesis of local adaptive filters is reduced to the determination of $(2M_1 + 1)(2M_2 + 1)C$ filter coefficients $\{ \eta_{r_1,r_2,\sigma}^{(k_1,k_2)} \}$. For optimal filter design in the domain of an orthogonal transform, one can, by virtue of Parceval's relation, reformulate the criterion of (4) in terms of signal spectra:

$$\mathbf{AVLOSS}(k_1, k_2)$$
$$= \mathbf{AV}_{\mathrm{imsys}} \mathbf{AV}_{\mathrm{obj}} \left\{ \sum_{r_1,r_2} \sum_{\sigma=1}^{C} \left| \eta_{r_1,r_2,\sigma}^{(k_1,k_2)} \beta_{r_1,r_2,\sigma}^{(k_1,k_2)} - \alpha_{r_1,r_2,\sigma}^{(k_1,k_2)} \right|^2 \right\} . \tag{6}$$

By minimizing $\mathbf{AVLOSS}(k_1, k_2)$ with respect to $\left\{\eta_{r_1,r_2,\sigma}^{(k_1,k_2)}\right\}$ one can find that the optimal values of the coefficients of the filter that minimize the filtration error as defined by (4) may be found from the following equation:

$$\eta_{r_1,r_2,\sigma}^{(k_1,k_2)} = \frac{\mathbf{AV}_{\mathrm{imsys}}\mathbf{AV}_{\mathrm{obj}}\left\{\alpha_{r_1,r_2,\sigma}^{(k_1,k_2)}\left(\beta_{r_1,r_2,\sigma}^{(k_1,k_2)}\right)^{\star}\right\}}{\mathbf{AV}_{\mathrm{imsys}}\mathbf{AV}_{\mathrm{obj}}\left\{\left|\beta_{r_1,r_2,\sigma}^{(k_1,k_2)}\right|^2\right\}} \tag{7}$$

with $\star$ denoting complex conjugate. The design of the local adaptive filter of (7) is therefore reduced to an estimation of the local power spectrum of the input image fragment and its mutual local spectrum with the "ideal" image.

2.2 Local Adaptive Filters for Image Restoration

Assume that image distortions can be modeled by the equation

$$\mathbf{b} = \mathbf{La} + \mathbf{n}, \tag{8}$$

where $\mathbf{L}$ is a linear operator of the imaging system and $\mathbf{n}$ is a random zero-mean signal-independent random vector that models the imaging system sensor's noise. Assume also that the imaging system operator $\mathbf{L}$ is such that the distorted image can be described in the domain of the chosen orthogonal transform by the following relationship:

$$\beta_{r_1,r_2,\sigma}^{(k_1,k_2)} = \lambda_{r_1,r_2,\sigma}^{(k_1,k_2)}\beta_{r_1,r_2,\sigma}^{(k_1,k_2)} + \nu_{r_1,r_2,\sigma}^{(k_1,k_2)} \tag{9}$$

where $\left\{\lambda_{r_1,r_2,\sigma}^{(k_1,k_2)}\right\}$ are running representation coefficients of the linear operator $\mathbf{L}$ in the domain of the orthogonal transform and $\left\{\nu_{r_1,r_2,\sigma}^{(k_1,k_2)}\right\}$ are zero-mean spectral coefficients of the realization of the noise interference. Then one can obtain from (7) that optimal restoration filter coefficients are defined as

$$\eta_{r_1,r_2,\sigma}^{(k_1,k_2)} = \frac{\left|\lambda_{r_1,r_2,\sigma}^{(k_1,k_2)}\right|^2\left|\alpha_{r_1,r_2,\sigma}^{(k_1,k_2)}\right|^2}{\lambda_{r_1,r_2,\sigma}^{(k_1,k_2)}\mathbf{AV}_{\mathrm{imsys}}\mathbf{AV}_{\mathrm{obj}}\left(\left|\beta_{r_1,r_2,\sigma}^{(k_1,k_2)}\right|^2\right)}. \tag{10}$$

The filter of (10) can be regarded as an "empirical" Wiener filter that assumes estimation, in each position of the window, of the parameters involved locally in its design from the observed distorted image fragments.

2.3 Local Adaptive Filters for Image Enhancement

Image enhancement is a process aimed at assisting visual image analysis. A reasonable basis for the design of linear filter parameters for image enhancement is to assume that the filtering should "restore" a "useful" signal, i.e. image details to be enhanced or extracted for the end user's convenience, against a "noise" background that obscures interpretation of the details. Following the above approach, one can, by analogy with the derivation of (10), design a filter that provides the minimum of the average squared modulus of the difference between this useful signal and the filtered signal:

$$\eta_{r_1,r_2,\sigma}^{(k_1,k_2)} = \frac{\mathbf{AV}_{\mathrm{obj}}\left(\left|\alpha_{r_1,r_2,\sigma}^{(k_1,k_2)}\right|^2\right)}{\mathbf{AV}_{\mathrm{imsys}}\mathbf{AV}_{\mathrm{obj}}\left(\left|\beta_{r_1,r_2,\sigma}^{(k_1,k_2)}\right|^2\right)} \tag{11}$$

where $\mathbf{AV}_{\mathrm{obj}}$ denotes averaging over such variations of the useful object(s) as object position, size, shape, etc.

Another useful criterion for image enhancement is signal spectrum restoration [12], which requires restoration of the power spectrum of the object signal. For this criterion, the following obvious filter results:

$$\eta_{r_1,r_2,\sigma}^{(k_1,k_2)} = \left(\frac{\mathbf{AV}_{\mathrm{obj}}\left(\left|\alpha_{r_1,r_2,\sigma}^{(k_1,k_2)}\right|^2\right)}{\left(\left|\beta_{r_1,r_2,\sigma}^{(k_1,k_2)}\right|^2\right)}\right)^{1/2}. \tag{12}$$

2.4 Local Adaptive Filters for Target Location

In target location, the target position is found by locating the maximum of the highest signal at the output of a linear filter [7,8]. Therefore the filter design has to be aimed at achieving the maximal ratio of the filter output signal at the target's position to the maximum of the highest signal in the field of search over the background component of the image fragment within the window. To all practical purposes, this requirement is equivalent to maximizing the ratio of the filter response to the target signal to the standard deviation of the signal over the field of search [8,10].

In this way one can arrive at the following optimal filter for target location:

$$\eta_{r_1,r_2,\sigma}^{(k_1,k_2)} - \frac{\left(\alpha_{r_1,r_2,\sigma}^{(k_1,k_2)}\right)^{\star}}{\mathbf{AV}_{\mathrm{imsys}}\mathbf{AV}_{\mathrm{obj}}\left(\left|\beta_{r_1,r_2,\sigma}^{(k_1,k_2)}\right|^2\right)}, \tag{13}$$

where $\left(\alpha_{r_1,r_2,\sigma}^{(k_1,k_2)}\right)^{\star}$ is the complex conjugate of the spectrum of the target signal, $\left|\beta_{r_1,r_2,\sigma}^{(k_1,k_2)}\right|^2$ is the power spectrum of the image background component, and $\mathbf{AV}_{\mathrm{obj}}$ in this particular formula denotes averaging over unknown positions of the target.

3 Selection of the Transform

The selection of orthogonal transforms for the implementation of the filters is governed by the convenience of formulating a priori knowledge regarding image spectra in the chosen base, by the accuracy of estimating the spectrum from the observed data required for the filter design, and by the computational complexity of implementing the filter.

The transforms that seem most appropriate in these respects are the discrete Fourier and discrete cosine transforms. DCT is more advantageous than

DFT in terms of the accuracy of spectral estimation because DCT, being the DFT of signals evenly extended outside their borders, substantially eliminates boundary effects that are characteristic of the DFT proper due to its periodicity. This is illustrated in Figs. 1 and 2, where one can easily see that local DCT spectra converge to zero much faster than those of DFT. An important computational advantage of DFT and DCT is that both can be computed recursively in a running window [2–5,7,11].

Therefore, the computational complexity of filtering in the base of DFT and DCT in a running window is of the order of O(size of the window) operations, which is the theoretical minimum for a general space-varying filtering. In the case of filtering in the DCT domain with an odd window size, in the computation of the inverse transform for the central pixel of the window only spectral coefficients with even indices are involved. Therefore, only these coefficients have to be computed and modified in the filtering in the DCT domain in a moving window, and the computational complexity of such a filtering can be asymptotically estimated as O(window size/4).

As for the component-wise transform, if we assume that it is separable from the transforms over spatial coordinates, it will not affect the computa-

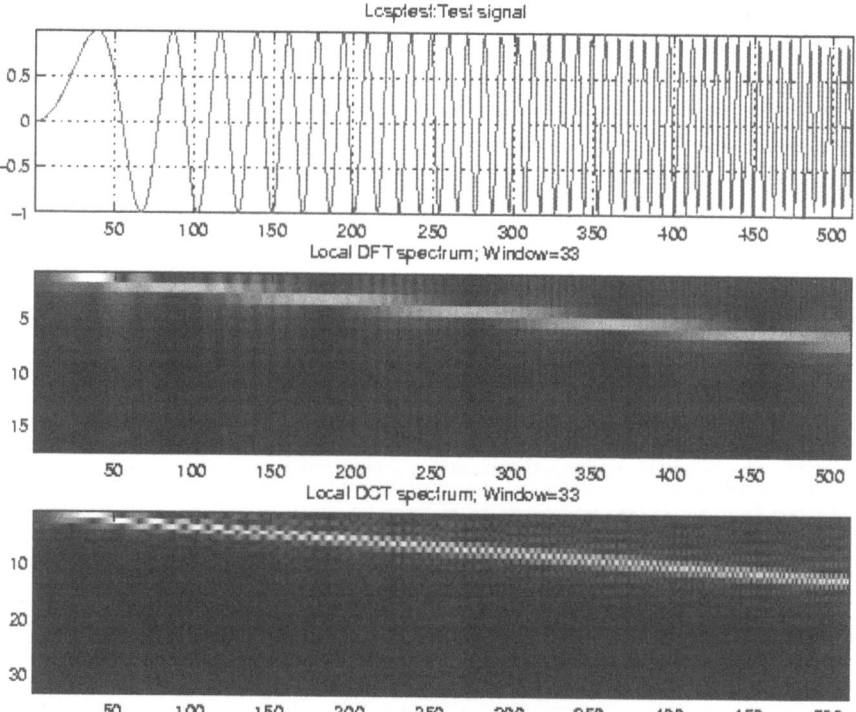

Fig. 1. Comparison of local DFT and DCT spectra shown in coordinates of spectral index (vertical) versus signal sample index (horizontal) for a test chirp signal

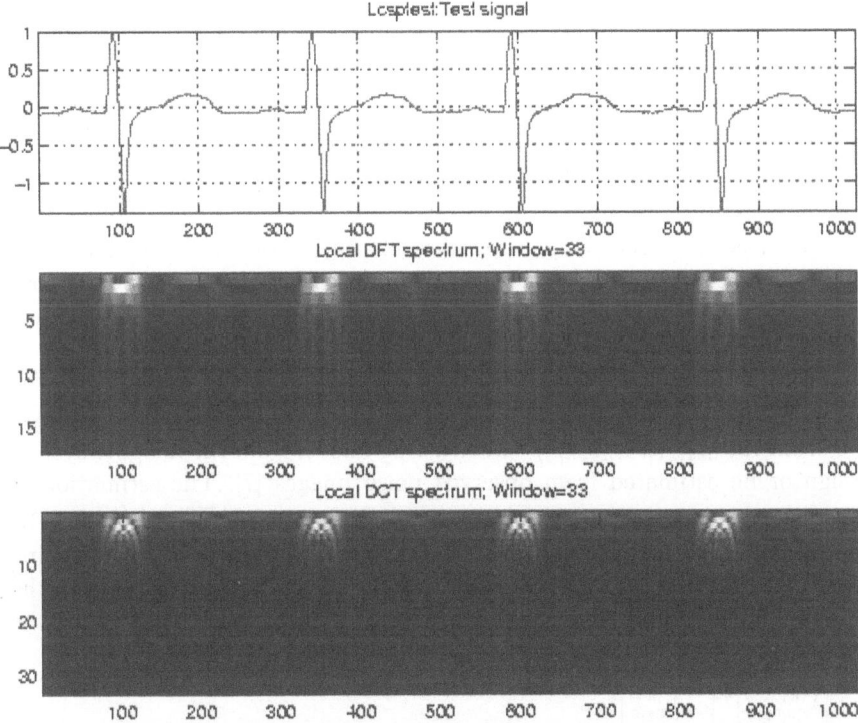

Fig. 2. Comparison of local DFT and DCT spectra for an electrocardiogram test signal

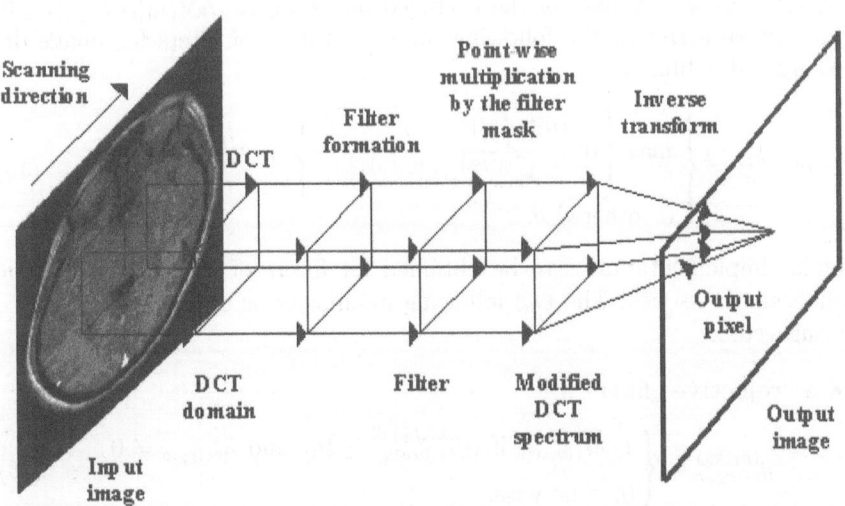

Fig. 3. Flow diagram of local adaptive filtering in the transform domain

tional complexity. We will not specify it here and leave the question of its selection open. An illustrative flow diagram of local adaptive filering in the DCT domain is shown in Fig. 3.

4   Filter Implementation: Local Adaptive Filters with Nonlinear Processing in Transform Domain

The main issue in the design and implementation of local adaptive filters in the transform domain is the estimation, in each position of the window, of averaged spectra of the observed image fragments, of the desired signal, and of the signal fraction regarded as additive signal-independent noise or as background noise. In this estimation, a priori knowledge regarding image and noise spectra should be utilized. In image restoration, the noise power spectrum $\mathbf{AV}_{\text{imsys}}\big(|\nu_{r_1,r_2,\sigma}^{(k_1,k_2)}|^2\big)$ can either be known from the imaging system design or be estimated from observed noisy images [7]. The estimation of the observed image fragment spectrum $\mathbf{AV}_{\text{imsys}}\mathbf{AV}_{\text{obj}}\big(|\beta_{r_1,r_2,\sigma}^{(k_1,k_2)}|^2\big)$ may be carried out by a smoothing of the observed spectrum of the fragment being processed. The estimation of the spectrum of "ideal" image fragments in the case of additive signal-independent noise can then be carried out using the relationship

$$
\begin{aligned}
&\big|\lambda_{r_1,r_2,\sigma}^{(k_1,k_2)}\big|^2 \big|\alpha_{r_1,r_2,\sigma}^{(k_1,k_2)}\big|^2 \\
&\simeq \max\left\{0,\ \mathbf{AV}_{\text{imgsys}}\mathbf{AV}_{\text{obj}}\big|\beta_{r_1,r_2,\sigma}^{(k_1,k_2)}\big|^2 - \mathbf{AV}_{\text{imgsys}}\big|\nu_{r_1,r_2,\sigma}^{(k_1,k_2)}\big|^2\right\}.
\end{aligned}
\tag{14}
$$

As a zero-order approximation, one can use the observed power spectrum $\big|\beta_{r_1,r_2,\sigma}^{(k_1,k_2)}\big|^2$ as an estimate of the averaged one $\mathbf{AV}_{\text{imgsys}}\mathbf{AV}_{\text{obj}}\big|\beta_{r_1,r_2,\sigma}^{(k_1,k_2)}\big|^2$. In this way we arrive at the following implementation of filters for image denoising and deblurring:

$$
\eta_{r_1,r_2,\sigma}^{(k_1,k_2)} =
\begin{cases}
\max\left\{0,\ \dfrac{\big|\beta_{r_1,r_2,\sigma}^{(k_1,k_2)}\big|^2 - \overline{\big|\nu_{r_1,r_2,\sigma}^{(k_1,k_2)}\big|^2}}{\lambda_{r_1,r_2,\sigma}^{(k_1,k_2)}\big|\beta_{r_1,r_2,\sigma}^{(k_1,k_2)}\big|^2}\right\} & \text{if } \lambda_{r_1,r_2,\sigma}^{(k_1,k_2)} \neq 0 \\
0, \text{ otherwise.}
\end{cases}
\tag{15}
$$

Similar implementations can be obtained for filters of (11)–(13) for image enhancement as well. The two following modifications of this filter can also be suggested:

- a **"rejective"** filter

$$
\eta_{r_1,r_2,\sigma}^{(k_1,k_2)} =
\begin{cases}
1/\lambda_{r_1,r_2,\sigma}, \text{ if } \big|\beta_{r_1,r_2,\sigma}^{(k_1,k_2)}\big|^2 \geq thr \text{ and } \lambda_{r_1,r_2,\sigma} \neq 0 \\
0, \text{ otherwise,}
\end{cases}
\tag{16}
$$

where the value of thr is associated with the variance of additive noise; and

- a **"fractional spectrum"** filter

$$\eta_{r_1,r_2,\sigma}^{(k_1,k_2)} = \begin{cases} G \left| \beta_{r_1,r_2,\sigma}^{(k_1,k_2)} \right|^{P-1}, & \text{if } \left| \beta_{r_1,r_2,\sigma}^{(k_1,k_2)} \right|^2 \geq thr_\sigma \\ 0, & \text{otherwise,} \end{cases} \qquad (17)$$

with P as a spectrum enhancement parameter and G as an energy normalization parameter. When $P = 1$, the filter is equivalent to that of (16). The selection $P \leq 1$ results in a signal energy redistribution in favor of weaker (most frequently, higher-frequency) components of local spectra. This modification is useful for image blind deblurring and image enhancement. Note that the idea of shrinkage of transform coefficients that are lower than a certain threshold has recently reappeared and obtained popularity in the form of wavelet shrinkage [13].

Note furthermore that if the threshold in the rejecting filter is made an appropriate function of the signal dc component (for instance, it is made proportional to it or to its square root), the filter can be used to suppress signal-dependent noise such as, for instance, speckle noise.

The design of filters for target location requires, as follows from (13), estimation of the power spectrum of the image fragment background component $\left| \beta_{r_1,r_2,\sigma}^{(k_1,k_2)} \right|^2$ averaged over the sensor's noise and over unknown coordinates of the target. As shown in [10], a reasonable approximation to this estimation is

$$\left| \beta_{r_1,r_2,\sigma}^{(k_1,k_2)} \right|^2 \cong \left| \beta_{r_1,r_2,\sigma}^{(k_1,k_2)} \right|^2 + \left| \alpha_{r_1,r_2,\sigma}^{(k_1,k_2)} \right|^2 \qquad (18)$$

such that the optimal local adaptive filter can be implemented as

$$\eta_{r_1,r_2,\sigma}^{(k_1,k_2)} \cong \frac{\left(\alpha_{r_1,r_2,\sigma}^{(k_1,k_2)} \right)^*}{\left| \beta_{r_1,r_2,\sigma}^{(k_1,k_2)} \right|^2 + \left| \alpha_{r_1,r_2,\sigma}^{(k_1,k_2)} \right|^2} . \qquad (19)$$

Image denoising, deblurring, and enhancement of the described local adaptive filters are illustrated in Figs. 4–9. Figure 4 and the graphs in Fig. 5 illustrate filtering to suppress additive Gaussian noise in a single-component of the constant image test piece. From the map of the local filter transparency (Fig. 4, bottom right corner) and from the graphs in Fig. 5, one can see that the filter is automatically context sensitive: within uniform areas it acts almost as a local mean filter which passes through practically only local dc components while in the vicinity of edges it is practically transparent to all local frequency components of the image and therefore preserves the edges (at the expense of not suppressing noise). Figure 6 illustrates the denoising of a color image; Figure 7 represents an example of blind restoration of a blurred color image; Fig. 8 shows an example of image enhancement; Fig. 9 illustrates the use of a local adaptive filter for target location in stereoscopic images.

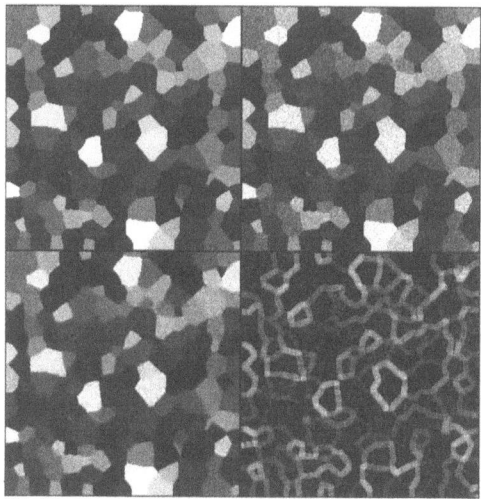

Fig. 4. Upper row: initial test piece constant and noisy test images (standard deviation of noise is 20 in the range 0–255). Bottom row: image filtered by the rejective filter of (16) with a window size of 7×7 pixels, $\lambda_{r_1,r_2,\sigma} = 1$, $thr = 20$, and a map of local filter transparency. Pixel gray level in this map is proportional to the number of signal spectral components preserved by the filter

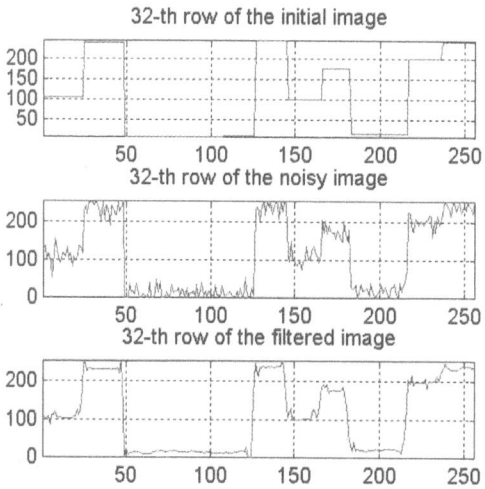

Fig. 5. Graphs of the video signal of the images in Fig. 4

5   The Use of Other Transforms

Although DCT and DFT are more advantageous than other transforms in terms of the convenience of incorporating a priori knowledge of images and their distortions into the filter design, and also in terms of the computational complexity of the filter implementation because they allow recursive

Fig. 6. Denoising additive Gaussian noise in a color image. *Upper row*: noisy RGB image components; standard deviation of noise in each of three components was 20 within signal dynamic range 0–255. *Bottom row*: corresponding denoised components. Filtering was performed by the filter of (16) in the window 7×7 pixels with $\lambda_{r_1,r_2,\sigma}^{(k_1,k_2)} = 1$, and $thr_r = thr_g = thr_b = 20$

spectral analysis in the moving window, some other transforms can nevertheless be used for local adaptive filtering in the transform domain [9]. The closest candidates are transforms with binary basis functions, such as the Walsh–Hadamard and Haar transforms, that are computationally the most simple ones. For moderate window sizes, basis functions of the DCT and Walsh–Hadamard transforms hardly differ from each other as one can see from Fig. 10. Figure 11 illustrates the denoising of a 1-D signal of an electrocardiogram by local adaptive rejective filtering in the Haar transform domain.

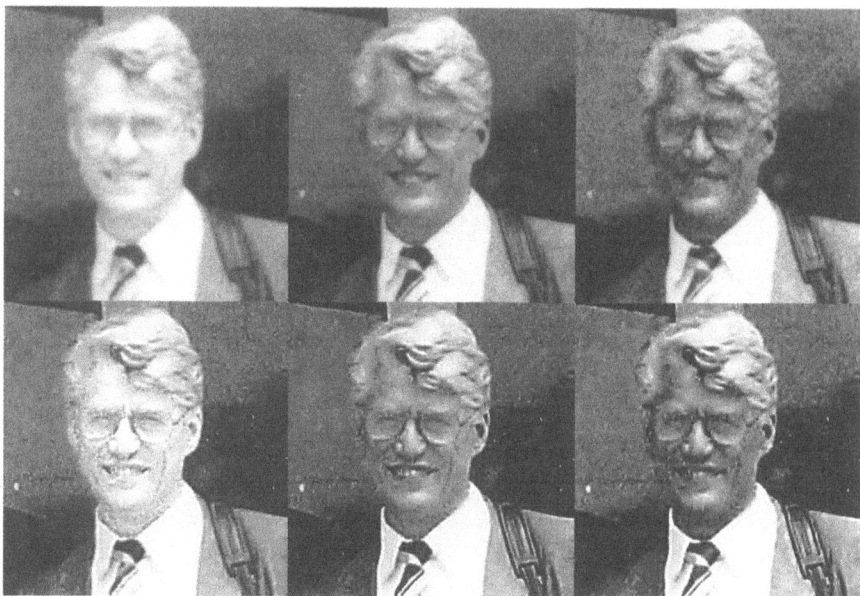

Fig. 7. Blind color image restoration. *Upper row*: RGB components of initial image. *Bottom row*: corresponding components of the restored image. Filtering was performed with the filter of (17) in the window 7×7 pixels with $P = 0.75$, $G = 2$ and $thr_r = 50$, $thr_g = 30$, and $thr_b = 60$.

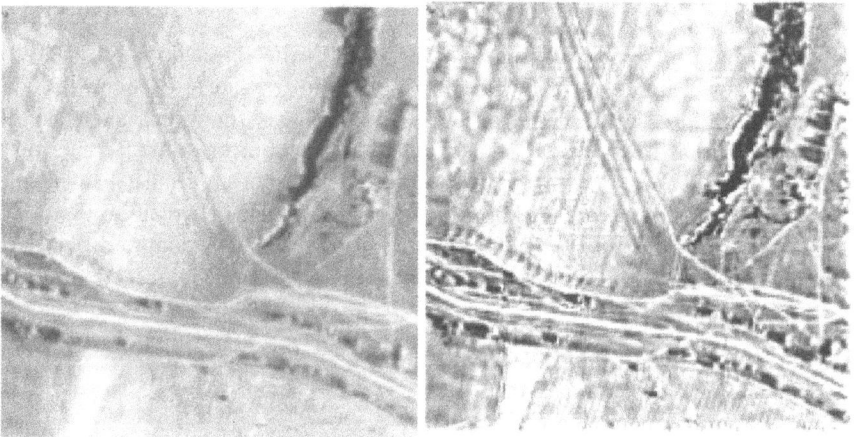

Fig. 8. Single-component image enhancement by filter of (17) in a window of 15×15 pixels with $P = 0.5$, $thr = 15$, and $G = 3$: initial (*left*) and filtered (*right*) images

The result of localization (marked with a cross)

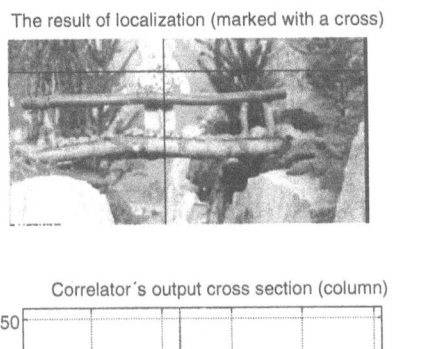

Target (highlighted)

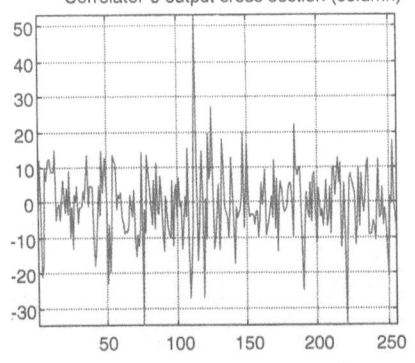

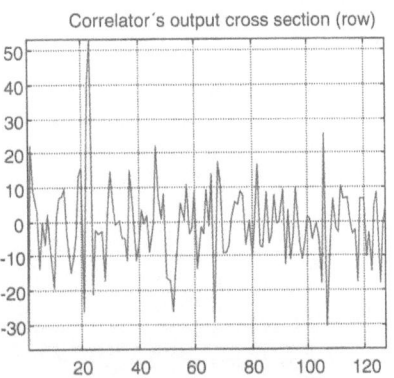

Fig. 9. Localization of a small (8 × 8 pixels, highlighted) fragment of the right of two stereoscopic images on the left image by the local adaptive filter of (19). The graphs illustrate the difference between the filter response to the target (sharp peak) and that to the rest of the image. Running window size is 32 × 32 pixels; image size is 128 × 256 pixels. The same filter applied globally rather than locally fails to locate the fragment

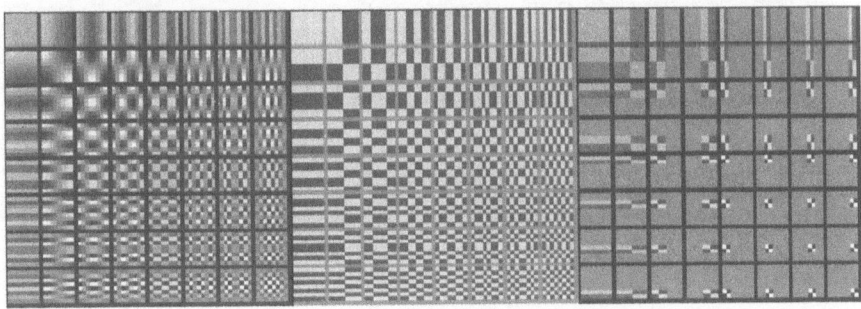

Fig. 10. Arrays of 8 × 8 basis functions of (*from left to right*) 2-D DCT, Walsh–Hadamard, and Haar transforms of order 8

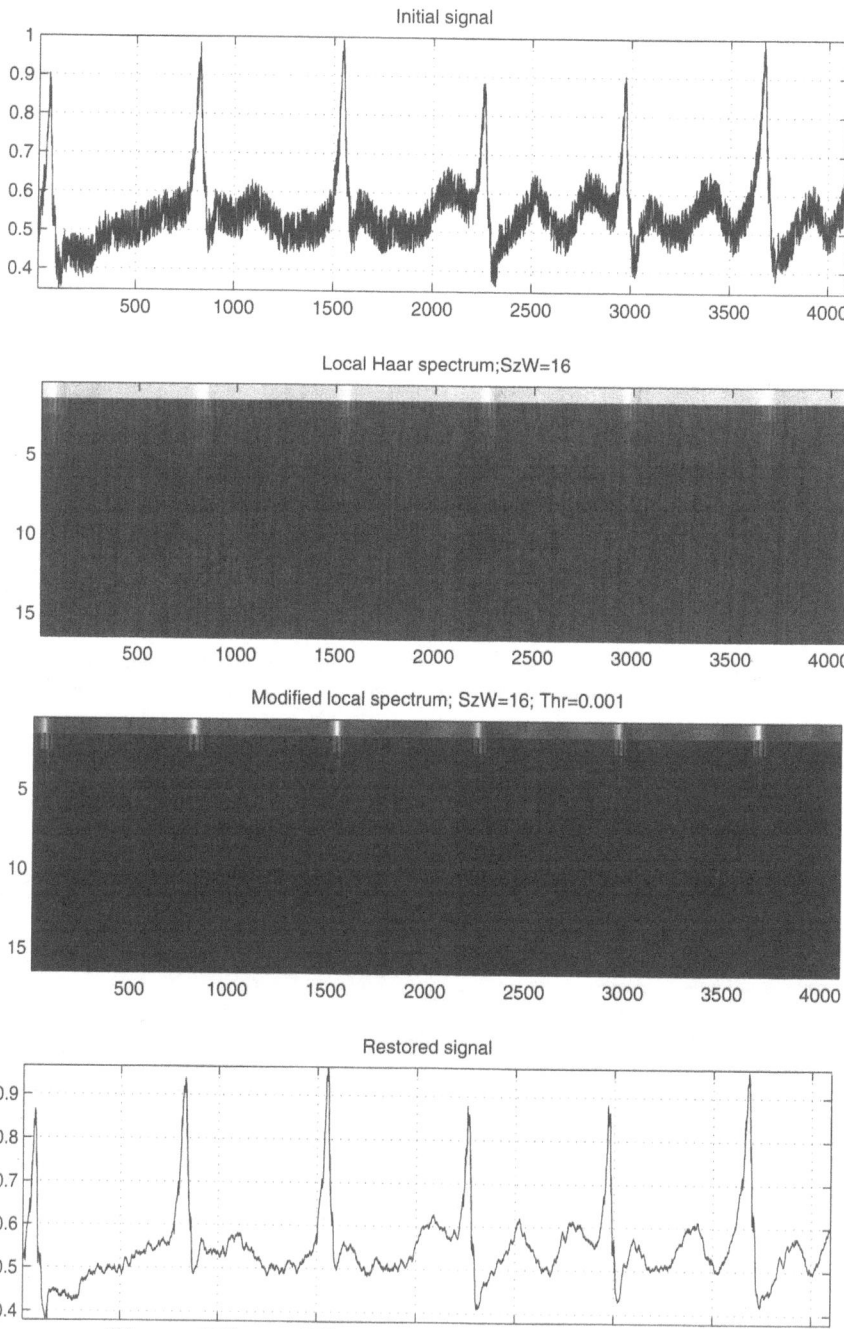

Fig. 11. Filtering a noisy electrocardiogram in the domain of the Haar transform

6   Modification of the Denoising Procedure: Thresholding the Directional Laplacians

The local adaptive filters described are context sensitive. They successfully suppress noise in areas where the video signal is changing slowly, while in the vicinity of edges and other heterogeneous areas they tend to keep the signal unchanged. Selection of the filter window size depends on what type of image details should be preserved and what can be neglected.

In the restoration of high-resolution images when objects of the size of about one pixel should be restored, the filter window size should be minimal in order to preserve these tiny details. The minimal window which preserves symmetry with respect to its central pixel is the 3×3 pixel window. Let us examine the basis functions of local DCT spectral analysis in such a window. The nine basis functions are

$$\begin{bmatrix} DCT00 & DCT10 & DCT20 \\ DCT01 & DCT11 & DCT21 \\ DCT02 & DCT12 & DCT22 \end{bmatrix}$$

$$\begin{bmatrix} \begin{bmatrix} 1 & 1 & 1 \\ 1 & 1 & 1 \\ 1 & 1 & 1 \end{bmatrix} & \begin{bmatrix} -\sqrt{3} & 0 & \sqrt{3} \\ -\sqrt{3} & 0 & \sqrt{3} \\ -\sqrt{3} & 0 & \sqrt{3} \end{bmatrix} & \begin{bmatrix} 1 & -2 & 1 \\ 1 & -2 & 1 \\ 1 & -2 & 1 \end{bmatrix} \\ \begin{bmatrix} \sqrt{3} & \sqrt{3} & \sqrt{3} \\ 0 & 0 & 0 \\ -\sqrt{3} & -\sqrt{3} & -\sqrt{3} \end{bmatrix} & \begin{bmatrix} 3 & 0 & -3 \\ 0 & 0 & 0 \\ -3 & 0 & 3 \end{bmatrix} & \begin{bmatrix} \sqrt{3} & -2\sqrt{3} & \sqrt{3} \\ 0 & 0 & 0 \\ -\sqrt{3} & 2\sqrt{3} & -\sqrt{3} \end{bmatrix} \\ \begin{bmatrix} 1 & 1 & 1 \\ -2 & -2 & -2 \\ 1 & 1 & 1 \end{bmatrix} & \begin{bmatrix} \sqrt{3} & 0 & -\sqrt{3} \\ -2\sqrt{3} & 0 & 2\sqrt{3} \\ \sqrt{3} & 0 & -\sqrt{3} \end{bmatrix} & \begin{bmatrix} 1 & -2 & 1 \\ -2 & 4 & -2 \\ 1 & -2 & 1 \end{bmatrix} \end{bmatrix}. \quad (20)$$

It follows from the definition of the 2-D DCT that from these functions, only DCT00, DCT20, DCT02, and DCT22 are involved in the computation of the window central pixel value. These functions, being applied in a running window, generate image local mean (function DCT00) and image horizontal (function DCT02), vertical (function DCT20) and isotropic (function DCT22) Laplacians. The last one can be decomposed into a sum of Laplacians taken at $\pm 45°$:

$$DCT22 = -(DCT220 + DCT221) \quad (21)$$

where

$$DCT220 = \begin{bmatrix} -2 & 1 & 1 \\ 1 & -2 & 1 \\ 1 & 1 & -2 \end{bmatrix} \quad \text{and} \quad DCT221 = \begin{bmatrix} 1 & 1 & -2 \\ 1 & -2 & 1 \\ -2 & 1 & 1 \end{bmatrix}. \quad (22)$$

This casts new light on the filter denoising capability and leads to different modifications of the filter. One of the possible modifications is a direct image five-channel convolution with window functions DCT00, DCT20, DCT02, DCT220, and DCT221, thresholding the convolution results and combining them into the output image.

The second possible modification is a "soft thresholding" of the convolution results:

$$OUTPUT = \begin{cases} INPUT, \text{ if } |INPUT| > thr \\ thr \cdot sign\,(INPUT) \cdot \left(\left| \dfrac{INPUT}{thr} \right|^{P} \right), \text{ otherwise,} \end{cases} \tag{23}$$

where thr is the same threshold parameter as that in (16) and P is a parameter that defines the degree of approximation of the "hard thresholding" of (16) (the "hard thresholding" is achieved when $P \to \infty$). The "soft thresholding" may help to reduce artifacts of denoising caused by the loss of low-contrast image details. Figure 12 illustrates image deblurring and denoising with the use of the above five-channel convolution with window functions DCT00, DCT20, DCT02, DCT220, and DCT221, and the subsequent "soft" thresholding.

Further natural modifications shape the Laplacian's masks DCT00, DCT20, DCT02, DCT220, and DCT221 by an appropriate window function and a directional smoothing of the convolution results. This shaping may help to reduce artifacts associated with the rectangular sampling raster. The directional smoothing may provide a better denoising capability than the simple element-wise thresholding.

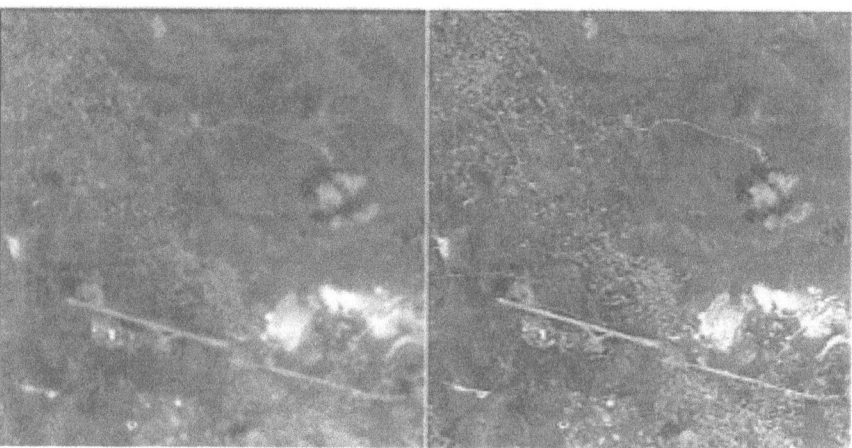

Fig. 12. Correction of image sensor scanning aperture by five-channel convolution with directional Laplacians: initial image and aperture-corrected images without and with noise suppression

7   Conclusion

We have described local adaptive filters for image restoration, enhancement, and target location that work in a moving window in the domain of the DCT and other transforms, outlined the filter design and implementation, and demonstrated their applications for processing 1-D signals, monochrome, and color images. We have also shown that filtering in the DCT domain by local adaptive filtering in a window of 3×3 is equivalent to image convolutions with uniform aperture and horizontal, vertical, and two diagonal ($\pm 45°$) 3×3 Laplacians.

References

1. Gabor, D. (1946) Theory of Communication. J Inst Elec Eng 93: 429–457
2. Yaroslavsky, L. (1987) Digital signal processing in optics and holography. Radio i Svyaz, Moscow (in Russian)
3. Vitkus, R.Yu., Yaroslavsky, L.P. (1988) Adaptive Linear Filters for Image Processing. In: Siforov, V.I., Yaroslavsky, L.P. (Eds.) Adaptive Methods for Image Processing. Nauka, Moscow, pp 6–35 (in Russian)
4. Yaroslavsky, L.P. (1991) Linear and Rank Adaptive Filters for Image Processing. In: Dimitrov, L., Wenger, E. (Eds.) Digital Image Processing and Computer Graphics. Theory and Applications. Oldenburg, Wien München, p 374
5. Yaroslavsky, L.P. (1996) Local Adaptive Filters for Image Restoration and Enhancement. In: Berger, M.-O., Deriche, R., Herlin, I., Jaffre, J., Morel, J.-M. (Eds.) 12th Int. Conf. on Analysis and Optimization of Systems. Images, Wavelets and PDE's, Paris, June 26–28. Lecture Notes in Control and Information Sciences, 219. Springer-Verlag, Heidelberg, pp 31–39
6. Yaroslavsky, L. (1996) Local Adaptive Image Restoration and Enhancement with the Use of DFT and DCT in a Running Window. In: Unser, A., Aldroubi, A., Laine, A.F. (Eds.) Wavelet Applications in Signal and Image Processing IV, SPIE Vol. 2825, pp 2–11
7. Yaroslavsky, L., Eden, M. (1996) Fundamentals of Digital Optics. Birkhauser, Boston
8. Yaroslavsky, L.P. (1993) The Theory of Optimal Methods for Localization of Objects in Pictures. In: Wolf, E. (Ed.) Progress in Optics, Vol. XXXII. Elsevier, Amsterdam
9. Yaroslavsky, L. (1997) Nonlinear Adaptive Image Processing in Transform Domain. In: IEEE Workshop on Nonlinear Signal and Image Processing, Sept. 7–11, Mckinac Island, Michigan
10. Yaroslavsky, L., Marom, E. (1997) Nonlinearity Optimization in Nonlinear Joint Transform Correlators. Appl Optics 36(July): 1–7
11. Vitkus, R.Y., Yaroslavsky, L.P. (1987) Recursive Algorithms for Local Adaptive Linear Filtration. In: Yaroslavskii, L.P., Rosenfeld, A., Wilhelmi, W. (Eds.) Mathematical Research. Academie Verlag, Berlin, pp 34–39
12. Pratt, W.K. (1978) Digital Image Processing. Wiley, New York
13. Donoho, D.L., Johnstone, I.M. (1994) Ideal spatial adaptation by wavelet shrinkage. Biometrica 81(3): 425–455

Fuzzy Problem for Correlation Recognition in Optical Digital Image Processing

G. Cheng, G. Jin, M. Wu, and Y. Yan

Summary. In this paper, the principle and technique of gray-tone image recognition using optical incoherent correlator is discussed with fuzzy theory. Several novel concepts (fuzzy relational matrix, changeable m-bit cycle-encoding method, uncertain pixels and fuzzy entropy segmentation) are proposed for improving the distortion-invariant ability and correctness of optical recognition. The optical experiments have proved the effectiveness of these new methods.

1 Introduction

Image recognition is the most important field in digital image processing, which includes classification, description, justification, recognition, and understanding. Using automatic machines to realize fast and accurate image recognition is one of the final aims of artificial intelligence. Optical pattern recognition has many merits in image processing, such as parallelism, high-speed, and huge capacity because optical methods can display, store, and process the images directly. Optical image recognition has become an important research field and involves character recognition, graph recognition, biomedical justification, industrial inspection, military reconnaissance, and so on, [1]

Model-based techniques, neural networks, and correlation pattern recognition (CPR) are three of the most popular approaches to machine vision and pattern recognition [1]. While the former two techniques are being widely pursued by the signal and image processing community, CPR has been investigated mostly by the optical pattern recognition community [2–9]. CPR offers several unique advantages in machine vision applications. Correlation is a "sensor independent" approach that generally does not require specific operations for data (such as segmentation, feature extraction, on-line model synthesis, or other processes). Consequently, correlation processors and algorithms can be applied in different types of sensors including infrared, synthetic aperture radar, CCDs, and laser radar. Thus CPR is a unifying approach to pattern recognition in a multi-sensor environment that does not require separate software to accommodate the different sensor phenomenology. Since the throughput of a correlator generally depends on the dimensions of the image, and the algorithm complexity depends on the scene size but not on the number of objects to be detected, CPR provides not only a precise estimate for throughput, but it also avoids the computational bottlenecks faced by some approaches due to scene complexity, clutter, and the number of objects presented.

Since optical correlators can realize pattern recognition, people have proposed many real-time optical correlators for this purpose [2–10]. Optical correlators may be roughly classified into three types: Vander Lugt matched filter, joint transform correlator, and incoherent optical correlator. The former two types, based on coherent optical sources, are known to be very sensitive to the misalignment of the system components and the noise generated by the optical scattering from defects and imperfections in the components [4–6,10,11]. Incoherent optical correlators have many advantages over their coherent counterparts. For example, they have less expensive light sources, a wider variety of input objects, fewer critical quality requirements for optical components, fewer critical alignment accuracy requirements between the input and the tested pattern, multi-channel redundancy easily exploited to suppress noise, and twice the spatial-bandwidth product of the coherent optical correlator for a given numerical aperture and input pattern size [7].

But incoherent correlators still have two main problems. One is the miscarriage of justice, and the other is that the correlation peaks are not as perfect as an ideal δ function. Many papers dealing with CPR, correlation and matching are about these same problems. In practice, the direct correlation of the input image with a reference mask would not reach the right justification in many cases. For example, it would be difficult to recognize the character **F** from characters **F** and **E** because the entire information of character **F** is included in character **E**. These situations also frequently happen in gray-scale image recognition. Many authors have introduced morphological hit-or-miss transform (HMT) to solve this problem for binary images [10–15], but optical HMT for gray-tone images is difficult to obtain [13]. In this kind of integral operation, the peaks of incoherent correlation are easily disturbed by noise so that the ability of recognition is largely reduced. In some previous papers, the results of matching were evaluated by using the intensity of the peaks, the ratio of peak to side-lobe, and the half-width parameters of the peaks, but all these parameters can only represent the local property of the optical correlation. F. Merkle proposed the method of statistical analysis and evaluation for optical correlation to improve the veracity of judgment, but for this method the computing complexity was increased and pre-knowledge of perfect matching was needed [16].

In this paper, the principles of optical gray-tone image recognition are discussed in Sect. 2. Section 3 proposes a fuzzy relationship matrix, which was constructed by a triangle fuzzy set and used to describe the characteristics of a gray-tone pixel matching. A variable m-bit cycle-encoding method was introduced to realize the fuzzy relation pixel matching on an optical correlator. We also discuss how to select m correctly for different distortions in gray-level object recognition. According to this method the optical experimental results are obtained by using an incoherent correlator.

The ordinary correlation's distortion-invariant ability for scaling and rotation is very limited. The calculation result will shut down rapidly when the input images contain noise [16]. Scientists have tried to improve the distortion-invariant ability by adjusting the threshold but this method is a statistical method for the results and is not a universal method for different kinds of distortion. Casasent has proposed a method of removing from the object the area that can be easily disturbed, which reduces the chance that the recognizing results will be disturbed and thus improves the distortion-invariant ability [13]. This preprocessing method, however, also loses some important information of the object, which may bring a miscarriage of justice. Another problem is that the correct choice of the removed area is very difficult in complex objects analyzing.

In Sect. 4, based on the fuzzy concepts, the concept of uncertain pixels is proposed to improve the distortion-invariant ability of optical correlation. This novel method reduces the effect of the pixels which are easily disturbed by the distortion of rotation and scaling. Without losing any information from the input images, the precise recognition between two images can be realized and a better distortion-invariant ability for scaling and rotation can be obtained.

In Sect. 5, based on the modern information theory and fuzzy concepts, a novel fuzzy entropy segmentation method for analyzing the optical computing results more simply, robustly, and accurately than the traditional principal or statistical character analyzing methods is presented. It has excellent ability in noise tolerance and resolution for identifying multi-objects and provides an approach for selecting the appropriate threshold quantitatively in a noisy environment.

2   Relationship Between Correlation and Matching

We define $X \Delta M$ as the recognition between input image X and reference mask M. Usually the size of X is larger than the size of M. We move M and X in any abitary direction. While M arrives at a position, pixel matching with the corresponding window of X will be performed. The output data will reflect how many pixels are fitted at this position. After M does the same processing with each window of X, the output image will indicate the matching relation between X and M. If the pixels of X contain all the pixels of M, the matching result will have a value corresponding to the number of the pixels in M. When the input image contains a noise which disturbs M, a perfect matching between X and M can not be obtained so that the rightness of recognition will be lowered. The matching process between X and M can be described as

$$[X \Delta M](m, n) = \sum_{i=1}^{p} \sum_{j=1}^{q} X(m+i, n+j) \# M(i, j), \qquad (1)$$

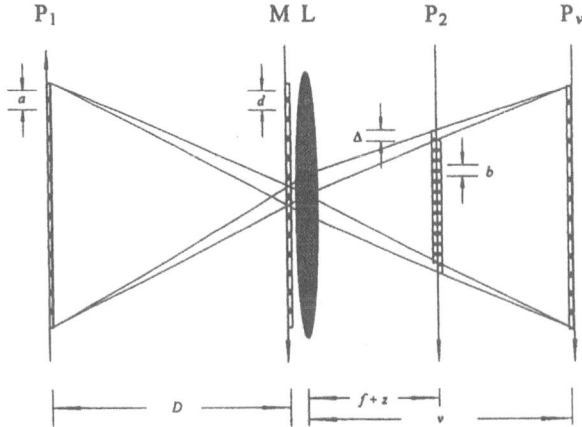

Fig. 1. Incoherent optical correlator for convolution

where $[X \Delta M](m, n)$ is the value of the position (m, n) of the output image, the symbol "#" indicates the pixel matching process between two pixels, and p and q are the width and height of M.

From (1), we know that the most important step of image matching is how to define the pixel matching "#". The pixel matching "#" can adopt different definitions according to the actual need, such as the NOR relationship used in binary images. To reflect the similarity between different pixels, the pixel matching "#" should satisfy the general signal distinguishing condition: If the function $\#(x_i, x_j)$ has the relationship $\#(x_i, x_j) \leq \#(x_i, x_i)$ for x_i ($i = 1, \ldots, n$), then function "#" can be used to distinguish the n singles. If a relational matrix can be used to show the cross-relationship between several variables, then the relational matrix could be described as

$$R_{X_{i=1,2,\ldots,n}} = \begin{bmatrix} R_{x_1,x_1} & \cdots & R_{x_1,x_n} \\ \cdots & \vdots & \cdots \\ R_{x_n,x_1} & \cdots & R_{x_n,x_n} \end{bmatrix} = \begin{bmatrix} \#(x_1, x_1) & \cdots & \#(x_1, x_n) \\ \cdots & \vdots & \cdots \\ \#(x_n, x_1) & \cdots & \#(x_n, x_n) \end{bmatrix}, \quad (2)$$

where R_{x_i,x_i} for $i = 1, \ldots, n$ will be the largest value in each vertical and horizontal line.

Figure 1 shows an incoherent optical correlator for performing correlations. The input plane P_1 is illuminated by a uniform incoherent light. Mask M is placed closely in front of an imaging lens L of a focal length f. The distance between plane P_1 and M is D. The distance between M and the lens is very small, so it can be neglected. The short distance between two near pixels (called the pixel sampling period) in P_1 is a. In M the pixel sampling period is d. Plane P_ν is the imaging plane in which a clear image of P_1 can be obtained. Symbol ν is the distance between the lens and P_ν.

If the convolution is obtained at

$$z = \frac{f^2(d-a)}{d(D-f)+fa},$$ (3)

then the intensity distribution I_2 of the output image in output plane P_2 has the form

$$I_2(mb, nb) = \sum_j \sum_k I_1\left[(m+j)a, (n+k)a\right] \times r(jd, kd),$$ (4)

where m, n, j, k are integers, and $r(jd, kd)$, the binary distribution on M. According to (4), the intensity distribution I_2 in P_2 is the correlated resultant image of I_1 with $r(jd, kd)$.

Comparing (1) and (4), we can see the different symbols used in these two equations. In fact, the multiplication "$\times$" cannot represent the pixel matching function "#" directly because multiplication does not satisfy the general signal distinguishing condition, which is the reason that ordinary optical correlation cannot obtain the right results in many cases.

Here we propose a fuzzy relationship to describe the pixel matching between two gray-scale pixels, which is defined as

$$t \# s = \begin{cases} m-n & \text{where } t-s = n \leq m \\ 0 & \text{where } t-s = n > m, \end{cases}$$ (5)

where t and s are two gray-level pixels and m is a variable which will be adjusted for different distortion-invariant abilities. According to (5), the relationship between different gray-level pixels constructs a fuzzy relational matrix C, whose elements on the diagonal are the largest values in each row and line. Each row or line in the fuzzy matrix C is a triangle function.

$$C_{iy} = \begin{cases} \dfrac{i}{m} + \dfrac{m-y}{m}, & y-m \leq i \leq y \\ \dfrac{-i}{m} + \dfrac{m+y}{m}, & y \leq i \leq y+m \\ 0 & \text{other} \end{cases}$$

$$C_{xj} = \begin{cases} \dfrac{i}{m} + \dfrac{m-x}{m}, & x-m \leq j \leq x \\ \dfrac{-i}{m} + \dfrac{m+x}{m}, & x \leq j \leq x+m \\ 0 & \text{other,} \end{cases}$$ (6)

where C_{iy} and Y_{xj} are the coefficients on the y line and x row of the relational matrix C. The main parameters of the triangle relational matrix are the gray-level of the image and the relational parameter m. As an example, a fuzzy relational matrix of pixels, whose gray-level is 16 and where m equals 5, is shown in Fig. 2.

	0	1	2	3	4	5	6	7	8	9	10	11	12	13	14	15
0	5	4	3	2	1	0	0	0	0	0	0	0	0	0	0	0
1	4	5	4	3	2	1	0	0	0	0	0	0	0	0	0	0
2	3	4	5	4	3	2	1	0	0	0	0	0	0	0	0	0
3	2	3	4	5	4	3	2	1	0	0	0	0	0	0	0	0
4	1	2	3	4	5	4	3	2	1	0	0	0	0	0	0	0
5	0	1	2	3	4	5	4	3	2	1	0	0	0	0	0	0
6	0	0	1	2	3	4	5	4	3	2	1	0	0	0	0	0
7	0	0	0	1	2	3	4	5	4	3	2	1	0	0	0	0
8	0	0	0	0	1	2	3	4	5	4	3	2	1	0	0	0
9	0	0	0	0	0	1	2	3	4	5	4	3	2	1	0	0
10	0	0	0	0	0	0	1	2	3	4	5	4	3	2	1	0
11	0	0	0	0	0	0	0	1	2	3	4	5	4	3	2	1
12	0	0	0	0	0	0	0	0	1	2	3	4	5	4	3	2
13	0	0	0	0	0	0	0	0	0	1	2	3	4	5	4	3
14	0	0	0	0	0	0	0	0	0	0	1	2	3	4	5	4
15	0	0	0	0	0	0	0	0	0	0	0	1	2	3	4	5

Fig. 2. The fuzzy relation matrix of pixels, whose gray-level is 16 and where m equals 5

The size of m can be chosen by trading off the ability of distortion-invariant and the space-bandwidth product of the optical system. Selecting a big m can obtain a better distortion-invariant ability, but a large space-bandwidth product is then needed and the distinguishing ability is limited. When two images are needed to match accurately and there is very little noise distortion, the value of m should be small. For example, when $m = 1$, the fuzzy relationship matrix C will transform into a Boolean relational matrix. In this condition the matching precision is very high but the distortion-invariant ability is very small. If there is a small amount of noise in the images we should enlarge the m to improve the distortion-invariant ability, and thus a small gray-level disturbance will be neglected. For example when $m = 3$, a gray-level disturbance less than 3 will be calculated into the result and a disturbance larger than 3 will be neglected. A larger m means a higher distortion-invariant ability with a lower matching precision.

The fuzzy relational measurement measures the similarity between images f and g.

$$FRM(f,g) = \int |f(x)\#g(x)|\,dx \tag{7}$$

where "#" is the fuzzy relational pixel matching processing between two gray-level pixels. When f is similar to g, the value of $FRM(f,g)$ will be larger. Using (1) and (7), the gray-scale image matching $X\Delta M$ can be defined as

$$[X\Delta M](i,j) = FRM(W(X,i,j,p,q),M). \tag{8}$$

A novel m-bit cycle-encoding method can be used to simply realize matching processing $X\Delta M$ in the optical correlator. Suppose the pixel of image I, whose gray-level is i $(0 \leq i \leq L - 1)$, can be expressed as $[b_{L+m-2}, b_{L+m-3}, \ldots, b_1, b_0]$ whose length is $L + m - 1$, and $b_i = \ldots = b_{i+m-1} = 1$, $b_j = 0$ $(j \neq i)$. For example, two pixels of 0 and 10 gray-level in an image, which has 16 gray levels and where m equals 5, can expressed as [0000000000000011111] [0000111110000000000]. "0" means the status of the light is dark and "1" means the status is bright. If two gray-level pixels t and s are represented using the m-bit cycle-encoding method, the codes T_e and S_e have the relation

$$[T_e] \times [S_e]^{\mathrm{T}} = \sum_{i=0}^{L+m-2} T_{ei} \times S_{ei} = \begin{cases} m - n & \text{where } |t - s| = n \leq m \\ 0 & \text{where } |t - s| = n \geq m. \end{cases} \tag{9}$$

From (9), (1), and (4), we can express the image X and mask M with the m-bit cycle-encoding method so that the fuzzy relational measurement between X and M can be realized with a simple correlation.

$$X\Delta M = (X_e \circ M_e)\big|_{E_d}, \tag{10}$$

where E_d is the decoding mask. In many cases, however, the decoding process using mask E_d is not necessary because the computing result always gives the largest value in the pixels corresponding to the pixels in the image E_d. So the FRM algorithm can be simplified to

$$X\Delta M = X_e \circ M_e. \tag{11}$$

An example of image matching based on FRM is given in Fig. 3. From the result of $X\Delta M$ we can find the information about the image matching of X with M. If $X\Delta M$ will be segmented with threshold T, the output image will reflect whether or not the image X is matching or not with mask M.

$$X = R \quad \text{if } (X\Delta M)\big|_{T=mN} > 0, \tag{12}$$

where N is the number of pixels in M. Figure 3f gives the result of segmentation of Fig. 3e. The central bright spot on the picture means that image X is matching with mask M.

Hence, by using FRM we can obtain high distortion-invariant ability for changing in illumination and noise, which has little distortion-invariant ability for rotation and scaling. In the next section, a fuzzy preprocessing method for the reference mask is proposed in order to improve the distortion-invariant ability for rotation and scaling.

3   Using Uncertain Pixels to Improve the Distortion-Invariant Ability

According to the degree they are disturbed by the noise, the pixels in a reference mask can be divided into two classes. In this paper the easily disturbed

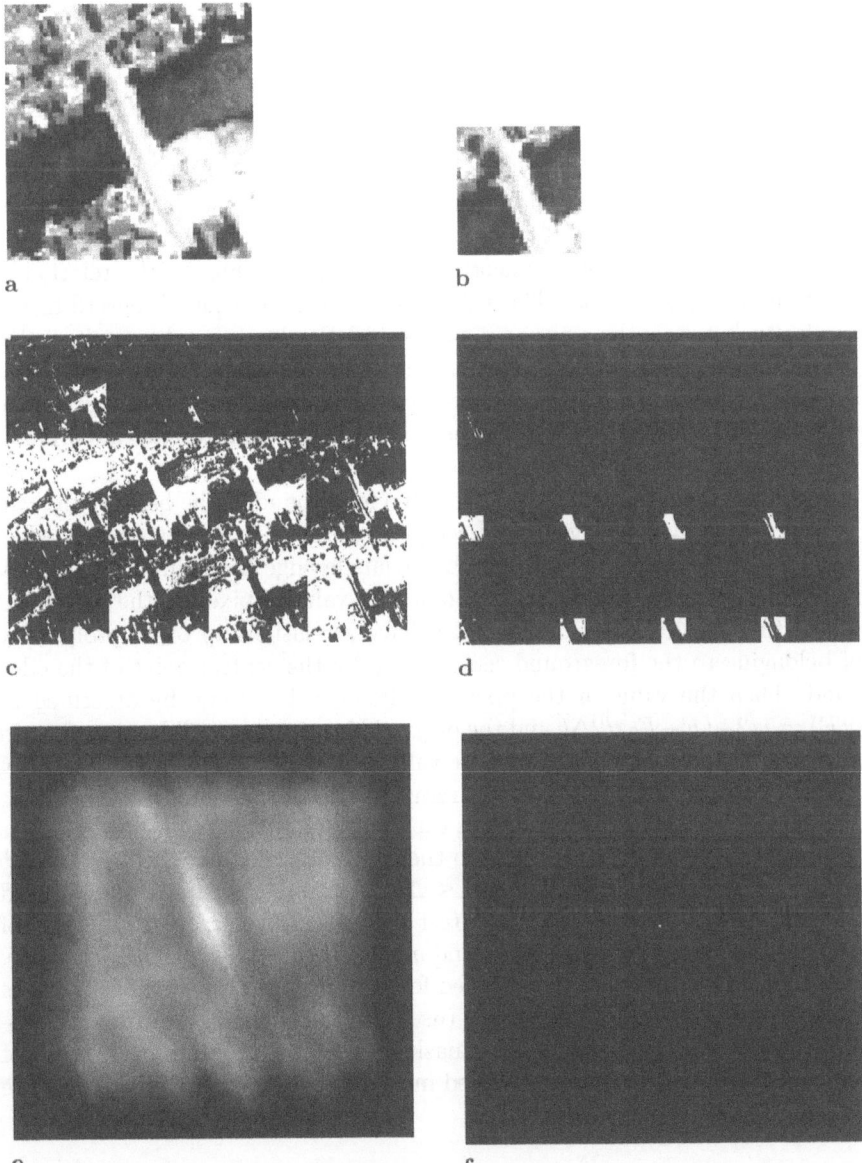

Fig. 3. Example of image matching based on *FRM*. (a) Input gray-scale image X. (b) Reference mask M. (c) Encoding image X_e. (d) Encoding image R_e. (e) Correlation of X_e and R_e. (f) Segmentation of (e)

pixels are called uncertain pixels because they have some fuzzy characteristics when we determine whether they belong to the foreground sets or the background sets. Selecting the right uncertain pixels is the key problem in processing and has an important relationship with the environment and the purpose of the image. For example, the pixels at the edge of the objects are easily disturbed by scale distortion and rotation distortion. If we want to improve the scale and rotation distortion-invariant abilities, we can reduce the certainty of these pixels to achieve such a target.

We can set the value of uncertain pixels T according to the relation of T to the foreground sets. The value t of uncertain pixels T describes the similarity between the uncertain pixels and the foreground sets; its value range is $0 \leq t \leq 1$. When the probability of T belonging to the foreground sets is large, the value of t is close to 1. That is, when T belongs to the foreground completely, $t = 1$. When T does not belong to the foreground, $t = 0$.

In binary images, the changes from the foreground to the background can be seen as rectangular edges. Here, we propose the concept of a triangle edge. The rectangle edge is expanded into a triangle edge by including a linearly changing edge band whose width is $2\Delta b$. The value of pixels in the edge band reflect their position in this band and have a relationship to its probability of belonging to the foreground sets. Let E_0 be the central point of the edge band. Then the value of the pixels in the edge band can be described as $p(E) = 1/2 + (E - E_0)/2\Delta b$ and the original binary image will be transformed into a multi-value data matrix whose value is limited as $p \rightarrow [0, 1]$. Increasing Δb will improve the distortion-invariant ability and have more uncertainty. In a real algorithm, if pixel $p(x, y)$ is the foreground pixel in mask R, and r is the shortest distance from $p(x, y)$ to the edge of R, then $p = 1$ when $r > \Delta b$ and $p = (\Delta b + r)/2\Delta b$ when $0 \leq r \leq \Delta b$. If pixel $q(x, y)$ is the background pixel in mask R, and r' is the shortest distance from $q(x, y)$ to the edge of mask R, then $q = 0$ when $r' > \Delta b$ and $q = (\Delta b - r')/2\Delta b$ when $0 \leq r' \leq \Delta b$.

In Fig. 4, a plane picture is needed for matching and three kinds of masks (an ordinary mask, an erased mask (used by Casasent), and a fuzzy preprocessed mask) are used. The erased mask loses the information in the edge of the object but the fuzzy preprocessed mask does the edge-softing around the

  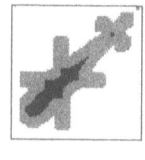 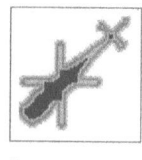

a   b   c   d

Fig. 4. Input image and three kinds of masks for correlation. (**a**) Input image. (**b**) Ordinary mask. (**c**) Erased mask (**d**) Fuzzy preprocessed mask

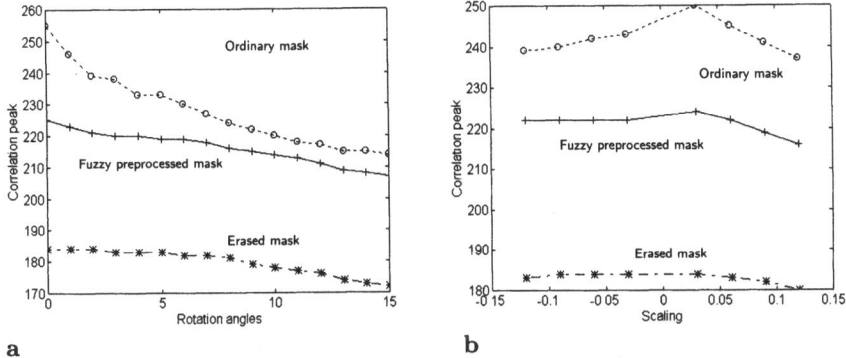

Fig. 5. The curves of the three methods in distortion-invariant ability. (a) Comparison of the robustness for rotation of the three methods. (b) Comparison of the robustness for scaling of the three methods

object. We compared the abilities of these methods for rotation distortion and scale distortion and the results are shown in Fig. 5.

The curves in Fig. 5a are the values of the correlation peaks of these three methods for a rotation distortion in the input image from 0 to 15 degrees. The curves in Fig. 5b are the values of the correlation peaks with a scaling distortion in the input image from 87.5–112.5%. From Fig. 5 we can see that the erased mask has better noise distortion ability than the ordinary methods. Meanwhile, the variance of the correlation peaks is small when noise disturbance existed but the values of the correlation peaks are low. Fuzzy preprocessed mask has both the merits of the former two methods; that is, it has a high correlation peak and well distortion-invariant ability without losing the important information of the images. Therefore, fuzzy preprocessing is the best one among these three methods.

4   Fuzzy Entropy Segmentation for Optical Correlation

First, a new matrix $\tilde{M}$ can be obtained by dividing the matching result M by the number of the pixels in mask R. Named as the pixel matching subjection matrix, the unitary matrix $\tilde{M}$ is a fuzzy matrix, whose value scope is from 0 to 1, varied with the perceptive characteristic of human beings. Then the fuzzy entropy of the fuzzy matrix $\tilde{M}$ is defined by

$$H_{\mathrm{f}}(\tilde{M}) = \frac{1}{W_M H_M} \sum_{m=1}^{W_M} \sum_{n=1}^{H_M} S(\tilde{M}(m,n)), \tag{13}$$

where W_M and H_M are the width and height of the matrix $\tilde{M}$, respectively. The function $S(\ )$ is a Shannon function $S(t) = -t \log_2 t - (1-t) \log_2 (1-t)$.

Apparently, the fuzzy entropy $H_{\mathrm{f}}(\tilde{M})$ is satisfied with $0 \leq H_{\mathrm{f}}(\tilde{M}) \leq 1$ and has the following mathematical properties:

1. Certainty: $H_f(\tilde{M}) = 0$, only when all the $\tilde{M}(m, n)$ equal 0 or 1
2. Extremum: $H_f(\tilde{M}) = 1$, only when all the $\tilde{M}(m, n)$ equal 0.5
3. Tendency: $H_f(\tilde{M})$ increases while $\tilde{M}(m, n) \to 0.5$

The relationships between fuzzy entropy and the matching result are:

1. When all $\tilde{M}(m, n)$ equal 0, which means that the input image does not match the reference mask anywhere, then the fuzzy entropy of the correlation is 0. The special case is that the input image is a white image and the reference mask is a black one.
2. When all $\tilde{M}(m, n)$ equal 1, which means that the input image matches the reference mask everywhere, then the fuzzy entropy of the correlation is 0. The special case is that both the input image and the reference mask are white images or black images.
3. The ideal matching between two images has the fuzzy entropy $H_f(\tilde{M}) = 0$ because the matrix $\tilde{M}$ is a δ function.
4. The non-ideal matching between two images has the fuzzy entropy $0 < H_f(\tilde{M}) \leq 1$ because the value scope of matrix $\tilde{M}$ is $[0, 1]$. Specifically when all $\tilde{M}(m, n)$ equal 0.5, the fuzzy entropy of the correlation is 1.

In fact, the first and second conditions will not happen in practical applications so that the input image and reference mask must satisfy the ideal matching while the fuzzy entropy of the correlation equals 0. In order to quantitatively estimate the inner information of the correlation, the fuzzy entropy is an important parameter reflecting the quality of optical recognition.

Here, a series of Gauss functions is used to simulate the correlation peaks with different heights and half-widths. These functions can be described as $f(m) = (\sqrt{2\pi}\,\sigma)^{-1}e^{(m-\sigma)^2/(2\sigma^2)}$, where $m \in [1, 512]$ and $\sigma \in [1, 100]$. Figure 6 shows the relationship between the fuzzy entropy and the half-width of the correlation peak. The logarithmic increased tendency of fuzzy entropy along

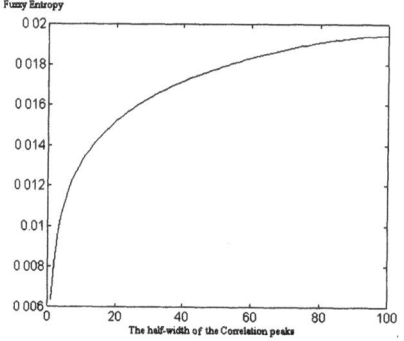

Fig. 6. The relationship between the fuzzy entropy and the half-width of the correlation peak

Table 1. Comparison of the three methods

	Fuzzy Entropy	Local Analysis	Statistical Analysis
Information Capacity	whole	local	local
Computing method	one-step serial	multi-step serial searching	multi-step nonlinear computation
Computing complexity	very little	large	very large
Processing speed	fast	normal	slow
Stability	high	low	normal
Pre-knowledge	no	no	yes

with the decrease in quality of the correlation peak indicates the inverse relationship between the fuzzy entropy and the quality of matching.

Compared with the former methods, using the concept of fuzzy entropy to describe the correlation peaks has many merits. Table 1 clearly shows the merits of the fuzzy entropy in optical recognition.

The justification is to segment the fuzzy matrix $\tilde{M}$ with an appropriate threshold,

$$\tilde{M}|_T(m,n) = \begin{cases} 1, & \text{while } \tilde{M}(m,n) \geq T \\ 0, & \text{while } \tilde{M}(m,n) < T, \end{cases} \tag{14}$$

where T is the threshold, and $\tilde{M}|_T$ is an ordinary Boolean matrix. $\tilde{M}|_T(m,n) = 1$ means that the reference mask is the same as the widow image of the image X at the position (m,n); and $\tilde{M}|_T(m,n) = 0$ means that the reference mask does not match the widow image of the image X at the position of (m,n).

Certainly, threshold T can be set as 1 when we only want to know whether there is absolute matching between two images. In practical image processing, there is little difference between input image and reference mask so the value scope of $\tilde{M}$ is $(0,1)$. If we select 1 as the threshold, the right recognition results cannot be obtained. Though the threshold should be set as $T < 1$, the right value cannot be ensured subjectively because there is a contradiction between getting better ability on the noise tolerance with a lower threshold and decreasing the probability of miscarriage of justice with a higher threshold. To solve this problem, a novel method, called fuzzy entropy segmentation, is proposed to compute the right threshold by the modern information theory:

1. The fuzzy entropy $H_f(\tilde{M})$ of the matrix $\tilde{M}$ is computed, where $0 \leq H_f(\tilde{M}) \leq 1$.
2. The mapping between the threshold and the fuzzy entropy is designed. The mapping can be linear or nonlinear functions. Here, a linear function is assumed:

Fig. 7. Testing Images. (a) Original Face Image. (b) Face Image with 5% noise. (c) Face Image with 10% noise. (d) Face Image with 15% noise. (e) Face Image with 20% noise. (f) Face Image with 25% noise. (g) Face Image with 30% noise

$$T = \frac{(1 - H_{\mathrm{f}}(\tilde{M}))}{2} + \frac{1}{2}, \tag{15}$$

where $T = 1$ while $H_{\mathrm{f}}(\tilde{M}) = 0$, and $T = 1/2$ while $H_{\mathrm{f}}(\tilde{M}) = 1$.

3. At the end, the matrix $\tilde{M}$ is segmented by the threshold. Checking the nonzero points in the matrix $\tilde{M}|_T$ will obtain the final recognition result.

Here, we use a series of human-face images with different noisy disturbances to test the distortion-invariant ability of fuzzy entropy segmentation. In Fig. 7a, a binary image with 64×64 pixels is considered as the reference mask. In Figs. 7b–g, the images with 5%, 10%, 15%, 20%, 25%, and 30% Gaussian noise are considered as input images. The curve of fuzzy entropy is shown in Fig. 8a. It is obvious that the fuzzy entropy increases proportionably with the noise of the images. Figure 8b shows the relationship between the threshold decided by the fuzzy entropy segmentation and the maximum value of the correlation peaks. (The solid line is the threshold and the dashed line is the maximum value of the peaks). In fact, people could not determine the similarity between the original image and the image with 30% Gaussian noise. Though the correlation still has a high peak, this peak does not indicate the similarity between two images. That is the most important meaning of the fuzzy entropy segmentation.

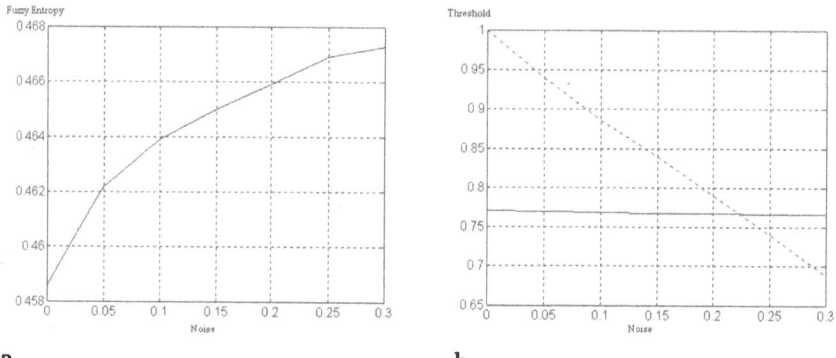

a   b

Fig. 8. The fuzzy entropy segmentation of correlation for the noise images. (a) The curve of fuzzy entropy. (b) The thresholds and the maximum values

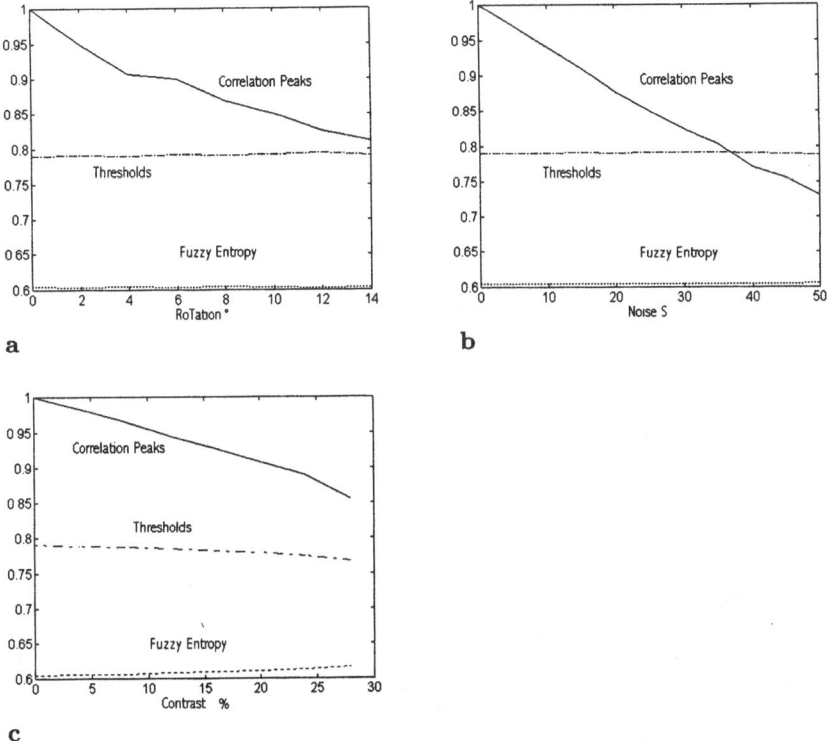

Fig. 9. Analysis of the triangle fuzzy rational measurement with fuzzy entropy segmentation. (**a**) Distortion-invariant ability for rotation. (**b**) Distortion-invariant ability for noise. (**c**) Distortion-invariant ability for contrast

We combined the triangle fuzzy rational measurement with the fuzzy entropy segmentation and tested the rightness and distortion-invariant ability of the algorithm. Figure 9 shows the test results, including the curves of the correlation peaks, fuzzy entropy and the thresholds.

Figure 9 shows that the novel algorithm has excellent distortion-invariant ability. The tolerance of distortion-invariant for rotation is larger than 14°, the tolerance of distortion-invariant for Gaussian noise is larger than $S = 37$, and the tolerance of distortion-invariant for rotation is larger than 28%, respectively.

5   Conclusion

In this paper, based on fuzzy theory, we discussed the principle and technique of gray-tone image recognition using optical correlators. In this novel method, a fuzzy relational matrix, which was constructed by triangle fuzzy set, was defined to describe the characteristics of gray-tone pixel matching. A changeable m-bit cycle-encoding method was introduced to realize the fuzzy relational measurement on a correlator structure. We also discussed how to select m correctly for different conditions of distortion in object recognition. This new method improved the performance of the conventional optical correlators and obtained the best balance between the spatial-bandwidth and the processing accuracy. Using the concept of uncertain pixels improves the distortion-invariant ability of optical correlation without losing information from the input images. Based on modern information theory, a novel fuzzy entropy segmentation method is proposed to analyze the optical computing results. It has excellent abilities regarding on the noise tolerance and resolution power on the identification of multi-objects and selecting the appropriate threshold quantitatively in the noisy environment. According to these algorithms the optical experimental results are obtained by using an incoherent correlator. The system has good fault-tolerance ability for rotation distortion, Gaussian noise disturbance, or information losing. The processing speed is 12 frames per second, which is so fast that it can be used in the field of machine vision.

Acknowledgements

We appreciate the support of the National Natural Science Foundation of China (69775008), the High Technology Research and Development Program of China, and Cao Guangbiao High Technology Development Foundation.

References

1. Mahalanobis, A. (1997) Correlation Pattern Recognition. Opt Engineering 36(10): 2631–2632
2. Goodman, J.W. (1968) Introduction to Fourier Optics. McGraw-Hill, New York
3. Weaver, C.S., Goodman, J.W. (1966) A Technique for Optically Convolving Two Functions. Appl Optics 5: 1248–1249
4. Caulfield, H.J., Maloney, W.T. (1987) Improved Discrimination in Optical Character Recognition. Appl Optics 26: 3633–3640
5. Javidi, B. (1989) Nonlinear Joint Transform Spectrum Based Optical Correlation. Appl Optics 28: 2358–2367
6. Gracht, J.V.D., Mait, J.N., Prather, D.W., Athale, R.A. (1994) Role of Coherence in Optical Pattern Recognition. Proc SPIE 2237: 152–163
7. Brasher, J.D., Johnson, E.G. (1997) Incoherent Optical Correlator and Phase Encoding of Identification Codes for Access Control or Authentication. Opt Engineering 36: 2409–2416
8. Kumar, B.V.K.V. (1992) Tutorial Survey of Composite Filter Designs for Optical Correlators. Appl Optics 31: 4773–4801
9. Chan, J., Braggins, D. (1996) Untiring Eyes. Manufacturing Engineer 75: 233–235
10. Serra, J. (1982) Image Analysis and Mathematical Morphology. Academic Press, New York

11. Maragos, P. (1987) Tutorial on Advances in Morphological Image Processing and Analysis. Opt Engineering 26(7): 623–632

12. Liu, L. (1994) Morphological Hit-or-Miss Transform for Binary and Gray-Tone Image Processing and Its Optical Implementation. Opt Engineering 33(10): 3447–3455

13. Schafer, R., Casasent, D. (1995) Nonlinear Optical Hit-Miss Transform for Detection. Appl Optics 34: 3869–3882

14. Yuan, S., Wu, M., Cheng, G., Jin, G. (1996) Optical Morphological Hit-Miss Transform for Pattern Recognition of Gray-Scale Image. Proc SPIE 2751: 271–277

15. Cheng, G., Yuan, S., Wu, M., Jin, G. (1997) Human-Face Recognition by Use of an Incoherent Optical System. High Technol Lett 3(2): 18–22

16. Merkle, F., Lorch, T. (1984) Hybrid Optical-Gigital Pattern Recognition. Appl Optics 23(10): 1509–1516

Part III

Optical Communication
(Photonics and Optoelectronics)

All-Optical Regeneration for Global-Distance Fiber-Optic Communications

E. Desurvire and O. Leclerc

Summary. As optical amplifiers have opened new perspectives for ultra-high-capacity transmission of lightwave signals over transoceanic distances (more than 100 Gbit/s over 10 000 km) fundamental limits are being felt. Such limits come from the combined effects of amplifier noise accumulation, fiber dispersion, fiber nonlinearities, and inter-channel interactions, contributing to various forms of signal degradation. In order to overcome these impairments and meet ever-growing transmission capacity needs, another technology revolution will soon be required. Promising developments concern in-line all-optical regeneration, which makes it possible to transmit optical data over virtually unlimited distances, without any electronic buffering. After recalling the basic principle of optical regeneration in lightwave systems, we describe the state of the art in experimental implementation. We also discuss the alternative offered by electronic regeneration, and highlight the advantages of the all-optical approach, with its technology challenges.

1 Introduction

Since the demonstration of the first low-loss optical fibers in 1970, followed by progress in single-frequency semiconductor lasers, and in the early 1990s, by the development of practical optical amplifiers, lightwave communications has now reached full maturity. Erbium-doped fiber amplifiers (EDFA) [1,2] introduced two fundamental improvements: first, optical signals can be periodically boosted along the line to compensate transmission loss, without electronic regeneration, thus expanding transmission distances as well as enabling transmission with significantly higher data rates; second, several optical channels can be transmitted at once through the same fiber, which opened the way to wavelength-division multiplexing (WDM), upon which future global networks of ultra-high capacities will be based [3]. The WDM technology benefited both from EDFAs, which opened further the usable fiber bandwidth, and progress in high-speed electronics, leading to multi-gigabit/s capacities for transmitters/receivers as well as the underlying multiplexing and switching circuitries.

The exploitation of the aforementionned technologies, which were unknown to the communication industry only a decade ago, made it possible in only a few years to transmit mult-100 Gbit/s over distances ranging from 1000 km to 10 000 km, breaking all previous transmission distance/capacity limits. In the meantime, global operator deregulation and the rise of new internet-driven services has opened a formidable market to this industry. It is now engaged in an accelerated capacity race, in view of providing ever-greater bandwidth to operators, regardless of geographical boundaries. Submarine fiber-optic cables, which carry 90% of the telephone traffic between continents, have superseded satellites for the long-haul transport, owing to their far greater bandwidth. New constellations of satellites are deployed to facilitate the access by an ever-increasing number of users to the world's communication network.

Submarine cables, which eventually concentrate all this traffic, thus represents the true 'information highway', as described below. In turn, the traffic is increasing exponentially, due to the rapid development of broadband services. With such an unprecedented market pull, long-haul communication technologies have entered a golden age, in which overcoming any capacity limit remains a key concern. Such a paradigm presses the R&D sector to rapidly propose, develop, qualify and standardize novel enabling technologies. Research in this field is now characterized by shorter conceptual-investigational overhead, gearing towards rapid technology evolutions, product definition and time-to-market cycles.

2   Transoceanic Systems and Related Technologies

It is useful to first consider the recent history and status of the world's undersea network, and then the related technologies. Figure 1 shows maps of the current amplifier-based network plant, as developed up to 1997, and of other projects through year 2000, including the first WDM generations. The figure does not convey the great diversity of smaller, short-haul undersea systems based on festoon and/or amplifier-free (or 'unrepeatered') links, which complete the network. Noteworthy is the network densification in both the Atlantic and the Pacific oceans. Increased network capacity is provided by the combined use of (1) ring topologies (two independent cables operating with mutual protection), (2) multiple-fiber pairs within the same cable (space-division multiplexing, or SDM), and (3) WDM, currently based on $N \times 2.5$ Gbit/s and $N \times 10$ Gbit/s configurations ($N = 8$, 16, 32 or 64). In such WDM systems, the implementation of submarine branching units (ADM), such as seen in Fig. 1(B) on the western side of Africa or North/South America, makes it possible to safely tap and feed the links from the coastlines, without use of submerged electronics.

This huge network plant still continues to develop in size, connectivity and capacity. Through 1988–1998, 300 000 km of undersea cables will have been installed! Year 1988 corresponds to the inauguration of the first submarine optical-fiber system, called TAT-8, a 2×280 Mbit/s SDM transatlantic link with 67-km-spaced electronic repeaters and $1.3 \mu m$ multi-longitudinal-mode signals [4]. Only in 1996 were the first optically amplified (or 'unrepeated') systems, called TAT-12/13 and TPC-5, installed in both the Atlantic and the

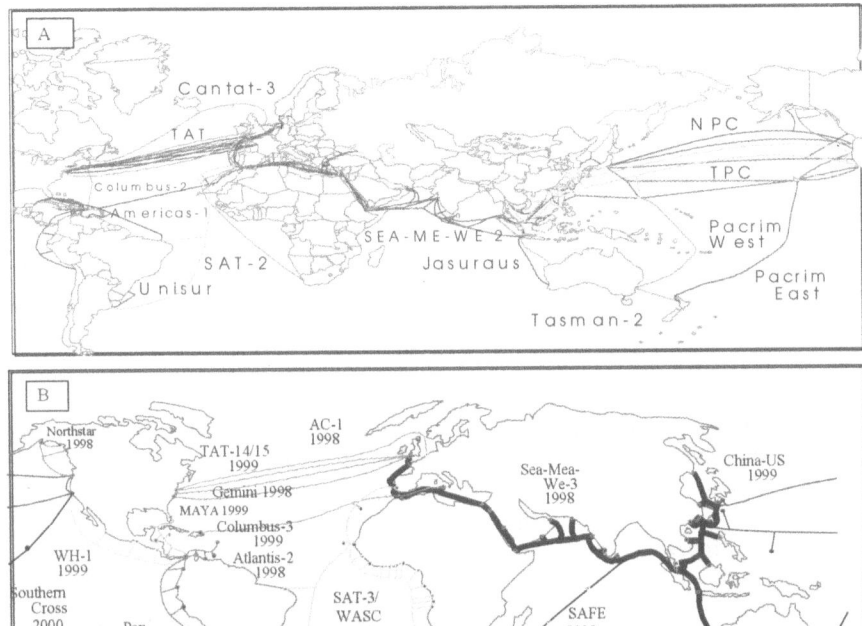

Fig. 1. Amplified submarine systems in service in 1997 (A) and other projects for 1998–2000 (B). Years indicate readiness for service; bold line corresponds to SeaMeWe-3 system, based on 4–8 × 2.5 Gbit/s WDM, with 36 000 km length and 40 landing points, and a longest unrepeated segment of 5100 km. Other projects in the Atlantic and Pacific areas (not shown in the figure) concern WDM systems based on $N \times 10$ Gbit/s capacities, with $N = 8$, 16 or 32 channels (courtesy Alcatel Submarine Networks)

Pacific oceans, as built for a 2×5 Gbit/s total ring capacity. Consider that TAT-8 was designed to meet all traffic needs up to the year 2001, but was actually 'filled to capacity' in 1990; as for TAT-12/13 and TPC5, planned for a 2008 fill year, they should be saturated in only two years from now [5]! Such a tidal wave of traffic demand of unpredictable proportions must be channeled through an ever-expanding SNS framework. The ambitious project Oxygen plans to have linked, by year 2003, not less than 175 countries through a 175 000-km-long, 262-landings cable mesh with capacities between 100 and 1000 Gbit/s [6]!

Such conditions of unquenched capacity demand requires intensification of research with shorter innovation-to-market cycles. Since the early 1990s, a lag of capacity per fiber pair (CPFP) of one order of magnitude is consistently observed between lab experiments and commercial systems. Figure 2 shows the five major technology cycles followed by long-haul transmission research since

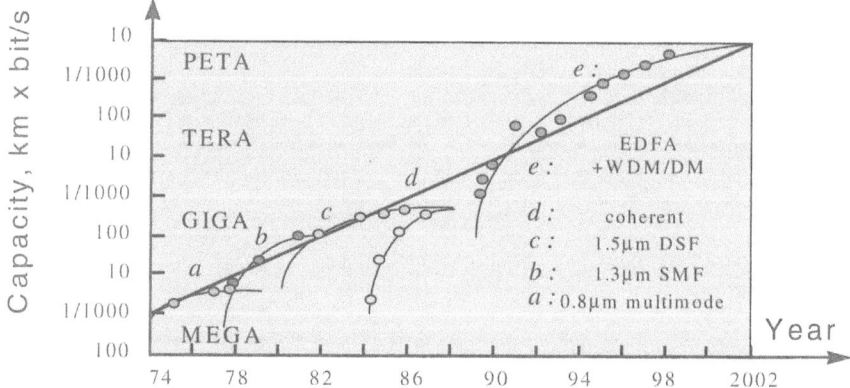

Fig. 2. Technology cycles in long-haul transmission research since 1974, showing consistent tenfold capacity increase every four years. Capacities of 10 km Pbit/s or 10 Mm Tbit/s are projected for year 2002 (EDFA, erbium-doped fiber amplifier; WDM, wavelength-division multiplexing; DM, dispersion management; DSF/SMF, dispersion-shifted/single-mode fiber)

1974 [2]. Our current cycle, the true 'golden age of optical fiber amplifiers' [1] is also that of WDM and dispersion-management (DM) technologies (which are discussed in this chapter as well). A tenfold increase of capacitydistance is observed every 4 years, predicting a value of 10 Mm Tbit/s (or 10 km Pbit/s) for the year 2002. Based on the previous observation, one concludes therefore that the SNS installed in the next four years should have CPFPs of at least 100 Gbit/s, meaning 200 Gbit/s for 'basic' rings. To express this performance in tangible 'customer' terms, a capacity of 100 Gbit/s permits the transmission of 1.5 million voice circuits at 64 kbit/s, and 50 000 high-resolution digital video communications at 2 Mbit/s. This potential represents a thousandfold improvement with respect to the first 100 Mbit/s lightwave systems, which existed twenty years ago. Also, the span of these systems was limited to a few tens of kilometers, while the current 'amplified system' generation concerns multi-thousand kilometer hauls.

Next, we review the enabling technologies for high-capacity transoceanic (> 100 Gbit/s, > 6 Mm) systems. The last two data in Fig. 2 correspond to recent lab experiments at 160 Gbit/s (9.3 Mm) and 320 Gbit/s (7.2 Mm), using 5 Gbit/s WDM signals [7]. Other transoceanic (9–10 Mm) WDM experiments with capacities greater than 160 Gbit/s have been carried out with 10 Gbit/s signals [8]. In addition to WDM, all these experiments used the 'return-to-zero' (RZ) format and the so-called dispersion management (DM) technique, which represent two major technology trends. These are worth explaining with simplified arguments, which is done in the following paragraph. This explanation also provides some background and introduction to the most recent developments in RZ transmission and, in particular, based upon *soliton* propagation.

The transition from the usual 'non-retun-to-zero' (NRZ) format to the RZ format, as the preferred choice for high-bit-rate signals in long-haul systems, can be simply explained by the combined and deleterious effects of self-induced Kerr nonlinearity and dispersion. Basically, the Kerr effect produces an instantaneous frequency change, which is proportional to the time derivative of the pulse envelope [9]. The higher the bit rate, the shorter the pulse rise and fall times, and therefore the greater the frequency change. Consider the NRZ format, as based ON/OFF intensity modulation. As long as there are symbols '1' to be transmitted in sequence, the light is kept at the constant 'ON' level; it is turned back to the 'OFF' level when '0' symbols must transmitted. Physically, the NRZ data stream is thus a sequence of rectangular pulses of variable durations and spacings, which represent strings of '1' symbols. Thus, the symbols '1' located at the edge of the strings experience different amounts of Kerr effect (i.e. frequency shifts of opposite values). As for the '1' symbols located within the string, they do not experience any frequency change, since the intensity is kept to a constant level. Different frequency components are associated with different velocities; upon propagation in the fiber, string distortion and inter-symbol interference occur, due to dispersion. The distorsion/interference effect is also a function of the string length (number of succeeding '1' symbols), which would make its compensation very difficult, if not intractable. In contrast, the '1' symbols of the RZ format are represented by individual light pulses. Physically, the RZ data stream is a comb of identical pulses, with some missing in the sequence, the corresponding gaps representing the '0' symbols. Thus, each RZ pulse experiences the same Kerr effect (i.e. up-/down-frequency shifts in the pulse's trailing/leading edges). Whether such pulses can propagate without distorsion is another question, but the distorsion is the same for each, regardless of their positions within the sequence. Clearly, the above considerations justify that, as the bit rate increases, signals should be transmitted in RZ format in order to minimize delterious Kerr nonlinearity. A second conclusion is that some propagation regime must be found where Kerr nonlinearity and dispersion could balance each other somewhat. Before analyzing this issue further, it is useful to describe the DM technique, which has introduced significant progress in long-haul RZ transmission.

Two most important developments in this field concern the so-called chirped RZ (CRZ) transmission, on one hand, and soliton transmission with various forms of DM approaches, on the other hand, which we briefly review here.

Soliton propagation, whose underlying principle is to use Kerr nonlinearity to balance fiber dispersion [9], has long been thought as the natural and unescapable solution to overcome the nonlinearity/dispersion limitations otherwise experienced with NRZ, especially in view of broadband WDM system implementation [10]. As it turned out, however, two major impairments in soliton transmission were identified: (a) the amplifier noise background, accumulating along the system, causes random frequency shifts and, hence, pulse timing jitter (called the Gordon–Haus jitter [10]), and (b) in WDM propagation, cross-phase modulation and nonlinear frequency mixing between colliding pulses from different channels also cause random frequency shifts (timing jitter) and inter-channel coupling (crosstalk). In an ideal, lossless transmission line, such deleterious interactions would not appear, since solitons act like independent particles (hence their name), which conserve their energy and integrity through nonlinear collisions. This picture was changed with the introduction of optical amplifiers. Initially, loss-compensated soliton propagation was thought to require *distributed* amplification, such as based upon Raman fiber amplifiers; the reason being that exact compensation of dispersion and nonlinearity requires the soliton power to be maintained constant along the transmission line. However, several groups showed the possibility of soliton propagation with *lumped* or periodic amplification, such as provided by practical in-line EDFAs. With the appropriate amount of power pre-emphasis at each amplifier's output, the resulting pulse behaves like the fundamental soliton (called the average or guiding-center soliton), as long as the amplification period is shorter that the characteristic soliton period (z_0). As previously mentioned, amplifier noise impairs transmission with Gordon–Haus jitter, in addition to the effect of the degradation due to the signal-to-noise ratio (SNR) observed in any system. In the WDM regime, the limitation becomes even more serious, because of collision jitter and crosstalk. Elegant countermeasures were actually discovered for both limitations; the first with sliding narrowband (frequency-guiding) filters, and the second with a dispersion-decreasing (DDF) fiber [10]. Such techniques made it possible to demonstrate the first $N \times 10$ Gbit/s transoceanic soliton transmission [11]. With sliding filters, the line becomes 'opaque' to amplifier noise after a few filtering spans, while the solitons pass through the filters, and SNR stabilization along the entire system eventually results. We note here, as it is central to the 'optical regeneration' topic of this chapter, that the sliding-filter technique provided a very important demonstration: *by exploiting the fundamental difference between (linear) amplifier noise and (nonlinear) soliton pulses, propagation without SNR degradation in amplified systems is actually possible.* The next section shows that synchronous intensity modulation produces a similar result, which represents a first step towards all-optical, in-line regeneration.

Despite the remarkable progress in WDM soliton systems, the proposed solution requiring mismatched narrowband filters and DDF cables went far to meet practical industry requirements. On the other hand, high-bit-rate NRZ transmission made significant advances by use of comparatively much less complex techniques. Overlooking the great improvements brought by 980 nm-pumped EDFAs (noise reduction), large effective-area fibers (nonlinearity reduction), and compensating filters (amplifier bandwidth enhancement), three fundamental changes were progressively introduced: (a) sinusoidal phase and intensity modulation is applied to the NRZ data at the transmitter level,

(b) accumulated dispersion is periodically compensated along the line by use of short segments of dispersion-compensating fiber (DCF), and (c) pre- and post-dispersion compensation are effected in the transmitter and receiver. The resulting pulse, called chirped RZ (CRZ), propagates in both positive and negative dispersion regimes. Progress in CRZ transmission by use of these different techniques has led to the aforementioned, multi-100 Gbit/s records [7,8], which overshadowed for the time being the competing 'soliton' approach.

In parallel to the development of transoceanic CRZ systems (which have now become commercial products for the year 2000), new types of soliton propagation regimes, based on dispersion management (DM), have been identified. It is beyond the scope of this chapter to provide a review of this very prolific field. Suffice it to state that CRZ and DM-soliton propagations conceptually share much in common, and that they cover a wide range of possibilities in balanced nonlinear/dispersive pulse behavior. The DM soliton, however, lends itself to a wide variety of 'stable' or 'periodic' solutions, which differ by the DM strategy. The strategy is determined by the type of 'dispersion map', which is based on alternating types of opposite-dispersion fibers (DSF, SMF, DCF, ...). Many novel properties of DM solitons were identified, e.g. soliton-like propagation in the negative and zero-dispersion regimes [12,13], or pulse energy enhancement enabling Gordon–Haus and WDM collision jitter suppression [14,15]. The DM-soliton technique made it possible to achieve, without any other form of in-line control, unprecedented system performance, such as 40 Gbit/s single-channel transmission over 8.6 Mm [16], or 640 Gbit/s (32×20 Gbit/s WDM) over 1.2 Mm of standard fiber (SMF) with 100 km amplifier spacing [17].

The aforementioned transmission technologies (CRZ and DM soliton) promise further developments in multi-100 Gbit/s amplified transoceanic systems, with the prospect of achieving 10 Mm Tbit/s CPFP in the early 2000s. A key question is whether the achieveable capacity will saturate past this point, to reach the end of a whole technology cycle (Fig. 2). The solution consisting in fitting more fiber pairs within submarine cables (SDM) is only a marginal one, let alone that the market *also* capitalizes on such capacity extensions. If a revolution is required at the end of this technology cycle, then new approaches for ultra-high-capacity WDM pulse transmission must be developed. The 'classical' CRZ and DM-soliton systems, which rely upon purely *passive* transmission techniques (including optical amplification), must leave place for novel forms of *active* in-line control. Ultimately, only such control could make possible to free future undersea systems from impairments caused by amplifier noise, timing jitter, pulse distorsion, WDM crosstalk and other signal degradation. In the following sections, we describe the possibilities offered by all-optical signal regeneration, and also discuss alternative types of regeneration which are based on purely electronic or hybrid opto-electronic approaches.

3   All-Optical Regeneration: Theory

In the literature, the label 'all-optical' usually refers to optical/optical (O/O) signal processing, as opposed to opto-electronic (O/E) approaches. By dubbing the principle of signal regeneration described in this chapter as 'all-optical', we reference not the control (which is electrical here), but the *signal carrier*, which remains unchanged through the process. This is in contrast with optical regeneration by interferometric wavelength conversion [18–20], where the optical data are transferred from one carrier to another (possibly at the same wavelength, but emitted from a different laser source), along with pulse reshaping and noise suppression. In this case, what is processed is only the intensity envelope of the input optical pulse, which acts as a driving signal for the emission of a 'new' pulse of suitable (noiseless) characteristics.

In all-optical regeneration (according to our definition), the input and output pulses physically represent the same electromagnetic wavepacket. The effect of all-optical regeneration is to suitably modify the pulse input characteristics such as: position (center of gravity in the time domain), frequency (center of gravity in the frequency domain), phase distribution (chirp), envelope (shape, pulse width, pedestal extinction, ...) and energy (photon contents). This whole signal processing can be implemented by combining active and passive functions such as synchronous modulation (SM) and narrowband filtering (NF), by using the nonlinear propagation properties of solitons [21], as explained below.

3.1   Principle

All-optical regeneration by SM/NF was first proposed in 1991 by Nakazawa et al., with the demonstration of 'unrepeatered' data transmission over one million kilometers [21]. Concerning the SM function, the possibility of using either intensity modulation (IM) as in [21], or phase modulation (PM), as in [22,23], or the combination of both with the extension of the technique to WDM systems [24] was also investigated, both theoretically and experimentally. In this subsection, we briefly describe the underlying principle of signal regeneration and background noise suppression through IM/PM/NF regeneration, by using both physical explanations and formalism. We consider, first, single-channel operation; possible solutions for implementation in WDM systems, and numerical simulation examples are described in the next two subsections.

The effects of IM and PM are qualitatively different; both act on the pulse's timing position (jitter suppression), but additionally, IM makes it possible to physically remove some finite amount of the noise background. Such a background is made of the accumulation of amplified spontaneous emission generated by the amplifiers in the line, and of the dispersive waves (continuum) generated by non-soliton components in the pulse envelope and other deleterious perturbations. This noise background manifests itself by the

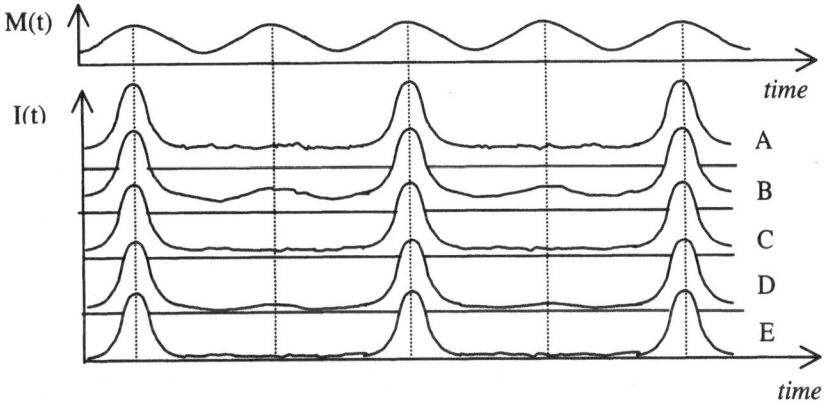

Fig. 3. Suppression of optical noise background intensity modulation (IM), as applied to a 10101 pulse sequence. A–B: Signal input/output at first modulator. C–D: Signal input/output at second modulator. E: Signal output after N modulator, showing asymptotic stabilization of noise background

unwanted presence of energy in the '0' slots and of a pulse pedestal in the '1' slots. Suppressing it with adequate periodicity leads to SNR stabilization throughout the system, as we should explain first. Figure 3 shows the effect of both periodic IM and soliton propagation, taking a jitter-free 10101 bit sequence as an example (A). Upon passing through the first modulator, the background noise in the '0' slots is shaped into pulses having 50% less energy (B). After propagation, the '1' soliton pulses remain unchanged, while the '0' noise spreads out in the time domain, due to fiber dispersion, filling up the gaps left out by the previous modulation (C); after several iterations, the noise background is stabilized to a constant, asymptotic level.

The previous description did not take into account the fact that between each modulation stage, some more noise is generated by the amplifiers included in that section, which accumulates with the previous one. Thus, the key issue is how often should the modulation occur in order to achieve convergence or stabilization of this overall background noise level. This can answered by a basic calculation [21], which we briefly recall here. The modulation function $M(t)$ with extinction ratio x takes the form

$$M(t) = x + \frac{1-x}{2}\left(1 + \cos\frac{2\pi}{T_{\text{bit}}}t\right), \tag{1}$$

where T_{bit} is the bit period. For constant signals (noise background), the IM transmittance is given by the time-averaged value of $M(T)$, i.e. $T_R = (1+x)/2$. For the '1' signal pulses, the transmittance is given by

$$T_{\text{m}} = \int I(t)M(t)\,\mathrm{d}t \Big/ \int M(t)\,\mathrm{d}t, \tag{2}$$

where $I(t)$ is the signal intensity ($I(0) = 1$). In the absence of timing jitter (perfect synchronization of each '1' pulse), the transmittance is less than unity (see below) and an excess gain $G_{ex} = 1/T_m$ must be introduced therefore for exact compensation. Let N be the noise generated by the sequence of amplifiers between each modulation stage; the total noise accumulated along a line containing k modulators is then:

$$N_{out} = N\left(1 + T_R G_{ex} + (T_R G_{ex})^2 + \ldots + (T_R G_{ex})^k\right)$$
$$= N\frac{1 - (T_R G_{ex})^{k+1}}{1 - T_R G_{ex}} \approx N\frac{1}{1 - T_R G_{ex}}, \tag{3}$$

where we used the property that $T_R G_{ex} = (T_R/T_m) < 1$ (as seen below). Thus, the total noise background effectively reaches an asymptotical value, as defined in (3). On the other hand, the transmission line is transparent to the '1' pulses ($G_{ex}T_m = 1$), making the SNR reach a constant value after some propagation distance. It is useful to define the modulator transmittance corresponding to a hyperbolic-secant-square pulse envelope (fundamental soliton) or Gaussian pulse envelope dispersion-managed soliton both with full width at half-maximum (FWHM) ΔT. Using either $I(t) = 1/\cosh^2(1/\tau)$ or $I(t) = \exp(-t^2/\tau^2)$, with $\tau = \Delta T/1.76$ or $\tau = \Delta T/2\sqrt{\log 2}$, respectively, integration in (2) yields a result [21,25] which can be put into the general form

$$T_m = x + \frac{1-x}{2}(1 + F(V)) \equiv T_R + \frac{1-x}{2}F(V). \tag{4}$$

In (4), the function $F(V)$ is equal either to $V/\sinh V$ or $\exp(-V^2)$, with the definitions $VT_{bit}/\Delta T = \pi^2/1.76$ or $VT_{bit}/\Delta T = \pi/2\sqrt{\log 2}$, respectively. Since $F(V)(1-x)/2$ is positive, it is seen indeed that $T_R G_{ex} = T_R/T_m < 1$, which is the convergence condition assumed in (3).

Next, we consider specifically the effect of IM and PM in terms of a 'pull-back' force in the time domain, which effectively results in jitter suppression. The principles of IM and PM are described in Fig. 4. The recovered RF clock, corresponding to the baseband signal modulation, drives the IM/PM modulator with the absolute phase reference. Timing jitter causes the '1' pulses to wander about the center of the bit frame, which is characterized by a random delay t_0. As seen in the figure, IM produces an asymmetric loss on jittered pulses, the loss being higher on the pulse edge most remote from the center of the bit frame. On the other hand, unjittered pulses ($t_0 = 0$) do not experience any loss since the modulator transmittance T_m has been exactly compensated by an excess gain G_{ex}, as described earlier. Soliton propagation immediately following IM causes the asymmetric pulses to recenter themselves about their new energy 'gravity center', as defined by

$$t_0' = \int tI(t - t_0)\Delta I(t)\,dt \Big/ \int I(t - t_0)\,dt, \tag{5}$$

while it is clear that due to the IM loss asymmetry we have $|t_0'| < |t_0|$, meaning that timing jitter decreases. Since additional timing jitter is introduced

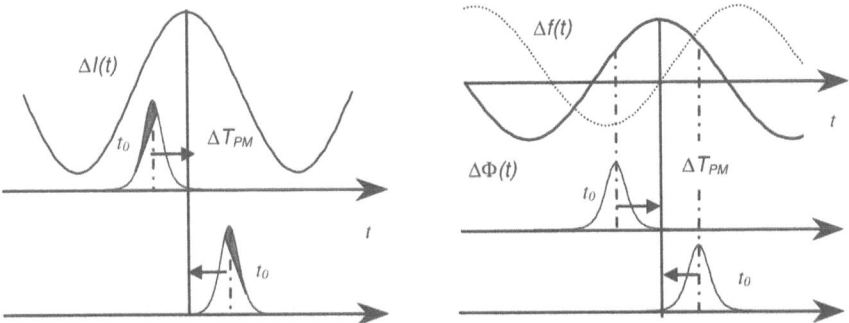

Fig. 4. Pull-back effect in the time domain resulting from synchronous IM (*left*) and synchronous PM (*right*). See text for description

between successive modulators, the issue of convergence towards $|t'_0| = 0$ after several IM stages is similar to that previously discussed for noise background stabilization. This point is developed further below with perturbation theory.

Next, we consider PM (Fig. 4). Unlike IM, PM does not introduce loss, therefore it is unable to remove dispersive waves. However, the frequency shift associated with PM, i.e. $2\pi\Delta f(t) = -\partial\Delta\Phi(t)/\partial t$, and subsequent soliton propagation together cause timing jitter suppresion. Assume the pulse to lag behind the center of the modulation (pulse on bottom axis, Fig. 4); it experiences a blue frequency shift ($\Delta f(t) > 0$); due to soliton properties, this shift becomes equally distributed across the pulse upon propagation. In the positive (anomalous) dispersion regime, the blue-shifted pulse propagates faster than the pulse would without a frequency shift; thus, PM converts timing delay into group velocity increase, which cancels the delay after some propagation distance. The same reasoning and reverse conclusion apply for jittered pulses arriving ahead of the PM window center (pulse on middle axis, Fig. 4); the resulting red shift corresponds to a group-velocity decrease, which delays the pulse. The issue of convergence in the jitter suppression process, in view of the additional jitter generated between each PM stage, is the same as in the IM case, and the conditions for convergence can be analyzed with perturbation theory (see next). To summarize, we can only say that both PM and IM act like 'pull-back' forces in the time domain, the magnitude of which is some monotonous function of timing jitter, which we need to specify at this point.

Soliton propagation in fibers is accurately described by the nonlinear Schrödinger equation (NLSE) [9], which is briefly discussed in Sect. 3.3. For constant dispersion, its most general form is

$$i\frac{\partial u}{\partial z} + \frac{1}{2}\frac{\partial^2 u}{\partial t^2} + |u|^2 u = f(u) + \varepsilon P(u), \tag{6}$$

where $u(t)$ is the soliton's dimensionless, electromagnetic field envelope, i.e. $|u(t)|^2 = A^2 \equiv I(t)$. The time derivative and the nonlinear term involved in (6) describe the effects of dispersion and Kerr nonlinearity, respectively. In the right-hand side (RHS) of this equation the function $f(u) = i(g - \alpha)u/2$ accounts for the effect of exponential loss and lumped amplification during propagation, while the function $\varepsilon P(u)$ describes all types of possible perturbations in the system, such as amplifier noise, narrowband filtering, and, in our case of interest, IM/PM [26]. The general solution of (6) takes the form

$$u(z,t) = A \operatorname{sech}[A(t - t_0)] \exp[i(\phi - \omega t)] \tag{7}$$

with $\operatorname{sech}(x) = 1/\cosh(x)$. The normalized soliton amplitude (A), timing position (t_0), phase (ϕ) and frequency (ω) obey the following coupled linear-equation system [26]:

$$\frac{dt_0}{dz} = -\omega, \quad \frac{d\phi}{dz} = \frac{A^2 - \omega^2}{2} + t_0 \frac{d\omega}{dz}. \tag{8}$$

The above relations describe the evolution of the pulse's timing position and chirp. By substituting the definitions in (7) and (8) into (6), it can be easily checked that $u(t)$ is the general solution of the NLSE with RHS = 0. The most interesting point arising from perturbation theory is the ability to predict the evolution of $u(z,t)$ in the presence of a perturbative function $\varepsilon P(u)$ in the RHS of the NLSE. The perturbation only affects the four soliton parameters (t_0, A, ω, ϕ) in the general solution $u(z,t)$ defined in (7) and (8). The effects of IM, PM, and narrowband filtering (NF) with center frequency ω_0 yield the differential system [26]

$$\frac{dt_0}{dz} = -\left\{ \frac{k_{\mathrm{IM}}}{A^3} t_0 \right\}_{\mathrm{IM}} - \omega, \tag{9a}$$

$$\frac{dA}{dz} = -A \left\{ \frac{k_{\mathrm{IM}}}{2A^2} \left(\frac{12}{\pi^2} t_0^2 + 1 \right) - g_{\mathrm{ex}}^{\mathrm{IM}} \right\}_{\mathrm{IM}}$$
$$- A \left\{ \frac{k_{\mathrm{NF}}}{2} (A^2 + 3(\omega - \omega_0)^2) - g_{\mathrm{ex}}^{\mathrm{NF}} \right\}_{\mathrm{NF}}, \tag{9b}$$

$$\frac{d\omega}{dz} = +\{k_{\mathrm{PM}} t_0\}_{\mathrm{PM}} - \{k_{\mathrm{NF}} A^2 (\omega - \omega_0)\}_{\mathrm{NF}}, \tag{9c}$$

$$\frac{d\phi}{dz} = -\left\{ \frac{k_{\mathrm{PM}}}{2} t_0^2 - \frac{\pi^2}{12A^2} \right\}_{\mathrm{PM}} + \frac{A^2 - \omega^2}{2} + t_0 \frac{d\omega}{dz}. \tag{9d}$$

In the above equations, terms relating to IM, PM and NF have been indexed in braces. The coupling constants k_{IM}, k_{PM} and k_{NF} relating to the IM, PM and NF strengths are proportional to the second derivative (curvature) of the modulators and filter transfer functions $\Delta I(t)$, $\Delta f(t)$ and $T(\omega)$, respectively; for the IM/PM, the curvature is proportional to the extinction ratio; for NF it is the reciprocal of the filter's squared FWHM [26]. In (9b), excess-gain coefficients $g_{\mathrm{ex}}^{\mathrm{IM,NF}}$ were introduced to compensate loss caused by IM and NF in conditions of perfect synchronicity and frequency matching ($t_0 = 0$ and $\omega = \omega_0$), as defined and discussed later on.

We consider first the effect of IM and PM on the soliton timing position t_0. Equation (9a) shows that IM causes the deviation (t_0) to exponentially vanish with the pull-back force k_{IM}/A^3; the force is thus modulated by amplitude, and it is stronger (weaker) for downward (upward) amplitude fluctuations. The second term ($-\omega$) reflects the soliton position change due to dispersive propagation. According to (9c), however, the frequency is modulated by PM, with a shift proportional to the position t_0. Combining (9a) and (9c) while overlooking NF and assuming constant amplitude ($A = 1$), we find that the position follows the equation of motion $t'' + k_{\text{IM}}t_0' + k_{\text{PM}}t_0 = 0$, whose general solution, specifically depending on the relative IM/PM strengths, is not physically meaningful. In the case of pure IM or PM, however, one can easily derive the following solutions:

$$t_0(z, \text{IM}) = t_0(0)e^{-k_{\text{IM}}z} - \frac{\omega(0)}{k_{\text{IM}}}(1 - e^{-k_{\text{IM}}z})$$

$$\approx t_0(0)e^{-k_{\text{IM}}z} - \omega(0)z, \tag{10a}$$

$$t_0(z, \text{PM}) = t_0(0)\cos\left(\sqrt{k_{\text{PM}}}\, z\right) - \frac{\omega(0)}{\sqrt{k_{\text{PM}}}}\sin\left(\sqrt{k_{\text{PM}}}\, z\right)$$

$$\approx t_0(0)e^{-k_{\text{PM}}z} - \omega(0)z. \tag{10b}$$

Thus, it is seen that both IM and PM cause random timing delays to exponentially vanish with propagation (while the soliton moves on with accumulated group delay $\omega(0)z$), as previously described in Fig. 4.

Next, we consider the effect of IM and PM on the soliton amplitude A. Equation (9b) shows that under IM, random timing delays, corresponding to a default of synchronicity, cause an amplitude loss proportional to $k_{\text{IM}}t_0^2$. Thus, timing jitter is converted by IM into amplitude noise, which is an important limitation in some non-optimized or overly perturbed conditions (see simulations in Sect. 4). Fortunately, such amplitude fluctuations can be (fully or partially) compensated by NF, as discussed hereafter. In conditions of perfect synchronicity ($t_0 = 0$) and frequency matching ($\omega = \omega_0$), the residual loss due to the finite curvatures of the modulation and filtering in the time/frequency domains is also compensated by the excess gains $g_{\text{ex}}^{\text{IM,NF}}$. Assuming a reference amplitude $A_{\text{ref}} = 1$ obtained under such conditions, and using $(dA/dz)_{\text{sync}} = 0$ in (9b), one gets the general amplitude equation:

$$\frac{dA}{dz} = \frac{k_{\text{IM}}}{2A}\{(A^2 - 1) - 12t_0^2/\pi^2\}_{\text{IM}}$$

$$- A\frac{k_{\text{NF}}}{2}\{(A^2 - 1) + 3(\omega - \omega_0)^2\}_{\text{NF}}. \tag{11}$$

It is seen from this result that upward or downward amplitude fluctuations ($A > 1$ or $A < 1$) reduce or increase the IM loss, while for NF the effect is reverse. These two effects can be physically (and simply) explained as follows. It is first recalled that, in the absence of chirp, the soliton temporal width ΔT varies like the reciprocal of both its spectral width $\Delta\omega$ (Fourier-transform limit) and its amplitude A (fundamental soliton relation) [9]. Thus, an increase in amplitude simultaneously corresponds to pulse narrowing in the time domain, and spectral broadening in the frequency domain. Conversely, a decrease in amplitude corresponds to pulse broadening and spectral narrowing. Thus, the filter is more lossy when the amplitude increases, and less lossy when the amplitude decreases, which is the aforementioned feedback, leading to power stabilization. As for IM transmittance, it is the reverse effect, albeit less intuitive. Assume a Gaussian pulse having $1/e$ duration σ, and passing through the intensity modulator at frequency Ω with a random delay t_0. The corresponding transmittance is then given by

$$T_{\text{IM}}(t_0) = \int_{-\infty}^{+\infty} \frac{\exp(-(t-t_0)^2/\sigma^2)}{\sigma\sqrt{\pi}} \cos(\Omega t)\, dt\,. \tag{12}$$

Elementary integration in (12) yields $T_{\text{IM}}(t_0) = \exp(-\Omega^2\sigma^2)\cos(\Omega t_0)$. We see that the transmittance exponentially decreases with the pulse duration σ (the maximum transmittance, corresponding to Dirac pulses being $\cos(\Omega t_0)$). Therefore, pulse narrowing (with amplitude increase) decreases the IM loss, while pulse broadening (with amplitude decrease) increases the IM loss; thus the power feedback is positive, leading (in the absence of other counter-effects) to unstable behavior.

While NF produces a beneficial counter-effect on timing-jitter/amplitude-noise conversion by IM (which is enhanced by the aforementioned instability), it also introduces an undesirable effect, which is frequency-jitter/amplitude-noise conversion. Indeed, timing jitter is a result of random frequency shifts about the nominal value ω_0, as discussed in the previous section. Thus, upon passing through the filter, pulses with random frequencies undergo random loss, which is described by the last term in the RHS of (11); for effective power stabilization by NF by negative feedback, the whole term between braces in this equation must remain positive, imposing the condition (in normalized units) $\delta\omega^2 < 2\delta A/3$ with $\delta A > 0$, which determines the maximum allowable frequency fluctuation (given the amplitude fluctuation). Another unwanted effect of the filter is to introduce excess gain ($g_{\text{ex}}^{\text{NF}}$). In the absence of IM, this excess gain causes the non-soliton components in the signal (dispersive waves) to increase exponentially with distance (as opposed to a linear increase when the net gain is zero). We have seen in (3) the IM locks the accumulated noise to a constant level (N_{out}). However, this noise level is higher when the excess gain increases, which enhances timing jitter due to noise/soliton interaction (Gordon–Haus jitter) and degrades the SNR.

Considering the combined action of PM and NF leads to similar conclusions. Under PM, timing jitter is converted into random frequency shifts (making jitter vanish upon propagation); however, NF converts these shifts into amplitude noise, as previously described. Although PM alone does not generate amplitude noise (which is advantageous over IM), the combination PM/NF does. Recall that PM does not remove the dispersive waves, unlike IM, thus the filter excess gain must remain compartively smaller, in order

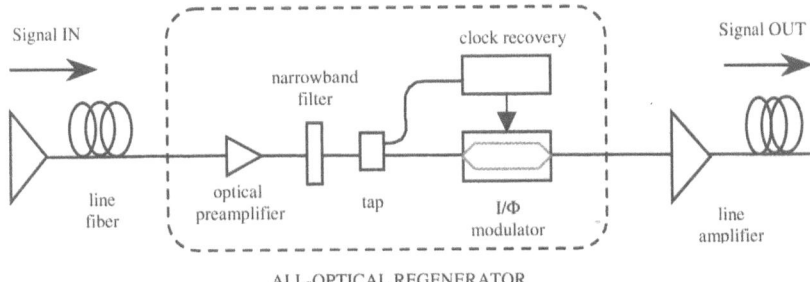

Fig. 5. Basic layout of the all-optical regenerator (see text for description)

to minimize SNR degradation. It is clear that combining IM and PM offers further flexibility for the most efficient implementation of signal regeneration.

To summarize the above analysis, we have seen that IM and NF generate different amplitude feedback mechanisms on the soliton signal; they are opposite, as long as the frequency fluctuation (timing jitter) remains below a certain limit. It can be concluded that the IM/NF strengths must be carefully optimized together in order to achieve the best operating conditions, i.e. maximum timing-jitter reduction by IM accompanied with minimal amplitude noise conversion, as balanced by NF, while amplitude noise generated by NF itself should also be minimal. The additional use of PM makes it possible to enhance jitter suppression, while also introducing amplitude noise conversion, which must also be compensated by NF. The multi-parameter optimization of the IM/PM/NF apparatus, given system assumptions (e.g. amplifier and regenerator spacing, bit rate, line dispersion, etc.) is a rather complex and non-intuitive task, and therefore, it must be reached through extensive numerical simulations. A few examples of regenerated system design optimization are described in the last subsection.

Having analyzed the principle of all-optical regeneration through SM/NF, the basic layout of the all-optical regenerator, shown in Fig. 5, is straightforward to understand. The regenerator module is inserted at periodic locations (spacing Z_R) in the amplified optical transmission line. An optical preamplifier provides compensation for the coupling/propagation loss of the various components involved, as well as the 'excess gain' required for nominal SM/NF operation, which corresponds to a power feedback bias. The narrowband filter is placed prior to the I/Φ modulator to control the pulse width characteristics, but it also acts as a power stabilizer with respect to the previous SM stage (as discussed above). The clock recovery circuit extracts the baseband RF component in the random data sequence, and generates a drive signal for the SM with the appropriate phase delay. The modulator can be either O/O or O/E (meaning driven by an optical or an RF clock signal), as discussed in further detail in Sect. 5.

3.2 Application to WDM Systems

In the previous section, we described all-optical regeneration through SM/NF for soliton signals charactrized by a frequency ω. High-capacity systems operate with the principle of wavelength-division multiplexing (WDM), which raises the issue of multiple-wavelength regeneration. There are three possible implemention schemes, as illustrated in Fig. 6. The first one consists in allocating a complete regenerator to each WDM channel, which requires both demultiplexing (DMUX) and re-multiplexing (MUX) stages; regeneration thus operates in parallel fashion, with no synchronicity between WDM channels. The two other possible schemes share the same modulator (and a periodic, Fabry–Pérot-type filter), and thus regenerate the channels in a serial fashion. This approach requires WDM synchronicity, meaning that all bits are (on average) time-coincident with the I/Φ modulation peak. Such synchronicity can be achieved either by use of appropriate time-delay lines located within a DMUX/MUX apparatus, or by making the WDM channels inherently time-coincident at the specific regenerator locations. How to achieve this last condition is discussed next.

The substantial reduction in WDM regenerator complexity obtained when all channels share a single I/Φ modulator makes the approach of periodic self-synchronization extremely attactive. It is beyond the scope of this chapter to discuss its feasibility in realistic systems, given the finite uncertainty in line parameters (such as group-velocity dispersion, or GVD) and their possible time/temperature-dependent fluctuations. It suffices to state here that this issue has been addressed to some extent, which leads to an optimistic conclusion [27]. The principle of what we refer to here as 'self-synchronous' WDM is based upon the property that in a dispersive transmission line, WDM channels have different GVDs and therefore they slide over each other upon propagation. Consider now two channels; their relative group delay increases until it is exactly equal to the bit period. At this location, the two channels are time-coincident or synchronous. In previous work [28], we have shown that through an appropriate choice of WDM comb parameters (i.e. wavelength spacing, zero-dispersion wavelength, dispersion slope), it is possible to achieve such a periodic synchonicity condition for all channels simultaneously. In this section, we briefly describe the associated requirements, based upon the most general analysis outlined in [24].

We consider first the case where the average dispersion slope (or third-order dispersion) of the transmission line, $D' = \partial D/\partial \lambda$, is constant and different from zero. The bit rate is $B = 1/T_{\text{bit}}$, where T_{bit} is the bit period, and the zero-dispersion wavelength is λ_0. At distance Z_R, the synchronization condition between any channel at wavelength λ_k and a reference channel at λ_1 is then

$$p(k)T_{\text{bit}} = \frac{Z_R}{2} D' \Delta\lambda_{k1}(\Delta\lambda_{k1} + 2\Delta\lambda_{10}), \tag{13}$$

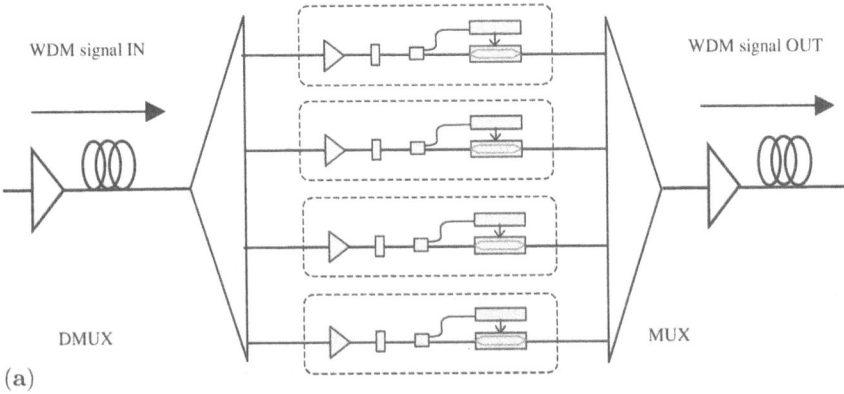

(a)

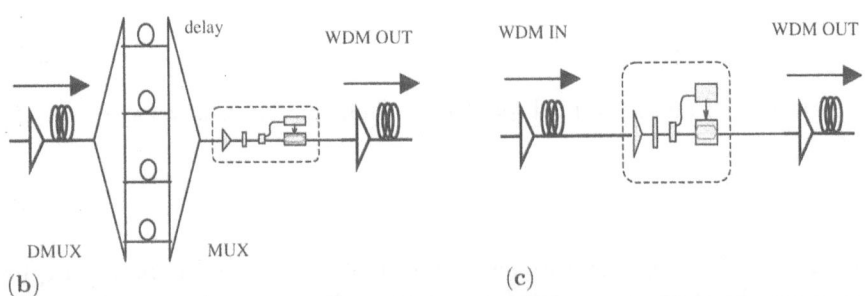

(b)   (c)

Fig. 6. Basic implementation schemes for multiple-wavelength all-optical regeneration: (a) parallel asynchronous; (b) serial re-synchronized; (c) serial self-synchronized. The dashed boxes correspond to the regenerator layout in Fig. 5

where $p(k)$ is an integer number, and $\Delta\lambda_{mn} = \lambda_m - \lambda_n$ with $\Delta\lambda_{m1} = (m-1)$ $\Delta\lambda$. The quantity $p(k)$ represents the number of times channel 1 has crossed channel k. It can be shown that (11) holds if one chooses the wavelength spacing $(\Delta\lambda = \lambda_{n+1} - \lambda_n)$ and the position of the WDM comb with respect to the zero dispersion $(\Delta\lambda_{10})$ according to the rule [24]

$$\Delta\lambda = \frac{2M}{2N - M}\Delta\lambda_{10}, \tag{14a}$$

$$\Delta\lambda_{10} = \frac{2N - M}{2\sqrt{M}}\sqrt{\frac{T_{\text{bit}}}{Z_R D}}, \tag{14b}$$

with M and N being any integers $1 \leq M < 2N$. Thus, periodic synchronous WDM propagation in the presence of third-order dispersion is governed by two 'quantum' numbers, which determine the pattern of collisions between channels. In particular, the number of collisions between channels k and k' is

$$p(k, k') = |k - k'| \left\{ N + \left(\frac{k + k'}{3} - 1 \right) M \right\}. \tag{15}$$

A detailed analysis of the implications of such 'quantization' of WDM propagation in system performance is made in [24]. Basically, channels experiencing the highest number of collisions, which are located on the extremities of the WDM comb, undergo the greatest number of frequency shifts and consequently accumulate the greatest timing jitter. In [24] it is shown that with finite third-order dispersion, the sum of all possible collisions between each synchronous location varies cubically with the number of WDM channels, which dramatically affects system performance.

Next, we assume a transmission line having constant average dispersion ($D' = 0$). This can be achieved either by use of a dispersion-flattened fiber (DFF), or by dispersion-slope compensation (DSC). This last approach can be implemented though periodic insertion of a dispersion-compensating fiber (DCF) of reverse slope, or by a DMUX/MUX apparatus containing fibers (or dispersive components) with adequate dispersion characteristics. In this case, the synchonicity condition takes the form

$$p(k)T_{\text{bit}} = Z_R D \Delta\lambda_{k1} \tag{16}$$

($p(k)$ integer) and is verified for WDM combs having a wavelength spacing of

$$\Delta\lambda = \frac{T_{\text{bit}}}{Z_R D} . \tag{17}$$

In [24] it is shown that the maximum number of collisions between syncchronous locations in a transmission line with constant dispersion varies quadratically with the number of WDM channels, in contrast to the cubic law previously mentioned. This feature reflects the importance of DSC for minimizing collision-jitter impairements in high-capacity transmission. It also suggests the interest of increasing the channel bit rate (as a means to reduce further the absolute number of WDM collisions), as simulations in the next section illustrate.

3.3 Numerical Simulation Examples

The simulation and design of WDM transmission systems based on dispersive/nonlinear RZ propagation require solving numerically the nonlinear Schrödinger equation (NLSE), as defined in (6). Such computer solving takes into account not only the proposed system configuration (e.g. amplifier and regenerator spacing) and parameters (e.g. transmission fiber and EDFA characteristics), but also the randomness involved in (a) amplifier noise (phase and amplitude of amplified spontaneous emission), (b) signal data (use of pseudo-random or PRBS bit sequences), and in some cases of interest (c) polarization states (e.g. use of polarization-interleaved bit sequences, effect of polarization-mode dispersion or PMD, and other polarization related impairments). The conventional NLSE resolution technique is based on the split-step Fourier transform [9]; this algorithm consists basically in toggling back and forth between time and frequency domains at each integration step to reduce the problem to linear integration [9]. The vector NLSE, which includes polarization effects, is a set of two equations similar to (6), each describing the u_x and u_y components with related cross-coupling terms (e.g. $u_y|u_x|^2$, $u_x|u_y|^2$). In most cases of interest, however, polarization effects introduce only secondary or negligible impairments, and solving the simpler scalar NLSE, (6), is sufficient. In WDM propagation, the signal has the form $U = U_1(\lambda_1) + U_2(\lambda_2) + \ldots + U_N(\lambda_N)$, with $U_k(\lambda_k) \approx u_k(z,t)\exp\{-\beta_k(\lambda_k)z\} + \mathrm{cc}.$, and the NLSE breaks into a system of N coupled equations, i.e. for two channels

$$i\left(\frac{\partial u_1}{\partial z} - \beta_1 \frac{\partial u_1}{\partial t}\right) + \frac{1}{2}\frac{\partial^2 u_1}{\partial t^2} + (|u_1|^2 + 2|u_2|^2)u_1 + u_1^* u_2^2 e^{2i\beta_{21}z}$$
$$= f(u_1), \tag{18a}$$

$$i\left(\frac{\partial u_2}{\partial z} - \beta_2 \frac{\partial u_2}{\partial t}\right) + \frac{1}{2}\frac{\partial^2 u_2}{\partial t^2} + (|u_2|^2 + 2|u_1|^2)u_2 + u_2^* u_1^2 e^{2i\beta_{21}z}$$
$$= f(u_2), \tag{18b}$$

with $\beta_{ij} = \beta_i - \beta_j$. In the left-hand side (LHS) of (18), terms in parenthesis represent self-phase (SPM) and cross-phase (XPM) modulations, respectively (in the last case a degeneracy factor "2" is introduced), while the last term with phase mismatch describes the effect of four-wave mixing (FWM); the function $f(u_i)$ in the right-hand side (RHS) takes into account effects of periodic gain/loss variations during transmission (amplifier spacing Z_a), the generation of amplified spontaneous emission (ASE) noise, and SM/NF processing at every distance Z_R. In the case of high dispersion, the phase mismatch varies rapidly with z, requiring small integration steps Δz for convergence, which significantly lengthens computation time.

The coupled-NLSE system computation involves several independent runs. At each amplifier stage, a random Gaussian-noise field U_{ASE} is added to the signal, using different noise seeds in each run. The WDM signal input is made of RZ (hyperbolic-secant) pulse trains coded as 128-bit PRBS. The resulting output signal sequences provide statistical information on the pulses' amplitude/timing noise characteristics in each WDFM channel, leading to bit-error-rate (BER) determination. Assuming Gaussian statistics, the BERs can be directly evaluated from the associated Q factors, which correspond to amplitude noise (Q_a) and timing jitter (Q_t), respectively. These are defined through

$$Q_a = \frac{\langle I_1 \rangle - \langle I_0 \rangle}{\sigma_1 - \sigma_0}, \qquad \mathrm{BER}_a = \frac{1}{Q_a\sqrt{2\pi}}\exp\left(-\frac{Q_a^2}{2}\right), \tag{19a}$$

$$Q_t = \frac{\chi T_{\mathrm{bit}}}{2\sigma_t}, \qquad \mathrm{BER}_t = \sqrt{\frac{2}{\pi}}\frac{1}{Q_t}\exp\left(-\frac{Q_t^2}{2}\right), \tag{19b}$$

with $\langle I_0 \rangle$ and $\langle I_1 \rangle$ being the mean values of the received energies in the "0" and "1" bits, respectively, and σ_0^2 and σ_1^2 the associated statistical variances. The parameter χT_{bit} corresponds to the receiver acceptance window ($\chi = 0.65$–0.75, typically). In (19b), σ_t^2 is the timing variance of the "1" pulses. The overall BER is eventually defined as the maximum value $\max(\text{BER}_a, \text{BER}_t)$. For $Q_{a,t} = 6$, the corresponding BERs take the value of 10^{-9}, which means the detection of a single error out of one billion transmitted bits.

The estimation of transmission quality, as based on this BER model, assumes Gaussian statistics for both energy and timing distributions of output signals. It is emphasized, however, that timing-jitter statistics of interacting solitons generally depart from Gaussian [29]. In previous work [30], we have shown that for all-optical SM regeneration, the Gaussian approximation for BER_t is somewhat optimistic, and in the PM case at high modulation index ($\Delta\phi > 3°$) it is rather inaccurate. However, since PM is generally used at low modulation index (in combination with high-index IM), the Gaussian statistics prediction remains dependable. Further work is required to analyze the evolution of non-Gaussian SM statistics through the system, taking into account all noise sources (Gordon–Haus effect, WDM collisions, soliton interactions, etc.), in order to evaluate the BER with improved accuracy. The main difficulty for direct numerical predictions of BERs is the prohibitive computation times involved in the acquisition of the actual probability distributions over a sufficient dynamic range. Therefore, new statistical models remain to be developed, although they will only be of indicative value given the complexity of the problem.

We describe next a few numerical simulations of all-optically regenerated WDM transmission. In high-capacity systems, a key issue is the WDM granularity, i.e. the choice of channel bit rate. A 160 Gbit/s system can be designed according to the following options: 64×2.5 Gbit/s, 32×5 Gbit/s, 16×10 Gbit/s, 8×20 Gbit/s, and 4×40 Gbit/s. Options based on bit rates that are compatible with the synchronous digital hierarchy (SDH), i.e. 2.5 Gbit/s, 10 Gbit/s and 40 Gbit/s are preferable, although the other solutions can be implemented by use of electronic bit-to-bit multiplexing (as was done in the first amplified submarine systems generation at 5 Gbit/s, see Sect. 2). The interest of moving to the highest bit rates (e.g. 20–40 Gbit/s) is to minimize the number of WDM channels, and thus terminal complexity and cost. As the bit rate increases, however, transmission impairments are greater, and system requirements become more stringent. Computer simulations help to identify the trade-offs and industrially sound options. The four examples described below concern (a) the study of WDM granularity in 80 Gbit/s systems, (b) the design of a 160 Gbit/s (8×20 Gbit/s) system, (c) the feasibility study of a 1 Tbit/s (25×40 Gbit/s) system, and (d) the implementation of an improved SM/NF regeneration technique, taking 160 Gbit/s (4×40 Gbit/s) transmission as an example.

Table 1. Optimized parameters corresponding to 8×10, 4×20, 2×40 and 8×20 Gbit/s SM/NF regenerated systems

Parameters	8×10 Gbit/s DSC system	4×20 Gbit/s DSC system	2×40 Gbit/s DSC system	8×20 Gbit/s DSC system
Pulse width Δt (ps)	20	10	5	10
Number of steps for DM	4	4	4	8
Mean dispersion (ps/nm km) at the center of multiplex	0.476	0.238	0.119	0.119
Channel spacing/filter FSR (nm)	0.6	1.2	2.4	1.2
Multiplex spectral width (nm)	4.2	3.6	2.4	8.4
Mean disp. slope (ps/nm^2 km)	0.0 (DSC)	0.0 (DSC)	0.0 (DSC)	0.0 (DSC)
Amplifier/filter spacing (km)	50	50	50	50
Regenerator spacing (km)	350	350	350	350
Shorter L_{coll}	20	23.33	35	20
Maximum N_{coll} over Z_R	28	12	4	28
Channel used for clock recovery	Channel D	Channel B	Channel B	Channel D
Filter 3 dB bandwidth (nm)	0.45	0.85	1.5	0.75
IM/PM depth	6 dB/6°	10 dB/6°	8 dB/6°	13 dB/6°

Considering first 80 Gbit/s regenerated systems, we investigate the potentials of 8×10, 4×20 and 2×40 Gbit/s, self-synchronous configurations with 350 km regenerator spacing [24]. Detailed description and parameters used for these three options are summarized in Table 1. Systems where the fiber dispersion slope is compensated for between two consecutive regenerators are referred to as DSC. Dispersion-decreasing fiber (DDF) is used in order to alleviate impairments generated by WDM collisions (the exponentially decreasing dispersion profile being approximated by a 4–8 segments step-wise DM).

Figure 7 shows Q-factor plots of the most degraded channel as a function of transmission distance in each of the three systems case. For comparison purposes, data concerning a similar 4×20 Gbit/s system without in-line regeneration are also shown.

While the Q factors decrease with distance, they reach asymptotic values of 8, 10 and 13 (8×10, 4×20 and 2×40 Gbit/s) after a few megameters. Such an asymptotic Q-factor convergence, which corresponds to 'infinite' error-free transmission potential, is due to the regenerative effect of in-line SM/NF. The transmission is ultimately limited by amplitude noise (Q_a) rather than timing jitter (Q_t), the latter being found higher than 40. This feature illustrates the

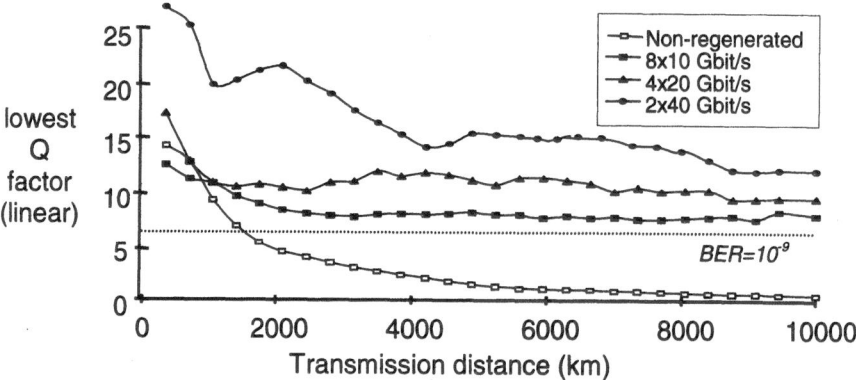

Fig. 7. Theoretical Q factor of most degraded channel as a function of propagation distance in regenerated 80 Gbit/s WDM transmission with different channel granularities of 10 Gbit/s, 20 Gbit/s and 40 Gbit/s (non-regenerated 4×20 Gbit/s system results are also shown for reference)

effect of jitter suppression and its conversion into amplitude noise through IM, as described in Sect. 3.1. The improved performance from 10 Gbit/s to 40 Gbit/s granularities is explained by the decrease, with bit rate, of the timing jitter generated between each SM/NF stage. Such jitter is a function of the number of WDM collisions experienced by each channel; the maximum number of collisions between modulators, N_{coll}, is far greater for systems having a large number of channels, i.e. 28, 12 and 4 for 8×10, 4×20 and 2×40 Gbit/s WDM configurations, respectively. It is concluded that for given WDM capacity, the best performance is achieved with the highest bit rate per channel (smallest WDM granularity); as the bit rate increases, however, other transmission impairments appear (e.g. interaction jitter, Raman self-frequency-shift). For this reason, it is clear that the best solution for achieving 160 Gbit/s capacities is not 2×80 Gbit/s or 1×160 Gbit/s, notwithstanding the extremely stringent requirements for system parameters (such as DSC). Therefore, trade-offs between performance and WDM granularity must eventually be found, with $N \times 40$ Gbit/s (SDH) solutions being the most appealing ones. Note that in the absence of regeneration (4×20 Gbit/s system), the error-free distance is limited to 2 Mm, which illustrates the performance improvement introduced by all-optical regeneration.

Next, we consider a self-synchronous 160 Gbit/s (8×20 Gbit/s) system with similar design parameters (Table 1) [24]. The main difference is the more stringent requirement for the DDF, which must include eight steps instead of four in the previous example. Figure 8 shows Q-factor plots corresponding to each WDM channel (A to H). As previously observed, the Q factors reach asymptotic values between 8 and 11 after a few megameters. The discrepancy in channel performance is attributed to the number of collisions experienced; as the analysis in [24] shows, channels located at the center of the multiplex

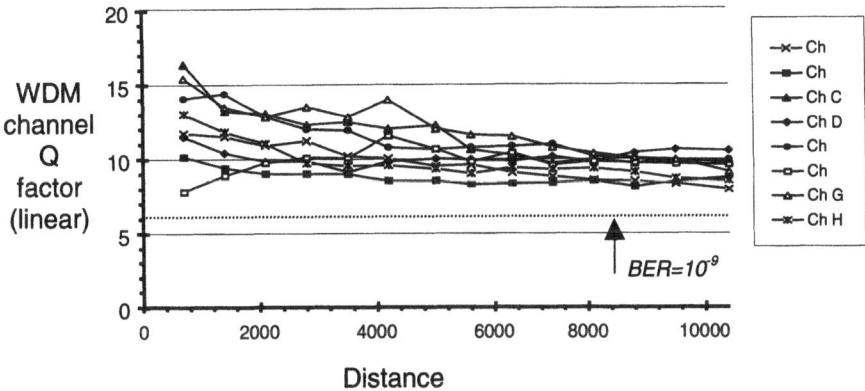

Fig. 8. Theoretical Q factor in 8×20 Gbit/s self-synchronous regenerated transmission

have less collisions and hence show a higher performance; the reverse is true for the outermost WDM channels.

The last two simulation examples concern regenerated WDM systems based on 40 Gbit/s line rates and having 1 Tbit/s aggregate capacity potential [31,32]. In this first feasibility demonstration stage, WDM self-synchronicity has not been required, meaning that there is a dedicated SM/NF regenerator for each channel, the signal being periodically demultiplexed along the transmission line at each regeneration stage (Fig. 6a). Improved designs to be investigated in the future will include serial WDM regeneration, where all channels are processed together in a single regenerator module, as shown in Figs. 6b,c.

In our first simulation [31], the impact of fiber dispersion slope and effective modal area in 25×40 Gbit/s transmission were investigated. It is found that Tbit/s capacities require precise control of the fiber dispersion ($D = 0.05$ ps/nm km) and associated slope ($D' = 7.5 \times 10^{-4}$ ps/(nm^2 km) through the WDM comb. It was also shown that (a) large effective-area (LEA) fibers substantially improve system performance, and (b) configurations with regenerator spacings of either 200 km (i.e. 5 amplifier spans) or 60 km (i.e. one amplifier span) are possible. Figure 9 shows plots of the Q_a factor as a function of channel number for three different distance (2, 6 and 10 Mm), corresponding to the case of an ideal system with (exactly) zero dispersion slope and 80 μm^2 effective-area fiber. The transmission performance is found to be nearly uniform across the multiplex. The Q factors rapidly reach asymptotic values, the lowest being 16, 12 and 10 at 2, 6 and 10 Mm, respectively. As previously observed, the associated timing jitter is found to be negligible, with Q_t factors higher than 40 for all channels at any transmission distance [corresponding to sub-picosecond timing jitter σ_t, according to (19b)].

To date (early 1999), these results represent the very first proof of existence of transoceanic soliton systems with 1 Tbit/s capacity per fiber. The

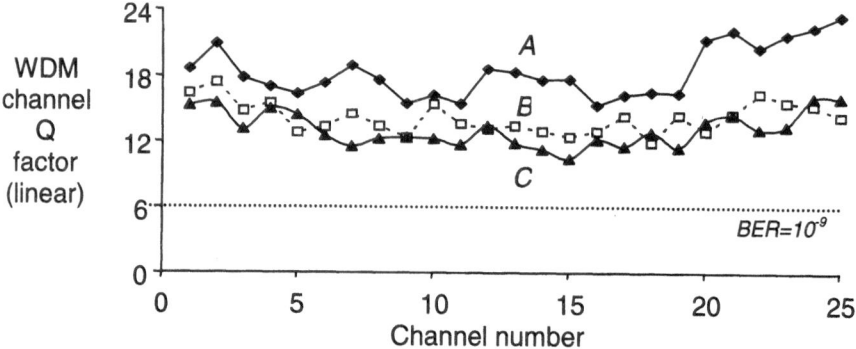

Fig. 9. Theoretical Q factors against WDM channel number (1 to 25) in 1 Tbit/s regenerated transmission, as calculated at (A) 2000 km, (B) 6000 km and (C) 10 000 km distances

relatively high system margin, corresponding to $Q > 10$ at 10 Mm (whereas practical requirements are $Q > 6$ at 6–8 Mm) reveals that even greater capacities (10Tbit/s) could be achieved with all-optical regeneration. This huge capacity could not tapped in without important developments in the field of ultra-wideband (100 nm) optical amplifiers, and ultra-flat dispersion fibers. The current objective is to demonstrate the feasibility of practical, realistic 1 Tbit/s systems for market applications in the early 2000s.

We conclude this section with recent simulation results concerning an improved technique of SM/NF regeneration. In Sect. 2, we have described the advantages of CRZ and DM-soliton transmissions, which present many similarities in the nonlinearity/dispersion trade-off. In particular, they offer the advantage of being far less sensitive to WDM collisions. On the other hand, these Gaussian-shaped, breathing-pulse formats do not exhibit the particle-like properties of the Schrödinger soliton, and therefore all-optical regeneration by SM/NF is not as efficient (monotonous Q-factor decrease, instead of asymptotic stabilization). In order to associate DM (CRZ or quasi-soliton propagation) with regeneration, we have proposed an improved SM/NF configuration [32]. This new concept, referred to as a black-box optical regeneration (BBOR), is based on the principle of periodic conversion of CRZ/DM solitons into actual Schrödinger solitons, by use of a short strand (1–5 km) of highly dispersive fiber (HDF); the converted signal is then regenerated according to the SM/NF principle, although PM and IM must be applied before and/or after HDF transmission, respectively [32]. Numerical simulations of a 160 Gbit/s (4 × 40 Gbit/s) system with amplifier/regenerator/channel spacings of 40 km/320 km/1.6 nm were performed by using optimal dispersion maps. The results are shown in Fig. 10. Without regeneration, the maximum error-free distance is seen to be around 2 Mm; with BBOR implementation, "infinite" transmission with $Q = 8$ at 10 Mm is achieved, with near-uniform performance between WDM channels.

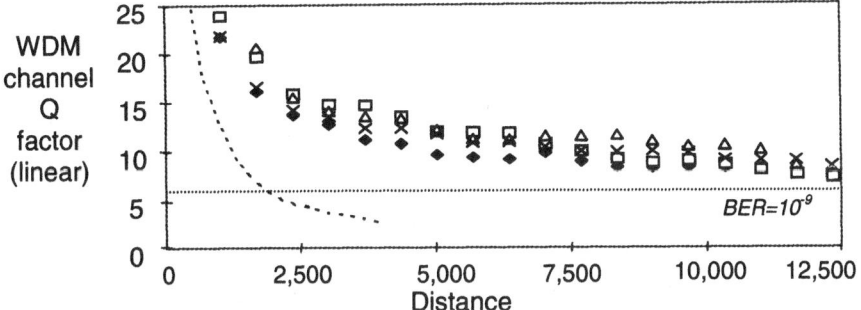

Fig. 10. Theoretical Q factors as a function of transmission distance in 4×40 Gbit/s system combining dispersion-managed soliton propagation and all-optical regeneration. The dashed line shows the worst channel performance obtained without regeneration

More-recent results concerning 640 Gbit/s (16×40 Gbit/s) capacities lead to similar conclusions, and it appears that Tbit/s transmission with high spectral efficiency (< 1.6 nm channel spacing) is potentially achievable. This first campaign of system design optimization confirms the interest of BBOR in future developments of regenerated WDM systems.

4 All-Optical Regeneration: Experiments

4.1 History and State of the Art

The principle of all-optical data regeneration by synchronous modulation and narrow-band filtering (SM/NF) was proposed and demonstrated by M. Nakazawa in 1991 [21]. In a 500 km-long loop experiment, 10 Gbit/s optical data (solitons) could propagate without any measurable degradation (later measured as error-free) over more than one million kilometers. Regeneration by IM was provided every 500 km through LiNbO$_3$ Mach–Zehnder modulators arranged in a polarization-independent configuration. Other types of intensity modulators, such as InP electro-absorption modulators (EAM), which provide IM with some amount of residual chirp (PM), have also been implemented; error-free loop transmissions at 10 or 20 Gbit/s over unlimited distances (> 20 Mm) were achieved [33]. Regeneration through EAM-IM made it possible to demonstrate 20 Gbit/s transmission with a substantial increase in amplifier span, i.e. up to 140 km [34] and even 200 km by using remotely-pumped amplifiers [35].

The transition to WDM was made later based on the parallel, in-line demultiplexing/regeneration approach (Fig. 6a). Several experiments were carried out with LiNbO$_3$ modulators, leading to 5×20 Gbit/s and 8×20 Gbit/s transmissions in 1998 [36,37]. In the last reference, only five parallel regenerators were used; based on the self-synchronization technique of [28], three pairs of WDM channels could be simultaneously regenerated in three SM modules, while the remaining channels were regenerated individually. The 240-km-long recirculating loop, defining the regenerator spacing, consisted of six spans of dispersion-decreasing fiber (DDF), which made it possible to reduce transmission impairments from WDM collisions.

In the meantime, all-optical regeneration by pure PM was proposed [22,23]. The first experimental demonstration of PM-regenerated transmission used a distributed Kerr fiber modulator (KFM) at 2.5 Gbit/s [38]. The retiming effect, dubbed "soliton shepherding" stems from XPM between the optical clock and signal data. In a more recent analysis, we have shown that retiming in KFM is possible despite the effect of clock/signal walk-off and fiber loss [39]. Other KFM-based transmission experiments at 20 Gbit/s led to the very first demonstration of a truly all-optical regenerator, including O/O modulation and all-optical clock recovery [40]. The principle of KFM regeneration was then extended to WDM with 2×20 Gbit/s error-free transmission over 3500 km; in this experiment, a single optical clock control was launched every 90 km, corresponding to the synchronicity period of the two channels [41].

The combination of PM and IM for more efficient regeneration was first analyzed in [42], showing the importance of applying PM with negative curvature (as illustrated in Fig. 4b, the opposite curvature resulting in repulsive timing control). We have experimentally verified this conclusion by implementing regeneration with a dual-control MZ interferometer [43]. In near push–pull operation, which enables proper IM/PM control (see next subsection), 20 Gbit/s error-free transmission over 40 000 km was achieved, to compare with 8000 km with single-electrode drive conditions. Finally, single-channel IM regeneration at 40 Gbit/s have been demonstrated by using either EAMs with 10 000 km transmission distance ($Z_R = 180$ km) [44] or LiNbO$_3$ MZ interferometers with 70 000 km distance ($Z_R = 240$ km) [45]. Significant advances in capacity are expected in the near future with the implementation of 40 Gbit/s WDM regeneration. The key to such progress is the development of low-loss, low-crosstalk, polarization-insensitive, high-IM-depth and accurately controllable IM/PM devices.

4.2   Some Recent Experimental Results

A polarization-independent InP Mach–Zehnder modulator with dual-electrode drive, fulfilling the aforementioned conditions, was developed in Alcatel-CIT's Corporate research centre laboratories. The deep-ridge InP waveguide structure provides very nearly identical TE/TM responses for the IM/PM transfer functions, while the dual drive provides independent control for IM/PM depths [43]. Figure 11 is an electron-scanning-microscope (ESM) picture of the modulator, in which one clearly sees one of the Y-coupler branch and the two-electrode drive layouts.

The InP modulator was tested in a single-channel 20 Gbit/s loop experiment ($Z_R = 90$ km), whose configuration is shown in Fig. 12 [43]. Transmis-

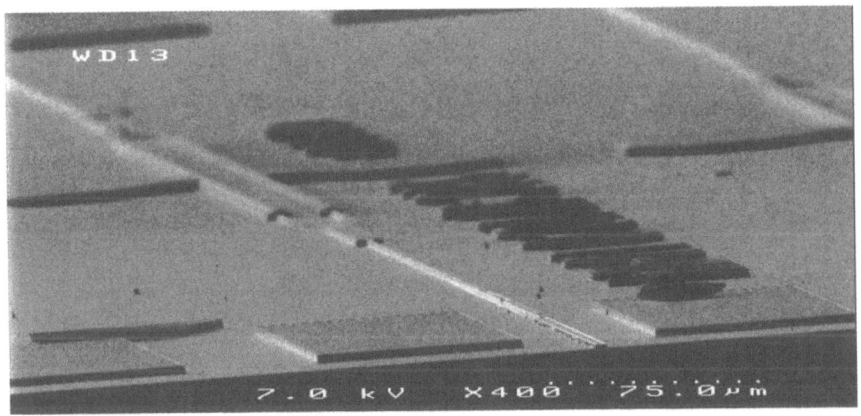

Fig. 11. ESM picture of polarization-independent InP Mach–Zehnder modulator for 20–40 Gbit/s all-optical regeneration, as developed by Alcatel-CIT's Corporate Research Centre

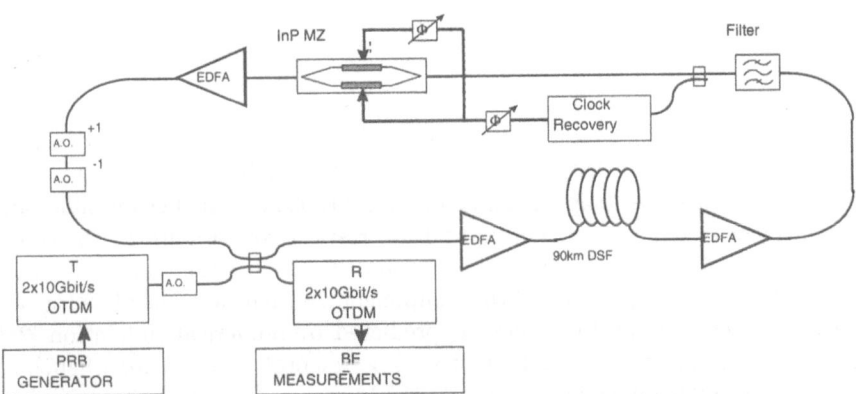

Fig. 12. Experimental loop setup for all-optically regenerated transmission at 20 Gbit/s, incorporating a polarization-independent, dual-drive InP Mach–Zehnder modulator

sion results using either single- or dual-control for IM/PM regeneration are summarized in Fig. 13. The figure shows the measured BER as a function of distance in both single and dual electrode drive configurations. For single drive, the associated IM/PM depths are 2 dB/1° and 2 dB/2°, respectively. Under dual-electrode driving (near push–pull operation) the PM depth and curvature are controllable via the relative RF phase of the electrical signals applied to the MZ electrodes (Fig. 12). Through optimal tuning of the RF phase, a 5°/4.5 dB PM/IM control could be achieved, representing near-optimal conditions. The measured error-free distance is in excess of 40 000 km, compared with 5000–10 000 km under single-drive conditions (Fig. 12).

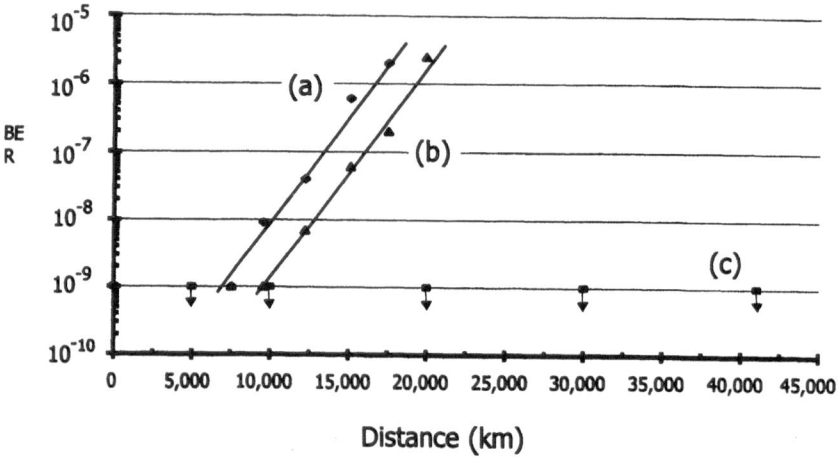

Fig. 13. Experimentally measured BER vs. transmission distance in three different operating conditions, corresponding to single-electrode driving (a,b), and near push–pull operation (c)

Very recently, we have demonstrated 40 Gbit/s operation of a polarisation-insensitive and wavelength-independent InP Mach–Zehnder regenerator through an error-free loop transmission over 20 000 km [44]. The amplifier and regenerator spacing are 45 km and 90 km, respectively. Figure 14 shows the evolution of Q factors measurements with distance in either single-electrode drive or push-pull operation of the component. In each case, the Q factors are seen to reach asymptotic values, as predicted by numerical simulation [24]; these measurements represent the first experimental confirmation of such an effect. The slight undershoot is attributed to a correction in the temporal interleaving of the four 10 Gbit/s OTDM tributaries. With single-electrode and dual-electrode drive, the asymptotic Q factors are 6.6 (BER = 10^{-11}) and 7.2 (BER = 10^{-12}), respectively. The improvement in the second case can be attributed to a regime of higher IM depth and optimized PM depth. Strictly identical results were obtained with a 40 Gbit/s OTDM signal having random polarizations in the four 10 Gbit/s tributaries, showing truly polarization-independent behaviour of the regenerator. Very stable operation of the loop was observed in either case, with no BER change over several hours.

Wavelength-independence and immunity to crosstalk was also measured, showing the potential of this 40 GHz regenerator for WDM transmission applications.

Further progress in 40 Gbit/s WDM implementation, focusing especially on self-synchronized propagation schemes, and on the InP component optimization (in particular the reduction of insertion loss) are now required to demonstrate and assess the high-capacity potential of SM/NF regeneration.

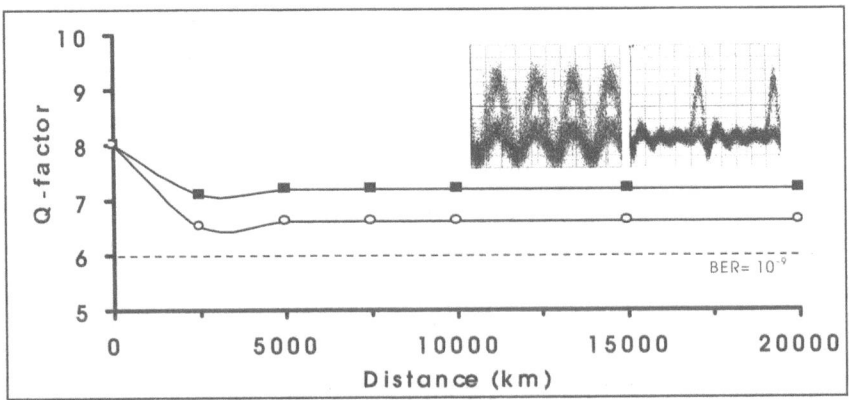

Fig. 14. Experimental Q factor vs. transmission distance with single-electrode (open circles) and push-pull (full squares) configurations. Insets shows 10 Mm eye diagrams at 40 Gbit/s and 10 Gbit/s monitored on 45 GHz photodiode before and after time-domain demultiplexing

5 Electronic and Opto-electronic Versus All-Optical Regeneration

The previous sections presented an overview of all-optical regeneration through SM/NF, which applies to the RZ format, as propagating either in the Schrödinger soliton or DM-soliton or CRZ regimes. So far, these two propagation techniques have provided the highest capacity/distance performance, using only passive transmission means (Sect. 2). Active in-line control by all-optical regeneration comes in as a development making it possible to push the limits further. Is all-optical regeneration the *only* solution for capacity enhancement? This is then a key question. With this increasing complexity, system designers wonder if it would not make sense to re-introduce in-line electronic regeneration. Long-haul systems could be made of shorter, unrepeated optical/amplified fiber trunks (500–1000 km), connected together by electronic repeaters. It is beyond the scope of this chapter to compare the relative merits of optical and electronic regeneration. This debate will surely develop in the forthcoming years, since it concerns all types of transmission systems. Here, we shall only recall the main differences between the two approaches and highlight the potential interest of the first. We will also briefly discuss the opto-electronic alternative offered by wavelength conversion.

The basic layout of the electronic repeater/regenerator is shown in Fig. 15. The optical signal is converted by a high-speed photodiode into a photocurrent; after RF pre-amplification, the baseband clock is extracted; after power amplification with RF equalization and automatic gain control, one-bit integration is performed by a flip–flop circuit controlled by the clock. The output voltage is compared to a 'decision' level, above/below which the bit symbol

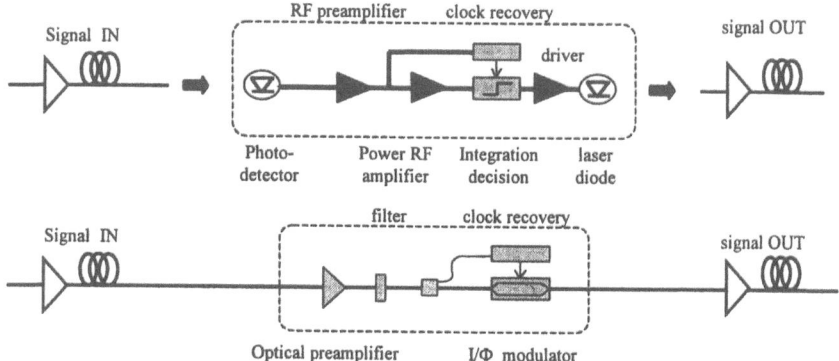

Fig. 15. Electronic (top) vs. all-optical (bottom) regenerator layouts, showing key components

is interpreted as 1/0, respectively. Any change from the previous cycle causes the circuit output to switch between ON and OFF states. The resulting signal is fed to a driver, which provides a current to an DFB laser for optical pulse re-emission; alternatively, laser modulation can be external.

In high-speed electronics, the main bandwidth limitation comes from the integration of hybrid components; upon assembly and optimization, significant performance degradation results from various factors (e.g. parasitic capacitance, impedance mismatch, propagation delays, dark current noise, imperfect equalization, low-frequency cut-off, dynamic range). For transistors operating at multi-ten-gigahertz frequencies, a key issue is the relatively low breakdown voltages (< 1–$2\,$V), which makes difficult to design the last decision/driver stage, corresponding to the E/O conversion function. While great advances have been made in 40 Gbit/s integrated photoreceivers, there is still a long way to go for the realization of fully operational O/E/O repeaters, which would include the different chain elements shown in Fig. 15a. Finally, another issue of decisive importance is cost, and that of electronic repeaters in dense WDM systems could be forbidding. Because of the number of parallel components required at each regeneration step, including the laser transmitters, and their duplication for securization purposes, the unit cost of the O/E/O repeater must be kept very low. In undersea systems applications, power consumption is also a key issue. Currently, power-feeding equipments (PFE) located at cable ends can provide about 1.5 A with 10 kV, which is sufficient to drive a 10 000-km-long system with 200 or 400 in-line EDFAs. The introduction of new active components in the line will substantially increase the PFE requirements, as well as impact the cable design itself (resistance minimization). These arguments show the importance of properly designing the in-line repeaters for minimal power consumption, complexity and cost.

In comparison, the all-optical regenerator appears simpler (Fig. 15b). In contrast with the electronic repeater version, its high-speed electronics is narrowband, which reduces the complexity of the RF amplifier circuitry (although the power consumption of the modulator driver could be as high as +30 dBm). The function of automatic power control is achieved optically, by the action of narrowband filtering (Sect. 3.1). Although the '0' symbols are reshaped by dispersive-wave elimination, there is no actual decision made regarding the symbol values. Rather, it is the pulse arrival time which determines the modulator's corrective action. The all-optical regenerator thus acts as an analog repeater with reshaping/retiming attributes; the third 'R' component of the '3R' regeneration is provided optically through EDFA re-amplification. Finally, a fundamental feature of all-optical SM/NF regeneration is that several WDM channels can be simultaneously processed through the same modulator (Figs. 6b,c), which introduces significant reduction in components, cost and power consumption. This approach, which we referred to as 'serial' regeneration, is conceptually similar to WDM amplification. Indeed, the tremendous capacity increase brought by WDM rests upon the three fundamental properties of EDFAs: (a) excess bandwidth, (b) bit-rate transparency, and (c) immunity to interchannel interference. Albeit bit-rate specific, serial WDM regeneration exploits similar properties. The associated constraints are threefold: the IM/PM modulator should be wavelength-independent and free from crosstalk, and the WDM frames should be synchronized upon reaching the modulator. Although the InP modulator is indeed immune to crosstalk [43,46], further work in SM/NF design and WDM frame synchronization is still required in order to meet the other conditions. If ever practical, serial WDM regeneration would then have an unquestionable advantage over parallel electronic regeneration.

To complete this chapter, it is important to mention the alternative optoelectronic solutions offered by interferometric semiconductor optical amplifier (SOA) gates [47–50]. The 2R/3R regenerative effect is based upon the nonlinear response of the SOA interferometer, which can be of the Mach–Zehnder or Michelson types. A most basic approach is the 2R regenerator (no retiming), based on a Michelson gate [50]. Wavelength conversion can be achieved by using the nonlinear gate as modulator for a cw probe source, which provides the output carrier (in this case the output data is logically inverted) [47,49]. The additional feature of retiming (3R regeneration) requires clock recovery; the clock is used to modulate the probe source, providing a noise/jitter-free pulsed signal. Data is then encoded onto the new carrier by the nonlinear gate. By using three probe sources at different wavelengths, 3R regeneration of NRZ signals could be recently demonstrated [48]. The corresponding layout is shown in Fig. 16. In this approach, NRZ data encoding is first performed, via an SOA, onto a twin-probe, two-wavelength input source (top path in the figure). The resulting signal is then converted to RZ through a modulator driven by the clock; such conversion performs retiming, since the clock is jitter-free. The signal passes then through a split-and-recombine wavelength-selective delay line, which broadens the pulse, thus achieving RZ-to-NRZ

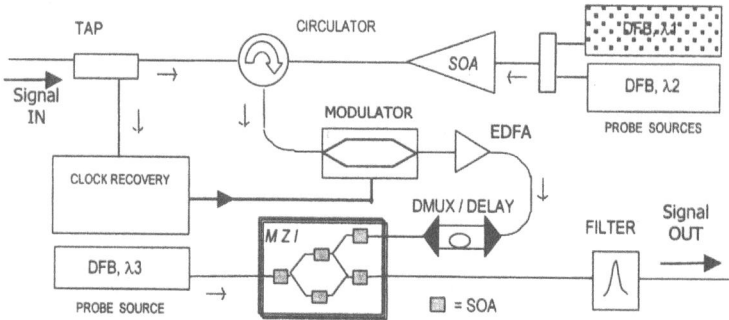

Fig. 16. Implementation of 3R regeneration and wavelength conversion for 10 Gbit/s NRZ signals, using interferometric SOA gates

conversion. Finally, a MZ interferometer (MZI) SOA gate encodes the information on a third probe signal (bottom path in the figure). This last operation results in amplitude noise suppression and pulse reshaping, due to the nonlinear transfer characteristics of the gate. Experimental implementation of this 3R regeneration scheme made possible to demonstrate error-free propagation of 10 Gbit/s NRZ signals over 200 000 km, as shown in Fig. 17 [48]. Such a huge transmission distance, corresponding to one light-second in a fiber path is not so representative of actual system requirements; however, the result fully illustrates the potential of 3R regeneration. The additional feature of wavelength conversion (whether accompanied with wavelength shift or not) is of central importance in WDM networking applications. In all-optical routing, optical data packets could actually propagate over distances significantly greater than the network size, and also experience optical storage/buffering over several millisecond durations.

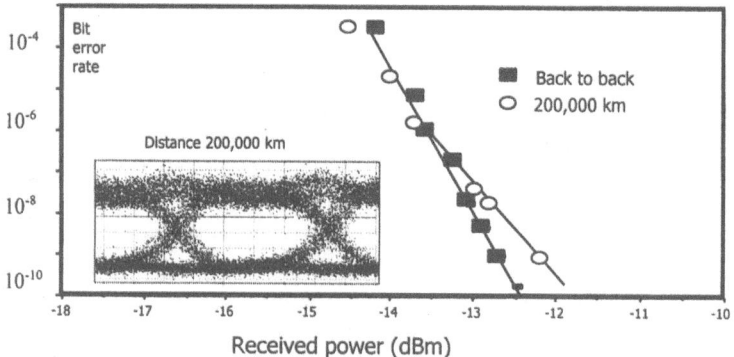

Fig. 17. Experimental BER measurements showing 200 000 km error-free propagation of 10 Gbit/s NRZ signals, based on 3R regeneration with the interferometric SOA apparatus shown in Fig. 15

The field of opto-electronic 2R/3R regeneration based on the interferometric SOA gate technology is progressing at rapid pace, due to their interest for wavelength-routing network applications. New developments with 40 Gbit/s operational capacities are expected soon. Other key improvements could concern the integration of the different subcomponents on a single chip, including the clock recovery function. Concerning this last issue, all-optical implementation based on self-pulsating DFB lasers [51] is of very promising potential.

6   Conclusion

In this chapter, we have presented a detailed review of optical signal regeneration for global communications applications. The forthcoming information age will be one of great demand for system capacity, which now pulls technology towards unprecedented performance levels. Signal transmission limitations in fibers have been successfully alleviated: loss by in-line optical amplification, dispersion and nonlinearity by new soliton-like RZ formats, and at our current stage, noise and distorsion by optical regeneration. In global communications, the optical layer only represents a small fraction of the network infrastructure, but its function for ultra-high-capacity data transport and routing is essential. We have highlighted the potentials of all-optical regeneration, while outlining the associated technology challenges and alternatives. The investigation of signal regeneration could represent one of the last building block in the history of optical communications technology.

Acknowledgments

The authors gratefully acknowledge Patrick Brindel, Elisabeth Brun-Maunand, Christian Duchet, Olivier Audouin, Erwan Pincemin, Bruno Dany, Dominique Chiaroni, Bruno Lavigne and Amaury Jourdan for their many contributions to the experimental and theoretical work described in this chapter.

References

1. Desurvire, E., The golden age of optical fiber amplifiers, Physics Today, 21–27, January 1994
2. Desurvire, E., Erbium-Doped Fiber Amplifiers, Principles and Applications, Wiley, New York, 1994
3. Hecht, J., Planned super-internet banks on wavelength-division multiplexing, Laser Focus World, p.103, May 1998
4. Runge, P., Undersea lightwave systems. Optics and Photonics News, Vol.1, N.11, 9 (1990); see also: Kerfoot, F., Marra, W., Undersea Fiber-optic networks: past, present and future, IEEE J. Select. Areas in Comm., Vol.16, N.7, 1220 (1998)

5. Submarine Fiber-Optic Comm. Syst., Vol.5, N.6, June 1997, Information Gatekeepers, Inc.

6. id.., Vol.5, N.11, November 1997

7. Bergano, N., et al., Proc. Conf. on Optical Fiber Communications (OFC'97)., paper PD16, Optical Society of America, Washington DC, 1997; ibid., (OFC'98), paper PD12

8. Taga, H., et al., Proc. Conf. on Optical Fiber Communications (OFC'98), paper PD16, Optical Society of America, Washington DC, 1998.; Suzuki, M., et al., ibid, paper PD17

9. Agrawal, G.P., Nonlinear fiber optics, Quantum electronics principles and applications, 2nd edn., Academic Press, San Diego, 1995

10. Mollenauer, L.F., Soliton transmission speeds greatly multiplied by sliding vfrequency-guiding filters, Optics and Photonics News, p.15, April 1994, and references therein

11. Mollenauer, L.F., et al., Demonstration of soliton WDM transmission at 6 and 7×10 Gbit/s, error-free over transoceanic distances, Electron. Lett., Vol. 32, N.5, 471 (1996)

12. Nijhof, J.H.B., et al., Stable soliton-like propagation in dispersion-managed systems with net anomalous, zero and normal dispersion, Electron. Lett., Vol.33, N.20, 1726 (1997)

13. Turytsin, S.K., et al., Dispersion-managed solitons and optimization of the dispersion management, Optical Fiber Technology, Vol.4, N.4, 384 (1998)

14. Nijhof, J.H.B., et al., Energy enhancement of dispersion-managed solitons and WDM, Electron. Lett., Vol.34, N.5, 481 (1998), and references therein

15. Devaney, J.F.L., et al., Soliton collisions in dispersion-managed wavelength-division-multiplexed systems, Optics Lett., Vol.22, N.22, 1695 (1997)

16. Morita, I., et al., 40 Gbit/s single-channel soliton transmission over 8600 km using periodic dispersion compensation, Electron. Lett., Vol.34, N.19, 1863 (1998)

17. LeGuen, D., et al., Narrowband 640 Gbit/s soliton DWDM trnsmission over 1200 km of standard fibre with 100 km, -21 dB amplifier spans, Electron. Lett., Vol.34, N.24, 2345 (1998)

18. Mikkelsen, B., et al., All-optical noise reduction capability of intrferometric wavelength converters, Electron. Lett., Vol.32, N.6, 566 (1996)

19. Chiaroni, D., et al., 10 Gbit/s optically regenerated NRZ transmission experiment over 20,000 kms with 140 km repeater spacing, Proc. Conference on Optical Fiber Communications, (OFC'98), paper PD15, Optical Society of America, Washington DC, 1998

20. Dupas, A., et al., 2R all-optical regenerator assessment at 2.5 Gbit/s over 3600 km using only standard fibre, Electron.Lett., Vol.34, N.25, 2424 (1998)

21. Kubota, H., Nakazawa, M., Soliton transmission control in time and frequency domains, IEEE J. Quantum Electron., Vol.29, N.7, 2189 (1993); see also (a) Nakazawa, M., et al., 10 Gbit/s soliton data transmission over one million kilometers, Electron. Lett., Vol.2, N.14, 1270 (1991), and (b) Nakazawa, M., et al., Experimental demonstration of soliton data transmission over unlimited distances with soliton control in time and frequency domains, Electron. Lett., Vol 29, N 9, 729 (1993)

22. Smith, N.J., et al., Soliton dynamics in the presence of phase modulators, Optics Comm., Vol.102, 324 (1993)

23. Smith, N.J., Doran, N.J., Evaluating the capacity of phase-modulator-controlled long-haul soliton transmission, Optical Fiber Technology, Vol.1, N.3, 218 (1995)

24. Leclerc, O., et al., Synchronous WDM soliton regeneration: towards 80–160 Gbit/s transoceanic systems, Optical Fiber Technology, Vol.3, N.2, 97 (1997)

25. Nakazawa, M., et al., Infinite-distance soliton transmission with soliton controls in time and frequency domains, Electron. Lett., Vol.28, N.12, 1099 (1992)

26. Georges, T., Perturbation theory for the assessment of soliton transmission control, Optical Fiber Technology, Vol.1, N.2, 97 (1995)

27. Leclerc, O., et al., Robustness of 80 Gbit/s (4 × 20 Gbit/s) regenerated WDM soliton transoceanic transmission to practical system implementation, Optical Fiber Technology, Vol.3, N.2, 117 (1997)

28. Desurvire, E., et al., Synchronous in-line regeneration of wavelength-division multiplexed soliton signals in optical fibers, Optics Lett., Vol.21, N.14, 1026 (1996)

29. Georges, T., Study of the non-Guassian timing jitter statistics induced by soliton interaction and filtering, Optics Comm., N.123, 617 (1996)

30. Leclerc, O., et al., Synchronously modulated soliton systems: a simple analysis of timing-jitter statistics and bit-error rate, Proc. Conf. on Optical Fiber Communications (OFC'98), paper ThI2, p.290, Optical Society of America, Washington DC, 1998

31. Pincemin, E., et al., Feasibility of 1 Tbit/s (25 × 40 Gbit/s) transoceanic optically-regenerated systems, submitted to Optics Lett.

32. Dany, B., et al., A transoceanic 4 × 40 Gbit/s system combining dispersion-managed soliton transmission and new "black-box" in-line optical regeneration, Electron. Lett., Vol.35, N.5, 418 (1999)

33. Widdowson, T., Ellis, A.D., 20 Gbit/s soliton transmission over 125 Mm, Electron. Lett., Vol.30, N.22, 1866 (1994)

34. Brun-Maunand,E., et al., Parametric study of chromatic dispersion influence in 20 Gbit/s, 20 Mm regenerated soliton systems having up to 140 km amplifier spacing, Electron. Lett., Vol.32, N.22, 1022 (1996)

35. Aubin, G., et al., Record 20 Gbit/s 200 km repeater span transoceanic soliton transmission using in-line remote pumping, IEEE Photonics Technology Lett., Vol.8, N.9, 1267 (1996)

36. Nakazawa, M., et al., 100 Gbit/s WDM (20 Gbit/s ×5 channels) soliton transmission over 10 000 km using in-line synchronous modulation and optical filtering, Electron. Lett., Vol.33, N.14, 1233 (1998)

37. Nakazawa, M., et al., 160 Gbit/s WDM (20 Gbit/s ×8 channels) soliton transmission over 10 000 km using in-line synchronous modulation and optical filtering, Electron. Lett., Vol.34, N.1, 103 (1998).

38. Widdowson, T., et al., Soliton shepherding: all-optical active soliton control over global distances, Electron. Lett., Vol.30, N.12, 990 (1994)

39. Bigo, S., et al., All-optical fiber signal processing for solitons communications, IEEE J. Select. Topics on Quantum Electron., Vol.3, N.5, 1208 (1997)

40. Bigo, S., et al., All-optical regenerator for 20 Gbit/s transoceanic transmission, Electron. Lett., Vol.33, N.11, 975 (1997)

41. Leclerc, O., et al., 2×20 Gbit/s, 3500 km regenerated WDM soliton transmission with all-optical Kerr fiber modulation, Electron. Lett., Vol.34, N.2, 199 (1998)

42. Harvey, H.J., An alternative derivation of soliton transmission control, Paper ThC2, p94, Proc. Top. Meeting on Optical Amplifiers and their Applications (OAA'94), Optical Society of America, Washington DC, 1994

43. Leclerc, O., et al., Polarisation-independent InP push-pull Mach-Zehnder modulator for 20 Gbit/s soliton regeneration, Electron. Lett., Vol.34, N.10, 1011 (1998)

44. Aubin, G., et al., 40 Gbit/s OTDM soliton transmission over transoceanic distances, Electron. Lett., Vol.32, N.24, 2188 (1996)

45. Suzuki, K., et al., 40 Gbit/s single channel optical soliton transmission over 70 000 km using in-line synchronous modulation and optical filtering, Electron. Lett., Vol.34, N.1, 98 (1998)

46. Leclerc, O., et al., 40 Gbit/s polarization-independent, push–pull InP Mach–Zehnder modulator for all-optical regeneration, Postdeadline Paper 35, Proc. Optical Fiber Communications (OFC'99), Optical Society of America, Washington DC, 1999

47. Mikkelsen, B., et al., All-optical noise reduction capability of interferometric wavelength converters, Electron. Lett., Vol.32, N.6, 566 (1996)

48. Jourdan, A., et al., Key building blocks for high-capacity WDM photonic transport networks, IEEE J. Select. Areas in Commun., Vol.16, N.7, 1 (1998); see also: Lavigne, B., et al., Performance and system margins at 10 Gbit/s of an optical repeater for long haul NRZ transmission, in Proc. European Conference on Optical Communications (ECOC'98), p.559

49. Dupas, A., et al., 2R all-optical regenerator assessment at 2.5 Gbit/s over 3600 km using only standard fibre, Electron. Lett., Vol.34, N.25, 2424 (1998)

50. Wolfson, D., et al., All-optical 2R regeneration based on interferometric structure incorporating semiconductor optical amplifiers, Electron. Lett., Vol.35, N.1, 59 (1999)

51. Sartorius, B., et al., All-optical clock recovery module based on self-pulsating DFB laser, Electron. Lett., Vol.34, N.17, 1664 (1998); see also: Sartorius, B., et al., Analysis and compression of pulses emitted from an all-optical clock recovery module, Electron. Lett., Vol.34, N.24, 2344 (1998)

Non-quantum Cryptography for Secure Optical Communications

J.P. Goedgebuer

Summary. Ongoing investigations on security in optical communications are reported. Methods and systems are sought that operate with data rates far beyond those of cryptographic equipment available at the present time. Emphasis is put on two novel methods, in which confidentiality is obtained using a key directly embedded in light.

1 Introduction

Along with the development of optical telecommunications came an interest in secure optical transmissions, especially in the case of fiber networks dedicated to confidential data transmissions between banks and their branches working well beyond national borders, between interconnected financial markets, between research and medical centers, between embassies, etc. With the great increase in operators, service providers, and users, proof of ownership, identity, payment, and billing become an increasingly demanding issue in which confidentiality is of key importance. Although optical fibers have been claimed to be tap-resistant, photonic communications links are vulnerable to eavesdropping. Clearly, a fiber tap is more difficult than an electric cable tap, for which a simple contact or inductive coupling suffices. However, fibers leak light when bent to a small radius, and the information can be taken and then retransmitted, perhaps after modification. To overcome this drawback inherent to fiber transmissions, confidentiality can be obtained with data encryption. In cryptography, the key is a pseudorandom binary number that controls the encryption algorithm. Encryption speed is currently limited to about 100 Mbits/s in the best cases. This is slow compared to data bit rates of several tens of Gbits/s encountered in optical communication systems.

The two methods we discuss insure a high bit rate and a degree of security that can be guaranteed to achieve a given level. In the first method, signal encryption is performed using the temporal coherence properties of light. Signals are encoded inside short coherence wavetrains of light. In the other scheme, the signals are encrypted within high-dimensional hyperchaotic waveforms.

2   Secure Communications by Coherence Modulation of Light

Coherence modulation of light (also termed path-difference modulation) is a method that utilizes the temporal coherence properties of light sources for imprinting a signal onto a light beam. The first articles on this subject were published in the 1980's in the field of communications and sensors [1–5]. The method relies on the general idea that an optical path-difference (OPD) between light fields can behave as an optical carrier under the condition that this OPD is larger than the coherence length of light. Unlike conventional modulation schemes, which require laser sources with narrow line-width, this condition implies the use of sources exhibiting short coherence length. The transmitter is formed by a two-beam interferometer that features an OPD greater than the coherence length, thus preventing any detectable constructive and destructive interference to occur. Decoding is carried out using a receiver formed by another two-beam interferometer with an OPD equal in length to that used in the transmitter. Security is based on the fact that the method requires OPD's of equal length to produce a detectable intensity modulation at the receiver output. The specific value of the optical delay used in the transmitter and receiver serves as the security key. Much work has been devoted to this method in the last years, especially for increasing the complexity of the key. The use of transmitters that combine several optical delays has been proposed to increase security, as well as optical delay lines featuring dispersive effects. With such modifications, the parameters forming the key (e.g. the values of a set of OPD's introduced in the transmitter, or a dispersion law) can be difficult for an eavesdropper to find. Moreover, the method can also be combined with code-division multiple-access (CDMA) coding techniques, thus providing a large family of schemes for secure communications that allow the degree of confidentiality required for a given application.

The next sections are intended as a review of the coherence modulation methods recently developed for security purposes. First we recall the fundamental aspects, placing major emphasis on system aspects. We also consider the vulnerability inherent of the method, should an eavesdropper attack the transmission link.

2.1   Background

The basic architecture of a coherence transmission system is shown in Fig. 1. The transmitter and receiver are Mach-Zehnder interferometers MZ_1 and MZ_2 with an identical path-difference D. In one arm of the transmitter, a phase modulator (PM) is included to enable a digital signal $s(t)$ to be encoded into light. Bit 0 corresponds to a phase $\Phi = 0$ (interferometers in phase) while bit 1 corresponds to a phase $\Phi - \pi$ (interferometers out of phase). The source (S) could be a continuous wave light emitting diode, a superluminescent diode, or a laser diode with a coherence length L. One of

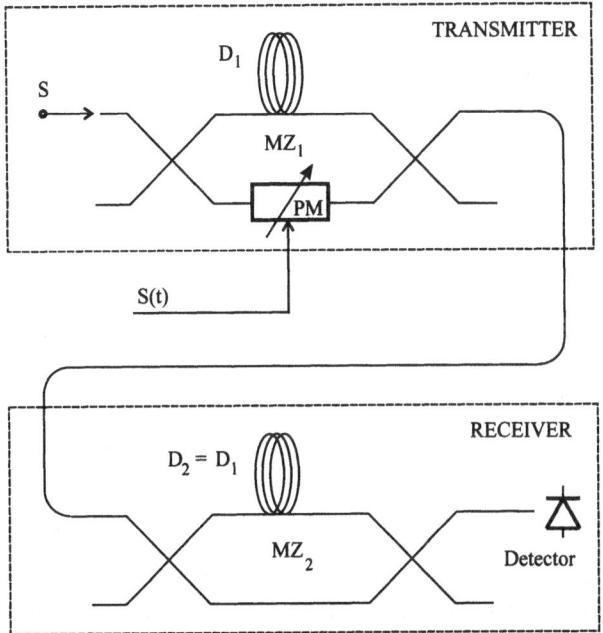

Fig. 1. Basic scheme for a coherence-modulated transmission system

the key requirements of the method is that D be greater than the coherence length of the source. Let $P(\sigma)$ be the power-spectrum of the source, where $\sigma = 1/\lambda$ is the wavenumber; and let $P_0 = \int_{\Delta\sigma} P(\sigma)\, d\sigma$ denote the optical power emitted by the source, where $\Delta\sigma$ is the line-width of the source, expressed as a wavenumber. Consider the light at the transmitter output. For clarity, we first assume the OPD of the transmitter is D_1 and that of the receiver is D_2. The power-spectrum $P_1(\sigma)$ obtained at the output of the transmitter is given by

$$P_1(\sigma) = \frac{1}{2}P(\sigma)\{1 + \cos(2\pi\sigma)[D_1 + Ks(t)]\}\,, \tag{1}$$

where K is a parameter of the phase modulator related to its half-wave voltage.

$P_1(\sigma)$ serves as the input power-spectrum for the receiver. Hence, the power-spectrum $P_2(\sigma)$ at the receiver output can be deduced by replacing $P(\sigma)$ with $P_1(\sigma)$ in (1), yielding

$$P_2(\sigma) = \frac{1}{4}P(\sigma)\{1 + \cos(2\pi\sigma)[D_1 + Ks(t)]\}[1 + \cos(2\pi\sigma)D_2]\,. \tag{2}$$

We obtain the intensity I_2 at the receiver output by integrating $P_2(\sigma)$ over the spectral range $\Delta\sigma$ emitted by the source:

$$I_2 = \int_{\Delta\sigma} P_2(\sigma)\, d\sigma. \tag{3}$$

Finally, letting $D_1 = D_2 = D$, and noting that

$$\int_{\Delta\sigma} P(\sigma)\cos(d\sigma 2\pi\sigma D)\,d\sigma \cong 0\,,$$

we obtain $I_2 = 3P_0/8$ for bit 0, and $I_2 = P_0/8$ for bit 1.

Consider now what happens if an eavesdropper, Eve, taps the link. We assume that she uses an interferometer with an OPD D_e. Then the intensity I_e detected at the output of the intercepting interferometer is given by (3) with $D_1 \neq D_e$:

$$I_e = \frac{1}{4}P_0 + \frac{1}{8}\int_{\Delta\sigma} P(\sigma)\cos(2\pi\sigma)(D_1 - D_e)\,d\sigma\,. \tag{4}$$

The second term in (4) is the cosine Fourier transform of the power-spectrum of the source, which is also the degree of temporal coherence of the source. This term is negligibly small if $|D_1 - D_e|$ is much greater than the coherence length, in which case the intensity detected by the eavesdropper remains constant, and the signal $s(t)$ cannot be detected. However if the eavesdropper uses an optical delay D_e that differs from D_1 by less than a coherence length L, i.e., $|D_1 - D_e| \leq L$, the signal can be detected. This condition sets an ultimate limit on the matching of D_1 and D_e to produce a detectable intensity modulation and to enable signal decoding.

2.2   Vulnerability

The previous paragraph shows that the OPD's D_1 and D_2 must have the same length within close tolerances, i.e., much smaller than a coherence length, for the transmitted signal to be decoded. Vulnerability is related to the amount of time required for the eavesdropper to find the correct value of D_1 to within an amount smaller than a coherence length. The eavesdropper can use two methods to defeat the transmission: a method derived from Fourier transform spectroscopy that uses a scanning Michelson interferometer with a variable OPD, or a method based on the channelled spectrum phenomena that uses a grating or a prism spectroscope.

Consider for instance the case of an OPD $D_1 = 200\,\text{mm}$, and a coherence length of 20 μm. This yields $200/20 \times 10^{-3} = 10^4$ keys. In the first method, an exhaustive key search requires a scanning interferometer and some minimum dwell time on each key, set either by the response time of the detector or by the velocity limit of the stepping motor of the scanning interferometer. Say the tapper's dwell time is $\tau = 100\,\mu\text{s}$, which represents a velocity of $20\,\mu\text{m}/100\,\mu\text{s} = 200\,\text{mm/s}$. Then, an exhaustive key search takes

$$(10^{-4}\text{s/key})\ 10^4\,\text{keys} = 1\,\text{s}\,.$$

This obviously is not sufficient time to guarantee a high degree of confidentiality.

Alternatively, the eavesdropper can use the channelled spectrum technique. The method consists of working in the spectral domain and using a spectroscope to detect the channelled spectrum (described by (1)) that exists at the transmitter output. The channelled spectrum exhibits fringes that are equispaced in wavenumbers by $\delta\sigma \cong 1/D_1$ and whose position changes with $s(t)$. Measuring the fringe spacing quickly yields the value of the optical delay used in the transmitter. A similar method, which is discussed in [6], is based on the use of a pass-band wavelength filter whose bandwidth matches the width of one channelled spectrum fringe. As the position of the fringe changes, the intensity transmitted by the filter (e.g. a Fabry–Pérot) varies accordingly. Therefore an eavesdropper is able to decode the secret signal without needing any matched receiver. Note however that these two methods are useable only if the channelled spectrum fringes can be resolved by the spectroscopic device. For a spectroscope operating around $\sigma = 1\,\mu m^{-1}$ and with a spectral resolution $R = \sigma/\delta\sigma = 10^6$, this method becomes impractical as the optical delay in the transmitter is greater than $D_1 = 1\,m$. The only solution is then to use a scanning interferometer to tap the line, but the technical problems associated with such a high value of D_1 are a real challenge.

It is thus seen that the basic scheme illustrated in Fig. 1 can exhibit high vulnerability. Other schemes featuring complex keys have been studied to enhance security. These are now discussed.

2.3   Enhancing Security

Coming back to the example given above, we saw that the basic scheme in Fig. 1 allows only modest security, since an exhaustive key search takes a few minutes. Improving the security can be carried out with more complex keys. Key complexity can be increased by means of the three following approaches.

To thwart a strong attack, we can employ a *variable delay line* in which the OPD $D(t)$ varies as a function of time and changes the key so quickly that the tapper cannot analyze it fast enough to keep up. A master key generator in the transmitter MZ_1 varies the delay line randomly, while a slave in the receiver tracks the keys. Conceptually the technique is simple, but there are many technological problems associated with fabricating such variable delay lines with a tuning capability of several coherence lengths (typically several tens of micrometers). Moreover, this method is still not secure enough because it is governed by a single parameter, the OPD of one delay line.

A better solution is provided by a transmitter in which *the delay line is formed by recirculating loops* [7]. Figure 2 shows a device in which L loops with different lengths are incorporated in both the transmitter and the receiver. The optical delay D_1, which describes the key in (1), (2), and (4), should now be replaced by a complex function D_1' that contains $L+1$ parameters, namely the original optical delay D_1 and the optical lengths of each loop: $D_1' = \{D_1, L_1, L_2, \ldots, L_L\}$. Let us now assume that the eavesdropper

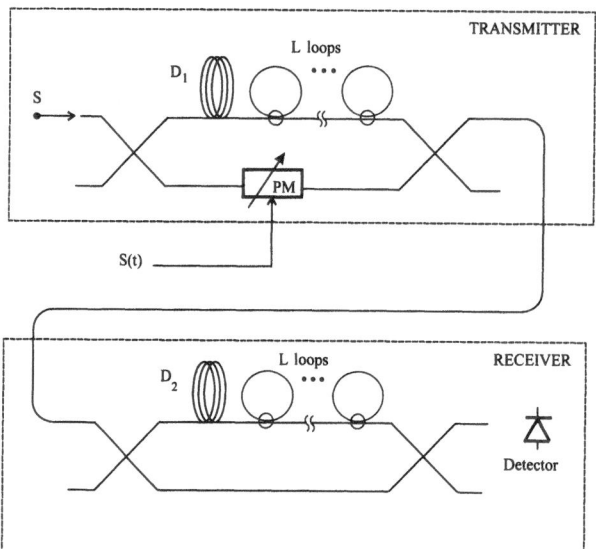

Fig. 2. Enhancing security in a coherence-modulated system

attempts to tap the transmission line using a scanning interferometer with no recirculating loops. It can be shown that she obtains an interferogram formed by fringe packets of very low contrast, making their detection difficult. As an example, calculations made in [7] in the simple case of a transmitter with a single recirculating loop and a 62% directional coupler show that the signal recovered by the eavesdropper drops to 38% (or $-4\,\mathrm{dB}$) down, relative to the authorized receiver (i.e. a receiver with a loop matched to that in the transmitter). For L loops, the signal drops exponentially with L, i.e., $-4L\,\mathrm{dB}$. Each loop operates as a coherence scrambler in which delays are the components of the key. Again following [7], let us assume a coherence scrambler formed by two loops with delays of about 200 mm (i.e., three delay lines), and a coherence length of 20 µm, a value typical of a light emitting diode. The key volume is then about 10^4 per key parameter, yielding 10^{12} since there are 3 parameters. An exhaustive key search then takes

$$(10^{-4}\,\mathrm{s/key})10^{12}\,\mathrm{keys} = 10^8\,\mathrm{s} = 3\,\mathrm{years}\ .$$

These figures are impressive. However, there are still many problems to solve for a practical system realization. One of these problems relates to the 4 dB power drop/loop mentioned previously. If the eavesdropper's receiver features ultra low-noise, it is still possible for her to retrieve the encoded signal. A way to enhance security still further is to use dispersive delay lines, as is now discussed.

A *dispersive delay line* can also be used as a coherence scrambler. It has three key parameters representing delay, linear dispersion, and non-linear dispersion. Its main advantage is to increase the power drop in the coherence function detected by an unauthorized receiver by about 20 dB, instead of the 4 dB achieved with a non-dispersive loop. Such a system was demonstrated recently using a pair of matched interferometers in which a branch is formed by a dispersive material [8]. Although the demonstration was performed with bulk optics, its implementation using fiberoptics technology is straightforward. In Fig. 3, it is realized with fiber interferometers, a non-dispersive fiber in one branch and a dispersive one in the other. Generally, the refractive index n of transparent optical materials can be approximated in the vicinity of a frequency ω_0 by

$$n(\omega) = n_0 + (\omega - \omega_0)n_1 + \frac{1}{2}(\omega - \omega_0)^2 n_2 \; , \tag{5}$$

where $n_1 = (dn/d\omega)_{\omega\,0}$ is the linear dispersion and $n_2 = (d^2n/d\omega^2)_{\omega\,0}$ the non-linear dispersion. In the case of fibers, the dispersion coefficient $D = -\lambda\, d^2n/c\, d\lambda^2$ (expressed in $\mathrm{ps\,nm^{-1}\,km^{-1}}$) is usually used, the relationship between n_1 and n_2 being

$$D = -\frac{\omega_0^2}{\pi c^2}\left(n_1 + \frac{\omega_0}{2}n_2\right) \; . \tag{6}$$

The OPD D_1 in the transmitter (1) then takes the form $D_1 = n(\omega)l_1 - n_0 l_2$ where l_1 and l_2 are the lengths of the two branches. If an eavesdropper attacks the line with a receiving interferometer without knowing the encoding dispersion law, the visibility of the fringe packet detected at his receiver output drops dramatically.

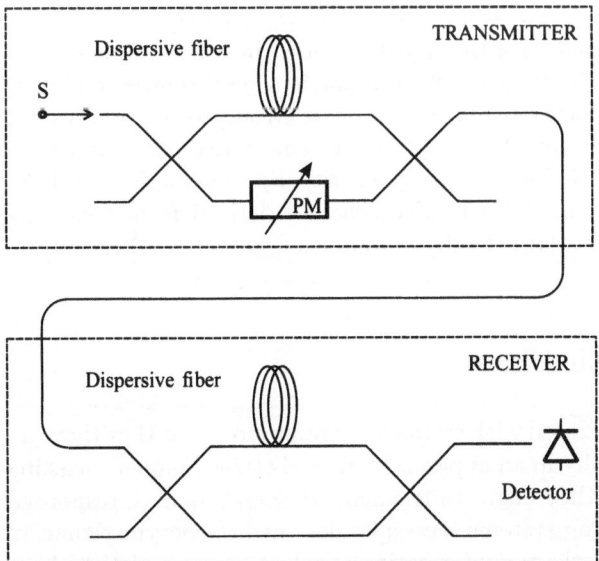

Fig. 3. An alternative way to increase security, based on the use of specific phase-encoding filters

Experimental demonstration has been carried out with a bulk Michelson interformeter containing a 1 cm thick flint-glass plate ($n_0 = 1.7756$, $n_1 = 2.4 \times 10^{-17}$, $n_2 = 9.39 \times 10^{-33}$ at $\lambda_0 = 660$ nm) and a light emitting diode with a 25 nm line-width as the source [8]. A visibility of 8% at the eavesdropper's receiver was obtained, as compared with 50% at the output of the authorized receiver. Recent results [9] indicate that the visibility drop can be reduced by up to -20 dB by using 1 km long dispersion-shifted optical fibers as the dispersive delay lines. Note that introducing a dispersive delay line is equivalent to operating with a spectral phase filter, i.e., with a filter that modifies the phase of each spectral component of light. This means that an eavesdropper using a spectroscope to find the key is unable to distinguish the phase parameter in the channelled spectrum. Finally, one can expect that this encoding method, when combined with the recirculating loops described previously, should provide strong security and should meet the previously predicted three year duration for an exhaustive key search. This expectation remains to be demonstrated experimentally.

2.4   CDMA Encoding

Finally let us note that the method can be combined with CDMA techniques for multiuser applications. Generally speaking, in CDMA techniques each user is assigned a unique digital code from a family of codes with zero or very low cross-correlation characteristics (such as the so-called Gold code). At the transmitter every "1" bit is encoded to a codeword, while every "0" bit is encoded in another codeword, both codewords having a bit rate higher than the original signal. At the receiver the total received signal (desired + co-users) is cross-correlated with the signal provided by a local code generator. If the code used by the receiver for decoding is a synchronized replica of the code used by the transmitter for encoding, a strong correlation occurs, which allows the original data to be recovered. This spread-spectrum technique can hence be regarded as providing security. Specific implementations in optics, especially in combination with schemes derived from coherence modulation, have yielded a large family of systems for secure multiple access communications [10,11].

3   Encrypting with Chaos

One of the problems associated with certain electronic circuits is that they can start to oscillate chaotically in an apparently unpredictable manner, masking entirely the information they ought to transmit. Researchers have responded to this problem by devising systems especially designed to encrypt signals in chaos. The chaos that masks the information is called "deterministic" chaos because it is ruled by an equation (or a set of equations) in which the parameters are known exactly; if exactly the same parameters are used, an identical chaotic signal can be theoretically reproduced. Secure communications based on chaos have been investigated especially in the area of electrical circuits and radiofrequency transmissions. Signal encryption and decryption are achieved using a carrier whose amplitude fluctuates chaotically. At least two classical methods have been demonstrated for communicating with chaos. The first method, due to Ott, Grebogi, and Yorke [12], utilizes controlling chaos. The dynamic behavior of a chaotic oscillator is made to follow prescribed orbits in the attractor by using small perturbations, thus allowing a message to be encoded in the chaotic signal [13]. However, this method is usually slow since it requires postprocessing that may be time consuming. Moreover most studies have been concerned with electrical circuits, a field out of the scope of this section. A very different concept developed by Pecora and Carroll [14] uses the idea of synchronized chaos. In the following we put emphasis on such a method since it allows potentially fast encryption of optical signals. We also report recent advances which show that signal encryption using optical chaos is on the verge of becoming a reality in telecommunications.

3.1   Chaos in Optics: Chaos Synchronization

The method developed by Pecora and Carroll uses the idea of synchronized chaos for achieving security in communications. In this case, a small information-bearing signal – the message – is masked by a large chaotic signal. The principle is illustrated in Fig. 4. The transmitter is a chaos generator that is divided into two subsystems, namely, the master and the slave. The slave is replicated at the receiver. The master subsystem is used to synchronize the two slave subsystems by means of a synchronization channel. At the transmitter output, the message signal $s(t)$ is added to the chaotic signal generated by the slave subsystem, and this composite signal is transmitted to the receiver. When the two subsystems are synchronized, the message can be reproduced by subtracting the chaotic part of the composite signal [15].

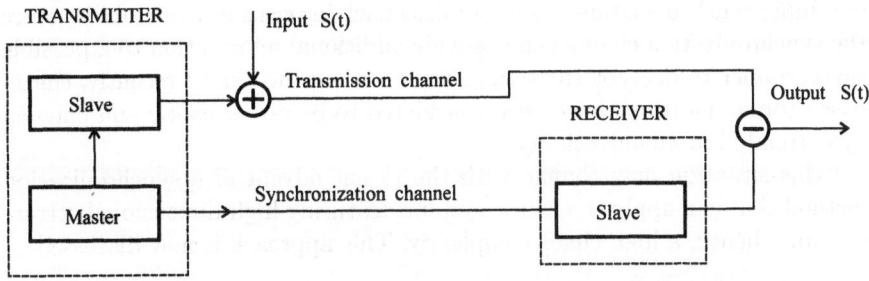

Fig. 4. Chaos synchronization using the Pecora–Carroll method. The message $s(t)$ is added at the output of the transmitter. Decryption is performed by substracting the chaos from the transmitter and the chaos from the receiver

So far systems based on this method have been implemented with electrical circuits featuring low-dimensional attractors, such as the double scroll or Chua's attractor, which has a single positive Lyapunov exponent [16]. But synchronization of chaos has also been studied in optics. Laser diodes can operate as chaotic oscillators in intensity and in polarization. Synchronizing optical chaos is a growing field of research that opens new perspectives in the study of very fast non-linear dynamics. The implementation of optical chaotic cryptosystems requires accurate control of the instabilities and non-linearities and poses severe practical problems. For this reason, most of the results reported in optics have been limited to numerical simulations [17–20], excepting some experimental demonstrations of control of laser chaos and digital encoded transmission [21]. Recently, a significant breakthrough has been achieved that allows these problems of synchronization and of low-dimensional chaos to be overcome. Successful optical transmitter-receiver systems have been developed and have been used to transmit encrypted data with very complex chaotic signals, sometimes termed hyperchaotic signals. In what follows, we first begin with some basic considerations on chaos synchronization.

3.2 Security Vulnerability

The Pecora–Carroll method for synchronizing chaos has been extensively studied with electrical circuits, such as Chua's circuit [16]. In most cases, these circuits are governed by a set of non-linear equations, termed Lorentz equations that yield attractors of low dimension. For instance, Chua's attractor (also termed the double scroll) has a single positive Lyapunov exponent. The chaotic process thus obtained is simple, i.e., chaos complexity is low. It can be defeated by an eavesdropper who has no a priori knowledge of the parameters governing the chaos and, hence, who has no synchronized receiver. The eavesdropper can use unmasking signal processing techniques such as Fourier analysis and correlation techniques [22], which work well to defeat simple chaotic processes. Moreover, the requirement of a second channel for obtaining synchronization is a major drawback in terms of vulnerability since the synchronization channel can provide additional information to a possible eavesdropper to decrypt the signal. As a consequence, until recently, chaos-based communications have been considered to be rather exotic and plagued by extremely low confidentiality.

This situation may change with the recent advent of a synchronization method that can apply to chaotic systems featuring high-dimensional attractors and, hence, a high chaos complexity. This approach is now discussed.

3.3 Increasing Chaos Complexity: Hyperchaos

One way to solve security problems is to use hyperchaotic systems with multiple positive Lyapunov exponents to mask the message. Hyperchaos usually

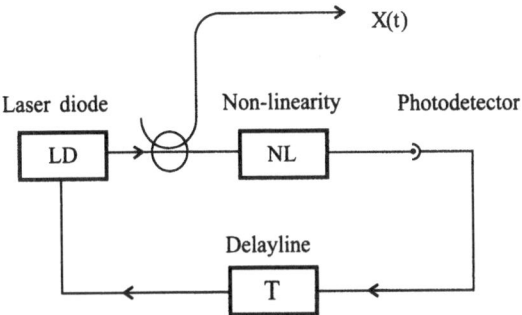

Fig. 5. Delayed-feedback generator used to produce hyperchaos $x(t)$

refers to chaos with more than five positive Lyapunov exponents. Such hyper-chaos can be achieved with a so-called delayed-feedback chaotic generator, which is known to produce high-dimensional chaos [23]. Figure 5 shows the principle of operation of the delayed-feedback chaotic generator described in [24,25] to produce hyperchaos. It is formed by a laser diode (LD), a feedback loop containing a non linear device (NL), and a delay line (T). In that specific example, the source is a wavelength-tunable laser diode, and the device NL is non-linear in wavelength. (Practically, it is formed by a bire-fringent plate set between parallel polarizers, yielding a $\sin^2$ spectral trans-mission curve.) The wavelength $x(t)$ emitted by the source is then shown to obey a delayed-differential equation (also termed Ikeda's equation) expressed as [24,25]

$$x(t) + \tau \, \mathrm{d}x(t)/\mathrm{d}t = \beta NL[x(t-T)] \, , \tag{7}$$

where τ is the time response of the detector, and β is a parameter – the bifurcation parameter – related to the gain of the feedback loop. By proper choice of the value of the bifurcation parameter, the solution for $x(t)$ of (7) is a high-dimensional chaotic function $ch(t)$. The dimension d of the hyperchaos is given by $d \approx 0.4\,\beta T/\tau$, and hence can be very high ($d \approx 5 \times 10^2$ for the system used). This value corresponds approximately to 250 zero or positive Lyapunov exponents, a number nearly 100 times greater than for Chua's circuit. Note that the scheme of the delayed-feedback generator of chaos in Fig. 5 is general and can be applied to chaos in intensity using suitable components.

3.4 Optical Cryptosystems Based on Hyperchaos

The Pecora–Carroll method cannot be used for synchronization of the pre-vious system since there is no rigorous analysis for synchronization of delay-differential equations. Another method, which seems to be specific to delayed-feedback systems, has recently been devised to obtain chaos synchronization. The method is illustrated in Fig. 6. The transmitter is formed by the hy-perchaos generator in Fig. 5. The receiver is a replica of the transmitter (in

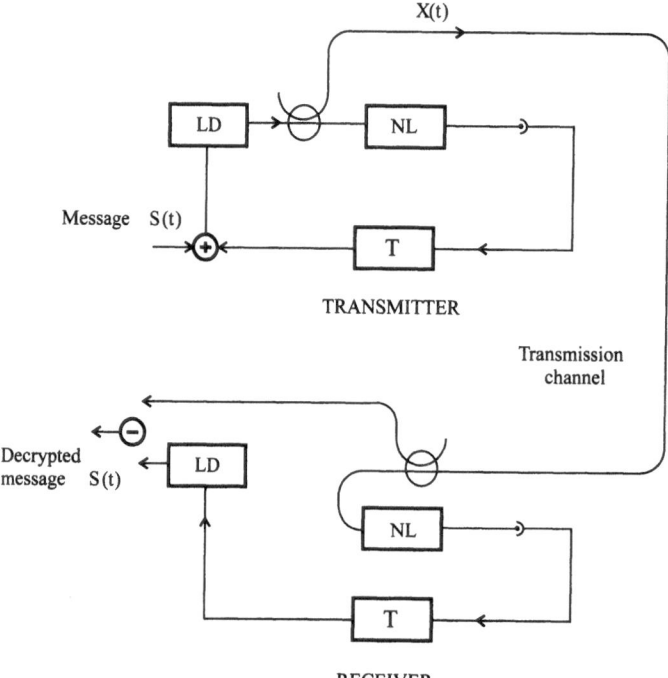

Fig. 6. Optical cryptosystem using optoelectronic generators of hyperchaos. At the transmitter, the message $s(t)$ drives the laser diode and becomes mixed with the dynamics of the whole transmitter, yielding a hyperchaotic encrypted signal $x(t)$. At the receiver, the signal is split into two. One part is fed into the receiver, which is identical to the transmitter except it operates with its feedback loop open. The signal at the receiver output is then synchronized with that of the transmitter, and is a duplicate of the pure laser signal from the transmitter. The other part is fed directly into a photodiode, which provides a duplicate of the laser-plus-information signal. The chaotic laser signal from the receiver is then subtracted from this laser-plus-information signal, yielding the initial message $s(t)$

the sense that it is formed by the same elements), but with its feedback loop open. The message $s(t)$ is injected in the feedback loop of the transmitter to drive the laser diode. The wavelength $x(t)$ emitted by the laser diode is then [26,27]

$$x(t) + \tau\, \mathrm{d}x(t)/\mathrm{d}t = \beta NL[x(t-T)] + f(t) , \qquad (8)$$

with $f(t) = A[s(t)+\tau\, \mathrm{d}s/\mathrm{d}t]$. A is an attenuation parameter; the term $\tau\, \mathrm{d}s/\mathrm{d}t$ arises from the low-pass filtering properties of the feedback loop whose time-response is τ. The chaotic signal $x(t)$ is sent into the feedback loop of the receiver at the correct point, namely into the non-linear element, to generate a signal $NL[x(t-T)]$. The wavelength $y(t)$ emitted by the laser diode in the receiver is given by

$$y(t) + \tau \, dy(t)/dt = \beta NL[x(t-T)] \ . \tag{9}$$

By subtracting (9) from (8), we obtain $x(t) - y(t) = As(t)$, meaning that the message recovery is performed by electronically subtracting the chaos at the emitter output from that at the emitter input. Such a system was successfully demonstrated at 1550 nm. Ongoing experiments show that bit rates of 1 Gbit/s can be obtained using inexpensive commercially available devices. Note that no external synchronization channel is required. Note also that the signal $s(t)$ is inserted into the feedback loop of the transmitter, in contrast to the case with the Pecora–Carroll method, where the signal $s(t)$ is simply superimposed on the chaos produced by the transmitter. Here, signal $s(t)$ is embedded in the structure of chaos, thus increasing confidentiality. Although a high confidentiality can be expected from those unique features, cryptoanalysis of such a system remains to be performed to evaluate accurately the performance in terms of security. A cryptosystem based on similar principles but operating with an all-optical feedback loop has been developed by Roy and colleagues [28,29]. The chaotic signals from the system are in 100 MHz frequency range, allowing a data rate comparable with those used in radiofrequency communications. At the heart of the system is a transmitter consisting of a ring laser made from an erbium-doped fiber amplifier. The message is converted into an optical signal and is injected into the laser. The receiver is identical to the transmitter, thereby ensuring that the signal is synchronized with the chaos of the ring-fiber laser in the transmitter. The information signal is decrypted electronically by subtracting the chaotic laser signal containing the information from the chaotic laser signal produced by the receiver. Several issues need to be resolved before such laser systems can be engineered into communications systems. A major concern is the information rate. Although these experiments have shown that data rates in the 10 MHz range can be used, in principle the method could be extended into at least the 10 GHz range to compete with current optical communication schemes. In more recent work, the researchers have increased the information rate to 1 Gbit/s [30].

4   Conclusion

The degree of security and the problems related to authentication are important issues to consider in the design of a cryptosystem. A low degree of security is sufficient for TV distribution to deter a casual tapper. In that case, the cost of the encrypting/decrypting system is also a key criterion. By way of contrast, a high degree of security is required for military applications. Investigations must still be carried out to determine whether encryption methods based on coherence modulation or hyperchaos can find effective industrial applications in the future. However, some conclusions can be drawn from the results obtained so far, the state-of-the-art of technology, and ongoing investigations. Coherence-modulation methods are at a more mature stage than chaos-based methods, especially for low-security systems. In that case, the devices required to operate coherence-encryption systems can be fabricated using present-day integrated optics technology. The maximum bit rate demonstrated so far is 8 Gbits/s [31]. The requirement of using low coherence sources limits the applicability of the method to local area networks. For a high degree of security, the large path differences required are a drawback, since they are difficult to implement using integrated optics technology. Fiber loops can be used, but they require piezoelectric fiber stretchers to compensate mechanical vibrations and thermal drift. Despite active compensation, transmissions have been reported to be limited to a few seconds in another type of application dealing with quantum cryptography involving interferometers with OPD's of about 1 m [31].

Chaos-encryption methods are only beginning to be investigated seriously. They seem to be very interesting since very simple systems can produce very complex chaos. Readily available devices can be used to generate high-dimensional chaos. Ongoing experiments indicate that signals up to 1 Gbit/s can be encrypted. However the vulnerability, which is known to be high for low-dimensional chaos, remains to be evaluated in the case of hyperchaos, even if a high confidentiality can be expected. Another point of concern that remains for investigation is the influence of noise due to the transmission channel, such as polarization and intensity noise. Despite such remaining investigations, the great simplicity of systems suitable for generating very complex chaotic dynamics is a unique feature for a cryptographic system operating with data rates far beyond the cryptographic equipment available at the present time. This reason explains why considerable interest has been shown for some years in secure transmissions based on optical chaos. Recent experiments have demonstrated bit rates of the order of 1 Gbit/s. This work is certain to continue since, in principle, all of the chaos-based encryption processing could be performed optically, with no limitation set by the use of electro–optical components. The technique would only be constrained by the dynamics of lasers, and signal encryption could be extended into at least the 10 GHz range.

References

1. Delisle, C., Cielo, P. (1975) Application de la Modulation Spectrale à la Transmission de l'Information. Can. J. Phys., **53**, 11, 1047–1053
2. Goedgebuer, J.P., Salcedo, J., Ferrière, R., Viénot, J.C. (1980) Communication via Electro–Optic Modulation of White Light. In: Macchado, M.A., Narducci, L.M. (Eds.) Proc. Conf. On Optics in Four Dimensions. American Institute of Physics, New York, **65**, 418–424

3. Goedgebuer, J.P., Salcedo, J., Viénot, J.C. (1982) Multiplex Communication via Electro–Optic Phase Modulation of White Light. Opt. Acta, **29**, 4, 471–477

4. Goedgebuer,J.P., Porte, H., Ferrière (1991) Recent Advances in Electrooptic Coherence Multiplexing. Int. J. of Optoelectronics, **6**, 4, 339–356

5. Brooks, J.L., Wentworth, R.H., Youngquist, R.C., Tur, M., Kim, B., Shaw, J. (1985) Coherence Multiplexing of Fiber Optic Interferometric Sensors. J. Lightwave Technol., **3**, 1062–1072

6. Wacogne, B., Jackson, D.A. (1996) Security Vulnerability in Coherence Modulation Communication Systems. IEEE Photonics Technol. Lett., **8**, 3, 470–472

7. Wells, W., Stone, R., Miles, E. (1993) Secure Communications by Optical Homodyne. IEEE J. on Selected Areas in Communications, **11**, 5, 770

8. Mazurenko, Y., Giust, R., Goedgebuer, J.P. (1997) Spectral Coding for Secure Optical Communications Using Refractive Index Dispersion. Opt. Commun., **133**, 1–6, 87–92

9. Goedgebuer, J.P., Poinsot, S. (to be published)

10. Griffin, R.A., Sampson, D.D., Jackson, D.A. (1992) Demonstration of Data Transmission Using Coherent Correlation to Reconstruct a Coded Pulse Sequence. IEEE Photonics Technol. Lett., **4**, 5, 513–515; Griffin, R.A., Sampson, D.D., Jackson, D.A. (1992) Optical Phase Coding for Code-Division Multiple Access Networks. IEEE Photonics Technol. Lett., **4**, 12, 1401–1404

11. Karafolas, N., Gupta, G.C., Uttamchandani, D. (1996) Combining Code Division Multiplexing and Coherence Multiplexing for Private Communications in Optical Fiber Multiple Access Networks. Opt. Commun., **123**, 1-3, 11–18

12. Ott, E., Grebogi, C., Yorke, J.A. (1990) Controlling Chaos. Phys. Rev. Lett., **64**, 11, 1196

13. Hayes, S., Grebogi, C., Ott, E. (1993) Communicating with Chaos. Phys. Rev. Lett., **70**, 20, 3031

14. Pecora, L.M., Carroll, T.L. (1990) Synchronization in Chaotic Systems. Phys. Rev. Lett., **64**, 8, 821

15. Cuomo, K.M., Oppenheim, A.V. (1993) Circuit Implementation of Synchronized Chaos with Applications to Communications. Phys. Rev. Lett., **71**, 65, 65

16. Matsumoto, T., Chua, L.O., Komura, M. (1985) The Double Scroll. IEEE Trans. Circuits and Systems, **32**, 8, 798

17. Annovazzi-Lodi, V., Donati, S., Scire, A. (1996) Synchronization of Chaotic Injected-laser Systems and Its Application to Optical Cryptography. IEEE J. Quantum Electron., **32**, 953

18. Mirasso, C.R., Colet, P., Garcia-Fernandez, P. (1996) Synchronization of Chaotic Semiconductor Lasers: Application to Encoded Communications. IEEE Photonics Technol. Lett., **8**, 299

19. Daisy, R., Fischer, B. (1997) Synchronization of Chaotic Nonlinear Optical Ring Oscillators. Opt. Commun., **133**, 282–286

20. Liu, Y., Ohtsubo, J. (1997) Dynamics and Chaos Stabilization of Semiconductor Lasers with Optical Feedback from an Interferometer. IEEE J. Quantum Electron., **33**, 1163

21. Colet, P., Roy, R. (1994) Digital Communication with Synchronized Chaotic Lasers. Opt. Lett., **19**, 2056

22. Beth, T., Lazic, D.E., Mathias, A. (1994) Lecture Notes in Computer Science, **839**, Springer, Berlin, 318–331

23. Dorizzi, B., Grammaticos, B., Leberre, M., Pomeau, Y., Ressayre, R., Tallet, A. (1987) Statistics and Dimension of Chaos in Differential Delay Systems. Phys. Rev. A, **35**, 1, 328

24. Larger, L., Goedgebuer, J.P., Mérolla, J.M. (1998) Chaotic Oscillator in Wavelength: A New Setup for Investigating Differential Difference Equations Describing Nonlinear Dynamics. IEEE J. of Quantum Electron., **34**, 4, 594–601

25. Goedgebuer, J.P., Larger, L., Porte, H. (1998) Chaos in Wavelength with a Feedback Tunable Laser Diode. Phys. Rev. E, **57**, 3, 2795–2798

26. Goedgebuer, J.P., Larger, L., Porte, H. (1998) Optical Cryptosystem Based on Synchronization of Hyperchaos Generated by a Delayed Feedback Tunable Laser Diode. Phys. Rev. Lett., **80**, 10, 2249–2252

27. Larger, L., Goedgebuer, J.P., Delorme, F. (1998) Optical Encryption System Using Hyperchaos Generated by an Optoelectronic Wavelength Oscillator. Phys. Rev. E, **57**, 6, 6618–6624

28. Van Wiggeren, G.D., Roy, R. (1998) Communication with Chaotic Lasers. Science, **279**, 1198–1200

29. Van Wiggeren, G.D., Roy, R. (1998) Optical Communication with Chaotic Waveforms. Phys. Rev. Lett., **81**, 16, 3547–3550

30. Guttierez, C., Porte, H., Goedgebuer, J.P., Sanchez, B., Hauden, J. (1997) A Microwave Coherence-Multiplexed Optical Transmission System on Ti:LiNbO$_3$ Integrated Optics Technology. Microwave and Optical Technol. Lett., **14**, 1, 64–69

31. Marand, C., Townsend, P.D. (1995) Quantum Key Distribution over Distances as Long as 30 km. Opt. Lett., **20**, 16, 1695–1697

Part IV

Optical Materials and Processing

Pulsed Laser Deposition: An Overview

I. N. Mihailescu and E. György

Summary. We present the historical background and the main development trends of a very rapidly growing research field: pulsed laser deposition for obtaining high-quality thin films for use in science and technology. We emphasize the main physical phenomena involved, such as the role of the plasma in the deposition process, the relation between the irradiation parameters and the plasma expansion, the overall deposition efficiency and the major shortcomings of this method. We discuss the influence upon the final performances of the films of the main deposition parameters, i.e. the incident laser wavelengths, duration and fluence, the ambient gas nature and pressure, the collector heating and the geometry of the experimental setup. We provide examples of highly adherent and uniform thin films, both stoichiometric and crystalline.

From the large number of definitions proposed in the literature for pulsed laser deposition (PLD) we mention here, without any pretence of a historical or other hierarchy, some of the more interesting synonyms: laser evaporation, laser-assisted deposition and annealing, laser flash evaporation, laser-assisted sputtering, laser MBE, hydrodynamic sputtering, laser ablation, laser ablation deposition, laser evaporation deposition or photonic sputtering. We notice as a common denominator of all these definitions the succession of two essential steps: a vaporization followed by a deposition onto a nearby collector. It is generally accepted nowadays that in comparison with other techniques applied for the deposition of high-quality thin films, PLD has at least the following main advantages: the ability to reproduce in thin films the stoichiometry and crystallographic status of very complex bulk materials; the relatively high growth rates of 1–5 Å/pulse and even higher; an energy source independent of the deposition environment; no ultrahigh vacuum requirements; rather high gas pressures (up to a few hundred pascals); a relative simplicity of the growth facility offering great experimental versatility (e.g. multilayers and doping); and a very reduced film contamination due to the use of light for promoting ablation. It is also generally accepted that the laser deposition can greatly reduce costs and allow more complex film growth. Following these considerations there is also interest in testing the potentialities of laser deposition to save time, energy and expensive materials with minimum pollution of the biosphere.

Very basically, the interactions between high-intensity laser radiation and the target in order to promote fine substance expulsion (ablation) result in subsequent target heating, melting, vaporization and ionization, followed by plasma generation, absorption, emission and optical regulation. The high temperature of the laser plasma (10^4 K and even higher) is expected to enhance chemical reactions among elements.

When PLD is conducted in the ambience of a chemically active gas, the ablated substance interacts with the gas molecules through linear and mostly non-linear channels and a film of a compound largely differing from the base material is eventually deposited. This method is usually known as reactive PLD (RPLD). The main criterion of RPLD is that the compound formed by the chemical reaction is more expensive and/or has better characteristics from the point of view of a precise application than those of the target material.

After the slow and somewhat disappointing start in the early 1960s, PLD and RPLD remained practically undeveloped for the next two decades. To our knowledge, the first report of an intentional deposition of the laser-ablated substance was by Smith and Turner [1]. It is interesting to observe that these authors were far from satisfied with the performance of the films they obtained. They used a free-running ruby-laser source and tried to deposit $PlCl_2$, MoO_2, CdTe, ZnTe and PbTe thin films. They state that the obtained films are non-uniform and poorly adherent to the substrate. One had to await the strong development in laser sources in the years 1970–1980. New high-intensity laser sources were developed which were able to deliver shorter laser pulses of nanosecond and later picosecond and even femtosecond pulse duration. One essential step was the progress of powerful pulsed UV laser sources using either excimer gases or frequency tripled or quadrupled solid state lasers for ablation. This made congruent vaporization possible, accompanied by ignition of the plasma. In this way possibilities arose for the stoichiometric transfer of the substance from the target or for controlled chemical reactions resulting in deposition of a very uniform and adherent thin layer.

Under the new premisses PLD and RPLD were successfully extended in 1986–1987 to obtain good-quality superconducting films [2,3]. It became evident from the first studies that the short nanosecond pulses generated by the new UV laser sources were more appropriate for obtaining excellent-quality thin films. In the particular case of the high-T_c superconductors it was soon observed that stoichiometric transfer was possible, but only in the ambience of a low-pressure oxygen atmosphere. One may consider this as the first consistent experiment of RPLD. There then followed a fast expansion of PLD and RPLD to metals [4,5] and semiconductors [6–8], binary compounds of the oxide type [9–14], nitrides and carbides [15–32], ternary and much more complicated compounds [33–44]. This is in our opinion a strong argument in favor of the quasi-universal potential of PLD and RPLD for obtaining thin films with the desired composition and structure in a precise place or a given application.

Typical apparatus for PLD and RPLD experiments is shown in Fig. 1. It consists of a vacuum chamber in which the target and substrate are placed.

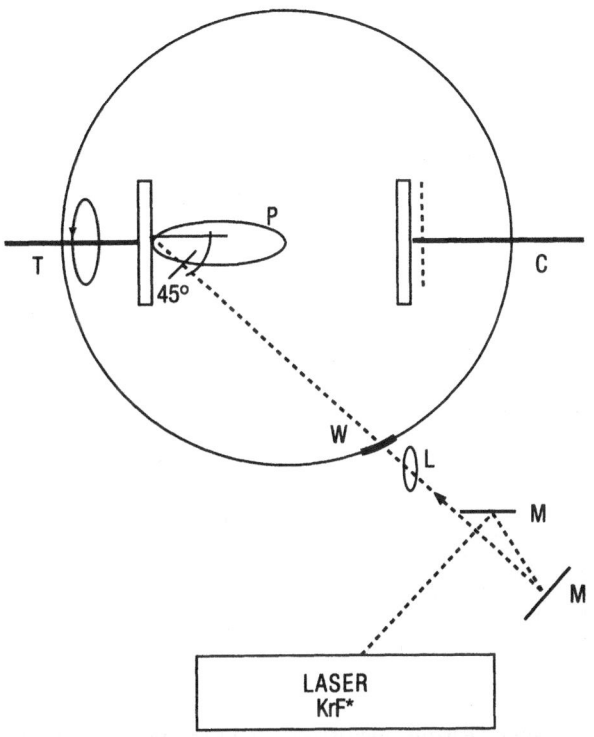

Fig. 1. Typical experimental set-up for PLD and RPLD: T, rotating target; C, collector; P, laser-generated plasma; M, mirrors; W, entrance window

The pulsed laser beam is usually incident at an angle of about 45°. The target is rotated at a frequency of a few hertz to avoid drilling. The ablated species are collected on a support usually placed plane-parallel to the target at a distance of a few centimeters. Heating of the substrate is usually not mandatory, except in those cases when epitaxial growth has to be promoted. The fact that PLD and RPLD need a low thermal budget is of key importance. Indeed, the traditional methods for obtaining crystalline thin films require very high temperatures (of up to 3000 K and even higher). The incident laser fluence in PLD and RPLD is usually set at a level of several joules per square centimeter.

One major advantage of PLD and RPLD is the deposition of highly uniform thin films, very adherent to any substrate. A typical illustration is given in Fig. 2. The film visible in the micrograph is uniform over a total area of a few square centimeters, with a thickness variation of less than 1%. The morphology of the film changes from the interface of the collector with the surface. The film is amorphous for the first 20 nm from the interface. It follows a transition zone extending over the next 10 nm. Finally we observe a columnar polycrystalline structure up to the surface. We consider that this

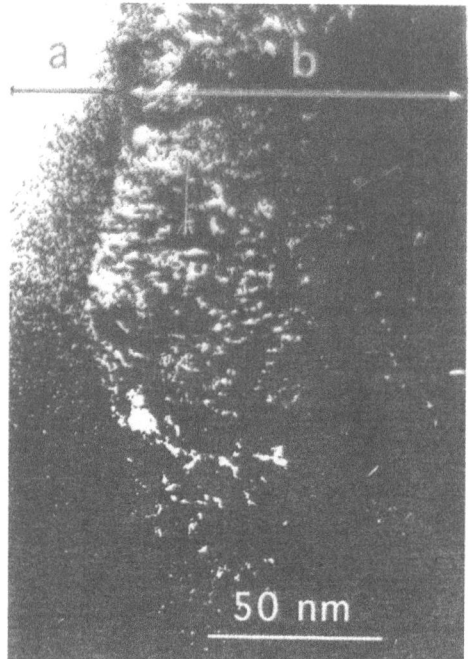

Fig. 2. Cross-section transmission electron microscopy (XTEM) micrograph of a tungsten carbide thin film, deposited by irradiation of a tungsten target with a KrF* excimer laser in a low-pressure (5 Pa) methane atmosphere

special structure is most probably responsible for the great hardness of this film of over 26 GPa, which exceeds any previously reported value for tungsten carbide.

An example of the successful stoichiometric transfer of the material, in the case of multicomponent targets, is given in Fig. 3. The films have been obtained by PLD from a hydroxyapatite HA ($Ca_{10}(PO_4)_6(OH)_2$) target. As is well known, HA is considered to be one of the best substitutes for human bone. It exhibits excellent biocompatibility and human tissues naturally "grow" around HA depositions. The protheses or pivots can be manufactured from Ti alloys, and simply protected by HA deposition before introduction into the human body. From Fig. 3 we observe that the deposition obtained by PLD has an identical composition and structure with the base material (powder), while mechanical tests have proved an increased microhardness and a higher adherence to the metal substrate.

The main parameters of PLD and RPLD determining the synthesis and the final characteristics of the deposited films are [45,46]: the ambient gas nature and pressure, the incident laser fluence, the deposition geometry, the

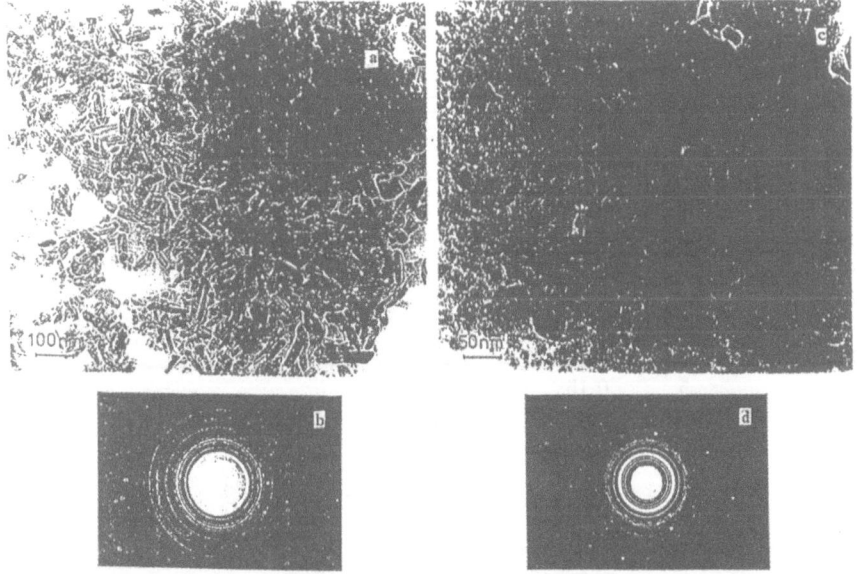

Fig. 3. TEM micrograph and selected area electron diffraction (SAED) patterns for the base HA powder (a,b) and the film obtained by PLD (c,d) in low-pressure oxygen. After deposition the films were heat treated in ambient air at 550°C

target–collector distance, the cleaning and heating of the collector, and the application of external electric or magnetic fields.

As one example we give in Table 1 the dependence on the pressure of the overall quality of the TiN thin films deposited by RPLD from a Ti target in N_2.

We notice in Table 1 the existence of a well-defined pressure range of 0.7–7 Pa, in which the synthesized TiN thin films are stoichiometric and polycrystalline with crystallites of 10–15 nm. For small pressures amorphous mixtures of Ti and TiN were deposited, while for pressures above 7 Pa and especially

Table 1. The main features of the films deposited by RPLD of TiN from titanium in nitrogen at different pressures, P_0

Pressure (Pa)	Description of the deposited film
0.06	Traces only of amorphous TiN embedded in amorphous silicon
0.2	Mixture of amorphous TiN and amorphous silicon
0.7, 1.3, 3.3, 7	Polycristalline f.c.c. TiN with crystallites of 10–15 nm
13	Oxidized polycrystalline TiN, most probably oxynitride
33	Mixture of amorphized TiN and amorphized TiO
100	Poorly crystallized TiO_2 with crystallites of a few nanometers

Table 2. The effect of the incident laser fluence in obtaining carbo-nitride (CN_x) thin films by RPLD from a graphite target in low-pressure nitrogen

E_s (J/cm^2)	P_0 (Pa)	D (Å/pulse)	N/C	H_v (GPa)	Crystalline status
3	1	0.4	0.06		a.
3	5	0.2	0.11		a.
3	10		0.10		a.
3	50		0.08		a.
3	100		0.10		a.
6	25	0.24	0.19		a.
6	50	0.12	0.20	11–12	a.
6	250	0.06	0.14		a.
12	1	0.22	0.20		p.c.
12	5	0.15	0.35		a.
12	10	0.06	0.30		a.
12	50	0.01	0.24	21	p.c.
16	1		0.20		p.c.
16	5		0.25		p.c.
16	10		0.70		p.c.

in excess of 10 Pa, the depositions consisted of mixtures of Ti oxynitrides, TiO_xN_y, and Ti oxides.

As for the effect of the incident laser fluence, E_s, we refer to one example of the synthesis of carbo-nitride (CN_x) thin films (Table 2) by RPLD from an electron-grade graphite target in low-pressure nitrogen.

The interest in synthesizing carbon nitride thin films was stimulated by the theoretical prediction of Liu and Cohen [47] concerning the extreme hardness of this material, comparable with or possibly greater than that of diamond. In the case of the stoichiometric carbo-nitride molecule (C_3N_4) the N/C atomic ratio is 1.33. The experiments were initially conducted with a laser fluence of 6 J/cm^2. The thin films obtained were characterized by an N/C atomic ratio up to 0.2. After that, the incident laser fluence was reduced in order to ablate less carbon and to increase the N/C atomic ratio. The results were quite unexpected: the atomic ratio decreased to below 0.1. Conversely, a significant increase in the N/C ratio up to 0.7 was possible when opposite to direct intuition the laser fluence was increased to 16 J/cm^2. We consider that such a behavior could be the result of changes in the evolution of the plasma initiated and evolved in front of the target in PLD and RPLD experiments. As proved by optical emission spectroscopic studies of the plasma [48], the C–N molecules are formed to a large extent in the gas phase, during transit through the ambient nitrogen of the ablated material. For higher incident laser fluences, the ablated molecules are more energetic and the probability of gas phase reactions between the carbon and ambient gas molecules is enhanced.

As for the role of plasma in PLD and RPLD, we further note that it was considered for a long time only as supplementary loss channel. Actually, the experiments have demonstrated that in the absence of plasma the deposition rates are very low, the ablated material is spread over the entire surface of the irradiation chamber, the deposited layers are non-stoichiometric and peel rather easily. On the contrary, for a well developed plasma the deposition rates arc largcr, thc contamination of the chamber is reduced, the deposited layers are adherent to the collector surface and the obtained films are usually stoichiometric. It was shown that the best compromise in order to obtain good-quality, adherent thin films is reached whenever the plasma column length is equal to the target–collector separation distance. We consider the following explanations of these evolutions: (i) the various species segregate during the post-plasma transit through the gas of the ablated substance from the target to the collector, while (ii) the plasma ensures a compact transit of substance along its trajectory.

The profile of the film produced by PLD and RPLD was theoretically predicted by analytical and numerical calculations [49,50]. The plasma expansion in vacuum was considered to be adiabatic. A schematic diagram of the plasma expansion is given in Fig. 4.

The focal spot has an elliptical shape with semiaxes X_0 and Y_0 while the expansion of the plasma has the geometrical shape of a triaxial ellipsoid, whose semiaxes are initially equal to X_0, Y_0 and Z_0. A characteristic ratio between the initial plasma dimensions is introduced as $k_0 = Z_0/X_0$. Here Z_0

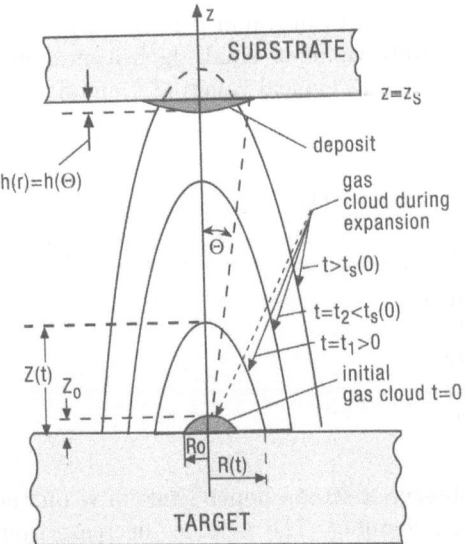

Fig. 4. Schematic for the vapour/plasma cloud expansion and deposition of a thin film

is the length of the initial plasma column, where X_0, Y_0 practically coincide with the dimensions of the focal spot on the surface. This ratio becomes $k(\infty) = Z(\infty)/X(\infty)$ at the moment of the impact of the plasma onto the collector. From numerical calculations it was found that an extended initial plasma ends in a spherical expansion, while a short initial plasma evoluates to a highly forwarded expansion (Table 3).

Table 3. The connection between the plasma parameters k_0 and $k(\infty)$ characteristic for the plasma expansion

k_0	0.001	0.003	0.01	0.03	0.1	0.3
$k(\infty)$	35.8	20.6	11.3	6.5	3.5	1.9

We point out that it is generally wrong to compare experimental results obtained with the same incident laser fluence but different focal spots (X_0, Y_0). Indeed, if the same incident laser fluence is obtained by stronger focusing (i.e. different value of X_0) the ratio k_0 is modified and the plasma expansion regime is changed from a spherical to a cylindrical (forwarded) expansion. Accordingly, the deposition rate is influenced, its value being significantly modified.

The same calculation shows a dependency of the thickness profile of the deposited films of the type

$$h(z_s, \theta) \sim z_s^{-2} \cos^n \theta \tag{1}$$

Here z_s is the target–collector separation distance and θ is the deposition angle (see Fig. 4). This result is entirely valid for small deposition angles when the applied series expansion is correct. Typical values of n are given in Table 4.

Table 4. Typical values for the exponent n from (1)

Target	Q (MW/cm^2)	P (Pa)	n
Gold	700	0.01	10.5
Graphite	700	0.01	11
Graphite	300	0.01	12
Graphite	700	10	2.5

By inspection of Table 4, we observe a strong dependence of n on the pressure of the gas in the irradiation chamber. The n values decrease from 10–12 to 2.5 only when the pressure in the irradiation chamber is increased from 0.01 to 10 Pa. The dependence of n on target nature is much weaker. By corroborating the predictions of the theoretical analyses with the results of many experimental studies we can mention several approaches to achieve very uniform films by PLD and RPLD:

(i)   to work with a large target–collector separation distance z_s;
(ii) to conduct the depositions in the ambience of an inert or chemically active gas;
(iii) to smoothly move the collector perpendicularly to the deposition axis during the multipulse ablation; and/or
(iv) to combine the previous three methods.

In response to the potential question of the present efficiency of PLD and RPLD we refer to the experimental results reported in [51]. We denote E_0 as the laser pulse energy, E_s the energy incident on the target surface, E_{pl} the energy absorbed in the plasma, E_a the energy effectively used for ablation, E_t the energy dissipated in the target (heat diffusion), E_r the energy which leaves the zone of the interaction, A the effective absorptivity of the target, q the optical thickness of the plasma and D_m the ablation rate. There exist the following simple relations between the parameters listed above:

$$E_s = E_0 e^{-q}, \quad A E_s = E_a + E_t, \quad E_r = (1 - A)E_0 e^{-2q},$$
$$E_0 = E_r + E_{pl} + E_a + E_t. \tag{2}$$

E_a can be estimated by heat calculations starting from the value of the ablation rate, D_m, E_r can be determined with the aid of the integrating sphere, while E_t can be measured with a thermocouple built into the target. The plasma radiation loss and plasma heat transfer to the target are neglected. We chose a typical example of PLD from an Al target in vacuum. An excimer laser source was used in the experiments ($\lambda = 308\,\mathrm{nm}$, $E_0 = 100\,\mathrm{mJ}$, $t \sim 25\,\mathrm{ns}$). The ablation rate, determined microscopically, was $D_m = 0.8 \times 10^{-3}\,\mathrm{mm/pulse}$. The experimental results were processed with the aid of (2). The main data obtained are given in Table 5.

From Table 5 we notice that the amount of energy effectively spent for ablation ($E_a + E_{pl}$) represents about 50% of the incident laser pulse energy – that is, the assertion that PLD and RPLD are efficient is therefore fully justified in our opinion.

The energy absorbed in the plasma, E_{pl}, is only 20% of E_0, in good agreement with the direct monitoring of the transit through plasma of a probe laser beam [52] (which indicates values of 20–30%). A plasma thickness of only 0.2 means that the plasma is fairly transparent to UV laser radiation.

Table 5. The main parameters characterizing the PLD process from an Al target in vacuum

E_t/E_0	E_r/E_0	E_a/E_0	E_{pl}/E_0	A	q
0.2	0.3	0.3	0.2	0.6	0.2

We note that in visible and IR light we typically have $q > 1$ (i.e. an opaque plasma). This makes the use of UV radiation indispensable for PLD and RPLD! Indeed, very strong absorption of radiation by plasma not only shields the target surface from the main part of the laser pulse, but also "explodes" the target surface with a detrimental effect upon the final quality of the deposited films [14].

The major drawback of PLD and RPLD processes is the presence of particulates of various shapes (round and irregular) and dimensions in the milimeter range on both the surface as well as in the depth of the obtained films (Fig. 5).

The presence of the particulates can spoil or limit the applications of the thin films obtained by PLD and RPLD in key technological fields such as VLSI microelectronics, optoelectronics, micro- and nanomachining and so on.

According to the present literature, the particulates are mainly formed and/or deposited as an effect of [45,46,53,54]: (i) explosive dislocation of the substance caused by the subsurface overheating of the target; (ii) gas phase condensation of the evaporated material; (iii) liquid phase expulsion under

Fig. 5. Typical SEM micrograph of a tungsten carbide thin film obtained by RPLD from a W target in CH_4 ($P_{CH_4} = 1.5\,Pa$)

the action of the recoil pressure of the ablated substance; and (iv) blast-wave explosion at the liquid (melt)–solid interface.

In our opinion, in order to decrease the density or even completely eliminate the particulates one has to avoid the presence of the liquid inside the crater during the laser–target interaction [55]. One possible solution to be applied in order to reach this goal is the proper choice of the laser wavelength. Indeed, when the attenuation coefficient is small, the absorbed laser energy is dissipated into the rather large zone. The target is melted in an extended volume and many particulates are formed and expelled. One then has to choose a target–laser combination for which the attenuation coefficient is high enough. According to the literature a practical criterion to be applied is an attenuation coefficient above 5×10^4 cm^{-1} [53]. Another solution is to set the incident laser fluence at a level high enough to vaporize (ablate) all the melted substance (e.g. the required fluence for Al and nanosecond laser pulses is 25–30 J/cm^2), or to use very short laser pulses (in the sub-picosecond range).

According to our experience and the literature some other efficient ways to eliminate particulates are: (i) employ a smooth target surface and uniform target ablation; (ii) combined rotation and translation of the target; (iii) variation of the substrate orientation with respect to the direction of the plasma expansion; (iv) fragmentation of particulates by means of an additional laser beam propagating parallel to the substrate surface; and (v) dual laser beam ablation from a single target. We emphasize that the experimental conditions for eliminating particulates can be optimized only for each particular system and for specific requirements.

To summarize we consider that:

(i)   PLD and RPLD are universal techniques. They can be applied to any compound existing in nature or predicted by theoretical models.
(ii) Droplets of various shapes and dimensions represent the major drawback in PLD and RPLD; however, there are now well-established methods to decrease their density down to complete elimination.
(iii) The plasma plays an essential role in PLD and RPLD, acts as a "piston" pushing the ablated substance from target to collector, intervenes in the chemical reaction, and provides hot particles which result in the formation of very adherent deposited layers.
(iv) PLD and RPLD are rather efficient processes: more than 50% of the incident laser energy is spent in ablation and plasma heating; the efficiency can be improved by using a target with poor thermal conductivity and/or external mirrors.
(v)   Progress is expected with shorter laser wavelengths and shorter laser pulses.

In conclusion, PLD and RPLD are rather simple, economic, low thermal budget and ambient preserving techniques. It can be foreseen that in the near future PLD and RPLD will be used in commercial thin film production.

References

1. Smith, H.M., Turner, A.F. (1965) Appl. Opt., 4(1), 147
2. Bednorz, J.G., Muller, K.A. (1986) J. Phys. B, 64, 189
3. Wu, M.K., Ashnuru, J.R., Torng, C.J., Hor, P.H., Meng, R.L., Gao, L., Huang, Z.J., Wang, Y.Q., Chu, C.W. (1987) Phys. Rev. Lett., 58, 908
4. Dupendant, H., Gavigan, J.P., Givord, D., Lienard, A., Rebouillat, J.P., Souche, Y. (1989) Appl. Surf. Sci., 43, 369
5. Pielmeier, R., Ballmann, D., Haberger, K. (1990) Appl. Surf. Sci., 46, 163
6. Biunno, N., Krishnaswamy, J., Sharan, S., Ganapathi, L., Narayan, J. (1990) Proc. Mater. Res. Soc., 158, 477
7. Afonso, C.N., Serna, R., Catalina, F., Bermejo, D. (1990) Appl. Surf. Sci., 46, 249
8. Marine, W., Scotto D'Aniello, J.M., Marfaing, J. (1990) Appl. Surf. Sci., 46, 239
9. Amirhaghi, S., Craciun, V., Craciun, D., Elders, J., Boyd, I.W. (1994) Microelectron. Eng., 25(2–4), 321
10. Craciun, V., Elders, J., Gardeniers, J.G.E., Boyd, I.W. (1994) Appl. Phys. Lett., 65(23), 2963
11. Craciun, V., Elders, J., Gardeniers, J.G.E., Geretovsky, J., Boyd, I.W. (1995) Thin Solid Films, 259(1), 1–4
12. Nanai, L., Vaitai, R., Hevesi, I., Jelski, D.A., George, T.F. (1993) Thin Solid Films, 227, 13
13. Fogarassy, E., Slaoui, A., Fuchs, C., Stoquet, J.P. (1990) Appl. Surf. Sci., 46, 195
14. Ozegowski, M., Metev, S., Sepold, G. (1998) Appl. Surf. Sci., 127, 514
15. Bulir, J., Jelinek, M., Vorlicek, V., Zemek, J., Perina, V. (1997) Thin Solid Films, 292, 318
16. Mihailescu, I.N., Chitica, N., Nistor, L.C., Popescu, M., Teodorescu, V.S., Ursu, I., Andrei, A., Barborica, A., Luches, A., Luisa de Giorgi, M., Perrone, A., Dubreuil, B., Hermann, J. (1993) J. Appl. Phys, 74(9), 5781
17. Mihailescu, I.N., Chitica, N., Teodorescu, V.S., Luisa De Giorgi, M., Leggieri, G., Martino, M., Perrone, A., Dubreuil, B. (1993) J. Vac. Sci. Technol. A, 11(5), 2577
18. Craciun, D., Craciun, V. (1992) Appl. Surf. Sci., 54, 75
19. Craciun, V., Craciun, D., Boyd, I.W. (1993) Mater. Sci. Eng. B, 18(2), 178
20. De Giorgi, M.I., Leggieri, G., Luches, A., Martino, M., Perrone, A., Majni, G., Mengucci, P., Zemek, J., Mihailescu, I.N. (1995) Appl. Phys., 60, 275
21. Luches, A., Leggieri, G., Martino, M., Perrone, A., Majni, G., Mengucci, P., Mihailescu, I.N. (1995) Thin Solid Films, 258, 40
22. D'Anna, A., Leggieri, G., Luches, A., Martino, M., Perrone, A., Majni, G., Mengucci, P., Alexandrescu, R., Mihailescu, I.N., Zemek, J. (1995) Appl. Surf. Sci., 86, 170
23. De Giorgi, M.L., Leggieri, G., Luches, A., Martino, M., Perrone, A., Majni, G., Mengucci, P., Zemek, J., Mihailescu, I.N. (1995) Appl. Phys. A, 60, 275
24. Mihailescu, I.N., György, E., Chitica, N., Teodorescu, V.S., Marin, G., Luches, A., Perrone, A., Martino, M., Neamtu, J. (1996) J. Mater. Sci., 31, 2909
25. Mihailescu, I.N., György, E., Popescu, M., Csutak, S., Marin, G., Ursu, I., Luches, A., Martino, M., Perrone, A., Hermann, J. (1996) Opt. Eng., 35(6), 1952

26. Mihailescu, I.N., Lita, A., Teodorescu, V.S., György, E., Alexandrescu, R., Luches, A., Martino, M., Barborica, A. (1996) J. Vac. Sci. Technol. A, 14(4), 1986

27. Leggieri, G., Luches, A., Martino, M., Perrone, A., Alexandrescu, R., Barborica, A., György, E., Mihailecu, I.N., Majni, G., Mengucci, P. (1996) Appl. Surf. Sci., 96, 866

28. Chitica, N., György, E., Lita, A., Marin, G., Mihailescu, I.N., Pantelica, D., Petrascu, M., Hadziapostolou, A., Grivas, C., Broll, N., Cornet, A., Mirica, C., Andrei, A. (1997) Thin Solid Films, 301, 71

29. Caricato, A.P., Leggieri, G., Luches, A., Perrone, A., György, E., Mihailescu, I.N., Popescu, M., Barucca, G., Mengucci, P., Zemek, J., Trchova, M. (1997) Thin Solid Films, 307, 54

30. De Giorgi, M.L., Leggieri, G., Luches, A., Martino, M., Perrone, A., Zocco, A., Barucca, G., Majni, G., György, E., Mihailescu, I.N., Popescu, M. (1998) Appl. Surf. Sci., 127–129, 481

31. Mihailescu, I.N., György, E., Alexandrescu, R., Luches, A., Perrone, A., Ghica, C., Werckmann, J., Cojocaru, I., Chumas, V. (1998) Thin Solid Films, 323(1,2), 72–78

32. Mihailescu, I.N., György, E., Marin, G., Popescu, M., Teodorescu, V. S., Van Landuyt, J., Grivas, C., Hadziapostolou (1999) J. Vac. Sci. Technol. A, 17(1), 245

33. Heitz, J., Wang, X.Z., Schwab, P., Bauerle, D., Schultz, L. (1990) J. Appl. Phys., 68, 2512

34. Proyer, S., Stangl, E., Schwab, P., Bauerle, D., Simon, P., Jordan, C. (1994) Appl. Phys. A, 58, 471

35. Proyer, S., Stangl, E., Borz, M., Hellebrand, B., Bauerle, D. (1996) Physica C, 257, 1

36. Tomov, R.I., Manolov, V.P., Atanasov, P.A., Tsaneva, V., Ouzounov, D.G., Tsanev, V.I. (1997) Physica C, 274, 187

37. Sutcliffe, E., Srinivasan, R. (1986) J. Appl. Phys., 60, 3315

38. Braren, B., Srinivasan, R. (1988) J. Vac. Sci. Technol. B, 6, 537

39. Basovich, A.J., Gaponov, S.V., Jastrabik, L., Jelinek, M., Kiselev, N.A., Kluenkov, E.B., Lebedev, O.I., Mazo, L.A., Soukup, L., Strikovskij, M.D., Talanov, V.V., Vasiliev, A.L. (1993) Thin Solid Films, 228, 193

40. Olsan, V., Jelinek, M. (1993) Physica C, 207, 391

41. Amin Sajjadi, A., Boyd, I.W. (1993) Appl. Phys. Lett., 63(24), 3373

42. Trtik, V., Jelinek, M., Kluenkov, E.B. (1994) J. Phys. D: Appl. Phys., 27, 384

43. Zhang, W., Boyd, I.W., Elliott, M., Herrenden-Harkerand, W. (1996) Appl. Phys. Lett., 69(23), 3599

44. Zhang, W., Boyd, I.W., Elliott, M., Herrenden-Harkerand, W. (1996) Appl. Phys. Lett., 69(25), 3929

45. Bauerle D. (1996) Laser Processing and Chemistry (2nd edn.), Springer-Verlag, Berlin Heidelberg

46. von Allmen, M., Blatter A. (1995) Laser-Beam Interactions with Materials (2nd edn.), Springer Ser. Mater. Sci., Berlin Heidelberg

47. Liu, A.Y., Cohen, M.L. (1989) Science, 245, 841

48. Vivien, C., Hermann, J., Perrone, A., Boulmer-Leborgne, C., Luches, A. (1998) J. Phys. D: Appl. Phys., 31, 1263

49. Anisimov, S.I., Luk'yanchuk, B.S., Luches, A. (1995) JETP, 81, 129

50. Anisimov, S.I., Bauerle, D., Luk'yanchuk, B. (1993) Phys. Rev. B, 48(16), 12076
51. Gorbunov, A., Konov, V. (1992) Proceedings of LAMP ' 9, Nagoya, 1061–1066
52. Mauro, H., Miyamoto, I., Ooie, T. (1992) Proceedings of LAMP ' 9, Nagoya, 293
53. Craciun, V., Craciun, D., Bunescu, M.C., Boulmer-Leborgne, C., Hermann, J. (1998) Phys. Rev. B, 58, 6787
54. Chrisey, D., Huber, G.K. (Eds.) (1994) Pulsed Laser Interactions with Materials, J. Wiley, New York
55. Mihailescu, I.N., Teodorescu, V.S., György, E., Luches, A., Perrone, A., Martino, M. (1998) J. Phys. D: Appl. Phys., 31, 2236

Absolute Scale
of Quadratic Nonlinear-Optical Susceptibilities

I. Shoji, T. Kondo, and R. Ito

Summary. An updated absolute scale of the quadratic nonlinear-optical susceptibilities is proposed. It is based on our recent redetermination for several important nonlinear-optical materials and the earlier reliable values. The problems that were unresolved or overlooked in earlier measurements are pointed out, and the guidelines for obtaining accurate values of the nonlinear-optical susceptibilities are presented through the detailed description of the measurement procedure applied in our determination.

1 Introduction

The quadratic nonlinear-optical susceptibility is a key parameter that determines the performance of nonlinear-optical devices such as harmonic generators and parametric oscillators. Knowledge of its absolute values is also essential for characterizing $\chi^{(3)}$ materials through the $\chi^{(2)}$ cascading process and for understanding the physics involved in the nonlinear-optical processes. Unfortunately, however, the absolute scale of the quadratic nonlinear-optical susceptibilities which should be referred to as standards for various materials at various wavelengths has not been available; significant discrepancies have been noted among the absolute values reported to date, even for such an important material as $LiNbO_3$.

The "mystery" that has caused confusion in establishing the absolute scale was the considerable difference between the obtained values in second-harmonic generation (SHG) and parametric-fluorescence (PF) measurements. Table 1 compares the reported values of the nonlinear-optical susceptibilities of several materials [1–8]. Although the two methods should yield essentially the same results from the theoretical point of view, the PF values are substantially larger than the SHG values. As a result, the absolute scale based on PF data [9] tends to be more than 50% larger than that based on SHG data [10,11].

Furthermore, our recent study has revealed two other factors that can be the source of considerable error in determination of the nonlinear-optical susceptibilities and has been overlooked by almost all the investigators. One is the multiple-reflection effect which occurs in measurements using (nearly)

Table 1. Summary of the nonlinear-optical susceptibilities determined by SHG and PF measurements[a]

Crystal	d_{ij} (pm/V)	SHG	PF
LiNbO$_3$	d_{31}	4.4 (0.532) [1]	5.8 (0.488) [2]
KTP	d_{15}	3.8 (0.440) [3]	4.1 (0.527) [4]
		2.5 (0.650) [5]	
LiIO$_3$	d_{31}	4.2 (0.532) [6]	7.3 (0.488) [2]
AgGaS$_2$	d_{36}	22 (0.532) [7]	31 (0.600) [8]

[a] SH wavelength or pump wavelength in μm is shown in parentheses.

plane-parallel-plate samples. The interference caused by the multiple reflections enhances the second-harmonic power in SHG or the signal power in PF, leading to overestimation of the nonlinear-optical susceptibilities when it is neglected, especially for materials with larger refractive indices. The other is the use of Miller's rule in wavelength scaling of nonlinear-optical susceptibilities. This rule assumes frequency independence of Miller's Δ [12], which is defined by

$$\Delta_{ijk} = \frac{d_{ijk}(-\omega_3; \omega_1, \omega_2)}{[n_i^2(\omega_3) - 1]\,[n_j^2(\omega_1) - 1]\,[n_k^2(\omega_2) - 1]}, \tag{1}$$

where $\omega_3 = \omega_1 + \omega_2$ and $n_i(\omega_3)$, etc., are the refractive indices. Theoretically, Miller's Δ should be independent of frequency if the system has a single anharmonic classical oscillator or a single resonant frequency [13,14], which may not be a good approximation in actual materials. The validity of Miller's rule, however, has not been experimentally examined in detail.

In this chapter, we propose an updated absolute scale of the quadratic nonlinear-optical susceptibilities. It is based on our recent redetermination for several important nonlinear-optical materials [15] and the previous reliable reported values. The measured materials in our redetermination include congruent LiNbO$_3$, MgO-doped LiNbO$_3$, LiTaO$_3$, KNbO$_3$, KTiOPO$_4$ (KTP), β-BaB$_2$O$_4$ (BBO), quartz, KH$_2$PO$_4$ (KDP), GaAs, GaP, α-ZnS, CdS, ZnSe, and CdTe. The important features of our measurement are as follows. First, three different nonlinear processes, SHG, PF, and difference-frequency generation (DFG), were used to determine the nonlinear-optical susceptibilities. The puzzling disparity between the values determined by SHG and PF methods has been rectified. Second, the multiple-reflection effect was fully taken into account in all the measurements. Moreover, we carried out measurements at several wavelengths to investigate the dispersion of the nonlinear-optical coefficients and assess Miller's rule quantitatively. In the SHG measurement, the fundamental wavelengths used were 0.532, 0.852, 1.064, 1.313, 1.533, and 1.548 μm. The PF measurement was made at pump wavelengths of 0.488 and

0.532 µm, and the DFG measurement was made at a pump wavelength of 0.532 µm.

In Sect. 2 the measurement methods are described in detail with typical data, and the values determined by the three methods are compared. Section 3 considers the multiple-reflection effect. The condition in which the multiple-reflection effect is significant is discussed. The dispersion of the nonlinear-optical susceptibilities is discussed in Sect. 4. Considerable frequency dependence of Miller's Δ is shown. Finally, in Sect. 5, the recommended standards of the second-order nonlinear-optical coefficients are presented.

2   SHG, PF, and DFG Determination

We used SHG, PF, and DFG methods to redetermine the nonlinear-optical coefficients. Absolute and relative SHG measurements were performed for all the materials, while PF measurements were carried out for congruent $LiNbO_3$ and 5%MgO-doped $LiNbO_3$, and DFG measurements were made for congruent $LiNbO_3$.

2.1   SHG Measurement

Absolute SHG measurements have usually been made using the phase-matching geometry. With the phase-matched SHG method, however, the diagonal components of a d tensor cannot be determined, measurable wavelengths are limited, differently cut samples are needed at different wavelengths, and the homogeneity over the whole length of the sample, as long as 10 mm, must be assured. On the other hand, we determined the absolute values of nonlinear-optical susceptibilities using the wedge technique, a non-phase-matched SHG method. It enabled us to determine the diagonal as well as non-diagonal components from visible to near infrared fundamental wavelengths without changing the samples. In addition, relative SHG measurements were also carried out using the wedge technique and the rotational Maker-fringe technique.

Wedge-Technique Measurement. The wedge samples used were typically 100 µm thick and had nearly parallel surfaces with an apex angle of 0.1°–0.2°. The quasi-parallel configuration caused distinct multiple-reflection oscillations superimposed on the Maker fringes. Earlier investigators have attempted to minimize the multiple-reflection effect by use of either thick samples or large-wedge-angle samples, or both. Thick samples, however, require a beam size large enough for the plane-wave approximation to be valid, thus making it impracticable to eliminate the overlap of the multiply reflected beams. Also, use of large-angle samples would make it difficult to take accurate Maker-fringe data, especially with materials of shorter coherence lengths. In the present study, we used thin, nearly parallel wedge samples and took full account of the multiple-reflection effect caused by both fundamental and second-harmonic waves.

Let us consider SHG from a sample with a wedge angle θ. The input fundamental wave is assumed to have a Gaussian distribution with the beam radii w_x and w_y (x is along the taper direction). Let both fundamental and second-harmonic waves be multiply reflected on the input and output surfaces of the sample. The transmitted second-harmonic power from the sample of thickness L is then given by [15]

$$P^{2\omega}(L) = \frac{2\omega^2 d^2 (P^\omega)^2}{\pi\varepsilon_0 c^3 n_\omega^2 n_{2\omega}} \left[\frac{2L^2}{\pi^{1/2} w_x^2 w_y (\Delta k L/2)^2} \right]$$

$$\times \int_{-\infty}^{\infty} \exp\left(-\frac{4x^2}{w_x^2}\right) F[l(x)] G[l(x)] \, dx, \tag{2}$$

where P^ω is the incident fundamental power, c is the light velocity in vacuum, n_ω and $n_{2\omega}$ are the refractive indices for the fundamental and the second-harmonic light, respectively, and $\Delta k = k_{2\omega} - 2k_\omega$ represents the wavevector mismatch. $l(x)$ is related to the sample thickness L at $x = 0$ and the wedge angle θ as $l(x) = L + x \tan\theta$, and l-dependent terms $F(l)$ and $G(l)$ are expressed using the refractive indices as given in [15].

The fundamental light sources used were laser diodes or diode-pumped solid-state lasers, both of which operated in a single-longitudinal mode. The cw-power stability, nearly ideal Gaussian transverse mode profiles, and well-behaved spectral characteristics of these lasers enabled us to obtain reproducible data with reasonable signal-to-noise ratios. The absolute magnitudes of the quadratic nonlinear-optical susceptibilities were determined by the measurement of the incident fundamental power P^ω, the beam radii w_x and w_y, and the generated SH power $P^{2\omega}$. The fundamental power was measured by calibrated powermeters. The beam radii of the focused elliptical Gaussian beam were determined by measuring the transverse beam distribution with a $\varnothing = 1\,\mu m$ pinhole. The SH light was detected with a photomultiplier tube, the signal from which was amplified by an I-V converter and processed with a lock-in amplifier. The absolute power of the SH light was obtained by calibrating the sensitivity of our SH detection system with a calibrated powermeter.

Figure 1 shows, as an example of the wedge-technique measurement, the SH power from d_{33} of congruent LiNbO$_3$ as a function of the sample thickness obtained at the fundamental wavelength of 1.313 μm. The incident fundamental power was 4.30 mW and the beam radii w_x, w_y were 20.81 μm and 24.35 μm, respectively. Figure 1a shows that the harmonic power is modulated by short-period oscillations superimposed on the sine-squared Maker fringes. These oscillations are caused by the interference effects resulting from the multiple reflections of both fundamental and second-harmonic waves. This is more clearly seen in Fig. 1b, which shows part of Fig. 1a around one of the

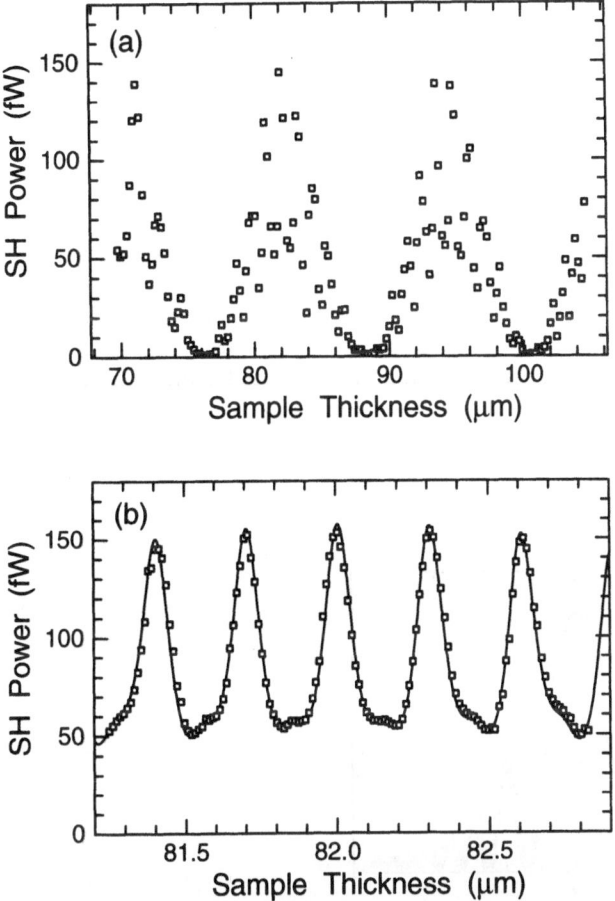

Fig. 1. Sample-thickness dependence of the SH power for d_{33}(congruent LiNbO$_3$) of the wedge sample at the fundamental wavelength of 1.313 μm. (b) Part of (a) around one of the Maker-fringe peaks plotted on an expanded horizontal scale. The fundamental power was 4.30 mW and the fundamental beam radii w_x and w_y were 20.81 μm and 24.35 μm, respectively. The open squares are experimental data, and the solid curve shows the theoretical fitting using (2)

Maker fringe peaks, plotted in an expanded horizontal scale. The solid line is the calculated result, which was least-squares-fitted to the experimental data by (2). The analysis yielded the absolute value of $d_{33} = 19.5$ pm/V at 1.313 μm. The overall accuracy is estimated to be better than ±5%.

Rotational Maker-Fringe Measurement. The wedge technique cannot be applied if the coherence length is too long to obtain even one period of Maker-fringe data. In such a case, e.g., the measurement of d_{31} of congruent LiNbO$_3$ at the fundamental wavelength of 1.064 μm, we performed relative measurements using the rotational Maker-fringe technique. The fundamental light source was a Q-switched Nd:YAG laser operating at 1.064 μm. The pulse width was 100 ns, the peak power 10 kW, and the repetition rate 1 kHz. To compensate for the fluctuations of the laser power, the measured SH power from the sample crystal was normalized by the SH power from a temperature-controlled LiNbO$_3$ plate placed just after the laser output aperture. The obtained ratio between the SH power from the measured material and that of the reference material gave the relative value of the nonlinear-optical susceptibilities.

The rotation-angle dependence of the SH power from d_{31} of congruent LiNbO$_3$ is shown in Fig. 2, as an example of the rotational Maker-fringe-technique measurement. An antireflection (AR) coated plane-parallel-plate sample was used. The relative value of d_{31} against d_{33} of congruent LiNbO$_3$ was 0.18, from which d_{31} was determined to be 4.5 pm/V when the absolute value $d_{33} = 25.2$ pm/V obtained in our absolute measurement at 1.064 μm was used.

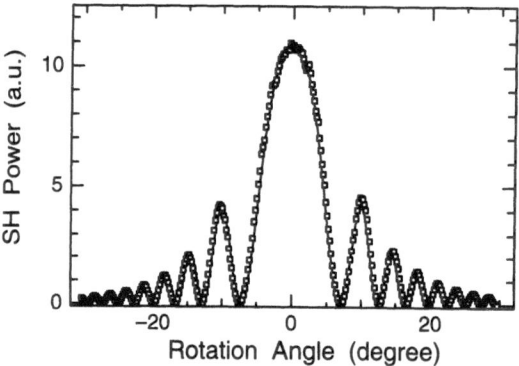

Fig. 2. Rotation angle versus the SH power for d_{31} of congruent LiNbO$_3$ at 1.064 μm. The sample was rotated around the X-axis and the fundamental wave was polarized in the Y-Z plane. The open squares are experimental data, and the solid curves are the theoretical fitting [15]

2.2   PF Measurement

The PF method we used is a type-I noncritically phase-matched process. A strong pump wave of frequency ω_p in a nonlinear-optical crystal spontaneously emits signal and idler waves of frequencies ω_s and ω_i, respectively, that satisfy the energy conservation equation, $\omega_p = \omega_s + \omega_i$. Of an infinite number of pairs of signal and idler waves, only those that satisfy the wavevector conservation or the phase-matching condition, $k_p = k_s + k_i$, grow to appreciable amplitudes.

The signal power around frequency ω_s radiated from the output end of a crystal of length L within a solid angle $\pi\theta^2$ is given by [16]

$$P_s = \frac{\hbar\omega_s^4\omega_i n_s}{2\pi^2\epsilon_0 c^5 n_i n_p |b|} d^2 L\pi\theta^2 P_p, \tag{3}$$

where $\hbar$ is the Planck constant, θ is the acceptance angle, and b is the dispersion factor defined by

$$b = \frac{\partial k_s}{\partial\omega_s} - \frac{\partial k_i}{\partial\omega_i}. \tag{4}$$

To deduce the nonlinear-optical coefficient d we have only to measure the ratio of the signal power P_s to the pump power P_p. Furthermore, because signal is proportional to pump, it does not depend on the diameter, focusing, or mode of the pump beam. Another advantage of PF measurement is that sample temperature is much less critical than in phase-matched SHG or DFG measurement because there usually exists a certain number of signal modes that satisfy the phase-matching condition.

Two pump sources were used. One was an argon-ion laser operating at $0.488\,\mu m$ which generated PF at $0.678\,\mu m$ (signal) and $1.741\,\mu m$ (idler). The other source was the SH radiation from a diode-pumped cw Nd:YAG laser. The $0.532\,\mu m$ pump generated fluorescence at $0.894\,\mu m$ (signal) and $1.314\,\mu m$ (idler). In the measurement for congruent LiNbO$_3$ we used Y-cut samples 16 mm long obtained from the same crystal as was used for our SHG measurement. Since the generation of the signal and the idler waves originates from the zero-point vibration state at the input end of the crystal, the existence of stray light has great influence on the output signal power. Much care was taken to keep out stray light. Luminescence generated from the optical components placed before the sample, which contributed as much as 20% of the detected signal, was blocked by a colored-glass band-pass filter. AR coating on both the input and output ends of the sample was made to ensure single pass of the pump, signal, and idler waves; without AR coating the nonlinear-optical susceptibility was overestimated by as much as 10%. Moreover, photoluminescence that was generated from the sample, probably due to unspecified impurities, was blocked by a spike filter placed just before the detector. The isolation of the signal from stray light realized the accurate measurements.

Figure 3 shows the dependence of the signal power on the pump power in the measurement of d_{31} for congruent LiNbO$_3$ at the pump wavelength of $0.488\,\mu m$. The analysis using (3) yielded the absolute value of $d_{31} = 4.8 \pm 0.5\,\mathrm{pm/V}$ at $\lambda_p = 0.488\,\mu m$.

2.3 DFG Measurement

DFG, which is a type-I noncritically phase-matched process as well, differs from PF only in the input condition; it is produced by two beams of comparable intensities. Unlike PF, it is not very sensitive to weak stray light. It

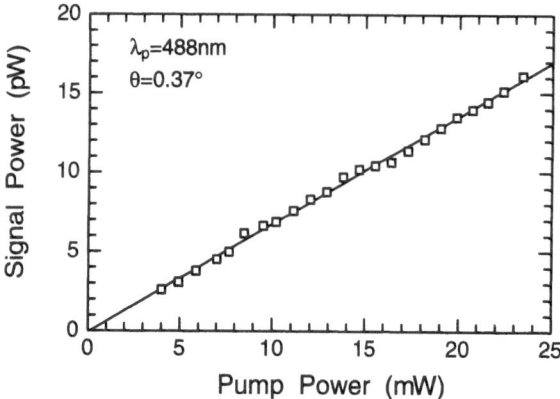

Fig. 3. PF signal power as a function of the pump power P_p. The pump wavelength was 0.488 μm. Open squares are experimental data, and the solid line represents the least-squares fitting

may, however, be more sensitive to the sample inhomogeneity and temperature variation because it involves a unique phase matching procedure. DFG is, therefore, expected to provide a good check on the quality of the sample and the consistency of the PF and DFG measurements.

The difference-frequency power is expressed, under the assumption of un-depleted pump and idler power, as [17]

$$P_s = \frac{4\omega_s^2 w_s^x w_s^y}{\pi\epsilon_0 c^3 n_p n_s n_i w_p^x w_p^y w_i^x w_i^y} P_p P_i L^2 d^2 \left(\frac{\sin(\Delta k L/2)}{\Delta k L/2}\right)^2, \tag{5}$$

where $\Delta k = k_p - k_i - k_s$ represents the wavevector mismatch. The effective radii of the elliptic Gaussian signal beam (w_s^x, w_s^y) are given, in terms of the beam radii of pump (w_p^x, w_p^y), and idler (w_i^x, w_i^y), as

$$\left(\frac{1}{w_s^{x,y}}\right)^2 = \left(\frac{1}{w_p^{x,y}}\right)^2 + \left(\frac{1}{w_i^{x,y}}\right)^2. \tag{6}$$

The pump beam was the 0.532 μm radiation from the YAG-SHG laser used for the PF measurement. The idler was the radiation from a single-mode distributed-feedback laser diode oscillating at 1.314 μm. The pump and idler produced a difference-frequency signal at 0.894 μm. The carefully aligned beams were made to impinge on the input end of the sample and propagate collinearly. The sample was the same Y-cut LiNbO$_3$ crystal that was used for the PF measurement. We achieved phase matching by changing the sample temperature around 28 °C. The pump and signal powers were measured by the calibrated powermeter. The beam radii of the pump and idler waves were determined by measuring the Gaussian spatial power distribution with a pinhole (∅ = 10 μm); the pump beam radii were $w_p^x = 0.96$ mm and $w_p^y = 0.53$ mm, and the idler beam radii were $w_i^x = 0.57$ mm and $w_i^y = 0.56$ mm.

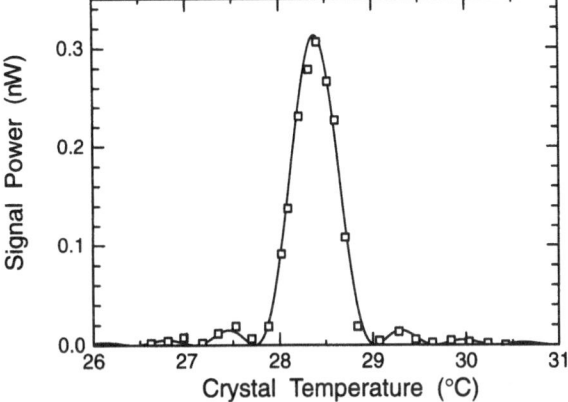

Fig. 4. Temperature tuning curve for DFG. Open squares represent experimental data, and the solid curve is the theoretical fitting

Figure 4 plots the signal power as a function of temperature. The solid curve is the theoretical fitting according to (5). The excellent fit between the experimental points and the theoretical curve is strong evidence for the homogeneity of the sample. The pump power of 7.4 mW and the idler power of 3.7 mW produced the signal power of 0.32 nW. The nonlinear-optical coefficient, d_{31}, of congruent $LiNbO_3$ determined from the DFG efficiency was $d_{31} = 4.3 \pm 0.6 \, \text{pm/V}$ for $\lambda_p = 0.532 \, \mu\text{m}$, $\lambda_i = 1.314 \, \mu\text{m}$, and $\lambda_s = 0.894 \, \mu\text{m}$, in exact agreement with our PF value.

2.4   Comparison of the Three Methods

Figure 5 shows the absolute values of d_{31} for congruent $LiNbO_3$ determined in our SHG, PF, and DFG measurement. Here the wavelength on the abscissa is either the fundamental wavelength of SHG measurement or twice the pump wavelength of the PF and DFG measurements. The values at 0.852 and 1.313 μm were obtained with the SHG method, the value at 0.976 μm with the PF method, and the value at 1.064 μm with the SHG, PF, and DFG methods. The three different processes provided consistent results. The earlier SHG value of Miller et al. [1] and the PF value of Choy and Byer [2] are also shown for comparison. Our value is in agreement with the SHG result by Miller et al., while Choy and Byer's PF value at 0.488 μm is more than 20% larger than our value. The previously reported PF values, which were significantly larger than the SHG values, were probably overestimated owing to insufficient isolation of the signal wave from the stray light and violation of the zero-point vibration condition at the input facet of the sample.

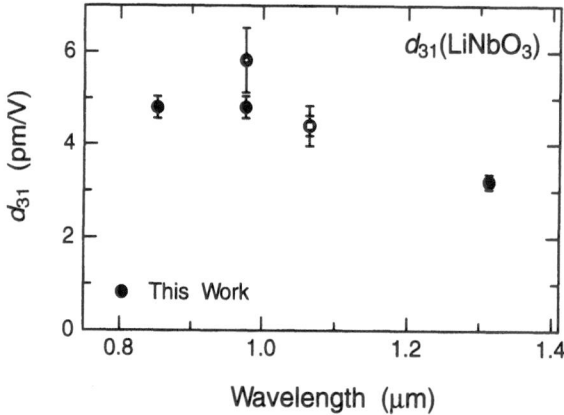

Fig. 5. Absolute magnitudes of d_{31} for congruent LiNbO$_3$ determined by the SHG, PF, and DFG methods (*filled circles*). The wavelength on the abscissa is either the fundamental wavelength of SHG measurement or twice the pump wavelength of PF and DFG measurements. The open square represents the SHG value of Miller et al. [1], and the thickly outlined circle is the PF value of Choy and Byer [2]

3 Multiple-Reflection Effect

As shown in Fig. 1, the multiple-reflection effect produces distinct interference oscillations superimposed on the Maker fringes in uncoated samples. This section considers the influence of the multiple-reflection effect on determination of the nonlinear-optical susceptibilities.

Figure 6 shows the sample-thickness dependence of SH power for d_{33} from the wedge sample of congruent LiNbO$_3$, using the Q-switched Nd:YAG laser as the fundamental light source. The incident beam radius was 0.95 mm, which was much larger than the typical beam radius of 20 μm used in the cw laser measurement. In contrast with Fig. 1, the interference apparently disappears. This is because the beam spot was large enough to extend over several periods of the interference oscillations, resulting in the smearing of the oscillations. However, the multiple-reflection effect does exist and enhances the second-harmonic power. We analyzed the obtained data using (2), taking full account of the multiple-reflection effect. The result of fitting is indicated by the solid curve in Fig. 6. Let us analyze the same data, this time neglecting the multiple-reflection effect; the conventional formula given by Boyd et al. [18] is used. Surprisingly, the fitting represented by the dotted curve is so good that the two fitting curves, with and without the multiple-reflection effect being taken into account, are apparently identical. Note that we plotted the dotted curve shifted upward by a small amount indicated by the vertical bar in Fig. 6. However, the d value thus estimated is 5.7% larger than the true d value based on (2). This arises because of the persistence of the multiple reflections despite their apparent disappearance. On the other hand, the earlier value of d_{33} (congruent LiNbO$_3$) frequently referred to as a standard

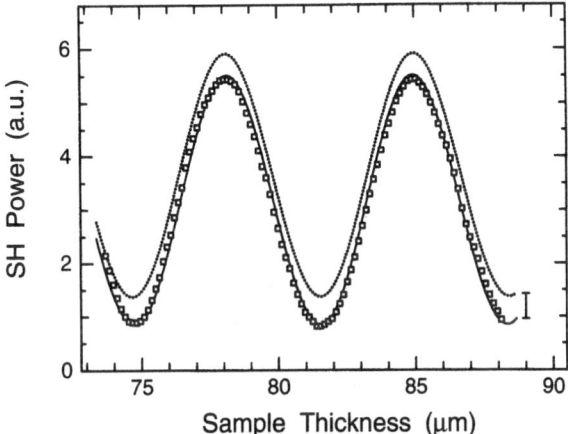

Fig. 6. Sample thickness versus the SH power for d_{33}(congruent LiNbO$_3$) of the wedge sample in the measurement using the Q-switched Nd:YAG laser. The fundamental beam radius was 0.95 mm. The open squares are experimental data, and the solid curve is the fitting according to (2). The theoretical fitting neglecting multiple-reflection effect (*dotted curve*) is shifted upward by a small amount indicated by the vertical bar

at 1.064 μm [11] is 27.2 pm/V, which was obtained by Miller et al. [1] with the rotational Maker-fringe technique. However, the multiple-reflection effect was not taken into account in their measurement. If their value is corrected by the overestimation factor of 5.7%, it reduces to 25.7 pm/V, which is in good agreement with our value at 1.064 μm, 25.2 pm/V.

The multiple-reflection effect is significant for materials with larger refractive indices. Especially for some semiconductors, whose refractive indices are as large as 3 in the near infrared, the overestimation factor can amount to 20%. Even the nonlinear-optical susceptibilities obtained from relative measurements are often overestimated if the multiple-reflection effect is neglected, because low-index materials such as KDP and quartz are usually used as the reference materials. We compared the nonlinear-optical susceptibilities of all the measured materials with previous reported values. As a result, almost all the previous values for the materials with refractive indices of more than 2 have been shown to be overestimated, probably owing to the neglect of the multiple-reflection effect.

4   Dispersion of Miller's Δ

Wavelength scaling of the nonlinear-optical susceptibilities has been conventionally made assuming the constancy of Miller's Δ. Our recent measurement has revealed, however, that the dispersion of the nonlinear-optical susceptibilities is not so simple as to be explained by Miller's rule. Miller's Δ's calculated from the nonlinear-optical susceptibilities determined at several wavelengths using (1) have been found to be hardly constant even if they are in the transparent wavelength region, and the wavelength dependence of Δ is different for different materials and different tensor components. For example, Δ_{15} of KTP at 1.06 µm is 0.35 pm/V, which is 35% larger than that at 1.31 µm, 0.26 pm/V. On the other hand, Δ_{33} of CdS at 1.06 µm, 0.156 pm/V, is 10% smaller than that at 1.31 µm, 0.173 pm/V. Therefore, the wavelength scaling using Miller's rule can lead to considerable errors. Although the constant Miller's Δ has been explained by the classical anharmonic oscillator model [13] and recently verified using the sum rules [14], the systems considered are assumed to have only a single resonant frequency. Since real systems have multiple resonant frequencies, even for the linear susceptibility or the refractive index, a three-oscillator model should be used to satisfactorily represent the experimentally obtained data [19]. A multiple-oscillator model must be constructed to explain the dispersion of the nonlinear-optical susceptibilities. Extended measurements using additional laser sources in the mid-IR spectral region as well as finely wavelength-scanning measurements would provide useful information for explaining the dispersion of the nonlinear-optical susceptibilities.

5   Recommended Standards of Nonlinear-Optical Susceptibilities

In Sects. 2–4 we pointed out the problems that were unresolved or overlooked in the previous measurements of the quadratic nonlinear-optical susceptibilities:

- SHG, PF, and DFG methods provide consistent results on the nonlinear-optical susceptibilities. The earlier PF measurements overestimated d values probably owing to insufficient isolation of the signal light from the stray light.
- The multiple-reflection effect should be taken into account in order to obtain accurate values of the nonlinear-optical susceptibilities. Almost all of the earlier measurement for the materials with large refractive indices ($n > 2$) overestimated d values because of neglecting the multiple-reflection effect.
- Miller's Δ is not so constant as to be used for the wavelength scaling of the nonlinear-optical susceptibilities. It can lead to considerable error in the obtained d values.

We have to clear all the problems above in order to obtain accurate quadratic nonlinear-optical susceptibilities. The previous careful SHG measurements for such standard materials as KDP, quartz, and BBO [6,20–22], are found to be consistent with our redetermination. This is because they have relatively low refractive indices. The absolute magnitudes of the quadratic nonlinear-optical susceptibilities based on these previous reliable measurements and our

Table 2. Recommended standards of quadratic nonlinear optical susceptibilities (pm/V)

Crystal	d_{il}	0.532 μm	0.852 μm	1.064 μm	1.313 μm	1.533 μm	1.548 μm
KDP	d_{36}			0.39			
quartz	d_{11}			0.30			
ADP	d_{36}			0.47			
Congruent	d_{33}		25.7	25.2	19.5		
LiNbO$_3$	d_{31}		4.8	4.4	3.2		
1%MgO:	d_{33}		27.5	24.9	20.3		
LiNbO$_3$	d_{31}		4.8	4.6	3.2		
5%MgO:	d_{33}		28.4	25.0	20.3		
LiNbO$_3$	d_{31}		4.9	4.4	3.4		
LiTaO$_3$	d_{33}		15.1	13.8	10.7		
	d_{31}			0.85			
KNbO$_3$	d_{33}		22.3	19.6	16.1		
	d_{31}		11.0	10.8	9.2		
	d_{15}			12.5			
KTP	d_{33}		16.6	14.6	11.1		
	d_{31}			3.7			
	d_{32}			2.2			
	d_{15}		3.9	3.7	2.6		
	d_{24}		1.9	1.9	1.4		
BBO	d_{22}	2.6	2.3	2.2	1.9		
	d_{33}			0.04			
	d_{31}			0.04			
	d_{15}			0.03			
GaAs	d_{36}			170		119	
GaP	d_{36}		159	70.6	36.8		
α-ZnS	d_{33}		17.0	12.5			9.0
	d_{31}		8.1	6.2			4.8
	d_{15}		8.0	5.8			4.3
CdS	d_{33}			19.1	16.8		14.2
	d_{31}			10.1	8.3		7.4
	d_{15}			10.7	8.8		8.0
ZnSe	d_{36}		53.8				
CdTe	d_{36}			109			73

determination are presented in Table 2. We recommend that these values be used as standards.

Acknowledgments

This work would not have been completed without the help of many people. In particular, we gratefully acknowledge the collaboration of our former colleagues and students: N. Ogasawara, A. Kitamoto, M. Shirane, K. Ohdaira,

H. Nakamura, K. Yashiki, and K. Tsuda. We also express our deep appreciation to our friends in industry who kindly supplied samples or lasers: M. Hirao of Hitachi, Ltd., T. Hirata of Yokogawa Electric, G. Hatagoshi and Y. Uematsu of Toshiba, M. Sakuta of Oki Electric Industry, H. Kuwatsuka of Fujitsu Laboratories, M. Oka, H. Kikuchi, T. Okamoto, K. Tatsuki, and S. Kubota of Sony, T. Kishimoto of Sumitomo Metal Mining, I. Nishino of Dowa Mining, and M. Omori of Japan Energy. We would also like to thank H. Kukimoto (formerly with the Tokyo Institute of Technology) for supplying some of the samples.

References

1. Miller, R.C., Nordland, W.A., Bridenbaugh, P.M. (1971) Dependence of Second-Harmonic-Generation Coefficients of $LiNbO_3$ on Melt Composition, J. Appl. Phys. **42**, 4145–4147

2. Choy, M.M., Byer, R.L. (1976) Accurate Second-order Susceptibility Measurements of Visible and Infrared Nonlinear Crystals, Phys. Rev. B **14**, 1693–1706

3. Vanherzeele, H., Bierlein, J.D. (1992) Magnitude of the Nonlinear-optical Coefficients of $KTiOPO_4$, Opt. Lett. **17**, 982–984

4. Cheung, E.C., Koch, K., Moore, G.T., Liu, J.M. (1994) Measurements of Second-order Nonlinear Optical Coefficients from the Spectral Brightness of Parametric Fluorescence, Opt. Lett. **19**, 168–170

5. Zondy, J.-J., Abed, M., Clairon, A. (1994) Type-II Frequency Doubling at $\lambda = 1.30\,\mu m$ and $\lambda = 2.53\,\mu m$ in Flux-grown Potassium Titanyl Phosphate, J. Opt. Soc. Am. B **11**, 2004–2015

6. Eckardt, R.C., Masuda, H., Fan, Y.X., Byer, R.L. (1990) Absolute and Relative Nonlinear Optical Coefficients of KDP, KD^*P, BaB_2O_4, $LiIO_3$, $MgO:LiNbO_3$, and KTP Measured by Phase-matched Second-harmonic Generation, IEEE J. Quantum Electron. **26**, 922–933

7. Kupecek, P.J., Schwartz, C.A., Chemla, D.S. (1974) Silver Thiogallate ($AgGaS_2$) – part I: Nonlinear Optical Properties, IEEE J. Quantum Electron. **QE-10**, 540–545

8. Canarelli, P., Benko, Z., Hielscher, A.H., Curl, R.F., Tittel, F.K. (1992) Measurement of Nonlinear Coefficient and Phase Matching Characteristics of $AgGaS_2$, IEEE J. Quantum Electron. **28**, 52–55

9. Kurtz, S.K., Jerphagnon, J., Choy, M.M. (1984) Nonlinear Dielectric Susceptibilities, In: Hellwege, K.H., Hellwege, A.M. (Eds.) Landolt-Bornstein, Numerical Data and Functional Relationships in Science and Technology, New Series, Springer, Berlin, **18**, Chap. S6

10. Singh, S. (1986) Nonlinear Optical Materials, In: Weber, M.J. (Ed.) Handbook of Laser Science and Technology, CRC, Boca Raton, FL, Vol. III, Part 1

11. Roberts, D.A. (1992) Simplified Characterization of Uniaxial and Biaxial Nonlinear Optical Crystals: A Plea for Standardization of Nomenclature and Conventions, IEEE J. Quantum Electron. **28**, 2057–2074

12. Miller, R.C. (1964) Optical Second Harmonic Generation in Piezoelectric Crystals, Appl. Phys. Lett. **5**, 17–19

13. Garrett, C.G.B., Robinson, F.N.H. (1966) Miller's Phenomenological Rule for Computing Nonlinear Susceptibilities, IEEE J. Quantum Electron. **QE-2**, 328–329

14. Scandolo, S., Bassani, F. (1955) Miller's Rule and the Static Limit for Second-harmonic Generation, Phys. Rev. B **51**, 6928–6931

15. Shoji, I., Kondo, T., Kitamoto, A., Shirane, M., Ito, R. (1997) Absolute Scale of Second-order Nonlinear-optical Coefficients, J. Opt. Soc. Am. B **14**, 2268–2294

16. Byer, R.L., Harris, S.E. (1968) Power and Bandwidth of Spontaneous Parametric Emission, Phys. Rev. **168**, 1064–1068

17. Canarelli, P., Benko, Z., Curl, R., Tittel, F.K. (1992) Continuous-wave Infrared Laser Spectrometer Based on Difference Frequency Generation in $AgGaS_2$ for High-resolution Spectroscopy, J. Opt. Soc. Am. B **9**, 197–202

18. Boyd, G.D., Kasper, H., Mcfee, J.H. (1971) Linear and Nonlinear Optical Properties of $AgGaS_2$, $CuGaS_2$, and $CuInS_2$, and Theory of the Wedge Technique for the Measurement of Nonlinear Coefficients, IEEE J. Quantum Electron. **QE-7**, 563–573

19. Zelmon, D.E., Small, D.L., Jundt, D. (1997) Infrared Corrected Sellmeier Coefficients for Congruently Grown Lithium Niobate and 5 mol.% Magnesium Oxide-doped Lithium Niobate, J. Opt. Soc. Am. B **14**, 3319–3322

20. Craxton, R.S. (1981) High Efficiency Frequency Tripling Schemes for High-power Nd:glass Lasers, IEEE J. Quantum Electron. **QE-17**, 1771–1782

21. Eimerl, D. (1987) Electro-optic, Linear, and Nonlinear Optical Properties of KDP and Its Isomorphs, Ferroelect. **72**, 95–139

22. Hagimoto, K., Mito, A. (1995) Determination of the Second-order Susceptibility of Ammonium Dihydrogen Phosphate and α-quartz at 633 and 1064 nm, Appl. Opt. **34**, 8276–8282

Part V

Optical Technologies

Femtosecond Fourier Optics: Shaping and Processing of Ultrashort Optical Pulses

A. M. Weiner

Summary. This chapter is structured as follows. In Sect. 2 we discuss femtosecond pulse shaping and waveform synthesis, and present some examples of applications of pulse shaping for ultrafast communications. In Sect. 3 we discuss the extension of pulse shaping to accomplish more sophisticated pulse processing operations by including holographic or nonlinear materials within the pulse-shaping apparatus. Finally, in Sect. 4, we briefly describe some applications of shaped femtosecond pulses in nonlinear optics and time-resolved materials studies.

1 Introduction

We discuss Fourier optics methods for ultrafast optical pulse shaping, waveform synthesis, and signal processing, which are achieved by spatial manipulation of the dispersed optical frequency spectra of femtosecond pulses. Using pulse-shaping techniques, one can now engineer femtosecond pulses into complex optical signals according to specification. A key point is that waveform synthesis is achieved by parallel modulation in the frequency domain, which is achieved by spatial modulation of the spatially dispersed optical frequency spectrum. Thus, waveforms with effective serial modulation bandwidths as high as terahertz can be generated without requiring any ultrafast modulators. Furthermore, by using an extension of pulse shaping called spectral holography, one can holographically record and then reconstruct such waveforms. During the reconstruction process, one can also perform interesting signal processing operations, such as time reversal, convolution, correlation, and matched filtering of femtosecond optical waveforms. Holographic and nonlinear methods also allow time-to-space conversion, where ultrafast time-domain signals are mapped (demultiplexed) into a spatial replica of the original ultrafast waveform.

These time/spectral-domain processing methods are in close analogy with traditional spatial-domain Fourier optics processing techniques. By applying such Fourier processing methods in the ultrafast time domain, one can achieve many new capabilities not available using other approaches.

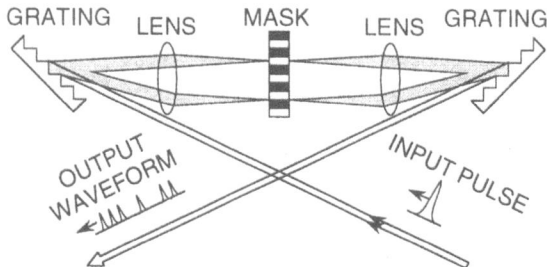

Fig. 1. Femtosecond pulse-shaping apparatus

2   Femtosecond Pulse Shaping

We first discuss femtosecond pulse shaping, in which powerful Fourier synthesis methods are utilized to generate almost arbitrarily shaped femtosecond optical waveforms [1,2]. As sketched in Fig. 1, in pulse shaping an incident femtosecond pulse is spread into its constituent spectral components by a grating and lens. A spatially patterned mask then modulates the phase and amplitude of the spatially dispersed spectral components. After the spectral components are recombined by a second lens and grating, a shaped output pulse is obtained, with the pulse shape given by the Fourier transform of the pattern transferred by the mask onto the spectrum. Pulse-shaping masks were originally implemented by using microlithographic patterning techniques and subsequently by using programmable spatial light modulators, holographic masks, and deformable mirrors. Pulse shaping has been or is currently being used in a number of laboratories for a broad range of applications, including coherent control over ultrafast physical processes, high-field physics, ultrafast nonlinear optics in fibers, and high-speed information networks. A review of some of these applications is given in [2].

The first use of the pulse-shaping apparatus shown in Fig. 1 was reported by Froehly et al. [3], who performed pulse-shaping experiments with input pulses 30 ps in duration. Related experiments demonstrating shaping of pulses a few picoseconds in duration by spatial masking within a fiber and grating pulse compressor were demonstrated independently by Heritage and Weiner [4,5]. The dispersion-free apparatus in Fig. 1 was subsequently adopted and popularized by Weiner et al. for manipulation of femtosecond optical pulses [1,6]. With minor modification, pulse-shaping operation has been successfully demonstrated for pulses below 20 fs in duration [7–9]. A fiber pigtailed pulse shaper with insertion loss as low as 5.3 dB has also been demonstrated for use in ultrafast optical communications experiments [10]. The apparatus of Fig. 1 (without the mask) can also be used for pulse stretching or compression by changing the grating–lens spacing. This idea was introduced and analyzed by Martinez [11] and is now used extensively in chirped pulse amplification.

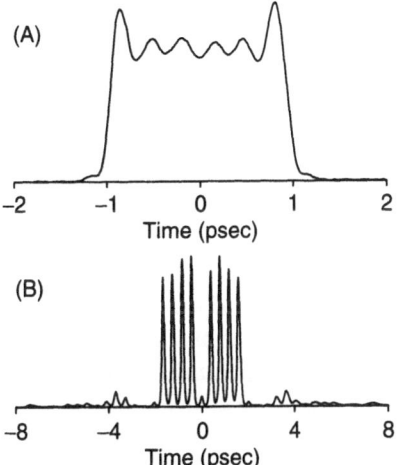

Fig. 2. Intensity cross-correlation traces of shaped pulses, measured using unshaped reference pulses directly from the laser. (**a**) Picosecond square pulse generated using microlithographically fabricated amplitude and phase masks. (**b**) Ultrafast optical pulse ("bit") sequence, generated using phase-only filtering

Figure 2 shows of examples of shaped pulses generated using fixed masks. Fig. 2a shows a 2 ps square pulse with $\sim 100\,\mathrm{fs}$ transition times (comparable to the duration of the input laser pulses) [1]. This pulse was generated by using both phase and amplitude masks to pattern the spectrum according to a truncated sinc function. Square pulses with flatter tops have also been demonstrated by appropriately apodizing the masking function [12]. Figure 2b shows an example of an ultrafast pulse sequence with an effective modulation rate of 2.5 THz [13]. Such sequences may be useful as ultrafast data packets for time-division multiplexed optical communications networks. It is worth noting that the pulse sequence in Fig. 2b was generated using a phase-only filter, which avoids the loss associated with amplitude filtering. Phase-only filters have been extensively explored in spatial optics. The pulse sequence shown here was actually generated using a phase-only filter known as a Dammann grating, which was originally designed for the generation of spot arrays for space-domain interconnect applications [14,15].

Currently most work on pulse shaping utilizes programmable spatial light modulators that allow computer control over the spatial masking pattern. The first demonstrations of programmable pulse shaping used one-dimensional (1-D) liquid crystal phase modulator arrays with up to 128 modulator pixels and millisecond reprogramming times [16]. In addition to computer programmability, these phase modulator arrays made possible gray-level phase control, which is difficult using fixed masks. Liquid crystal arrays allowing independent gray-level phase and amplitude control were subsequently developed [17] and are now available commercially. Faster liquid crystal modulators allowing binary spectral phase modulation with 100 μs reprogramming times have also been reported [18]. Pulse shaping using an acoustooptic modulator as a programmable mask has also been demonstrated [19] and is seeing application with amplified femtosecond systems. The development of programmable pulse shaping using commercially available spatial light modulators was a key step enabling widespread adoption of this technique, since it eliminated the need to fabricate custom microlithographic masks for each new pulse-shaping experiment.

Figure 3 shows examples of pulse-shaping data obtained using a programmable liquid crystal modulator (LCM) array at 1.55 μm. Both of these examples have relevance to ultrafast optical communications. In the top two traces, the LCM is programmed to provide a cubic spectral phase function [20]. This corresponds to a quadratic frequency-dependent delay, which leads to the observed asymmetric distortion and interference structure in the measured intensity profile. Cubic spectral phase is important in the transmission of ultrashort pulses over optical fibers, especially in systems (such as dispersion-shifted fibers) where the lowest order dispersion can be made zero. The sign of the cubic phase is changed between the two traces, leading to a change in the direction of the asymmetric tail. The LCM can be used to compensate for the dispersion slope of optical fibers (see below), as well as for third-order phase effects in femtosecond amplifier systems. The bottom trace shows a pseudonoise burst generated by applying a pseudorandom phase code onto the optical spectrum [10]. The phase code converts the few hundred femtosecond input pulse into a pseudonoise waveform several tens of picoseconds in duration with substantially reduced peak intensity. By decoding using a subsequent pulse shaper with the inverse of the original phase code, coded waveforms can be restored to their original femtosecond duration (not shown). This spectral coding–decoding process enables ultrashort pulse code-division multiple-access (CDMA) communications, where different users share a common fiber-optic medium based on the use of different spectral codes. The CDMA receiver recognizes the desired data in the presence of multi-access interference on the basis of the strong intensity contrast between correctly and incorrectly decoded pulses. Femtosecond encoding and decoding was first demonstrated using fixed phase masks in the late 1980s [6], and a theoretical analysis of femtosecond CDMA was reported shortly thereafter [21]. Taking advantage of advances in fiber-optic technology in the early 1990s, at Purdue we have now constructed a femtosecond CDMA testbed, including encoding and decoding, dispersion-compensated transmission of encoded pulses over kilometer lengths of fiber, and nonlinear optical thresholding to distinguish between correctly and incorrectly decoded pulses. Detailed discussion of our femtosecond CDMA experimental results is given in [10].

We discuss the fiber dispersion compensation application in more detail. In our group we have recently demonstrated transmission of sub-500 femtosecond pulses over a 3 km link consisting of lengths of standard single-mode

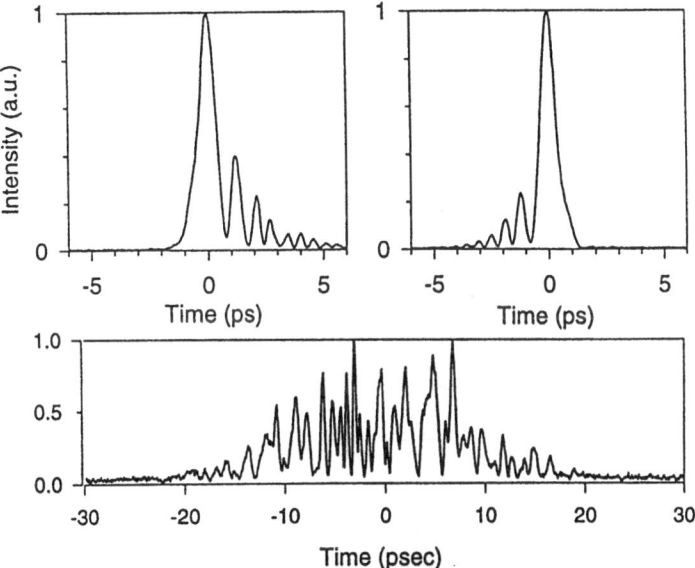

Fig. 3. Programmable pulse-shaping data at 1.5 μm using a liquid crystal modulator array. *Top*: distorted pulses resulting from programmable cubic spectral phase. *Bottom*: pseudonoise burst resulting from pseudorandom spectral phase coding

fiber (SMF) and dispersion-compensating fiber (DCF). By carefully matching both the dispersions and dispersion slopes of the fibers, we obtain less than a factor of two pulse broadening, despite the fact that the pulse first broadens by several hundred times in the SMF part of the link. An example of our data is shown in Fig. 4. Due to small residual dispersion and dispersion slope in our link, the output pulse in Fig. 4b has an asymmetric distortion as discussed above. We can compensate for this phase distortion by programming a pulse shaper for an equal and opposite phase characteristic (Fig. 4d). The resulting pulse, shown in Fig. 4c, is completely recompressed with no observable distortion [20]. This technique for fine tuning out small amounts of residual dispersion from approximately dispersion-compensated links should be applicable to both CDMA and ultrahigh-speed TDM optical transmission and networking.

In our current research, we use pulsing shaping as a core technology enabling novel investigations into ultrafast communications and femtosecond spectroscopy. We are also seeking to advance the state of the art, e.g. by incorporating optoelectronic modulator arrays into pulse shapers with the goal of demonstrating pulse sequence (data packet) generation with subnanosecond reprogramming times. Finally, pulse shaping can be extended to accomplish more sophisticated pulse processing operations by including holographic or nonlinear materials in place of a mask or spatial light modulator within the pulse-shaping apparatus. This is the subject of the next section.

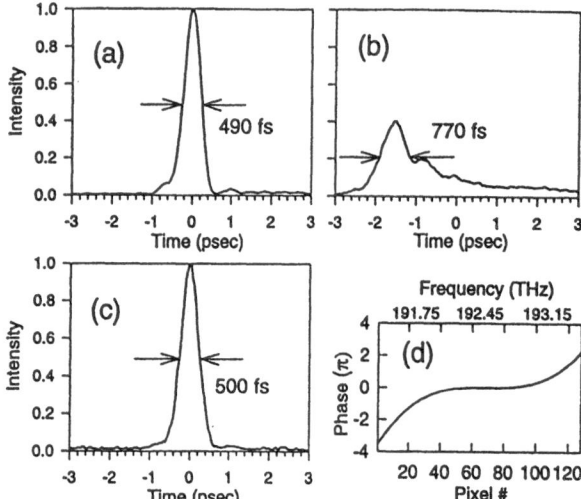

Fig. 4. (a) Input pulse to the 3 km dispersion compensated fiber link and output pulse from the fiber link when (**b**) constant-phase or (**c**) cubic- and quadratic-phase correction is applied to the LCM. (**d**) Phase-correction function

3   Holographic and Nonlinear Fourier Pulse Processing

3.1   Time-domain Pulse Processing

Holographic and nonlinear Fourier processing techniques can be used for pure time-domain processing as well as time–space conversions. We first discuss pure time-domain processing using an extension of pulse shaping called spectral holography, in which the pulse-shaping mask is replaced by a holographic material. Spectral holography was first proposed theoretically [22] in the Russian literature, and subsequent experiments demonstrating the principles of time-domain processing via spectral holography were performed by Weiner et al. [23]. In analogy with off-axis spatial holography, two beams are incident: an unshaped femtosecond reference pulse with a uniform spectrum, and a temporally shaped signal waveform with information patterned onto the spectrum. The spectral components making up the reference and signal pulses are spatially dispersed and interfere at the Fourier plane. The resulting fringe pattern is stored by using a holographic recording medium. During readout with a short test pulse, each spectral component from the test pulse diffracts off that part of the hologram containing phase and amplitude information corresponding to the same frequency component from the signal beam. The diffracted frequencies are then recombined into a pair of output beams, corresponding to +1 and −1 order diffraction, respectively. Assuming linear holographic recording and sufficient spectral resolution, the reconstructed field $E_{\text{out}}(\omega)$ can be written as follows:

$$E_{\text{out}}(\omega) \approx E_{\text{t}}(\omega) E_{\text{r}}^*(\omega) E_{\text{s}}(\omega) e^{i\boldsymbol{K}_1 \cdot \boldsymbol{r}} + E_{\text{t}}(\omega) E_{\text{r}}(\omega) E_{\text{s}}^*(\omega) e^{i\boldsymbol{K}_2 \cdot \boldsymbol{r}} . \tag{1}$$

Here, $E_t(\omega)$, $E_r(\omega)$ and $E_s(\omega)$ are the complex spectral amplitudes of the test, reference, and signal fields, respectively, and $\boldsymbol{K}_1$ and $\boldsymbol{K}_2$ are the propagation vectors of the diffracted output beams. When both test and reference beams consist of unshaped pulses with durations short compared to the duration of the shaped signal pulse, the output pulse is either a real or a time-reversed reconstruction of the original signal pulse, depending on the diffraction direction. This is in analogy to the reconstruction of real or conjugate versions of a stored image in spatial holography. Furthermore, if the test beam itself is a shaped pulse, then one can generate the convolution or the correlation of the signal and test electric field envelopes. In the special case where the test and signal waveforms are identical, the correlation becomes a matched filtering operation, which is useful for chirp compensation and pulse compression and for ultrafast pattern matching. This is the time-domain analog of holographic matched filtering used for pattern matching of spatial images in a van der Lugt correlator [24], for example.

The first demonstrations of spectral holography were performed using femtosecond pulses from a CPM dye laser to record fixed spectral holograms using thermoplastic plates as the holographic material [23]. The full range of signal processing operations enumerated above were successfully demonstrated. As one example, Fig. 5 shows output intensity profile data corresponding to matched filtering operation. The short output pulse shown in Fig. 5a results when all three pulses (signal, reference, and test) are ultrashort pulses with no distortion. In Fig. 5b, pulse shaping is used to encode the signal pulse only into a low intensity pseudonoise burst. Finally, in Fig. 5c, both signal and test pulses are encoded in the same way; the identical distortions cancel, and an intense bandwidth-limited output pulse is restored. These data illustrate the coding–decoding process that forms the basis for femtosecond CDMA communications. It is worth noting that such holographic matched filtering is a self-aligned process, so that one can decode or process incoming signals without having to precisely specify those signals beforehand.

Together with Profs. D. Nolte and M. Melloch, we have recently demonstrated a series of dynamic spectral holography experiments [25]. These experiments use pulses at $\sim 850\,\mathrm{nm}$ from a femtosecond Ti:sapphire laser and photorefractive quantum wells (PRQWs) [26] as an extremely sensitive, dynamic holographic medium. PRQWs are able to dynamically track changes in the input beams occurring on time scales as fast as microseconds. We have utilized dynamic spectral holography to demonstrate the ability to remove slow timing jitter from an input ultrashort pulse signal.

3.2   Space-to-time Conversion

Holographic space-to-time conversion can be achieved by placing a holographic mask at the Fourier plane of a pulse shaper. This can be achieved using either a fixed computer-generated hologram [27] or a dynamic optically addressed hologram [28–30]. We have recently performed space-to-time con-

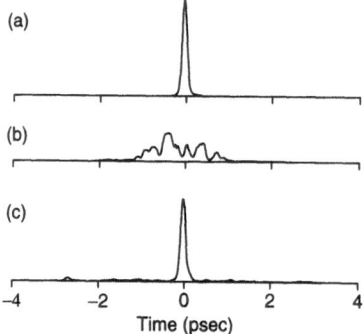

Fig. 5. Holographic matched filtering of coded ultrafast waveforms. (a) None of the pulses are coded. (b) The signal is coded using pulse shaping. (c) Both signal and test pulses are coded, resulting in matched filtering operation and a restored output pulse

version experiments using PRQWs as the dynamic holographic material [25]. The temporal profile resulting when a spectrally dispersed femtosecond pulse is diffracted from the hologram can be either a direct or Fourier-transformed version of an input spatial image, depending on whether a direct or Fourier transform hologram of the image is recorded. Figure 6 shows data for the direct space-to-time conversion case. The input image is a simple slit, resulting in a square-like output pulse (the electric field cross-correlation measurement is shown as Fig. 6b). The corresponding power spectrum is the well-known sinc function squared (Fig. 6a). An interesting effect occurs when the input signal beam amplitude is increased above the reference beam amplitude, leading to nonlinear recording conditions. In this case the central peak of the spectrum is strongly saturated, while the sidelocks continue to increase (Fig. 7a). This amounts to a high-pass filtering operation, which enhances the edges in the time-domain waveform, as is clearly evident in Fig. 7b. Such edge enhancement could potentially be useful for distinguishing between closely spaced pulses in an ultrafast communications link. We note also that this time-domain processing once again has close analogies with spatial Fourier optics, where holographic edge enhancement was proposed for identification of breaks in photolithography masks [31].

3.3   Time-to-space Conversion

Holographic and nonlinear Fourier pulse processing can also be used for time-to-space conversion of femtosecond pulses. One method for time-to-space mapping is based on the recording of a spectral hologram using time-domain signal and reference pulses, as in Sect. 3.1, and then reading out using a monochromatic, continuous-wave (CW) laser. Nuss et al. [32] used the arrangement sketched in Fig. 8 for experiments demonstrating such time-to-space mapping. Femtosecond signal and reference pulses from a mode-locked

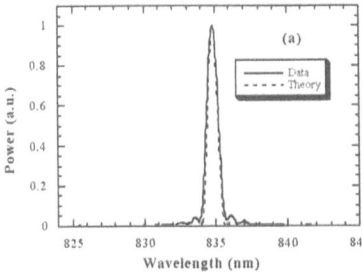 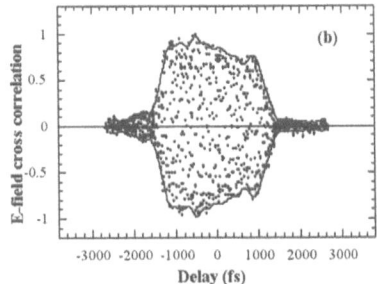

Fig. 6. Space-to-time conversion: time-domain image of a slit: (**a**) power spectrum; (**b**) electric field cross-correlation data

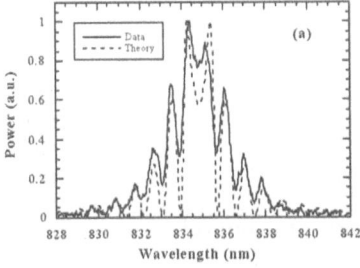 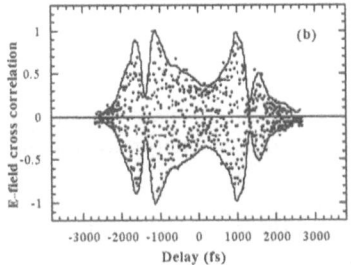

Fig. 7. Edge-enhanced time-domain image of a slit: (**a**) power spectrum; (**b**) electric field cross-correlation data

Ti:sapphire laser are used in conjunction with a PRQW as a dynamic holographic recording material. Readout is accomplished by using a CW laser diode. By passing the diffracted output beam through a Fourier transform lens, the original temporal information is converted into spatial information. The spatial profile of the resulting output beam is given by the electric field cross-correlation between signal and reference pulses. For a short-bandwidth-limited reference pulse, this yields a spatial replica of the electric field amplitude of the original time-domain signal pulse.

In addition to PRQWs, bulk holographic crystals have also been employed for time-to-space conversion experiments [30]. However, both PRQWs and bulk holographic materials are too slow for use in most communication applications requiring Gb/s frame rates. A related technique for time-to-space conversion of picosecond pulses was demonstrated by Ema, who utilized the exciton resonance in ZnSe for the nonlinearity required for holography [33]. Although this material has the advantage of fast response (~ 13 ps), its operation wavelength is 442 nm and it must be cooled to cryogenic temperatures. A third scheme, first demonstrated by Mazurenko, Fainman, and coworkers [34,35], relies on second-harmonic generation (SHG) using a nonlinear optical crystal within a pulse shaper. The setup is similar to that of Fig. 8, except

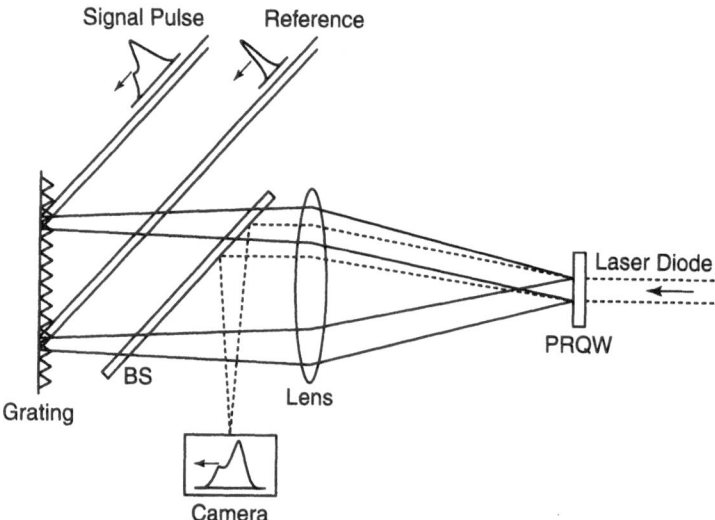

Fig. 8. Apparatus for space-to-time mapping using spectral holography and dynamic semiconductor photorefractive films

that the reference and signal beams are brought in from different directions such that the spatial dispersions are equal but opposite. This leads to the generation of a quasimonochromatic second-harmonic beam which can be transformed through a subsequent Fourier transform lens, resulting in the desired time-to-space conversion. This scheme is an important advance due to the combination of fast response and operation at convenient wavelengths and temperatures. However, in the original experiments where SHG was performed using angle-tuned type-I phase matching in an LBO crystal [35], the conversion efficiency was rather low ($\sim 0.1\%$). We subsequently analyzed the factors governing the efficiency in SHG-based time-to-space conversion and have recently achieved femtosecond optical time-to-space mapping with more than 50% conversion efficiency using temperature-tuned noncritical phase matching in a thick $KNbO_3$ nonlinear crystal [36,37]. This increase in efficiency by more than 500 times may contribute to the realization of systems performing sophisticated ultrafast pulse processing operations repeatable at communication rates with realistic power budgets.

3.4   Generalized Time–Space Processing

By exploiting the ability to convert between space and time in pulse shaping and spectral holography, one can envision new opportunities for data processing and manipulation of extremely broadband optical signals. Figure 9 illustrates the cascaded time-space systems concept [37]. Here, an ultrafast time-domain data stream (or possibly a multiwavelength data signal) would first be converted into the space domain. Various processing operations could

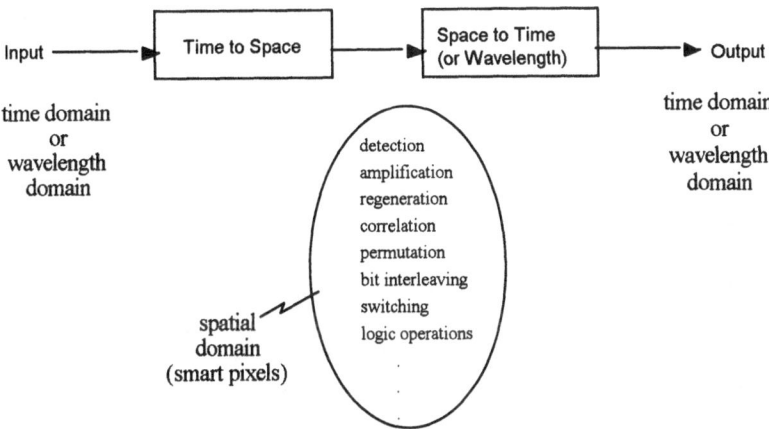

Fig. 9. Block diagram of generalized spacetime pulse processing systems. By converting interchangeably between time, space, and wavelength domains, sophisticated data manipulation applications may be possible

then be performed in parallel using high-speed smart pixel optoelectronic arrays. The smart pixel array would include detectors for reading the space-domain data, VLSI electronics for processing, and modulators for converting back to the ultrafast time (or the wavelength) domain using an appropriate pulse-shaping setup. Key to this concept is the use of arrays of optoelectronic "smart pixels" to implement the spatial-domain processing function. Smart pixel device arrays, in which optoelectronic transceivers such as detectors and modulators are intimately coupled to electronic processing circuitry at every pixel of the array, have recently emerged as a key technology for traditional (i.e. spatial-domain) optical signal processing and interconnect systems [38,39]. Here, our aim is to employ smart pixel technology in an application for which it has previously received little attention – namely, for manipulation and processing of ultrafast time-domain optical signals. Experiments demonstrating cascaded space-to-time and time-to-space conversion based on spectral holography were reported previously [30], but no further processing was performed. By incorporating high-performance smart pixel device arrays into ultrafast pulse processing systems, it should be possible to accelerate processing times substantially and achieve completely new functionalities. We are currently pursuing proof-of-concept experiments using hybrid SEED smart pixel technology [40] as a demonstration vehicle.

4   Selected Applications of Shaped Pulses

Femtosecond pulse shaping is now used in a number of laboratories worldwide for a variety of applications in technology and ultrafast science. Rather than review the entire field, here we briefly survey pulse-shaping applications that have been pursued by the author. Generally, these fall into three categories – namely, optical communications, nonlinear optics, and ultrast interactions with materials.

We have already touched on applications in ultrafast communications, such as dispersion compensation and CDMA, earlier in this chapter. It is also worth noting that pulse-shaping techniques are being extended by others for applications involving filtering of multiple wavelength optical signals in WDM communications. Examples include construction of WDM cross-connect switches with flat-topped frequency response [41] and multichannel WDM gain equalizers [42]. Finally, pulse shaping can be applied for phase filtering of broadband incoherent light for coherence coding applications [43,44].

In ultrafast nonlinear guided wave optics, pulse shaping has been used to obtain femtosecond square pulses for all-optical switching experiments with a fiber nonlinear coupler, where square pulses gave improved switching characteristics compared to normal (unshaped) pulses by avoiding averaging over the pulse intensity profile [12]. Specially shaped and phase-modulated femtosecond "dark pulses" were also used for the first clear demonstration of fundamental dark soliton propagation in optical fibers [45]. Pulse shaping has also been applied by others for dark soliton experiments on a picosecond time scale [46].

Finally, the author has pursued several experiments in which femtosecond pulse sequences are used to enhance control in femtosecond spectroscopy and ultrafast laser–materials interactions. In the first such studies, performed in collaboration with Keith Nelson at MIT, a train of approximately a dozen evenly spaced pulses (generated via pulse shaping) were used to excite molecular crystals through impulsive stimulated Raman scattering (ISRS) [47]. By tuning the pulse repetition frequency to match a phonon frequency, it was possible to selectively amplify coherent optical phonons. This was one of the first experiments in the coherent control field [48] where specially engineered femtosecond waveforms are used to manipulate quantum mechanical motions. Later, in experiments performed with Martin Nuss at Bell Laboratories, similar pulse sequences were used to excite coherent quantum mechanical charge oscillations in coupled GaAs/GaAlAs quantum wells, revealing the sensitivity of these oscillations to the optical phase [49].

Recently we have been investigating the use of shaped femtosecond pulses for manipulation and enhancement of optically excited terahertz radiation. Terahertz waveform synthesis has been achieved by using shaped pulses to excite ultrafast photoconductive dipole antennas, resulting in the generation of quasi-narrowband terahertz tone bursts and terahertz pulse trains with internal phase and amplitude modulations [50]. Through these experiments we discovered that multiple pulse excitation could result in an enhanced power spectral density at a selected terahertz frequency through avoidance of saturation effects. These multiple pulse excitation experiments originally performed on dipole antennas have now been extended to large-aperture photoconductors driven by an amplified femtosecond source, where the enhancement gives prospects for substantially higher peak terahertz electric fields [51].

References

1. Weiner, A.M., Heritage, J.P., Kirschner, E.M. (1988) J Opt Soc Am B 5: 1563–1572
2. Weiner, A.M. (1995) Prog Quantum Electron 19(3): 161–238
3. Froehly, C., Colombeau, B., Vampouille, M. (1983) Shaping and Analysis of Picosecond Light Pulses. In: Wolf, E. (Ed.) Progress in Optics, Vol. 20. North-Holland, Amsterdam, pp 65–153
4. Heritage, J.P., Weiner, A.M., Thurston, R.N. (1985) Opt Lett 10: 609
5. Weiner, A.M., Heritage, J.P. (1987) Rev Phys Appl 22: 1619
6. Weiner, A.M., Heritage, J.P., Salehi, J.A. (1988) Opt Lett 13: 300–302
7. Reitze, D.H., Weiner, A.M., Leaird, D.E. (1992) Appl Phys Lett 61: 1260–1262
8. Efimov, A., Schaffer, C., Reitze, D.H. (1995) J Opt Soc Am B 12(10): 1968–1980
9. Yelin, D., Meshulach, D., Silberberg, Y. (1997) Opt Lett 22(23): 1793–1795
10. Sardesai, H.P., Chang, C.-C., Weiner, A.M. (1998) J Lightwave Technol 16: 1953–1964
11. Martinez, O.E. (1987) IEEE J Quantum Electron 23: 59
12. Weiner, A.M., et al. (1989) IEEE J Quantum Electron 25: 2648
13. Weiner, A.M., Leaird, D.E. (1990) Opt Lett 15: 51–53
14. Dammann, H., Gortler, K. (1971) Opt Commun 3: 312–315
15. Killat, U., Rabe, G., Rave, W. (1982) Fiber Integr Opt 4: 159–167
16. Weiner, A.M., Leaird, D.E., Patel, J.S., Wullert, J.R. (1992) IEEE J Quantum Electron 28: 908–920
17. Wefers, M.M., Nelson, K.A. (1995) Opt Lett 20: 1047–1049
18. Ratsep, M., Tian, M., Lorgere, I., Grelet, F., Le Gouet, J.-L. (1996) Opt Lett 21: 83–85
19. Hillegas, C.W., Tull, J.X., Goswami, D., Strickland, D., Warren, W.S. (1994) Opt Lett 19: 737–739
20. Chang, C.-C., Sardesai, H.P., Weiner, A.M. (1998) Opt Lett 23: 283–285
21. Salehi, J.A., Weiner, A.M., Heritage, J.P. (1990) J Lightwave Technol 8: 478
22. Mazurenko, Y.T. (1990) Appl Phys B 50: 101–114
23. Weiner, A.M., Leaird, D.E., Reitze, D.H., Paek, E.G. (1992) IEEE J Quantum Electron 28: 2251–2261
24. Goodman, J.W. (1968) Introduction to Fourier Optics. McGraw-Hill, New York
25. Ding, Y., Nolte, D.D., Melloch, M.R., Weiner, A.M. (1998) Paper presented at Ultrafast Phenomena XI, Garmisch-Partenkirchen, Germany (unpublished)
26. Nolte, D., Melloch, M. (1994) MRS Bulletin 19: 44–49
27. Nuss, M.C., Morrison, R.L. (1995) Opt Lett 20: 740–742
28. Ding, Y., Nolte, D.D., Mellcoh, M.R., Weiner, A.M. (1998) IEEE J Sel Top Quantum Electron 4: 332–341
29. Ema, K., Shimizu, F. (1990) Jpn J Appl Phys 29: 1458
30. Sun, P.C., Mazurenko, Y.T., Chang, W.S.C., Yu, P.K.L., Fainman, Y. (1995) Opt Lett 20(16): 1728–1730
31. Ochoa, E., Goodman, J.W., Hesselink, L. (1985) Opt Lett 10: 430–432

32. Nuss, M.C., Li, M., Chiu, T.H., Weiner, A.M., Partovi, A. (1994) Opt Lett 19: 664–666
33. Ema, K., Kuwata-Gonokami, M., Shimizu, F. (1991) Appl Phys Lett 59(22): 2799–2801
34. Mazurenko, Y.T., Putilin, S.E., Spiro, A.G., Beliaev, A.G., Yashin, V.E., Chizhov, S.A. (1996) Opt Lett 21: 1753–1755
35. Sun, P.C., Mazurenko, Y.T., Fainman, Y. (1997) J Opt Soc Am A 14: 1159
36. Kan'an, A.M., Weiner, A.M. (1998) J Opt Soc Am B 15: 1242–1245
37. Weiner, A.M., Kanan, A.M. (1998) IEEE J Sel Top Quantum Electron 4: 317–331
38. Forrest, S.R., Hinton, H.S. (Eds.) (1993) IEEE J Quantum Electron 29: 598–813
39. Lentine, A.L., Miller, D.A.B. (1993) IEEE J Quantum Electron 29: 655–669
40. Goossen, K.W., et al. (1995) IEEE Phot Tech Lett 7: 360–362
41. Patel, J.S., Silberberg, Y. (1995) IEEE Phot Tech Lett 7: 514–516
42. Ford, J.E., Walker, J.A. (1998) IEEE Phot Tech Lett 10: 1440–1442
43. Binjrajka, V., Chang, C.-C., Emanuel, A.W.R., Leaird, D.E., Weiner, A.M. (1996) Opt Lett 21: 1756–1758
44. Griffin, R.A., Sampson, D.D., Jackson, D.A. (1995) IEEE J Lightwave Technol 13: 1826–1837
45. Weiner, A.M., et al. (1988) Phys Rev Lett 61: 2445
46. Emplit, P., Haelterman, M., Hamaide, J.-P. (1993) Opt Lett 18: 1047–1049
47. Weiner, A.M., Leaird, D.E., Wiederrecht, G.P., Nelson, K.A. (1990) Science 247: 1317
48. Warren, W.S., Rabitz, R., Dahleh, M. (1993) Science 259: 1581
49. Brener, I., Planken, P.C.M., Nuss, M.C., Pfeiffer, L., Leaird, D.E., Weiner, A.M. (1993) Appl Phys Lett 63: 2213
50. Liu, Y., Park, S.-G., Weiner, A.M. (1996) IEEE J Sel Top Quantum Electron 2: 709–719
51. Siders, C.W., Siders, J.L.W., Taylor, A.J., Park, S.-G., Melloch, M.R., Weiner, A.M. (1999) Opt Lett 24: 241–243

Aperture-modulated Diffusers (AMDs)

H. P. Herzig and P. Kipfer

Summary. In this chapter we discuss the design of optical diffusers for the generation of flat-top intensity distributions, as required for 248 nm or 193 nm excimer-laser applications [4,5]. We investigate the generation of far-field distributions with various shapes, uniform intensity and high efficiency.

1 Introduction

Lenses are among the oldest and best known optical elements besides mirrors. For example, the Vikings [1] made lenses nearly a thousand years ago even with elliptical surfaces. They used them for focusing and imaging. These applications are still in use till today in ordinary life. Also the invention of the Fresnel lens and the Fresnel zone plate did not change the general span of applications.

With the introduction of micro-optics a powerful tool was created to make various optical elements that inaugurate new applications. Refractive and diffractive elements with arbitrary phase profiles are used for laser beam shaping, optical interconnects, or for illumination systems [2,3]. The fabrication technologies for micro-optical elements permit flexibility, high accuracy and reproducibility, especially for the periodicity of diffractive elements.

In this chapter we discuss the design of optical diffusers for the generation of flat-top intensity distributions, as required for 248 nm or 193 nm excimer-laser applications [4,5].

For applications in illumination systems often a space-invariant response of the diffractive elements is required. In detail, the diffusers have to be almost independent of the size, shape and homogeneity of the illumination. For a practical application in flexible systems, the elements should furthermore be insensitive to small alignment errors.

The utilization of arrays of micro-lenses with adapted geometry is straightforward for this application. The lenses generate the desired angular spectrum, while the array property warrants the space invariance of the element. We call these elements aperture-modulated diffusers (AMDs), because the geometry of the aperture determines the shape of the generated far field (angular spectrum).

Section 2 introduces the design of AMDs. These elements are basically focusing (or diverging) lenses with arbitrary aperture shapes. However, we are interested in a uniform far-field distribution of accurate width and not in an optimum focus spot. Section 3 compares these two cases with respect to the optimum optical function, the fabrication tolerances and the scaling behavior. Such elements can be realized as refractive or diffractive components. Diffractive optics has the advantage of higher flexibility and the deflection angles can be controlled very accurately. The drawback is the amount of light at the zeroth order. In Sect. 4 we briefly summarize these problems and present some fabricated diffusers.

2   Design of AMDs

A diffuser generates a well-defined diffraction pattern in the far field when illuminated with an incident light beam. The following setup is considered: a diffuser is illuminated by a multi-mode excimer laser with a given beam divergency in the x and y direction at the wavelength $\lambda = 248$ nm. The required intensity distribution is then obtained in the Fourier plane (far field) of the second lens (Fig. 1).

The basic principle of AMDs is very simple. The phase structure to generate the flat-top intensity distribution is designed as an array of lenses of hexagonal, rectangular or arbitrary shape. For a given focal length, the aperture shape of a single lens defines the shape and the width L of the far-field distribution. Due to the divergent illumination with an incoherent source, the far field of the lenses is added incoherently and with small displacements. This can also be described in conventional Fourier optics by the use of an extended source $U_0(x_0, y_0)$ in the Fourier domain, where the extension S of the source is defined by the divergence angle (see Fig. 1b) [4].

The AMD with an aperture shape $a(x_1, y_1)$ is considered in the thin-element approximation by a complex transmission function $t(x_1, y_1)$ [6]:

$$t(x_1, y_1) = a(x_1, y_1) \exp[i\Phi(x_1, y_1)] , \tag{1}$$

with

$$
\begin{aligned}
a &= 0 \quad \text{for} \quad (x_1, y_1) \text{ outside the aperture} \\
a &= 1 \quad \text{for} \quad (x_1, y_1) \text{ inside the aperture.}
\end{aligned}
\tag{2}
$$

In the case of coherent illumination, the intensity distribution I_2 in the far field is given by a convolution of the amplitudes

$$
\begin{aligned}
I_2(x_2, y_2) &= |U_2(x_2, y_2)|^2 = |\mathrm{FT}\{U_1(x_1, y_1)\}|^2 \\
&= |\mathrm{FT}\{\mathrm{FT}^{-1}\{U_0(x_0, y_0)\} \cdot t(x_1, y_1)\}|^2 \\
&= |U_0(x_0, y_0) \otimes \mathrm{FT}\{t(x_1, y_1)\}|^2 .
\end{aligned}
\tag{3}
$$

In the case of incoherent illumination, however, the intensity distribution I_2 in the far field is given by a convolution of the intensities

$$I_2(x_2, y_2) = |U_0(x_0, y_0)|^2 \otimes |\mathrm{FT}\{t(x_1, y_1)\}|^2 . \tag{4}$$

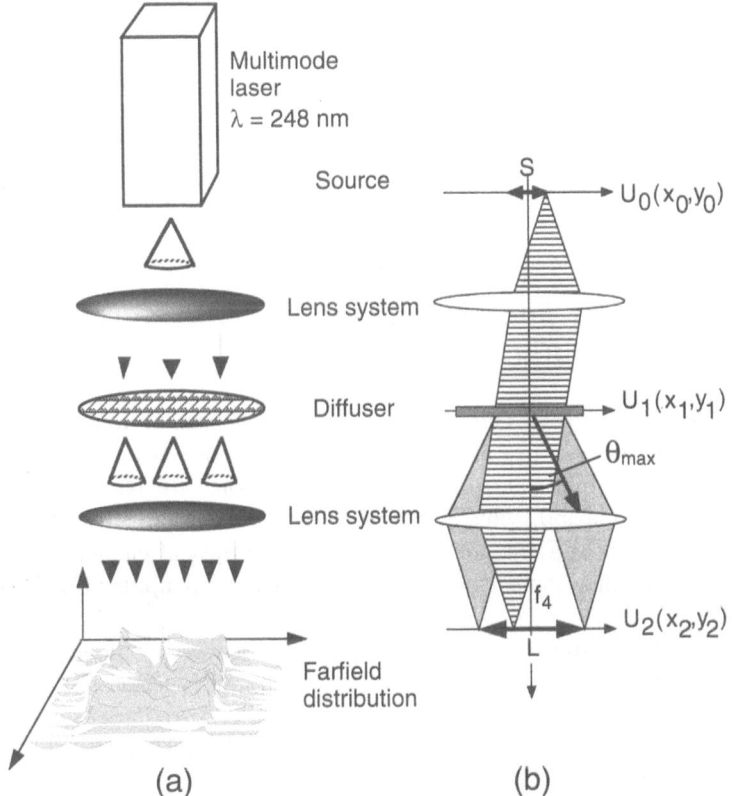

(a)   (b)

Fig. 1. Setup for the generation of flat-top distributions

In the following we assume that the source is ideally incoherent. The modal distribution of the laser is approximated very roughly by a rect-function. The convolution of the intensity distribution with a rect-function corresponds to detection of the intensity distribution with a common detector with finite pixel size. This has to be taken into account for a comparison of the far-field distribution with measurements.

Henceforth, we will use the deflection angle θ instead of the far-field co-ordinates (x_2, y_2). Consequently the maximum diameter of the far field is described by (for small angles)

$$2\theta_{max} = L/f_4 . \tag{5}$$

f_4 is the focal length of the second lens system, as shown in Fig. 1b.

The principle of AMDs works well for rectangular, hexagonal and triangular shapes, where the entire surface of the diffuser can be covered without any dead space. Note that in the far field the generated pattern will be increased by convolution with the source size and also by diffraction at each aperture. The influence of diffraction at the aperture is given by

$$\sin \varepsilon = \lambda/A \ , \tag{6}$$

where ε is the deflection angle, λ is the wavelength and A is the diameter of the aperture. Thus, AMDs are arrays of elements, where each element is typically larger than $100\,\mu$m to reduce the influence of diffraction at the aperture.

3   Scaling Law for Focusing Lenses and Far-field Diffusers

In principle, an AMD can be considered as a focusing lens. However, there are important differences concerning the optimum structure, the fabrication tolerances and the scaling behavior. Our considerations are based on the thin-element approximation, which assumes a linear relation between the wavefront deformation and the surface profile of the optical element. This approximation is justified, since the deflection angles in our applications are less than $10°$.

In the thin-element approximation a lens is described by its transmission function $t_1(x_1, y_1)$, as shown in (1). The optimum phase function $\Phi(r)$ of a focusing lens is given by

$$\Phi(r) = \frac{2\pi}{\lambda} \left(\sqrt{r^2 + f^2} - f \right) . \tag{7}$$

On the other hand, a diffuser that generates a flat-top far field is described by

$$\Phi(r) = \frac{2\pi}{\lambda} \frac{r^2}{2f} \ , \tag{8}$$

with $r^2 = x_1^2 + y_1^2$, wavelength λ and focal length f. The far field is uniform because the sine of the local deflection angle θ depends linearly on the position r:

$$\sin \theta(r) = \frac{\lambda}{2\pi} \frac{\partial \Phi}{\partial r} = \frac{r}{f} \ . \tag{9}$$

For further discussions below we use $\theta \approx \sin \theta$.

For small apertures, (7) and (8) give the same results. The main difference is the scaling behavior.

It is well known that for a given aperture and lens shape the geometrical aberrations scale with the size of the lens. Thus a smaller lens has fewer aberrations. AMDs behave differently, because the angular spectrum generated by the lens remains unchanged (as long as the influence of diffraction is negligible).

The wavefront error ΔW is an important value to characterize the quality of an optical system. A lens is considered as diffraction limited if $\Delta W < \lambda/4$.

Here, we consider deviations ΔW that are caused by fabrication errors. Assuming a lens with diameter D, which generates a wavefront error of maximum ΔW_m (over the distance $D/2$), then the outgoing beam deviates by the angle

$$\Delta\theta = 2\Delta W_m/D . \tag{10}$$

In the case of a focusing lens, the spot size is increased by

$$\Delta s = f\Delta\theta = \frac{2f}{D}\Delta W_m = 2f\#\Delta W_m , \tag{11}$$

where f is the focal length. Equation (11) shows that for a given f-number ($f\#$), we get the same error Δs, independently of the scaling factor. On the other hand, the angular spectrum (far field) is enlarged by $\Delta\theta$ (equation (10)). $\Delta\theta$ increases for smaller lenses (smaller D).

In order to illustrate the influence of lens errors (e.g. etch depth errors in micro-optics) on the far field, we calculate some examples.

We consider a rotationally symmetric lens (focal length f) which generates a wavefront deformation $\Delta W(r)$. The wavefront W can now be written as

$$W(r) \approx \frac{\Phi(r)\lambda}{2\pi} = \frac{1}{2f}r^2 + \Delta W(r) \tag{12}$$

and

$$\Delta W(r) = a_2 r^2 + a_4 r^4 , \tag{13}$$

with $r^2 = x_1^2 + y_1^2$. The coefficient a_2 of the quadratic term describes the defocusing error and the coefficient a_4 describes spherical aberration. The error $\Delta\theta$ in the far field can be obtained by differentiating (13):

$$\Delta\theta = 2a_2 r + 4a_4 r^3 . \tag{14}$$

Equations (13) and (14) show another important difference between a focusing element and a far-field element. That is, the size of the focus spot can be reduced by defocusing, whereas in the case of far-field elements defocusing generates an error $\Delta\theta$ in the width of the angular spectrum. This error can be estimated from (14). If the error is only defocusing (i.e. $a_4 = 0$), we get a relative error of

$$\frac{\Delta\theta}{\theta_{max}} = \frac{4\Delta W_m}{D\theta_{max}} . \tag{15}$$

ΔW_m is the maximum deformation of the wavefront and $2\theta_{max}$ ($= 2\times$ numerical aperture) is the width of the far field generated by the lens; D is the lens diameter. The relative error becomes two times larger if spherical aberration is the origin of ΔW_m.

From (12) we can also estimate the uniformity error. For the deflection angle $\theta(r)$

$$\theta(r) = \frac{r}{f} + 2a_2r + 4a_4r^3 . \tag{16}$$

The defocusing term can be included in an effective focal length f^*, i.e.

$$\frac{1}{f^*} = \frac{1}{f} + 2a_2 . \tag{17}$$

The geometrical uniformity error in the far-field distribution can now be found by differentiation:

$$\frac{dr}{d\theta} = \frac{1}{1/f^* + 12a_4r^2} \approx f^*(1 - 12a_4r^2f^*) = I_{mean} + \Delta I . \tag{18}$$

ΔI is the deviation from the mean intensity I_{mean}.

Considering the influence of spherical aberration with a maximum wavefront deviation of ΔW_m, we get a uniformity error of

$$\frac{\Delta I}{I_{mean}} = \frac{24\Delta W_m}{\theta_{max} D} . \tag{19}$$

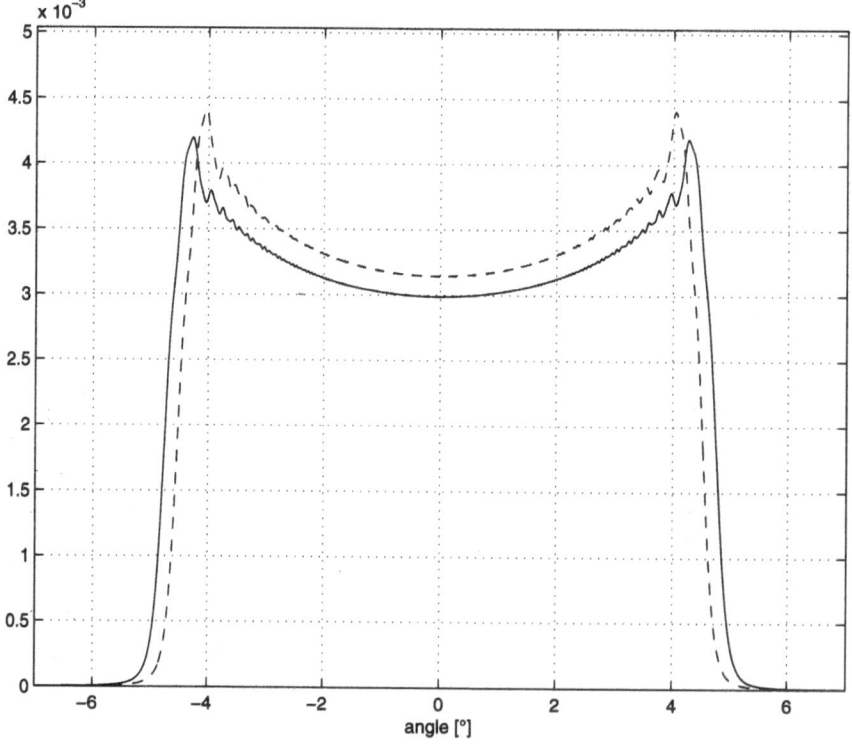

Fig. 2. Influence of defocusing on the far field; the dashed line corresponds to the defocused lens

The influence of defocusing (equation (15)) and spherical aberration (equation (19)) are illustrated in Figs. 2 and 3, respectively. Note that (15) and (19) are geometrical estimations, whereas Figs. 2 and 3 show calculations according to (4). For all calculations the wavelength is $\lambda = 1\,\mu\text{m}$ and the divergence is $s = 5\,\text{mrad}$.

Figure 2 shows the influence of defocusing. In one curve (solid line), defocusing compensates a wavefront deformation of $\Delta W_m = 1.5\lambda$ at the rim. The lens diameter is $D = 2\,\text{mm}$. Both curves are similar, which indicates that the defocusing affects the maximum angle but not the uniformity. This is in good agreement with the prediction by (18) in which the deviation from the uniformity is independent of the defocusing terms. The relative change of the maximum angle corresponds to (15).

The far field of a lens which has only spherical aberration ($\Delta W_m = 1.5\lambda$) is shown in Fig. 3. The lens has a diameter of $D = 500\,\mu\text{m}$ (dashed line) and $2\,\text{mm}$ (solid line). The diffuser generates an angular spectrum of $\theta_{max} = \pm 5°$.

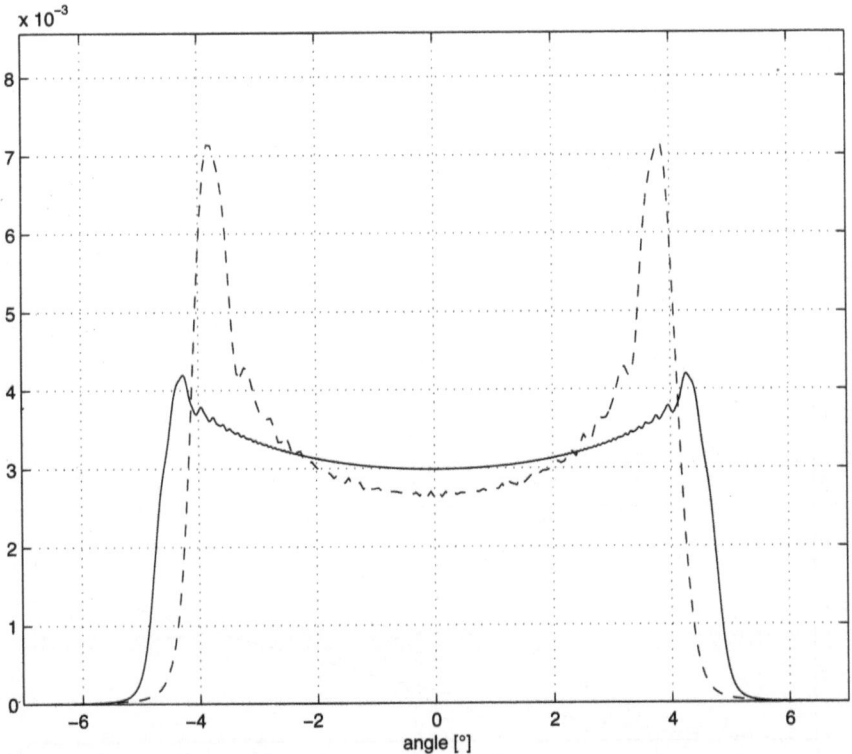

Fig. 3. Far field of two diffusers with maximum spherical aberration of $\Delta W_m = 1.5\lambda$, but with two different lens diameters of $0.5\,\text{mm}$ (dashed line) and $2\,\text{mm}$ (solid line)

4   Examples of Realized Diffractive AMDs

The phase function $\Phi(x, y)$ can be implemented as a refractive or diffractive element. A refractive element generates the phase distribution Φ by varying the optical path length through a phase plate. In the case of a diffractive element, the phase function is mainly generated by the position and the period of a local grating [2]. The shape of the grating period determines the efficiency of the element, which is the amount of light that goes into a particular diffraction order.

Diffractive optics has the advantage of higher flexibility for arbitrary structures and the deflection angles can be controlled very accurately. The drawback is the amount of light at the zeroth order. Figure 4 shows the amount of light at the zeroth order versus the etch depth error. We have used (20) for the calculation [7]:

$$\eta_0 = \left[\frac{\sin(\Delta\varphi/2)}{M \sin(\Delta\varphi/2M)} \right]^2 . \tag{20}$$

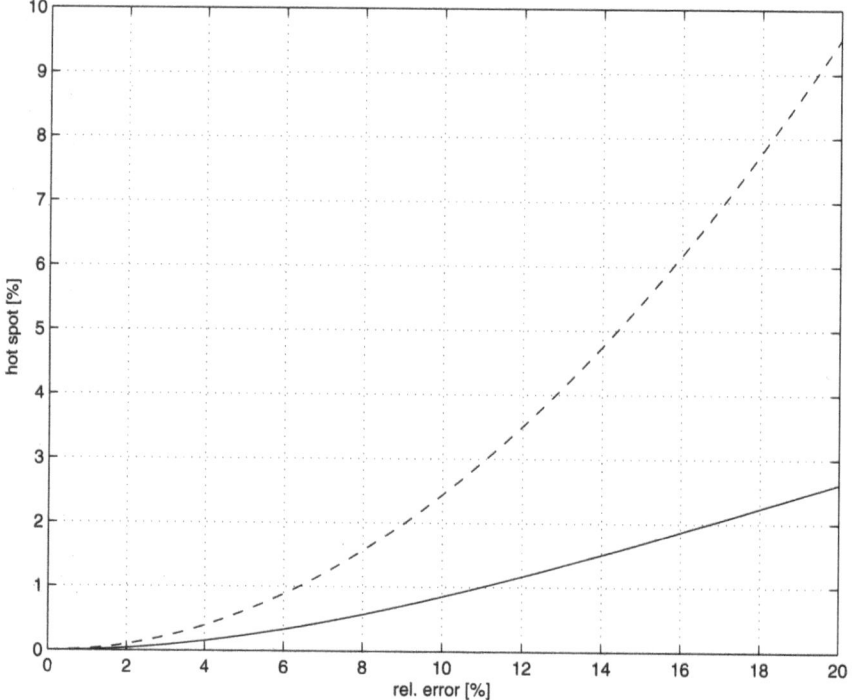

Fig. 4. Efficiency of the zeroth order as a function of the relative etch depth error for a two-level (*dashed line*) and eight-level (*solid line*) diffractive diffuser

η_0 is the diffraction efficiency of zeroth order, $\Delta\varphi$ is the phase error and M is the number of phase steps. The amount of light at zeroth order should not exceed 0.5%.

Binary diffractive elements with only two phase levels usually have low diffraction efficiency (40.5%). For the generation of symmetrical intensity distributions, however, both the plus and the minus first diffraction order contribute to the desired intensity distribution, increasing the diffraction efficiency ($\sim 80\%$). To improve the efficiency, multi-level elements have to be employed.

Different diffusers have been fabricated and tested. The far-field distribution measurements were made on fused silica elements at a wavelength of $\lambda = 248\,\mathrm{nm}$ using a KrF excimer laser. All elements were coated with an antireflection layer for this wavelength. Characteristic magnitudes are the homogeneity and the efficiency. The shapes of the far field were rectangular, annular, hexagonal and trapezoidal. The last two were also made as multipol elements. The rectangular-shaped elements were fabricated as two-level elements because of the desired angular spectrum of $\pm 7°$; the other elements were eight-level diffractive elements. The minimum feature size of all elements is $1\,\mu\mathrm{m}$. All elements had a parabolic phase profile, except the annular element. The multipol elements have to be considered as off-axis lenses. Since there is more scientific interest in non-conventional-shaped diffusers, only measurements on eight-level diffusers are presented here.

Eight-level elements are required if the influence of the zeroth order on the far field is very critical and if the desired efficiency should exceed 90%.

A hexagonal-shaped diffuser is the best compromise for adapting a circular field in combination with a high filling ratio. The generated angular spectrum is shown in Fig. 5. The efficiency is 92% which is close to the theoretical limit of 95% for an eight-level diffractive phase element.

In some applications illumination at an oblique angle is advantageous. For this case off- axis Fresnel zone lenses were designed and fabricated. As an example, we present here a ring diffuser which has a phase profile deviating from a parabola.

In general, toric lenses are used for focusing light into a ring. Their phase function is represented by

$$\Phi(r) = \frac{\pi}{f\lambda}(r - r_0)^2 \tag{21}$$

with focal length f and lateral shift r_0.

Due to the rotational symmetry the amount of light in the different angle regions scales with the radius, leading to a linearly increasing instead of a flat-top intensity distribution in the far field. The parts of the ring which are generated by inner parts of the toric lens have less intensity than the outer parts. Adding a cubic phase term to (21) corrects this non-uniformity. Figure 6 shows the well-achieved flat-top distribution in the angular spectrum of the ring diffuser.

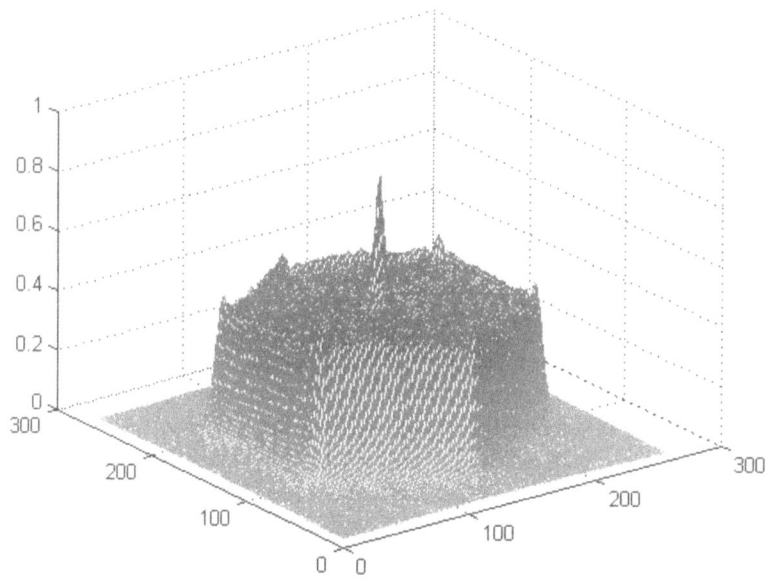

Fig. 5. Far field of a hexagonal-shaped diffuser. The element has been fabricated at CSEM Neuchâtel, Switzerland

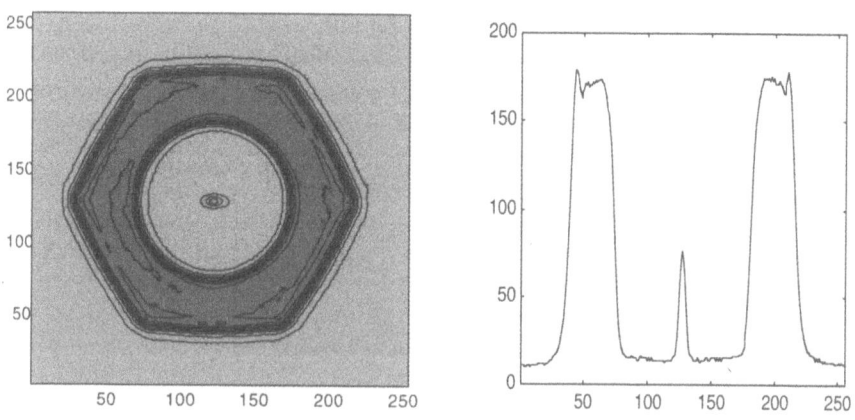

Fig. 6. Far-field and line scan of diffuser that generates a ring pattern. The elements have been fabricated at CSEM Neuchâtel, Switzerland

5   Conclusion

AMDs are phase elements for the generation of flat-top intensity distributions, in our case for excimer laser applications ($\lambda = 248\,\text{nm}$). The phase elements are arrays of micro-lenses with arbitrary aperture shape. We investigated the influence of surface deviations like defocusing and spherical aberration on the far field. Spherical aberration causes errors in the width of the far field and additional deviations from a uniform flat-top distribution. Defocusing has no influence on the uniformity. Since we are interested in arbitrary far-field patterns and highly accurate deflection angles, we fabricated the diffusers as diffractive elements. Sophisticated diffractive AMDs were designed at our Institute (IMT-NE), fabricated at CSEM Neuchâtel and characterized at Carl Zeiss Oberkochen. The eight-level elements always showed more than 90% efficiency. The energy content at the zeroth order was less than 0.5%.

Acknowlegdements

We are indebted to Peter Blattner and Wolfgang Singer for fruitful discussions. For the fabrication and measurements Philippe Regnault from CSEM Neuchâtel is also acknowledged. We thank Johannes Wangler and Manfred Maul from the Carl Zeiss Oberkochen Company for the characterization.

This work was supported by Carl Zeiss Oberkochen, Germany, and the Swiss Priority Program 'Optique'.

References

1. Schmitz, E.-H. (1981) Handbuch zur Geschichte der Optik, Bd. 1: von der Antike bis Newton. Verlag J.P. Wayenborgh, Bonn
2. Herzig, H.P. (Ed.) (1997) Micro-Optics: Elements, Systems, and Applications. Taylor & Francis, London
3. Turunen, J., Wyrowski, F. (Eds.) (1997) Diffractive Optics for Industrial and Commercial Applications. Akademie Verlag, Berlin
4. Singer, W., Herzig, H.P., Kuittinen, M., Piper, E., Wangler, J. (1996) Diffractive Beamshaping Elements at the Fabrication Limit. Opt. Eng. 35: 2779–2787
5. Ogawa, T., Uematsu, M., Uesawa, F., Kimura, M., Shimizu, H., Oda, T. (1995) Sub-quarter Micron Optical Lithography with Practical Super Resolution Technique. Proc. SPIE 2440: 772–783
6. Goodman, J.W. (1996) Introduction to Fourier Optics. McGraw-Hill, New York
7. Ricks, D.W. (1993) Scattering from Diffractive Optics. In: Lee, S.H. (Ed.) Diffractive and Miniaturized Optics. Critical Reviews of Optical Science and Technology, Vol. CR49, pp. 187–211

Optical Properties of Quasiperiodic Structures: Linear and Nonlinear Analysis

M. Bertolotti and C. Sibilia

Summary. The transmission properties of self-similar optical multilayer structures are discussed in both the c.w. and the pulsed time domain, when the input light intensity is low enough for any nonlinear effect to be neglected. The considered structures are obtained by alternating two dielectric layers of different refractive indices such that the highest refractive index layers belong to some fractal set. The triadic Cantor and the Fibonacci sets are considered as examples. The transfer matrix method is used and some of its properties are discussed. Then the nonlinear transmission properties are discussed when the input level intensity is so high to induce third-order nonlinear polarization in the dielectric media constituting the structures. A mesoscopic model is also introduced.

1 Introduction

During the past few years, two- and three-dimensional periodic dielectric structures, usually referred to as photonic band gap (PBG) crystals, have been intensively studied, theoretically as well as experimentally [1, 3]. An essential property of photonic crystals is the existence of forbidden frequency bands, from which propagating modes, spontaneous emission, and zero-point fluctuations are all absent. At the same time, the electric-field intensity strongly increases near the PBG edges in the frequency domain [4].

Many possible applications in optical devices can be obtained because of the field localization. Field localization in periodical structures can be described through the density of modes (DOM) [5]. The density of modes is strongly increased if defects are introduced inside the structure or it is made quasiperiodic.

One of the peculiar aspects of the fractal structures is the spatial field localization, so it is of interest to investigate the behaviour of quasiperiodic structures, which follow a fractal code, such as, for example, the Cantor code. Recently, many theoretical studies of one-dimensional (1D) quasiperiodic structures realizing a Fibonacci or a Cantor sequence have been performed [6–13], and interesting experimental work has been done [6–9]. In [10] a comparison of the DOM of these structures suggest Cantor multilayer structures, which have a maximum density of modes at the band edge, are preferable over other structures for all applications which can profit from a high density of modes, such as in delay lines or optical limiting [4, 5].

The interesting properties of the structures studied in [6–13] are linked to the properties of self-similar spaces and also related to the possibility of weak localization of photons. The treatment is so general that it can be applied to any kind of waves propagating in a self-similar medium. Localized photons in fractal structures have been called "fractons". The existence of fracton modes [14] has been experimentally proven for acoustic waves in one-dimensional Cantor composites [15]. In optics, scattering and diffraction of fractal objects have been studied by many researchers [16].

In what follows we describe some general properties of self-similar fractal structures. Then we discuss some optical linear and nonlinear properties of specific layered systems.

2   What Are Fractals?

The term "fractal" was introduced by Mandelbrot [17] to describe geometrical objects with no integer dimension. The definition given by Mandelbrot states that a fractal is a self-similar set whose dimension is different from the topological dimension; a self-similar set being an invariant set with respect to a scale change.

Self-similar fractals are generated mathematically by means of a recursive operation of *generators* and *initiators* [17]. A process is defined on an object, called the *initiator*. In the case of the Koch fractal the initiator is a unitary line. A straight-line segment of length 1/3 is erased in the middle of the initiator and an equilateral triangle, without basis, is build on the segment. This operation can be repeated again at the smallest scale: a line of 1/3 is erased again in the middle of each of the 4 segments. This fractal has a scale factor of 3 (Fig. 1).

The triadic Cantor set has a generation procedure very similar to the Kock fractal, the difference is that one removes n segments of length $l_n = (1/3)^n$, without adding any more: we start from a straight-line segment of unit length. Then we "wipe away" the open middle third and we repeat the process on the remaining two segments of length 1/3. Repeating the middle-third wiping-out process over and over again leaves not a single connected segment. In Fig. 2 the first three levels of the Cantor set generation are shown [19]. The optical structures that we will discuss in the following can be suitably constructed by using the criteria used to define the set. For example, a 1D structure can be realized with a multilayered stack of different materials assembled following the fractal code.

One of the interesting phenomena appearing in fractal structures of this type is wave localization such that the field (acoustic, electromagnetic, or other) becomes spatially confined in some suitable regions, or delocalized in some other parts. Many theoretical works have been written on this interesting subject [13, 18]. The fractal properties of the structure lead to a transmission spectrum that exhibits isolated peaks in the middle of frequency band

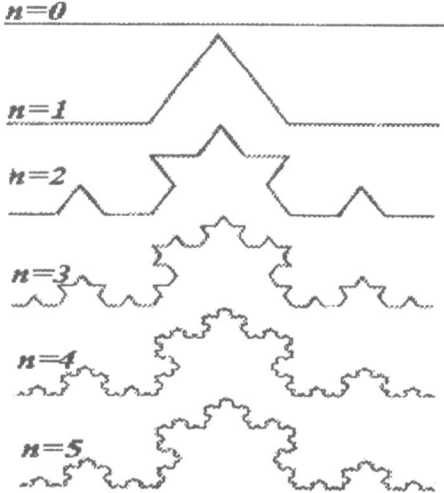

Fig. 1. Example of a self-similar fractal: the Koch fractal

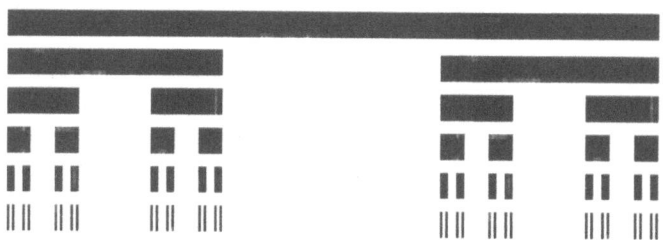

Fig. 2. Example of the generation sequence of a Cantor set

gaps, and it is possible to have a field localization for the mode pattern [20]. All these properties allow fractal structures to appear very attractive from the optical point of view; also in the framework of their nonlinear response.

In what follows, we discuss in more detail the linear and nonlinear optical properties of a multilayer material realized following a Fibonacci or a Cantor code [21–23]. Two different regimes for the optical properties can be distinguished related to the layer thicknesses: one in which interference is predominant (layer thickness larger than a wavelength) and the other (mesoscopic) in which, when the thicknesses are less than a quarter of a wavelength, a simplified approach to the optical transmission properties of the layers can be performed, introducing an effective refractive index which takes care of the optical properties of the structure as a whole [24].

3   Transmission Properties of Filters Realized with a Fractal Code

Let us consider a structure realized by alternating two dielectric layers of different refractive indices such that the highest refractive index layers belong to a triadic Cantor set, as described in Figs. 2 and 3. This is obtained by alternating two nondispersive, planar, dielectric layers of refractive index n_2 and n_1 ($n_2 > n_1$) and of such thickness that their optical paths are the same. Let us take the layer of refractive index n_2 as the initiator. If L is the optical thickness of the initiator, the generator is obtained by substituting the central part of the initiator, having an optical thickness of $L/3$, with a layer of refractive index n_1 and optical thickness $L/3$. The layered structure is obtained by iterating the operation up and down, and stopping the iteration at the Nth step (see Fig. 3). The incident light is assumed to be a plane wave propagating in the direction having an angle θ with respect to the normal at the interfaces planes.

To discuss the optical properties of the structure, the transfer matrix method [19] can be used. In this way it is very easy to evaluate the transmission spectrum for both TE or TM polarization. Examples of transmission spectra for normal incidence are shown in Fig. 4, where the magnitude of the transmission as a function of φ ($\varphi = k_0 L$, where $k_0 = 2\pi/\lambda$ and L is the optical path of the generator) for the first four levels of the Cantor se-

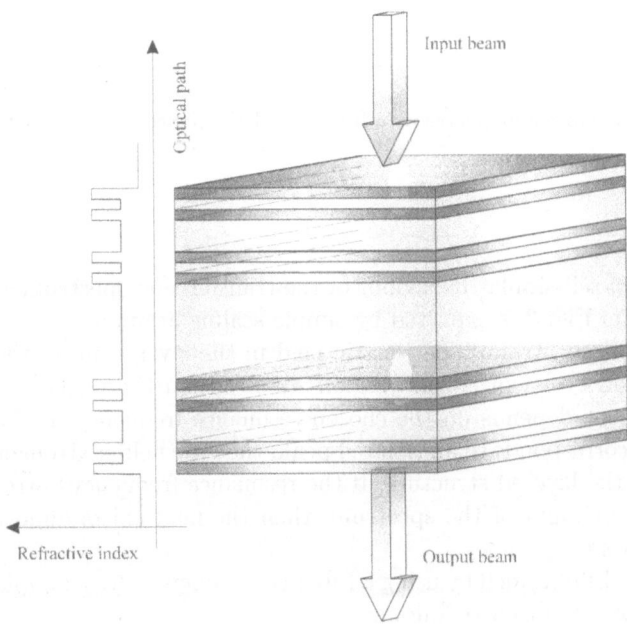

Fig. 3. Layered structure which follows the Cantor set

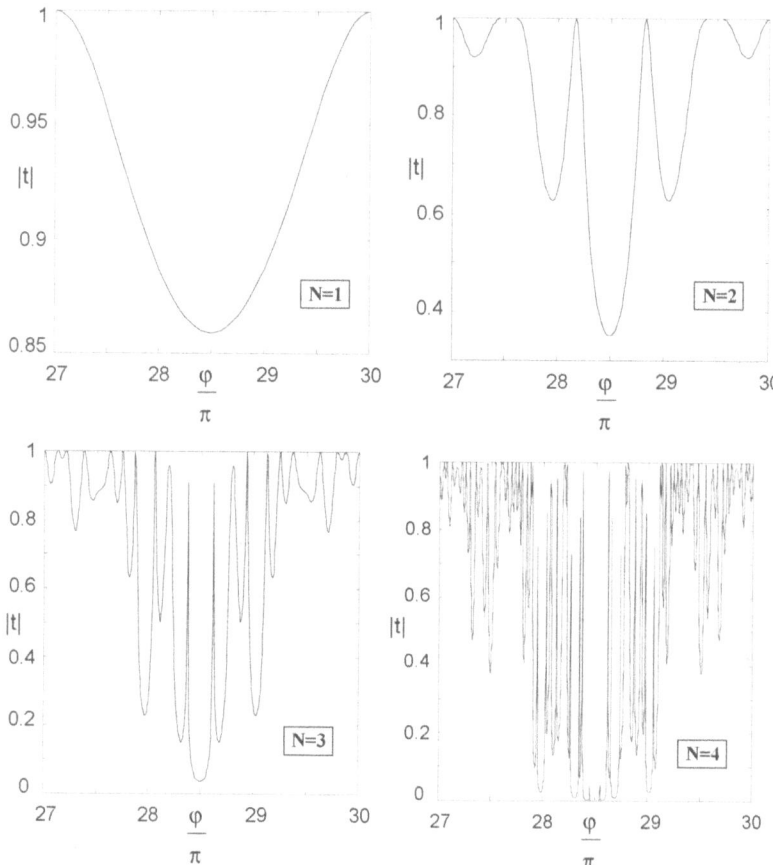

Fig. 4. Examples of transmission spectra as a function of the phase φ for different levels of the sequence

quence is given. Quasi-self-similar behaviour of the transmission spectrum of the fractal structure of Fig. 3 is expected by simple scaling arguments.

Let us consider the spectrum. The electric field in the layers can fill the whole structure or can be localized in a region smaller than the resonator. This different behaviour depends on the chosen resonance frequency. If the resonance frequency corresponds to an isolated peak, then the field is stronger in a selected part of the layered structure. If the resonance frequency corresponds to a broad maximum of the spectrum, then the field exists almost everywhere in the layers.

Similar results can be obtained by using a Fibonacci sequence. An example of Fibonacci sequence is realized as follows:

external medium / BAABAABAABAABAABABAABABAABAABABAABA / substrate,

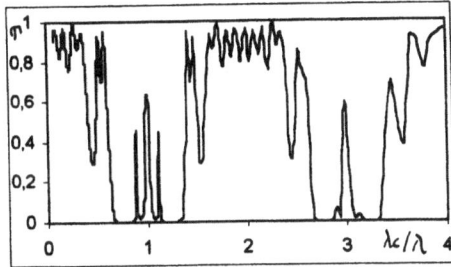

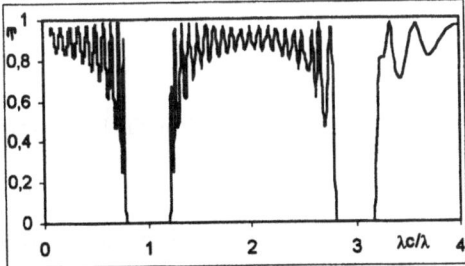

Fig. 5. Example of a transmission spectrum for a layered structure which follows the Fibonacci code (**a**) and periodical structure (**b**)

where each layer has an optical thickness of a quarter of a wavelength; the symbol A means a layer of refractive index n_1 and B means a layer of refractive index n_2. In a Fibonacci multilayer S_j there are F_j layers given recursively by $F_{j+1} = F_j + F_{j-1}$ ($j \geq 1$, with $F_0 = A$, $F_1 = B$). Figure 5 shows the transmission spectrum of such a layered structure, compared with a periodic one. In [23], a detailed description of the Fibonacci filter transmission properties is performed for TE and TM waves, including a comparison with an equivalent Cantor code structure. A more detailed discussion on the localization problem in a Fibonacci multilayer in which a nonlinearity is taken into account is discussed in [13].

4   Properties of a Fractal Filter. Dynamical Map

The dynamical properties of multilayer structures can be studied through the construction of the dynamical map and the research into the invariants of the structure. This method has been applied in [6] to Fibonacci multilayers. The basis of the method is the following. Let us consider a multilayer in which two types of layers A and B are arranged in some sequence. In order to understand the light propagation in this media, first consider an interface of two layers. The electric field for the light in layer A is given by

$$E = E_A^{(1)} \exp[i(k_A^{(1)} x - \omega t)] + E_A^{(2)} \exp[i(k_A^{(2)} x - \omega t)] , \qquad (1)$$

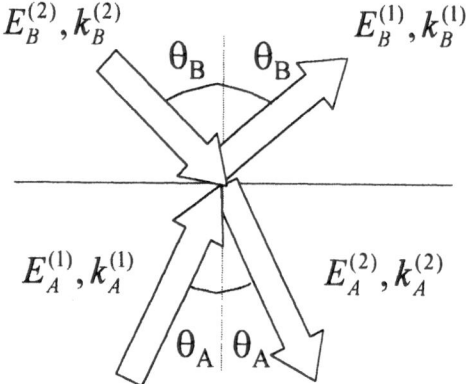

Fig. 6. Scheme of the transmitted and reflected electric field through the interface between two dielectrics

where the superscripts (1) and (2) identify the reflected and incident waves, respectively (Fig. 6). The electric field in layer B is given by the same expression with subscript A replaced by B.

We consider a polarization which is perpendicular to the plane of the light path (TE wave). The appropriate boundary conditions at the interface give

$$
\begin{aligned}
E_A^{(1)} + E_A^{(2)} &= E_B^{(1)} + E_B^{(2)} , \\
n_A \cos\theta_A (E_A^{(1)} - E_A^{(2)}) &= n_B \cos\theta_B (E_B^{(1)} - E_B^{(2)}) ,
\end{aligned}
\tag{2}
$$

where n_A and n_B are the refractive indices of A and B, respectively, and the angles θ_A and θ_B are shown in Fig. 6. Snell's law means $\sin\theta_A / \sin\theta_B = n_B/n_A$.

It is convenient to choose the two independent variables for the light as

$$
E_+ = E^{(1)} + E^{(2)}, \quad E_- = (E^{(1)} - E^{(2)})/i .
\tag{3}
$$

Then (2) gives

$$
\begin{bmatrix} E_+ \\ E_- \end{bmatrix}_B = T_{BA} \begin{bmatrix} E_+ \\ E_- \end{bmatrix}_A ,
\tag{4}
$$

where T_{BA} is given by

$$
T_{BA} = \begin{bmatrix} 1 & 0 \\ 0 & n_A \cos\theta_A / n_B \cos\theta_B \end{bmatrix} .
\tag{5}
$$

Also we define

$$
T_{AB} = T_{BA}^{-1} = \begin{bmatrix} 1 & 0 \\ 0 & n_B \cos\theta_B / n_B \cos\theta_A \end{bmatrix} .
\tag{6}
$$

The matrices T_{BA} and T_{AB} represent the light propagation across interfaces $B \leftarrow A$ and $A \leftarrow B$, respectively. Propagation within one layer is represented by

$$T_A = \begin{bmatrix} \cos \delta_A & -\sin \delta_A \\ \sin \delta_A & \cos \delta_A \end{bmatrix} , \tag{7}$$

for a layer of type A, and the same expression for T_B in which δ_A is replaced by δ_B. The phases are given by

$$\delta_A = n_A k d_A / \cos \theta_A$$

and

$$\delta_B = n_B k d_B / \cos \theta_B , \tag{8}$$

where k is the wave number in vacuum, d_A and d_B are the thicknesses of the layers.

Now the structure is completely determined by the proper combination of the matrices describing the light propagation through the interfaces and its transfer from one interface to the other. At any level the total matrix M_j describing the structure is easily calculated with knowledge of the matrices T_A, T_B, T_{AB}, T_{BA}. If we call u the diagonal element and w the outdiagonal element, the transfer matrix has two useful properties:

$$M_j = \begin{pmatrix} u_j & -w_j \left(\dfrac{n_A}{n_B} \right) \\ w_j \left(\dfrac{n_B}{n_A} \right) & u_j \end{pmatrix} , \tag{9}$$

$$\det(M_j) = 1 , \tag{10}$$

irrespective of j. Property (9) allows us to determine all the matrices to any order from only u_j and w_j. Equation (10) is the field conservation in the passage from one material to the other and during field propagation.

The dynamical map of the structure is defined by the relation between $M_j(x)$ and $M_{j-1}(x)$. Considering the different situations, an invariant can be constructed, with the elements of the matrix M_j, which is characteristic of the particular structure. Moreover, the condition of periodicity of the transmission spectrum as a function of $\varphi = kL$ that we have seen, for example, in Figs. 4a–d, gives a limiting condition on the trace of the matrix M_j: the absolute value of the trace must be less or equal to 2, namely, $|\mathrm{Tr}(M)/2| = |u| \leq 1$.

We will now show practical applications of these considerations to the cases of Cantor or Fibonacci structures. Let us first consider a Cantor set made with two materials A and B. We choose thicknesses so as to realize a perfect triadic Cantor set with the optical paths of different layers made the same length:

$$l_A = l_B = l ,$$
$$l_A = n_A d_A / \cos \theta_A . \tag{11}$$

Assuming the external layer is made with material A, we find the matrices M_N for different levels of Cantor code sequence ($N = 1, 2, 3, \ldots$) are

$$
\begin{aligned}
M_1(x) &= T_{AB}T_B(x)T_{BA} \ , \\
M_2(x) &= T_{AB}T_B(x)T_{BA}T_A(x)T_{AB}T_B(x)T_{BA} = M_1(x)T_A(x)M_1(x) \ , \\
M_3(x) &= M_2(x)T_A(x3)M_2(x) \ , \\
M_j(x) &= M_{j-1}(x)T_A(x3^{j-2})M_{j-1}(x)
\end{aligned}
\tag{12}
$$

where $x = lk$.

The dynamical map of the structure is defined by

$$
M_j(x) = M_{j-1}(x) \begin{pmatrix} \cos(x) & -\sin(x) \\ \sin(x) & \cos(x) \end{pmatrix}, M_{j-1}(x)
\tag{13}
$$

with M_1 given by (12) as the initial condition. Performing the calculation we find

$$
w_1(n_A/n_B) = [n_A \cos(\theta_A)/n_B \cos(\theta_B)] \sin(x)
\tag{14}
$$

and an analogous relation for $w_1(n_B/n_A)$. Therefore we define

$$
\begin{aligned}
&\mathrm{Tr}(M_j(x)) = 2u_j(x) \ , \\
&\Delta w_j = w_j(n_A/n_B) + w_j(n_B/n_A) \ , \\
&\Delta w_1 = w_1(n_A/n_B) + w_1(n_B/n_A) \\
&\quad\quad = \sin(x)[n_B \cos(\theta_B)/n_A \cos(\delta_A) + n_A \cos(\theta_A)/n_B \cos(\theta_B)] \ , \\
&\mathrm{Tr}(M_1(x)) = 2\cos(x)n_A \ .
\end{aligned}
\tag{15}
$$

For the Cantor set the phases x_j follow the rule

$$
x3^{j-2} = x_j \ .
\tag{16}
$$

By substituting we obtain a simple dynamical map

$$
\begin{aligned}
u_j(x) &= (2u_{j-1}^2(x) - 1)\cos(x_1) - \sin(x_1)u_{j-1}(x)\Delta w_{j-1}(x) \ , \\
w_j(x) &= 2u_{j-1}(x)\Delta w_{j-1}(x)\cos(x_1) - \sin(x_1)[\Delta w_{j-1}^2(x) - 2] \ .
\end{aligned}
\tag{17}
$$

We can replace the element of jth order with the first derivative of the element of $(j-1)$th order. In this way (17) become a system of differential equations which give the map and its eigenvalues, the constant of motions are the so-called invariants.

We can also start from different initial conditions defined through the following matrices

$$
\begin{aligned}
M_1(x) &= T_B(x) \ , \\
M_2(x) &= T_B(x)T_{BA}T_A(x)T_{AB}T_B(x) = M_1(x)T_{BA}T_A(x)T_{AB}M_1(x) \ , \\
M_3(x) &= M_2(x)T_{BA}T_A(x3)T_{AB}M_2(x) \ , \\
M_j(x) &= M_{j-1}(x)T_{BA}T_A(x3^{j-2})T_{AB}M_{j-1}(x) \ .
\end{aligned}
\tag{18}
$$

This representation does not change the properties of the map and its initial conditions; we should only replace $\Delta w_1 = w_1(n_A/n_B) + w_1(n_B/n_A)$ with $\Delta w_1 = w_1(n_A/n_B)n_B \cos(\theta_B)/n_A \cos(\theta_A) + w_1(n_B/n_A)n_A \cos(\theta_A)/n_B \cos(\theta_B)$. The characteristics of the map are still u_j (half trace) and Δw_j. u_j is an independent quantity from the specific definition; it is unchanged because of the conservation of the flow through the structure (Pointing vector). The value of the traces for any j is constant and it is an important parameter of the multilayer. The structure has an invariant [25]

$$I = 2u_{j-1}(x)\Delta w_{j-1}(x)\cos(x_1) - \sin(x_1)(\Delta w_{j-1}^2(x) - 2) , \tag{19}$$

which may be explicitly written, replacing the initial data, as

$$I_c = \sin(x)2\cos^2(x)[n_A \cos(\theta_A)/n_B \cos(\theta_B) + n_B \cos(\theta_B)/n_A \cos(\theta_A)] \\ + 2 - y\sin^2(x) , \tag{20}$$
$$y = [n_A \cos(\theta_A)/n_B \cos(\theta_B) + n_B \cos(\theta_B)/n_A \cos(\theta_A)]^2 .$$

For the case of a Fibonacci structure with [6]

$$M_1 = T_A , \quad M_2 = T_{AB}T_B T_{BA} T_A \tag{21}$$

one has

$$M_j = M_{j-2}M_{j-1} \tag{22}$$

with the initial conditions (21). The invariant in this case is [6]

$$I = \frac{1}{4}\sin^2\delta_A \sin^2\delta_B \left[\frac{n_A \cos\delta_A}{n_B \cos\delta_B} - \frac{n_B \cos\delta_B}{n_A \cos\delta_A}\right]^2 .$$

The transmission coefficient T is given in terms of the matrix M_j as

$$T = 4/(|M_j|^2 + 2) , \tag{23}$$

where $|M_j|^2$ is the sum of the squares of the four elements of M_j. Equation (23) is valid in general and not only for a Fibonacci structure. This is a quantity measured experimentally and has a rich structure with respect to a variation of either the wavelength of the light or the number of layers.

Going back to the Cantor multilayer, extreme cases are $u_j = 0$ and $u_j = 1$. In the first case

$$u_j = 0 \quad u_{j-1} = 0 \quad \forall j \Rightarrow \\ u_j(x) = (2u_{j-1}^2(x) - 1)\cos(k_1) - \sin(k_1)u_{j-1}(x)\Delta w_{j-1}(x) \tag{24} \\ = -\cos(k_j) = 0 .$$

From (11) and (15), we obtain

$$\cos\left(3^{j-2}\frac{knd}{\cos(\theta)}\right) = 0 \Rightarrow 3^{j-2}\frac{knd}{\cos(\theta)} = 3^{j-2}\frac{2\pi nd}{\lambda \cos(\theta)} = \pi(m + 1/2) , \tag{25}$$

where m is an integer and j is the recursive order of the structure. Equation (25) can be translated into a relation on the optical path, which for normal incidence, is

$$(m + 1/2)\lambda/2 = 3^{j-2}nd \ . \tag{26}$$

Fixing j, we always have an integer m that satisfies relation (26) when $nd = 3^i\lambda/4$, where i is any integer. Then the structure fulfils the condition (26) if the layers of the constituent materials are selected to be $\lambda/4$, $3\lambda/4$, $3^i\lambda/4$, where λ is the incident radiation. With condition (26) the quasiperiodicity is most effective, and the absolute value of the invariant (20) is maximum. For a normally incident wave ($\theta_A = \theta_B = 0$) and $nd = \lambda/4$, then $\cos(\theta_A) = \cos(\theta_B) = 1$ and $\cos(x_j) = 0$ and $\sin(x_j) = 1$. Equation (20) becomes $I_c = 2 - (n_A/n_B + n_B/n_A)^2$.

In the opposite limit, $|u| = 1$,

$$u_j = 1 \quad u_{j-1} = 1 \quad \forall j \Rightarrow |u| = 1 = |-\sin(x_j)\Delta w + \cos(x_j)| \ . \tag{27}$$

Equation (27) is effective for $x_j = m\pi$, with m integer, that is for layers built so that their optical lengths satisfy $\lambda/2$, $3\lambda/2$, $3^i\lambda/2$. With condition (27), $I_c = 0$, the transmission is maximum and the periodicity is very bad. Condition (24) offers the best situation for the construction of the layers if we want the quasiperiodicity to be the most effective.

Similar results can be found for a Fibonacci multilayer [6]. The map for Fibonacci, if we define $y = \mathrm{Tr}(M_j)/2$, is

$$y_{j+1} = 2j_j j_{j-1} - y_{j-2} \tag{28}$$

and the invariant is

$$I_f = y_{j+1}^2 + y_j^2 + y_{j-1}^2 - 2y_{j+1}y_j y_{j-1} \ . \tag{29}$$

For $\lambda/4$ layers the invariant I_f is maximum and the quasiperiodicity is the most effective. The dynamical map of the Fibonacci structure is a one-dimensional noninvertible map. The quasilocalization of the light wave has been demonstrate in the Fibonacci multilayer by the self-similarity of the transmission coefficient under the given boundary conditions [6].

If we consider a Cantor sequence and we compare the spectra of the optimized structures, the fractal nature appears not only in the sequences but also in the resultant spectra. The spectrum for some N order appears quasi-self-similar in itself (Figs. 4a–d). We can differentiate the figure relative to the order $N = 4$ from the one relative to $N = 3$ only by the value of k_0/π (Fig. 4), that is by a scaling factor. Localization is connected to the similarity between one figure and the others, and inside spectra. The localization follows a power law: this is the only possible explanation for the spectral scale invariance [25].

These advantageous fractal properties of the Cantor multilayer are not present in the Fibonacci multilayer. In the Cantor structure the fractal characteristic increases progressively from the recurrence 1 to 2, 3, 4, with only a few layers. In the Fibonacci structure it is necessary to increase greatly the number of layers to get similar results.

5   Time-Domain Response of the Filter

In the previous section some peculiarities of Cantor and Fibonacci filters have been discussed in terms of the dynamical map, and the transmission spectra have been written as a function of the vacuum wave number $k_0 = \omega/c$. The incident and the transmitted beams can be seen as the Fourier transform of a time-dependent input signal $x(t)$ and a time-dependent output signal $y(t)$, respectively. In what follows we describe the properties of the filters connected to a frequency response when a pulse of finite bandwidth is propagating through them. The signals' Fourier transform will be represented by capital letters, while the small letters will be used to represent signals in the time domain.

The transmission represents the transfer function of a linear and permanent filter [26],

$$H(\omega) = \frac{2}{u_j(\omega) - w_j(\omega, n_A/n_B) - w_j(\omega, n_B/n_A) + u_j(\omega)} , \tag{30}$$

defined in terms of the elements of the transfer matrix the structure (we restrict our consideration to TE polarization). Therefore, the output (transmitted) signal is given by

$$Y(\omega) = H(\omega)X(\omega) , \tag{31a}$$

which can be expressed in the time domain as

$$y(t) = h(t) * x(t) . \tag{31b}$$

The operator $*$ means a convolution integral between the operands. The function $h(t)$ is the filter impulsive response and corresponds to the inverse Fourier transform of the transfer function (30).

The input signal may be considered as an amplitude modulation of the carrier at some optical frequency ω_0:

$$x(t) = x_m(t) \cos(\omega_0 t + \varphi) , \tag{32}$$

in which $x_m(t)$ represents the pulse shape (envelope) and φ is a constant temporal phase shift. It is useful to note some definitions about signal theory. The input signal can be expressed in terms of the analytical signal

$$x^+(t) = \frac{1}{2\pi} \int_0^{+\infty} X(\omega)e^{i\omega t} \, d\omega ,$$

$$x(t) = 2\,\mathrm{Re}\{x^+(t)\} , \tag{33}$$

or in terms of the complex envelope $\underline{x(t)} = x_m(t)e^{-i\varphi}$,

$$x(t) = \mathrm{Re}\{\underline{x(t)}e^{-i\omega_0 t}\} . \tag{34}$$

From (33) and (34), we have

$$x(t) = 2x^+(t)e^{-i\omega_0 t} , \tag{35}$$

which in the frequency domain becomes

$$X(\omega) = 2X^+(\omega + \omega_0) . \tag{36}$$

This equation can be applied to each signal, even to the impulsive response $h(t)$. Therefore, with the help of (32), we have

$$Y(\omega) = \frac{1}{2} H(\omega) X(\omega) . \tag{37}$$

This means that also the output (transmitted) signal is an amplitude modulation of the carrier at the optical frequency ω_0,

$$y(t) = \text{Re}\{y(t)e^{-i\omega_0 t}\} , \tag{38}$$

in which the Fourier transform of the complex envelope is given by (37).

In order to study the pulse compression through, for example, a Cantor Fabry–Perot filter, we consider a gaussian imput pulse shape

$$x_m(t) = Ae^{-(1/\tau)^2} \tag{39}$$

and the constant phase shift $\varphi = 0$ so that the input complex envelope is

$$x(t) = Ae^{-(1/\tau)^2} , \tag{40}$$

where A is the pulse peak amplitude.

When the pulse passes through the filter, its shape remains unchanged only if the filter has a bandwidth as large as the frequency spectrum of the pulse. The narrower the bandwidth is with respect to the spectrum of the pulse the larger the pulse is, because the higher spectral components are filtered out. If the filter is characterized by a more complex transmission function, the shape of the pulse can dramatically change because of the attenuation of some spectral components that are not necessarily the higher ones. There can even be a modulation of the original pulse if most frequencies are attenuated except a particular frequency or a group of frequencies centred around a narrow transmission peak.

The results obtained with a Cantor structure show peculiar compression behaviour that is not present in a more traditional periodic structure [26]. To evaluate the compression capacity of the Cantor device with respect to the length of the pulse, it is necessary to introduce some new parameters. The first parameter is the compression ratio (CR), which is the ratio between the width at half height of the input pulse and the width at half height of the output pulse. If compression behaviour is present, the output pulse is obviously narrower than the input pulse and the CR is greater than unity. In Fig. 7 the CR as a function of the input pulse value of $1/\tau$ is shown (for a level

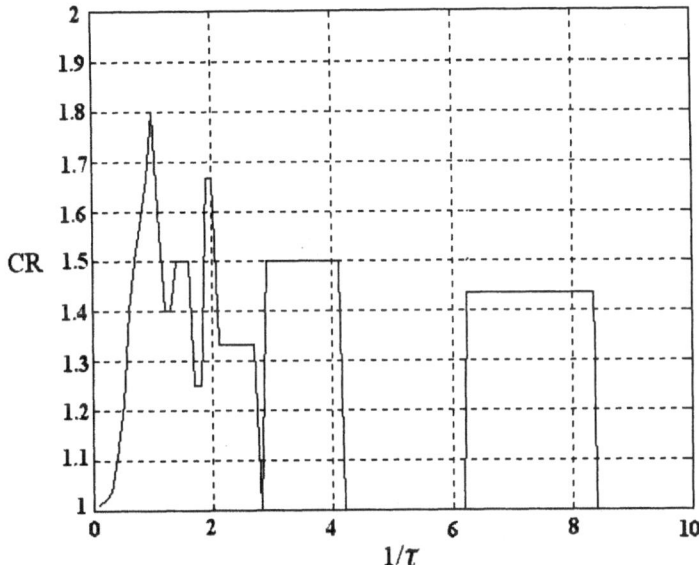

Fig. 7. Compression ratio (CR) as a function of the input pulse of bandwidth $1/\tau$

$N = 3$ of the Cantor set). It is possible to see that the CR is maximum when $1/\tau = 1$. The irregular behaviour of the curve is due to the fact that when the input pulse becomes narrower, its transform becomes wider, and effects different peaks of the transmission spectrum of the device. This behaviour results in sudden transmission of the pulse.

6   Nonlinear Model of the Filter

One of the advantages coming from field localization in fractal structures is the sensitivity to input intensity when nonlinear materials are considered. To describe nonlinear propagation in the Cantor filter when a third-order material without saturation is considered we may use a formalism based on the application of the slowly varying approximation (SVEA) and the omission of the third harmonic generated in the structure and of nonlinear terms appearing in the boundary conditions [27]. Studies of optical nonlinear properties of a single-layer nonlinear Fabry–Perot resonator have shown that this standard approach is valid provided the following two conditions are both met: $n_2 I_{\mathrm{cav}} \ll n_0$ and $D > \lambda/n_0$, where n_0 and n_2 are the linear and nonlinear parts of the refractive index of the considered structure, I_{cav} is the cavity irradiance level, and D is the length of the resonator [27]. In what follows we should take into account this limit of validity of the standard approach. The problem is conveniently treated by using the transfer matrix method.

To describe the behaviour of each single nonlinear layer, in the following we separate the fields written in (2) and (3) into modulus and phase, and if we assume a plane wave with a TE polarization, we write the fields as a superposition of forward and backward waves, starting from E and H components at each interface, as

$$Ey = a + b ,$$
$$Hx = \frac{1}{\eta}(a - b) ,$$

(41)

and, for the hth layer, the waves can be expressed as

$$a_h(z) = \hat{a}_h e^{i(k_0 n_h z + \phi_{ah})} ,$$
$$b_h(z) = \hat{b}_h e^{-i(k_0 n_h z + \phi_{ah})} ,$$

(42)

where η is the characteristic impedance of the medium, k_0 is the free-space wave numbers, n_h is the complex index of refraction (to include losses), and ϕ_a and ϕ_b are additional phase shifts. The link between the field in the hth medium and the fields into the $(h + 1)$th is given by applying the suitable boundaries conditions for the electric and magnetic field at the interface $z = \delta_h$:

$$\hat{a}_h e^{i(k_0 n_h z + \phi_{ah})} + \hat{b}_h e^{-i(k_0 n_h z + \phi_{bh})} = \hat{a}_{h+1} + \hat{b}_{h+1} ,$$
$$\hat{a}_h e^{i(k_0 n_h z + \phi_{ah})} - \hat{b}_h e^{-i(k_0 n_h z + \phi_{bh})} = (\hat{a}_{h+1} + \hat{b}_{h+1})\frac{n_{h+1}}{n_h} ,$$

(43)

and in a matrix form

$$\begin{pmatrix} \hat{a}_h \\ \hat{b}_h \end{pmatrix} = M_h \begin{pmatrix} \hat{a}_{h+1} \\ \hat{b}_{h+1} \end{pmatrix} ,$$

(44)

where M_h is the scattering matrix between the hth and the $(h+1)$th medium, given by

$$M_h = \frac{e^{-i\phi_{bh}}}{t_h} \begin{pmatrix} \exp[-i(\phi_h^l + \phi_h^{nl})] & \Gamma_h \exp[-i(\phi_h^l + \phi_h^{nl})] \\ \Gamma_h \exp[-i\phi_h^l] & \exp[-i(\phi_h^l)] \end{pmatrix}$$

(45)

with

$$t_h = \frac{2n_h}{n_h + n_{h+1}} , \qquad r_h = \frac{n_h - n_{h+1}}{n_h + n_{h+1}} .$$

The linear phase is

$$\varphi_{Lh} = k_0 n_h \delta_h .$$

(46)

The nonlinear phase shift ϕ_h^{nl} is obtained, as given in more detail in [21,27], by starting from the nonlinear Maxwell equations. It turns out to be proportional to the intensity inside each layer:

$$\phi_h^{nl} = \phi_{ah} - \phi_{bh} = 3k_0 n_{2h}^{nl} \int_0^{\delta_h} (|\hat{a}_h|^2 + |\hat{b}_h|^2)\, dz .$$

(47)

The intensity present in each layer is defined as

$$I_h(z) = \frac{1}{2}\varepsilon_0 c n_h (|a_h(z)|^2 + |b_h(z)|^2) , \tag{48}$$

and the average intensity is given by

$$\bar{I}_h(z) = \frac{1}{\delta_h} \int_0^{\delta_h} I_h(z)\, dz = \frac{\varepsilon_0 c n_h}{2\delta_h} \int_0^{\delta_h} (|\hat{a}_h(z)|^2 + |\hat{b}_h(z)|^2)\, dz .$$

It is also useful to define the effective intensity of each layer as

$$I_{\text{eff}} = \bar{I}_h(z)[2\alpha\delta_h] \tag{49}$$

such that the nonlinear phase shift ϕ_h^{nl} may be written as a function of this variable:

$$\phi_h^{\text{nl}} = 3k_0 n_{2h}^{\text{nl}} \left[|\hat{a}_h|^2 \frac{1 - e^{-2\alpha_h \delta_h}}{2\alpha_h} \right] + \left[|\hat{b}_h|^2 \frac{-1 + e^{2\alpha_h \delta_h}}{2\alpha_h} \right] . \tag{50}$$

To find the nonlinear response of the Cantor filter, a generalization of the dummy-variable method [27] can be used, which is a useful technique for obtaining plots in a nonlinear Fabry–Perot cavity. It consists of the definition of a dummy variable in relation to the input and output intensity.

Examples of nonlinear transmission for fixed spectral values are given in Fig. 8. Examples of the output versus the input intensity is shown for the level $N = 2$ (Fig. 8a) and $N = 3$ (Fig. 8c). The corresponding spectral positions are shown in Figs. 8b,d. A more detailed discussion is presented in [21,23] where a comparison has also been performed with a traditional layered structure, finding a reduction of the input threshold intensity for bistability. Multistable behaviour can also be present depending on the spectral position. These properties make the nonlinear quasiperiodic structures very interesting for the filtering properties.

7   Mesoscopic Layered Structures

In a stratified structure, the fields must be determined by the solution of a set of transfer- matrix equations, with the field calculated in any given layer for the specific incident field. Such an approach is necessary if the layer thicknesses are of the order of the light wavelength; true interference effects then become important. If the layer thicknesses are much less than the wavelength, the multilayer structure can be considered a uniform effective medium. It is then possible to introduce a great simplification in all calculations concerning the layer optical properties. In fact, in this case the propagation of the light through the structure can be described in terms of effective linear and nonlinear optical susceptibilities. We assume that the thickness of each layer is much larger than an atomic dimension but much smaller than the incident wavelength [28].

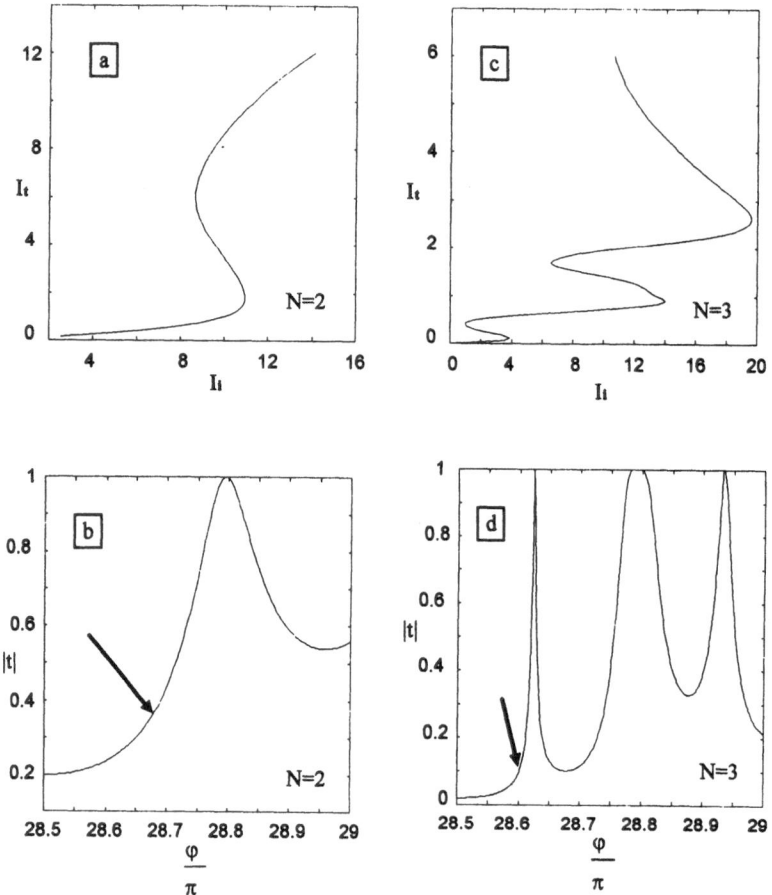

Fig. 8. Nonlinear transmission, at fixed frequency value, as a function of the input intensity assuming a third-order nonlinear medium

Results of the analysis depend critically on the directions of polarization of the incident beam. In particular, if the electric field is TE polarized, then it is spatially uniform within the composite material (because of the boundary condition that states that the tangential component of the electric field must be continuous at an interface); consequently, the optical constants of the composite become simple averages of those of the constituent materials [28].

On the other hand, if the electric field is TM polarized, then the electric field becomes nonuniformly distributed inside the layers of the composite and, taking advantage of boundary conditions at each layer, the effective linear optical constants are given by [28]

$$\frac{1}{n_{\text{eff}}^2} = \frac{f_A}{n_A^2} + \frac{f_B}{n_B^2} . \tag{51}$$

If we have a periodic layered distribution, the volume fraction f_A and f_B of each material is given by

$$f_A = \frac{d_{tot}|_A}{d_{tot}} , \quad f_B = \frac{d_{tot}|_B}{d_{tot}} , \tag{52}$$

where d_{tot} is the total thickness of the structure and $d_{tot}|_j$ $(j = A, B)$ is the total thickness of the structure if the initiator is the material A or B.

If a triadic Cantor structure is considered, then the volume fractions f_A and f_B of each material are given by

$$f_A = \frac{d_{tot}|_A}{d_{tot}} = \frac{2^N d_A}{2^N d_A + (3^N - 2^N) d_B} ,$$
$$f_B = \frac{(3^N - 2^N) d_B}{(3^N - 2^N) d_B + 2^N d_A} , \tag{53}$$

where d_A and d_B are the thicknesses of the smallest layers of the structure. We observe that we have an additional parameter compared with a periodic structure that is the Cantor level N.

The layering produces a large enhancement of the effective refractive index. An example is given in Fig. 9, where the ratio n_{eff}/n_A is presented for several values of the Cantor level as a function of $n_A(\omega)/n_B(\omega)$. In this example, the multilayer is realized with two different materials whose optical paths follow the triadic Cantor code. The same can be done with a Fibonacci code, as discussed in [24], where it is shown that an enhancement of the effective index is found when a nonlinear material is taken into account in one of the layers constituting the structure.

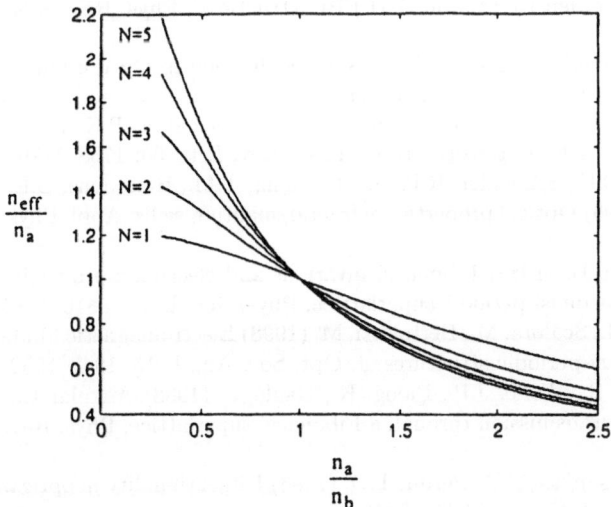

Fig. 9. Ratio of n_{eff}/n_A as a function of n_A/n_B for several values of the Cantor level

8   Conclusions

Among layered composites, fractal materials exhibit very interesting and flexible properties. We have here discussed layered structures following some fractal code. Considering the Cantor layered structure, the resulting structures seem to offer a larger flexibility in the handling of spectral transmission, when compared to periodic structures, which may turn out useful for unique filtering properties, mainly when the nonlinear response of the structure is taken into account. A mesoscopic treatment of the layers can be performed when their thickness is less than the wavelength, giving rise to an interesting enhancement of the effective refractive index. The treatment through an effective index describing macroscopically the whole structure can be extended also to structures of any layer thickness. A recent study is discussing this approach, which looks very interesting [28].

References

1. Joannopulos, J.D., Villeneuve, P.R., Fan, S. (1997) Putting a new twist on light, Nature 386: 143–149
2. Yablonovich, E. (1987) Inhibited spontaneous emission in solid-state physics and electronics, Phys. Rev. Lett. 58: 2059–2062
3. Robertson, W. (1992) Measurement of photonic band structures in a two-dimensional periodic dielectric array, Phys. Rev. Lett. 68: 2023–2026
4. Scalora, M., Dowling, J.P., Bowden, C.M., Bloemer, M. (1994) Optical limiting and switching of ultrashort pulses in Nonlinear PBG Materials, Phys. Rev. Lett. 73: 136–137
5. Benedickson, J.M., Dowling, J.P., Scalora, M. (1996) Analytic expression for electromagnetic mode density of finite 1-D PBG structures, Phys. Rev. E 53: 4107–4121
6. Komoto, M., Sutherland, B., Iguchi, K. (1987) Localization in Optics: Quasi-periodic media, Phys. Rev. Lett. 58: 2436–2438
7. Merlin, A., Bajema, K., Clarke, R., Juang, F.Y., Bhattacharya, P.K. (1985) Quasi-periodic GaAs-AlAs Heterostructures, Phys. Rev. Lett. 55: 1768–1770
8. Gourley, L., Tiggs, C.P., Schneider, R.P., Jr., Brennan, T.M., Hammons, B.E., McDonald, A.E. (1993) Optical properties of fractal quantum wells, Appl. Phys. Lett. 62: 1736–1738
9. Laruelle, F., Etienne, B. (1988) Fibonacci invariant and electronic properties of GaAs/Ga$_{1-x}$Al$_x$As quasi-periodic superlattice, Phys. Rev. B. 37: 4816–4819
10. Sibilia, C., Nefedov, I., Scalora, M., Bertolotti, M. (1998) Electromagnetic Mode density for finite quasi-periodic structures, J. Opt. Soc. Am. B 15: 1947–1952
11. Hurley, C., Tamura, S., Wolfe, J.P., Ploog, K., Nagle, J. (1988) Angular Dependence of phonon transmission through a Fibonacci superlattice, Phys. Rev. B 37: 2354–2365
12. Angelsky, O.V., Maksimyak, P.P., Perun, T.O. (1993) Dimensionality in optical fields and signals, Appl. Opt. 32: 6066–6071
13. Dutta Gupta, S., Ray, D.S. (1990) Localization problem in optics: Nonlinear quasi-periodic media, Phy. Rev. B 41: 8047–8063

14. Rammal, R., Toulouse, G. (1983) Random walks on fractal structures and percolation clusters, Jour. Phys. Lett. 44: 412–421

15. Craciun, F., Bettucci, A., Molinari, E., Petri, A., Alippi, A. (1992) Direct Experimental Observation of Fracton Mode Patterns in one dimensional Cantor Composite, Phys. Rev. Lett. 68: 342–347

16. Sakurada, Y., Uozumi, J., Asakura, T. (1992) Fresnel diffraction by one-dimensional regular fractals, Pure Appl. Opt. 1: 29–40

    Berry, M.V. (1979) Diffractals, J. Phys. A 12: 781–797

    Uozumi, J., Asakura, T. (1994) Current Trends in Optics, Vol. 6, ed. by J.C. Dainty, Academic Press, London, pp. 83–93

    Allain C., Cloitre, M. (1986) Optical diffraction on fractals, Phys. Rev. B 33: 3566–3569

    Zimnyakov, D.A., Tuchin, V.V. (1997) Scale properties of the diffraction fields induced by pre-fractal random screens, in Fractal Frontiers, ed. by M.M. Novak, T.G. Dewey, World Scientific, Singapore, pp. 281–290

    Jakeman, E. (1982) Fresnel scattering by a corrugated random surface with fractal slope, J. Opt. Soc. Am. 72: 1032–1041

    Elson, J.M., Bennett, J.M. (1979) Relation between the angular dependence of scattering and the statistical properties of optical surfaces, J. Opt. Soc. Am. 69: 31–49

17. Mandelbrot, B.B. (1992) The Fractal Geometry of Nature, W.H. Freeman, San Francisco, CA

18. de Vries, P., De Raedt, H., Lagendijk, A. (1989) Localization of waves in Fractals: spatial behaviour, Phys. Rev. Lett. 62: 2515–2518

19. Bertolotti, M., Masciulli, P., Sibilia, C. (1994) Spectral transmission properties of a self-similar optical Fabry-Perot resonator, Optics Lett. 19: 777–780

20. Bertolotti, M., Masciulli, P., Sibilia, C., in Linear and Nonlinear Integrated Optics, eds. Righini, G.C., Yevick, D. (1994) Transmission properties of integrated resonator realized with Cantor-like Code, SPIE 2212: 607–612

21. Bertolotti, M., Masciulli, P., Ranieri, P., Sibilia, C. (1996) Optical bistability in a nonlinear Cantor corrugated waveguide, J. Opt. Soc. Am. B 13: 1512–1519

22. Bertolotti, M., Masciulli, P., Sibilia, C., Wijnands, F., Hoekstra, H. (1996) Transmission properties of a Cantor corrugated waveguide, J. Opt. Soc. Am. B 13: 628

23. Panajotov, K., Sibilia, C., Bertolotti, M., Scalora, M. (–) Optical limiting with Fractal multilayers, to be published

24. Sibilia, C., Tropea, F., Bertolotti, M. (1998) Enhanced nonlinear optical response of a Cantor-like and Fibonacci-like quasi-periodic structures, J. of Mod. Optics 45: 2255–2267

25. Milillo, E., Sibilia, C., Rusu, V., Gravè, I., Bertolotti, M. (–) Properties of Cantor multilayers, to be published

26. Garzia, F., Masciulli, P., Sibilia, C., Bertolotti, M. (1998) Temporal pulse propagation in a Cantor multilayer filter, Opt. Comm. 142: 333–340

27. Danckaert, J., Fobelets K., Veretennicoff, I., Vitrant, G., Reinish, R. (1991) Dispersive optical bistability in stratified structures, Phys. Rev. B 44: 8214–8225

28. Centini, M., Scalora, M., Sibilia, C., Bertolotti, M., Bloemer, M., Bowden, C. (–) High Efficient Parametric Interaction in 1-D PBG structures, submitted for publ. in Phys. Rev. Lett.

Part VI

Optical Metrology
(Optical Systems)

Diffractive Optical Elements in Materials Inspection

R. Silvennoinen, K.-E. Peiponen, and T. Asakura

Summary. Theory and applications of diffractive-element-based sensors for materials inspections are considered. The diffractive element is computer-generated hologram which has been fabricated using either conventional or advanced electron beam lithography techniques. The sensor applications include metal, transparent, and porous material inspection for quality assessment.

1 Introduction

Optical metrology can provide new means for product inspection and process control in various fields of modern industry. In principle, in industry there is usually interest in optical or other new metrology, provided that such a new metrology can assist, for instance, in material savings, product quality inspection or optimal process control.

We have experience of joint projects with process and other industries in developing optical devices for such purposes. Usually these projects have involved laboratory experiments and also the development of prototypes, including testing in industrial environments. Recently, we investigated the application of diffractive elements for sensing a diversity of material properties.

In this chapter we consider first the imaging theory of diffractive optical elements (DOEs). The fabrication of DOEs and application in optical information processing are nicely described by Herzig and Dändliker [1], and Frisem and Amitai [2]. We then present various sensor applications. These include, for instance, surface roughness and curvature detection of machined metal surfaces, quality inspection of transparent objects such as security windows, ferroelectric ceramics and laser active crystals and surface quality inspection of porous materials, which include pharmaceutical compacts and paper. In other words, we deal with materials that have great importance in the engineering sciences and in the process industry.

2 Theory of Diffractive Elements

The diffractive elements, which in our case are computer-generated holograms, obey the formulas of hologram imagery [3,4]. The Fresnel number of the element is much larger than one [5]. For that reason the stigmatism [6] of the images is lower than in the case where the Fresnel number of the element is less than one. The efficiency of the DOE was maximized by adjusting the threshold level of the element to zero [7]. The effect of degradation caused by the possible wavelength shift is also taken into account in the planning of the DOE [4,8].

For the present hologram sensors the reconstructing wavefront can be guided in two ways to an aperture of the DOE depending on the sensor application. In the first case, the reconstructing wavefront is first scattered from the surface under test and, thereafter, is diffracted from the hologram towards the focal plane of the DOE. In the second case, the reconstructing wavefront is first diffracted towards the surface under test and, thereafter, is scattered from the test surface towards the focal plane of the DOE. One can easily observe that these two geometric setups differ from each other, and thus it is necessary to present the imaging properties for each setup separately.

2.1 Wavefront from a Test Surface to an Aperture of the DOE

First we consider the planarity properties of the surface. It is well known that a plane wavefront reflected from a curved surface with known radius will obey the curvature with the radius divided by two. If this wavefront is used for the reconstruction of the DOE, the location of the image will change. Detecting these changes from the original location of the image makes it possible to extract information from the reconstructing wavefront. To analyse the radius of the curvature of the wavefront, we need to write the mean distance variables, R_r, R_o, R_c, R_i, and the respective angular variables of the hologram imagery, α_r, β_r, α_o, β_o, α_c, β_c, α_i and β_i, as stated in [3] and Fig. 1, to obey the case of reconstruction. The subscripts r, o, c and i denote the respective variables for reference, object, reconstruction and image.

Let us denote the optical axis and the hologram plane by the z-axis and the xy-plane, respectively. The variable α is an angle between the mean position vector and its projection located on the xz-plane, whereas the variable β describes the angle between the z-axis and the projection of the mean position vector of the reference, object, reconstruction or image point onto the xz-plane. For the present sensor, the reference source was at infinity and the respective angular variables α_r and β_r were zero. The scaling factor was assumed to be unity. After these substitutions the formulae of the mean distance and angular variables for a reconstructed image reduce to the following forms:

$$\frac{1}{R_c} = \frac{1}{R_i} - \frac{\mu}{m^2 R_o}, \tag{1a}$$

$$\alpha_c = \sin^{-1}\left(\sin\alpha_i - \frac{\mu}{m^2}\sin\alpha_o\right), \tag{1b}$$

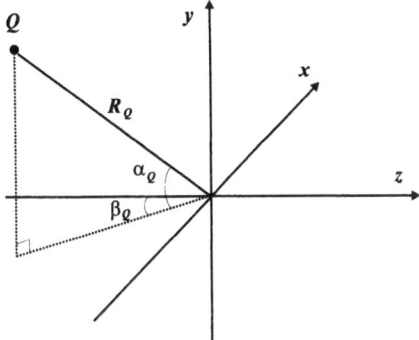

Fig. 1. The mean distance variable R_Q and the respective angular variables α_Q and β_Q of source point Q in hologram imagery. Note that Q denotes the subscripts r, o, c and i mentioned in the text for reference, object, reconstruction and image, respectively

$$\beta_c = \sin^{-1} \frac{\cos \alpha_i \sin \beta_i - \mu/m^2 \cos \alpha_o \sin \beta_o}{\cos \alpha_c}, \tag{1c}$$

where $\mu = \pm\lambda_c/\lambda_o$ is the ratio of the reconstructing and recording wavelengths and m is the scaling factor. The $\pm$ signs normally denote $+$ for a virtual image and $-$ for a real image. For the changes in the mean image distance $R_i = R_o \pm \Delta R_i$ caused by the curved reconstructing wavefront, the distance R_c of (1a) can now be written in the form

$$R_c = \frac{R_o(R_o \pm \Delta R_i)}{R_o - \mu(R_o \mp \Delta R_i)}. \tag{2}$$

To investigate the behaviour of the angular variables α_c, β_c, α_i and β_i we encoded a ring pattern in the DOE. We define a ratio given by

$$\gamma = \varrho/R_i, \tag{3}$$

where ϱ is the radius of magnified ring (due to the aberrations of the inspected surface) in the image plane and R_i is the mean image distance mentioned above. By using the above definition and the knowledge that $\varrho \cos \delta \ll R_i$, the α- and β-variables in the image and object domains can be written as follows:

$$\alpha_i = \tan^{-1}(\gamma \cos \delta), \tag{4a}$$

$$\beta_i = \tan^{-1} \frac{\varrho \sin \delta}{\sqrt{R_i^2 + (\varrho \cos \delta)^2}} \approx \tan^{-1}(\gamma \sin \delta), \tag{4b}$$

and

$$\alpha_o = \tan^{-1} \frac{\varrho_o \cos \delta}{R_i} = \tan^{-1}(\gamma_o \cos \delta), \tag{5a}$$

$$\beta_o = \tan^{-1} \frac{\varrho_o \sin \delta}{\sqrt{R_i^2 + (\varrho_o \cos \delta)^2}} \approx \tan^{-1}(\gamma_o \sin \delta), \tag{5b}$$

where $\gamma_o = \varrho_o/R_i$ is the ratio in the angular direction (α_o, β_o) from the optical axis to the point located on the original ring at the azimuth angle δ.

Next we consider the relation of the radius of surface curvature and the shifts of Gaussian image variables detected by the sensor. First, we suppose that the curvature of the surface is spherical with a known radius. For such a surface the radius of curvature can be calculated in the x- and y-directions from the formulae

$$\rho_x = \frac{S_x}{\Delta\alpha_c}, \quad \rho_y = \frac{S_y}{\Delta\beta_c}, \tag{6}$$

where S_x and S_y are the lengths of the surface segment under test and $\Delta\alpha_c$ and $\Delta\beta_c$ are the respective angles. By using the imaging properties of the sensor, we can first calculate the values of angular variables α_i and β_i from the data of the ring observed by the CCD camera. If the observed values coincide with those used in the recording of the DOE, both angular variables α_c and β_c in the reconstruction domain become zero. In practice, such a situation holds for the normal incident light reflected from the planar surface. Otherwise, at least one of the α_c and β_c values will be non-zero. By observing the possible non-zero values of the angular variables in the image domain with the sensor, we can calculate the shifts $\Delta\alpha_c$ and $\Delta\beta_c$ for the reconstruction angle caused by a surface of spherical or other non-planar type.

The respective maxima of height h of the curved surface segment can be estimated from

$$h_x = \frac{S_x^2}{8\rho_x}; \quad h_y = \frac{S_y^2}{8\rho_y}. \tag{7}$$

Controlled height maxima can be produced by bending a mirror beam clamped on both edges and loaded at the centre. Then the profile of the beam will obey the formula

$$h = \frac{3t}{l} + \frac{4t^3}{l^3}; \quad 0 \le t \le l/2, \tag{8}$$

where l is the length of the beam. To get a measure for the radius of curvature, we define an integrated radius of curvature ρ_{Int} by

$$\rho_{\text{Int}} = \frac{1}{N} \sum_{i=1}^{N} \frac{(\frac{l}{2} - t_i)\Delta_{t_i}}{h_{t_{i+1}} - h_{t_i}}, \tag{9}$$

where $\Delta_{t_i} = t_{i+1} - t_i$ is the sampling interval for the integration.

2.2   Wavefront from an Aperture of the DOE to a Test Surface

For the second case of the sensor, the reconstructing wavefront is first diffracted through the DOE aperture, and scattered from the test surface towards its image plane. Owing to the curvature of a test surface that is situated between the DOE and its image plane, the location of the reconstructed image is changed. Let us first consider the planarity properties of the surface as a function of distance of the test surface and the geometry of the DOE. After these geometrical assumptions the formulae of the mean distance and angular variables for a reconstructed image reduce to the following forms:

$$\frac{1}{R_i} = \frac{\mu}{m^2} \frac{1}{R_o}, \tag{10a}$$

$$\alpha_i = \sin^{-1}(\sin \alpha_c - \frac{\mu}{m^2} \sin \alpha_o), \tag{10b}$$

$$\beta_i = \sin^{-1} \frac{\cos \alpha_c \sin \beta_c - \mu/m^2 \cos \alpha_o \sin \beta_o}{\cos \alpha_i}. \tag{10c}$$

If the curved surface with the radius of R_s ($R_s = \infty$ for a flat surface) is located between the aperture plane of the DOE and its image plane, the image plane changes from its ideal position. This can also be understood as a change in the focal length of the sensor. If the change of the image plane from its original position is denoted by $\pm \Delta R_i$ (+ denotes the concave and − the convex surface), the respective change in mean distances can now be written in the form

$$\frac{\mu}{R_o \pm \Delta R_i} = \frac{\mu}{R_o} + \frac{2}{\Delta R_s} - \frac{2\mu s}{R_o \Delta R_s}, \tag{11}$$

where s is the distance from the DOE to the curved test surface with the radius of curvature change equal to ΔR_s. From (11) one can solve the radius of curvature change ΔR_s to read

$$\Delta R_s = \frac{2R_o(R_o \pm \Delta R_i - \mu s) \mp 2\mu s \Delta R_i}{\mp \mu \Delta R_i}. \tag{12}$$

For all of the measurements, $R_c = \infty$ and $\alpha_c = \beta_c = 0$. The behaviour of the angular variables α_i and β_i as presented in (6) can be calculated from (3)–(5). Owing to the small values of γ, the variables α_i, β_i, α_c and β_c were observed to show rather linear responses. Finally, we can consider the relation between the radius of test surface curvature and the shifts of Gaussian image variables detected by the ring encoded in the sensor. Because of the fact that the α_r, β_r, α_o, β_o, α_c, β_c, α_i and β_i variables remain constant for all of the wavefront that diffracts from the hologram aperture, the only reason for the change in the curvature of the diffracted wavefront is the curvature of the test surface itself. Thus we remark here that the image shift ΔR_i and the radius of curvature change ΔR_s can be rewritten as

$$\Delta R_i = \gamma_N (R_i - s)/2, \tag{13a}$$

$$\Delta R_s = \frac{2R_o[R_o \pm \gamma_N(R_o/\mu - s)/2 - \mu s] \mp \mu s(R_o/\mu - s)}{\mp \mu \gamma_N(R_o/\mu - s)/2}, \tag{13b}$$

where $\gamma_N = |(\gamma - \gamma_o)/\gamma_o|$, $\gamma = \varrho/R_i$ and $\gamma_o = \varrho_o/R_i$ as defined in (3)–(6).

3   Applications

Here we consider some examples that show the applicability of the DOE sensor for material inspection. First we present some data related to the surface roughness and curvature measurement of machined metal surfaces. In addition we show how image data from transparent objects can be achieved for quality assessment. Also we deal with cases where the surface quality of porous materials is examined.

3.1   Metal Surface Roughness and Curvature Measurement

Various kinds of test patterns can be encoded to act as a DOE sensor. For metal surface inspection we have used, for example, a 4×4 light spot matrix [8,9], prefractal [10,11], and a specific test pattern [12] as shown in Fig. 2. The element of Fig. 2 was designed in a manner where four 4×4 light spot matrices and a ring element were stagged together.

The complex amplitudes of the designed subelements are first summed and the element is coded using an on-axis carrier wave. The size of the entire element is $6 \times 6\,\mathrm{mm}^2$ with a 3 mm diameter ring. The focal length of the DOE is 100 mm. The purpose of the four light spot matrix subelements is to measure the local roughness of four subareas of the inspected surface, whereas the purpose of the ring is to measure the overall curvature of the inspected area. The operation of the sensor was tested using commercial surface roughness metal standards. For surface curvature measurement we bent the surface mirror in a controlled manner. Experimental results and calculations, using (1)–(9), are in relatively good agreement as shown in Fig. 3.

To avoid of reflection ability of the metal surface and instrument factors, we have used a measure for optical surface roughness, called the contrast parameter C. It is defined as

$$C = \frac{\Gamma_{\mathrm{surface}}}{\Gamma_{\mathrm{DOE}}},\tag{14}$$

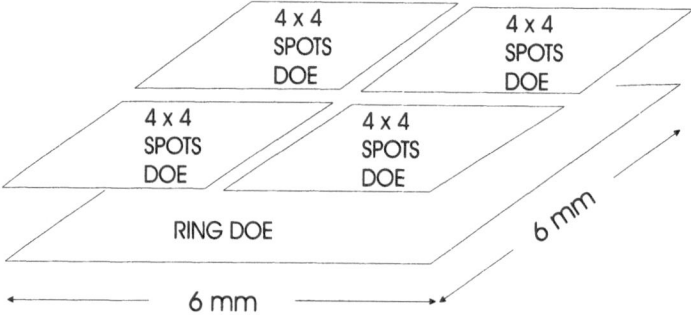

Fig. 2. Schematic diagram of construction of the DOE for the purpose of simultaneous measurement of surface roughness and curvature

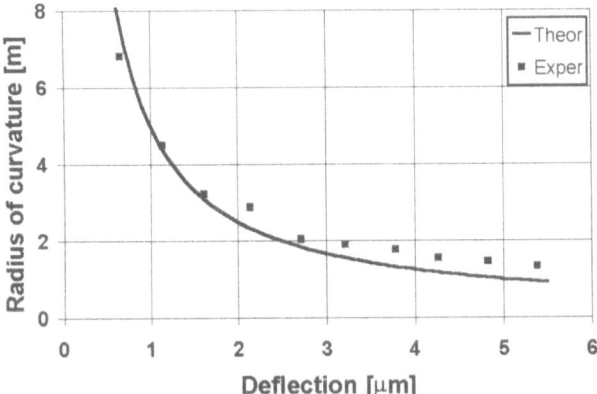

Fig. 3. Radius of curvature of mirror as a function of the maximum deflection with respect to the bending profile in the three-point loading domain: integrated radius of curvature (*line*); experimental values (*squares*)

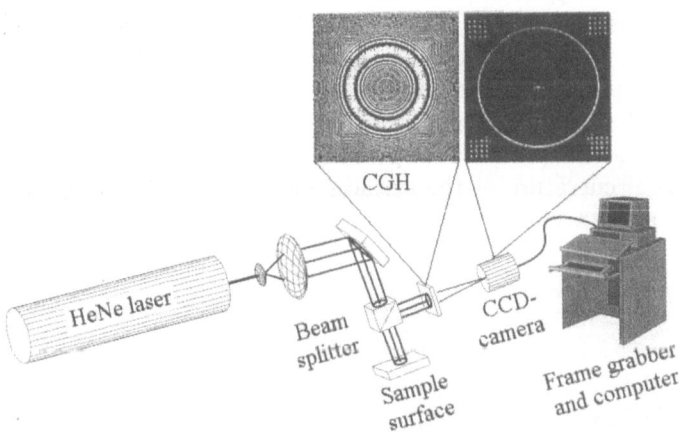

Fig. 4. Schematic diagram of inspection system for metal surface roughness and flatness

where Γ is the energy ratio $E_{peaks}/(E_{total} - E_{peaks})$ at the detection plane, in which E_{peaks} and E_{total} are the energy of the light spots and the total energy detected, respectively. The value of Γ_{DOE} is constant for a fixed sensor setup.

In the inspection of metal flatness we can measure diameters of the ring pattern, say, in the x- and y-directions after detecting the ring by a CCD camera. If the surface is flat then the ratio of diameters is equal to unity. For non-flat surface deviation from unity can be observed. In Fig. 4 we show the schematic diagram of a measurement system for metal surface roughness and flatness inspection. The device was applied for off-line inspection of a mass flow metal surface related to a paper machine [13].

3.2   Inspection of Transparent Objects

In the previous section we described a sensor where the DOE is acting as an analyser of scattered laser light. For the inspection of transparent objects, a useful measurement configuration is based on using the DOE as an image-producing element. That is to say, the DOE diffracts the light which is incident on the object. The image is then detected by the CCD camera located at the focal plane of the system. Now, the mathematical modelling of the system is based on (10)–(13). For instance, the change in curvature of the transparent object can be estimated using (13).

A typical measurement configuration is shown in Fig. 5. We have applied this techniques for the inspection of commercial NaCl IR windows [14], laser active crystal face quality [15], defects of multilayered security glasses [16], and ferroelectric ceramics [17].

The latter materials are polycrystalline in nature, being composed of a large number of tiny crystallites bonded together but randomly orientated. On a macroscopic scale, these crystallites are indistinguishable from their grain boundaries and hence both appear and act as one entity. Such materials have found applications as eye protective devices, TV displays, reflective displays, optical shutters, linear page composers and optical sensors. The main representative of transparent ferroelectric ceramics is lead zirconate titanate (PLZT) modified by La. Application of an external electric field induces an optically active, ferroelectric phase which persists only as long as the electric field is applied. Local curvature of the surfaces and the wedgeness has an impact on the electro-optic effect.

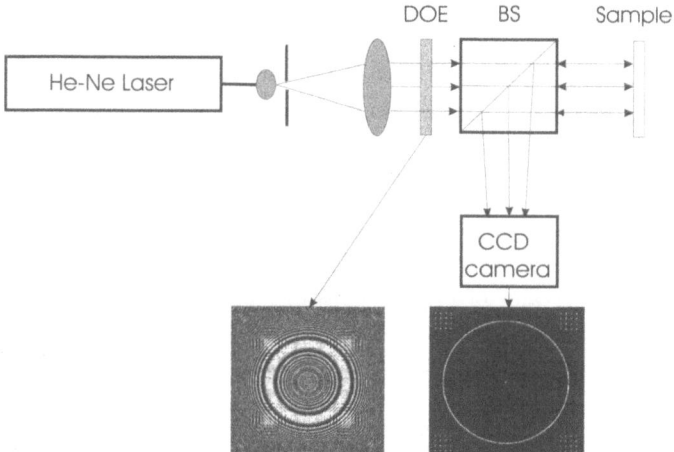

Fig. 5. Schematic diagram of measurement system, where the DOE diffracts the wavefront via the transparent test surface to the CCD camera

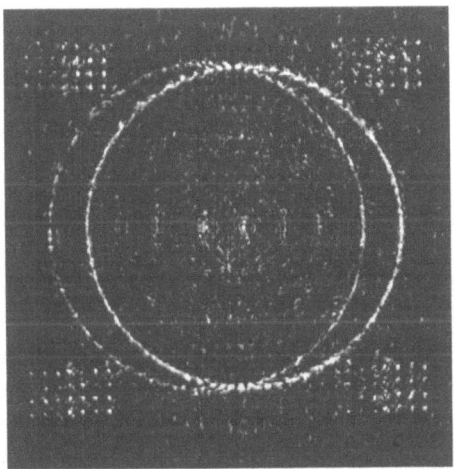

Fig. 6. Shifted ring images observed by the sensor indicating a local wedgeness of 3.62 mrad for the 0.398 mm thick $Pb_{0.9}La_{0.1}Zr_{0.65}Ti_{0.35}O_3$ test plate

In Fig. 6 we show an example of hot-pressed PLZT sample inspection of local wedgeness, $\Delta\alpha$, which can be calculated using the formulae

$$\Delta\alpha = \tan^{-1}(1-F)\varrho_o\frac{\Delta w_i}{\varrho_{o_i}}, \quad F = \frac{2n_oL_o}{\mu(R_o - s)}, \tag{15}$$

where Δw_i is the maximum distance deviation of the frame-grabbed images of rings in the image plane, ρ_{o_i} the radius of the frame-grabbed image of the ring reflected from the first interface, n_o the refractive index of ferroelectric ceramics (in the present example, $n_o = 2.5$), L_o the thickness of the sample, and other symbols are as defined previously.

3.3   Surface Quality Inspection of Porous Products

Optical inspection of porous materials is rather complicated owing to the strong scattering of light in most cases. This means, for instance, that we cannot usually apply in data analysis Fresnel's formulas for reflection. Generally speaking, classical models for the interaction of light with porous materials like that of Kubelka–Munk [18] are usually insufficient in interpreting measured data. However, some qualitative picture of light scattering from the surfaces of porous materials can be obtained using the DOE for analysing the scattered light. We have applied such techniques especially in the inspection of pharmaceutical compacts [19–21] and paper from a paper mill [22].

In the case of pharmaceutical compacts the surface porosity is important since it can affect the rate of release and absorption of the drug. With optimal surface porosity it is possible to help in administering correct medical treatment. It was observed that the quality of the upper and lower surfaces of tablets can differ depending on the powder material and bulk porosity. This was achieved by comparing the surface quality of the tablets and calculating the correlation matrices for each bulk porosity. For this purpose we exploited the integrated average correlation (IAC), defined by the formula

$$\text{IAC} = \sqrt{\sum_{i=1}^{N}\sum_{j=1}^{M} \text{CM}_{ij}^2}\,, \tag{16}$$

where CM_{ij} is ith and jth element of the correlation matrix. The higher the value of IAC, the smoother or more homogeneous is the inspected surface. IAC values as a function of bulk porosity for some tablets are shown in Fig. 7a). It was observed that plastically deformed starch tablets had their lower surfaces optically more homogeneous than those of upper surfaces unlike dicalcium phosphate tablets (Emco). The tablets were produced by a hydraulic compaction simulator.

In some cases the tablet has a thin coating which is used for the physical or chemical protection of a drug and to protect a drug from the gastric environment of the stomach. The quality of the coating is important for quality control. The variation (or lack) of coating thickness can be observed in the image data. This can be seen for instance in Fig. 7b where the intensity profiles of a 4×4 matrix are shown as projections onto a plane. The change of the thin film has an effect on the intensity map due to the interference phenomenon. The data of Fig. 7b was detected using a DOE that was produced by an electron beam writer.

Paper is another example of a porous material. The purpose of testing paper quality is to provide reliable data on the physical properties and characteristics of the paper products.

In the paper industry, many of the testing methods are based on the requirements of older user technology, and some may no longer be relevant. More effort needs to be directed towards the development of new test meth-

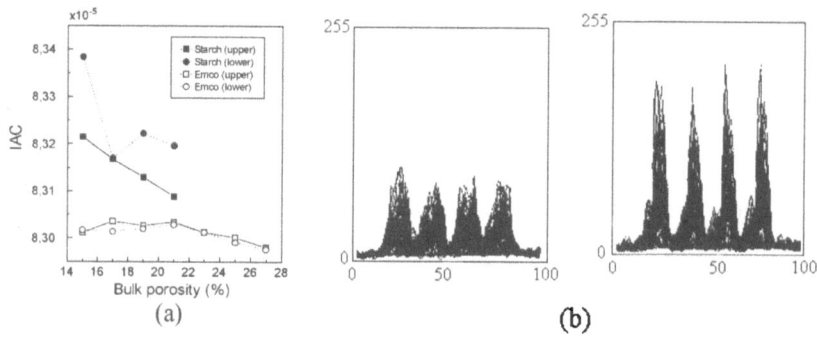

Fig. 7. (a) IAC values as a function of bulk porosity of tablets and (b) intensity plots of coated tablet. Numbers shown on the axes are pixel numbers

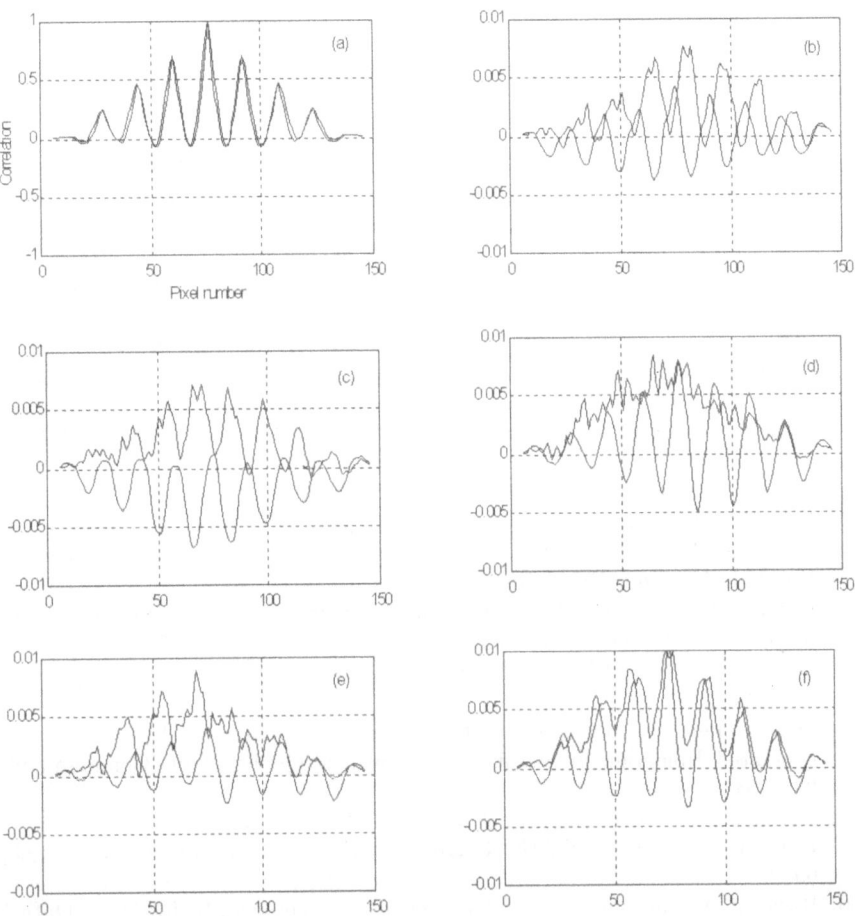

Fig. 8. Cross sections in x and y directions obtained from the two-dimensional correlation images of the sensor: (**a**) autocorrelations from a smooth surface; (**b**) to (**f**) consecutive cross correlations from paper when the sample locations are changed at 10 µm intervals

ods. For that purpose we have been investigating the potential advantages of the DOE for paper surface quality inspection. There is a need, for example, for on-line and also off-line laboratory devices for the surface roughness inspection of paper. Anisotropy of the surface texture of paper (there is a preferential fibre orientation along the machine direction) could be observed by using the DOE and image cross correlation analysis. Some results of such a study are presented in Fig. 8. In Fig. 8d it can be seen that relatively strong anisotropy of the surface texture, whereas in Fig. 8f a rather isotropic paper surface is present.

Acknowledgements

RS and K-EP wish to express their gratitude to the Academy of Finland and the Technology Development Center (TEKES) for financial support.

References

1. Herzig, H.P., Dändliker, R. (1991) Holographic Optical Elements for Use with Semiconductor Lasers, In: Goodman, J.W. (Ed.), International Trends in Optics, Academic Press, San Diego
2. Friesem, A.A., Amitai, Y. (1996) Planar Diffractive Elements for Compact Optics, In: Consortini, A. (Ed.), Trends in Optics, Academic Press, San Diego
3. Latta, J.N. (1971) Computer-based Analysis of Hologram Imagery and Aberrations. I. Hologram Types and Their Nonchromatic Aberration, Appl. Opt., **10**, 599–608
4. Latta, J.N. (1971) Computer-based Analysis of Hologram Imagery and Aberrations. II. Aberrations Induced by a Wavelength Shift, Appl. Opt., **10**, 609–618
5. Shannon, R.R., Wyant, J.C. (1979) Applied Optics and Optical Engineering Vol. VII. Academic Press, New York
6. Hazra, L.N., Han, Y., Delisle, C. (1994) Imaging by Zone Plates: Axial Stigmatism at a Particular Order, J. Opt. Soc. Am., **11**, 2750–2754
7. Roux, F.S. (1994) Wavelength Dependence of Thin Diffractive Lenses, Opt. Eng., **33**, 2843–2848
8. Silvennoinen, R., Räsänen, J., Peiponen, K.-E., Gu, C. (1993) Computer-generated Hologram in Sensing Metal Surface Quality, Opt. Commun., **95**, 231–234
9. Räsänen, J., Silvennoinen, R., Peiponen, K.-E., Asakura, T. (1994) On Surface Damage Detection of Slightly Rough Metal Surfaces, Opt. Laser Eng., **20**, 65–69
10. Räsänen, J., Savolainen, M., Silvennoinen, R., Peiponen, K.-E. (1995) Optical Sensing of Surface Roughness and Waviness by a Computer-generated Hologram, Opt. Eng., **34**, 2574–2580
11. Peiponen, K.-E., Silvennoinen, R., Uozumi, J., Savolainen, M., Asakura, T. (1994) Prefractals in Optical Information Coding, Optik, **97**, 127–128
12. Silvennoinen, R., Räsänen, J., Savolainen, M., Peiponen, K.-E., Uozumi, J., Asakura, T. (1996) On Simultaneous Optical Sensing of Local Curvature and Roughness of Metal Surfaces, Sensors & Actuators A, **51**, 117–123
13. Savolainen, M., Räsänen, J., Peiponen, K.-E., Silvennoinen, R. (1996) Off-line System for Simultaneous Inspection of Metal Surface Roughness and Flatness by Computer-generated Holograms, Opt. Eng., **35**, 3637–3639
14. Räsänen, J., Peiponen, K.-E., Silvennoinen, R., Silfsten, P. (1992) Crystal Surface Quality Inspection Using Computer-generated Hologram and Digital Image Processing, Phys. Status Solidi (a), **134**, K53–K55.
15. Ketolainen, P., Silvennoinen, R., Peiponen, K.-E., Raerinne, P. (1994) On Optical Sensing of Alkali Halide Crystal Surfaces, Opt. Rev., bf1, 94–96
16. Savolainen, M., Peiponen, K.-E., Savander, P., Silvennoinen, R., Vehviläinen, H. (1995) Novel Optical Techniques for Window Glass Inspection, Meas. Sci. Technol., **6**, 1016–1021

17. Silvennoinen, R., Peiponen, K.-E., Krumins, A., Räsänen, J. (1997) Optical Quality Inspection of PLZT Ceramics by Using CGH-element, Opt. Mater., **7**, 145–152

18. Kubelka, P. (1948) New Contributions to the Optics of Intensely Light-scattering Materials. Part I, J. Opt. Soc. Am., **38**, 448–457

19. Silvennoinen, R., Peiponen, K.-E., Laakkonen, P., Ketolainen, J., Suihko, E., Paronen, P., Räsänen, J., Matsuda, K. (1997) On Optical Inspection of the Surface Quality of Pharmaceutical Tablets, Meas. Sci. Technol., **8**, 550–554

20. Peiponen, K.-E., Silvennoinen, R., Räsänen, J., Matsuda, K., Tanninen, V.P. (1997) Optical Coating Inspection of Pharmacaeutical Tablets by Diffractive Element, Meas. Sci. Technol., **8**, 815–818

21. Silvennoinen, R., Peiponen, K.-E., Sorjonen, M., Ketolainen, J., Suihko, E., Paronen, P. (1998) Diffractive Optical Element Based Sensor for Surface Quality of Pharmaceutical Tablets, J. Mod. Opt., **45**, 1507–1512

22. Silvennoinen, R., Peiponen, K.-E., Räsänen, J., Sorjonen, M., Keränen, E.J., Eiju, T., Tenjimbayashi, K., Matsuda, K. (1998) Diffractive Element in Optical Inspection of Paper, Opt. Eng., **37**, 1482–1487

Multiple-Wavelength Interferometry for Absolute Distance Measurement

R. Dändliker and Y. Salvadé

Summary. Multiple-wavelength interferometry enables an increase in the range of nonambiguity and a reduction in the sensitivity of classical interferometry. It can also be operated on rough surfaces. The accuracy depends on the stability and the calibration of the different wavelengths. An electronically calibrated three-wavelength source for synthetic wavelengths in the millimeter range with an accuracy better than 10^{-5} was demonstrated. Absolute distance measurements were performed up to 200 mm with a resolution better than 10 μm. Distance measurements to non-cooperative targets, using a custom designed lock-in CCD and appropriate signal processing, was also demonstrated.

1 Introduction

Since the advent of lasers in 1960, distances or displacement measurements by optical techniques became very attractive in a wide range of applications. For instance, they are currently used for the calibration of machine tools, for geodesy, for surface inspection in aeronautics, for surveying systems, for robotics, and for space applications. The main advantage of optical metrology is that the object is probed without contact. Triangulation techniques, as well as time-of-flight systems, are noncoherent methods for distance measurements. The measurement accuracy of such techniques is typically larger than one millimeter while coherent methods, based on interferometry, enable high precision measurements. Classical interferometry is commonly used for high-resolution displacement measurements. Resolution better than 100 nm is obtained by using commercially available interferometers. However, the main drawback of this technique is the incremental manner of measuring, resulting from counting optical fringes. Several alternative interferometric methods have been developed in order to perform absolute distance measurements, based on multiple-wavelength interferometry or white-light interferometry.

In 1895, A. A. Michelson and J. R. Benoît managed to determine the number of cadmium red line wavelengths in the international meter prototype [1]. The cadmium red line was the most coherent source in those days, and interferometric measurement above 10 cm was a difficult task. They determined, therefore, the number of wavelengths in an etalon 10 cm in length, with a resolution of 1/50 of a fringe, and they compared this etalon with the meter prototype. For this task, they used eight intermediate etalons with lengths of

10×2^{-1} cm, 10×2^{-2} cm, ..., 10×2^{-8} cm. They first counted the number of wavelengths in the smallest etalon and compared it with the second etalon which is two times longer. This comparison was performed by moving the smallest one over a distance equal to its own length. Observation of white-light interference allowed them to measure the optical path difference between the two rear sides of the etalons. Quasi-monochromatic light was then used to determine the excess fraction of the interferometric fringe. Comparisons between different etalons were carried out similarly in order to determine the number of wavelengths in the 10 cm long etalon. After each comparison, the excess fractions were measured with red, green, and blue emission lines of the cadmium. Michelson and Benoît concluded that the use of different colors allows the direct measurement of the number of wavelengths in the etalons, without another operation. Evidently, principles of classical, white-light, and multiple-wavelength interferometry were known as early as 1895.

Today, the use of highly coherent lasers allows the measurement of displacements or distances up to at least 10 m by interferometric techniques. Absolute distance measurement with a resolution of better than 0.1 mm over several meters cannot, however, be covered by classical interferometry or by current time-of-flight metrology. Multiple-wavelength interferometry (MWI) is, as classical interferometry, a coherent method, but it offers greater flexibility in sensitivity by appropriately choosing the two different wavelengths [2,3]. Indeed, the use of two different wavelengths, λ_1 and λ_2, permits the generation of a synthetic wavelength $\Lambda = \lambda_1 \lambda_2 / |\lambda_1 - \lambda_2|$, much longer than the two individual optical wavelengths. This method thus makes it possible to increase the range of non-ambiguity for interferometry and to reduce the sensitivity of the measurement. Moreover, this technique is also applicable to rough surfaces.

In order to obtain this new synthetic wavelength, the different optical wavelengths have to be interrelated. Real-time electronic signal processing is mandatory for practical applications. The absolute accuracy of a distance measurement made by MWI depends essentially on the properties of the source (coherence, stability, power) and on the calibration of the synthetic wavelength. For highly accurate measurements, that is for $\delta L/L < 10^{-5}$, where L is the working distance and δL the resolution, the synthetic wavelength has to be known with at least the same accuracy. Therefore, the two laser sources must be stabilized and the synthetic wavelength has to be calibrated [3,4].

This paper presents some solutions for signal processing and for the calibration of multiple-wavelength sources. Experimental results of absolute distance measurements with cooperative (reflecting) targets and non-cooperative targets (rough surfaces) will be presented.

2   Multiple-Wavelength Interferometry

2.1   Basic Concepts

Let us consider two-wavelength interferometry using the optical wavelengths λ_1 and λ_2. For an interferometric path difference L, the phases ϕ_1 and ϕ_2 corresponding to the wavelengths λ_1 and λ_2 for a refraction index $n = 1$ are given by

$$\Delta\phi_1 = \frac{2\pi}{\lambda_1}2L \quad \text{and} \quad \Delta\phi_2 = \frac{2\pi}{\lambda_2}2L. \tag{1}$$

The phase difference between ϕ_1 and ϕ_2 is then given by

$$\Delta\phi_{12} = \Delta\phi_1 - \Delta\phi_2 = 2\pi\left(\frac{1}{\lambda_1} - \frac{1}{\lambda_2}\right)2L = \frac{2\pi}{\Lambda}2L. \tag{2}$$

The phase difference is thus sensitive to a new synthetic wavelength Λ which can be expressed as

$$\Lambda = \frac{\lambda_1\lambda_2}{|\lambda_1 - \lambda_2|} = \frac{c}{|\nu_1 - \nu_2|}. \tag{3}$$

Therefore, the use of two slightly different wavelengths permits the generation of a new synthetic wavelength much longer than the individual optical wavelengths. The range of non-ambiguity of the phase difference $\Delta\phi_{12}$, which is also known as the synthetic phase, is increased compared to the range of non-ambiguity for classical interferometry. Moreover, the sensitivity of the measurement is reduced.

Two-wavelength interferometry can be accomplished by injecting two wavelengths simultaneously into the interferometer and optically separating them at the output using a prism or a grating [5,6], as shown in Fig. 1. Both interference signals are then detected individually. The synthetic phase can

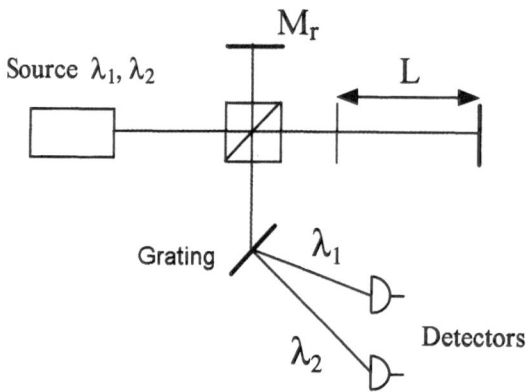

Fig. 1. Two-wavelength interferometer with optical separation of both wavelengths

be determined by measuring the interferometric phases at both wavelengths and computing the difference. However, this method works only for relatively large wavelength differences and thus small synthetic wavelengths (<1 mm), since both wavelengths have to be separated by means of a prism or a grating. In addition, this method requires interferometric stability at the optical wavelength, which is difficult to achieve in applications where reduced sensitivity is desired. Alternative methods [2,3,7–9] can be used to solve this problem by detecting the total interference signal without any optical separation of the two wavelengths. The complex amplitudes are given by two contributions at the wavelengths λ_1 and λ_2, namely,

$$U(t) = U_{\lambda_1}(t) + U_{\lambda_2}(t), \tag{4}$$

with

$$U_{\lambda_1}(t) = V_{\lambda_1} \exp(\mathrm{i}2\pi\nu_1 t) \quad \text{and} \quad U_{\lambda_2}(t) = V_{\lambda_2} \exp(\mathrm{i}2\pi\nu_2 t), \tag{5}$$

where ν_1 and ν_2 are the optical frequencies corresponding to λ_1 and λ_2.

After beam recombination, the superposition of the reference wave and the measuring wave is delayed by the time $\tau = 2L/c$. If the two waves are of equal intensity, the averaged interference signal becomes

$$I(\tau) = \langle |U(t) + U(t-\tau)|^2 \rangle = 2(I_{\lambda_1} + I_{\lambda_2}) + 2\mathrm{Re}\big[\langle U(t)U^*(t-\tau)\rangle\big], \tag{6}$$

where $I_{\lambda_1} = |V_{\lambda_1}|^2$ and $I_{\lambda_2} = |V_{\lambda_2}|^2$. The brackets $\langle\,\rangle$ denote the time average. The interference term is then given by the real part of the autocorrelation function of $U(t)$. Using (4), the autocorrelation function becomes

$$\begin{aligned}
\langle U(t)U^*(t-\tau)\rangle &= \langle U_{\lambda_1}(t)U_{\lambda_1}^*(t-\tau)\rangle + \langle U_{\lambda_2}(t)U_{\lambda_2}^*(t-\tau)\rangle \\
&\quad + \langle U_{\lambda_1}(t)U_{\lambda_2}^*(t-\tau)\rangle + \langle U_{\lambda_2}(t)U_{\lambda_1}^*(t-\tau)\rangle.
\end{aligned} \tag{7}$$

The autocorrelation function of $U(t)$ depends not only on the individual autocorrelation functions of $U_{\lambda_1}(t)$ and $U_{\lambda_2}(t)$, but also, through the cross-correlation functions, on the statistical relation between these two contributions. However, assuming that the integration time is longer than the period of the beat frequency ($\nu_1 - \nu_2$) of the two optical frequencies ν_1 and ν_2, we see that

$$\begin{aligned}
\langle U_{\lambda_1}(t)U_{\lambda_2}^*(t-\tau)\rangle &= V_{\lambda_1}V_{\lambda_2}^* \exp(\mathrm{i}2\pi\nu_2\tau)\langle \exp[\mathrm{i}2\pi(\nu_1-\nu_2)t]\rangle = 0 \\
\langle U_{\lambda_2}(t)U_{\lambda_1}^*(t-\tau)\rangle &= V_{\lambda_2}V_{\lambda_1}^* \exp(\mathrm{i}2\pi\nu_1\tau)\langle \exp[\mathrm{i}2\pi(\nu_2-\nu_1)t]\rangle = 0.
\end{aligned} \tag{8}$$

Using (5)–(8), the interference signal is

$$I(\tau) = 2(I_{\lambda_1} + I_{\lambda_2}) + 2I_{\lambda_1}\cos(2\pi\nu_1\tau) + 2I_{\lambda_2}\cos(2\pi\nu_2\tau), \tag{9}$$

which is simply the incoherent superposition of the individual interference signals for λ_1 and λ_2. This interference signal is depicted in the upper part of Fig. 2 as a function of the optical path difference $2L = c\tau$. We note that

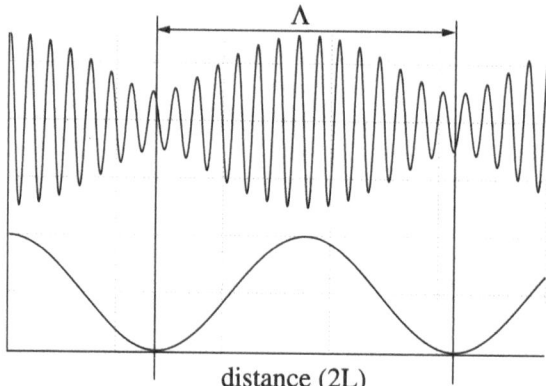

Fig. 2. Interference signal obtained by using two wavelengths simultaneously (*upper part*), and the square of its corresponding fringe visibility (*lower part*)

the modulation depth of the fringes varies periodically along the optical path difference. The fringe visibility, first introduced by Michelson, is defined by

$$|g(\tau)| = \frac{I_{\max}(\tau) - I_{\min}(\tau)}{I_{\max}(\tau) + I_{\min}(\tau)}, \tag{10}$$

where $I_{\max}(\tau)$ and $I_{\min}(\tau)$ are the intensities at the maximum and minimum of the fringes next to the interferometric delay τ, and $g(\tau)$ is the normalized autocorrelation function

$$g(\tau) = \frac{\langle U(t)U^*(t-\tau)\rangle}{\langle |U(t)|^2\rangle}, \tag{11}$$

also known as the complex degree of temporal coherence. Using (4), (5) and (11), it can be shown that the square of the fringe visibility becomes

$$|g(\tau)|^2 = 1 + 2\frac{I_{\lambda_1}I_{\lambda_2}}{I_{\lambda_1}^2 + I_{\lambda_2}^2}\cos[2\pi(\nu_2 - \nu_1)\tau]. \tag{12}$$

The lower part of Fig. 2 shows $|g(\tau)|^2$ as a function of the optical path difference. It looks like a typical interference signal, but now rather for the synthetic wavelength Λ than for the optical wavelength λ.

So far we have only considered perfectly monochromatic waves. In this ideal case, displacements or distances can be measured at arbitrary large optical path differences. However, single-mode laser sources also have a finite spectral width. Depending on the laser type, the linewidth can vary from a few kHz (gas lasers) to a few 10 MHz (standard diode lasers). The linewidth $\Delta\nu$ and the coherence time τ_c are inversely related. The exact relation between $\Delta\nu$ and τ_c depends on the lineshape function. For a Lorentzian lineshape the relation is $\tau_c = 1/\pi\Delta\nu$, where $\Delta\nu$ is the full width at half the maximum, and at τ_c the interference contrast becomes $|g(\tau_c)| = 1/e$. Highly coherent lasers are thus required for the measurement of long distances by interferometric techniques.

2.2   Signal Processing

Application of heterodyne technique to MWI was first reported by Fercher et al. [5]. The synthetic phase is determined by measuring the interferometric phases at both wavelengths and by computing the difference (see Fig. 1). This method provides a fast measurement and also works for rough surfaces. However, as already mentioned, the technique can be used only for relatively large wavelength differences and thus small synthetic wavelengths Λ.

Superheterodyne detection, introduced by Dändliker et al. [7,8], allows high-resolution measurements at arbitrary synthetic wavelengths Λ without the need for interferometric stability at the optical wavelengths λ_1 and λ_2 or the optical separation of these wavelengths. This is of great importance for range finders and for measuring large industrial distances with sub-millimeter resolution. Both wavelengths are used to simultaneously illuminate a Michelson interferometer. Two different heterodyne frequencies f_1 and f_2 are generated for each wavelength. These frequency differences can be produced by acousto-optical modulators and are typically $f_1 = 40.0\,\mathrm{MHz}$ and $f_2 = 40.1\,\mathrm{MHz}$. The heterodyne signals corresponding to the individual wavelengths λ_1 and λ_2 are of the form

$$
\begin{aligned}
I_{\lambda_1}(t) &= C_0 + C_1 \cos(2\pi f_1 t + \Delta\phi_1) \quad \text{and} \\
I_{\lambda_2}(t) &= C_2 + C_3 \cos(2\pi f_2 t + \Delta\phi_2),
\end{aligned}
\tag{13}
$$

where the interferometric phases $\Delta\phi_1$ and $\Delta\phi_2$ are given by (1). Since the wavelengths are not separated optically, the interference signal is given by the incoherent superposition of I_{λ_1} and I_{λ_2}; namely,

$$
I(t) = A_0 + A_1 \cos(2\pi f_1 t + \Delta\phi_1) + A_2 \cos(2\pi f_2 t + \Delta\phi_2).
\tag{14}
$$

Because $f_1 - f_2$ is chosen to be small compared with f_1 and f_2, the detector output has the form of a carrier-suppressed amplitude-modulated signal with carrier $(f_1 + f_2)/2$ and modulation frequency $(f_1 - f_2)/2$. After amplitude demodulation, the result is

$$
I_{\mathrm{dem}}(t) = A_{12} \cos[2\pi(f_1 - f_2)t + (\Delta\phi_1 - \Delta\phi_2)].
\tag{15}
$$

This signal at $f = f_1 - f_2$ makes it possible to measure directly the phase difference $\Delta\phi_1 - \Delta\phi_2 = 4\pi L/\Lambda$ which is now only sensitive to the synthetic wavelength Λ.

Successful application of superheterodyne detection has been reported for MWI with different types of sources; namely, two detuned single frequency Ar lasers ($\Lambda = 60\,\mathrm{mm}$) [8], a diode laser and an acousto-optic modulator for a 500 MHz frequency shift ($\Lambda = 0.6\,\mathrm{m}$) [10], two-wavelength He:Ne laser ($\Lambda = 55.5\,\mu\mathrm{m}$) [11], and tunable Nd:YAG lasers ($\Lambda = 0.12\ldots 1.5\,\mathrm{m}$) [12,13].

Simpler detection methods, which do not need separate modulation of the two wavelengths, might be of interest [2,3,9]. If the two heterodyne frequencies f_1 and f_2 are chosen to be equal $(f_1 = f_2)$, (14) becomes

$$I(t) = A_0 + A_1 \cos(2\pi f_1 t + \Delta\phi_1) + A_2 \cos(2\pi f_1 t + \Delta\phi_2). \tag{16}$$

The interference fringe function for the synthetic wavelength Λ is then obtained by detecting the electrical power of the ac part of this signal, which is

$$P_{ac} = \frac{1}{2} \left[A_1^2 + A_2^2 + 2A_1 A_2 \cos(\Delta\phi_{12}) \right]. \tag{17}$$

Equation (17) looks like a typical interference signal, but now for the synthetic wavelength Λ rather than for the optical wavelength λ (see Fig. 2). The interference phase $\Delta\phi_{12} = \Delta\phi_1 - \Delta\phi_2$ of the synthetic wavelength can now be determined by techniques similar to those used for phase interpolation in interferometry, such as phase-stepping [14,15].

2.3   Two-Dimensional Detection

In the previous sections, we showed the advantage of heterodyne detection applied to MWI. However, this technique requires synchronous detection of the heterodyne signal, which is not compatible with a standard CCD. In view of the application of MWI to non-cooperative targets (rough surfaces), a new type of CCD image sensor for two-dimensional synchronous detection at the heterodyne frequency was proposed and developed [16]. The so-called lock-in CCD is based on photocharge detection and storage. For each pixel, modulated light is detected with four or more samples per period. The corresponding photo-charges are stored at different locations (buckets). The samples are then read out by charge transfer. Amplitude, phase, or power of the modulation (beat frequency) can then be determined for each pixel. Improved versions of the lock-in CCD have recently been developed at CSEM Zürich [17]. Arrays with 5×12 pixels have been designed and fabricated with a standard CMOS process. The layout is shown in Fig. 3. Each pixel allows detection of up to eight samples per modulation period. The fill-factor is about 0.7%. This device has been used for measurements on non-cooperative targets in order to improve the probability of detecting speckles (see Sect. 4.2).

To further simplify the optical setup, we now use a moving mirror on a magnetic translator (loud-speaker) to provide the phase stepping at the synthetic wavelength and to generate the frequency shift for the heterodyne detection (Fig. 4). This mirror moves at a nearly constant speed of about 20 mm/s over a distance of about 3 mm, which produces a heterodyne frequency of $2v/\lambda \approx 50$ kHz. The control signals for the lock-in CCD and the phase-shifting algorithm [18] are obtained from the interference signal I_{ref}.

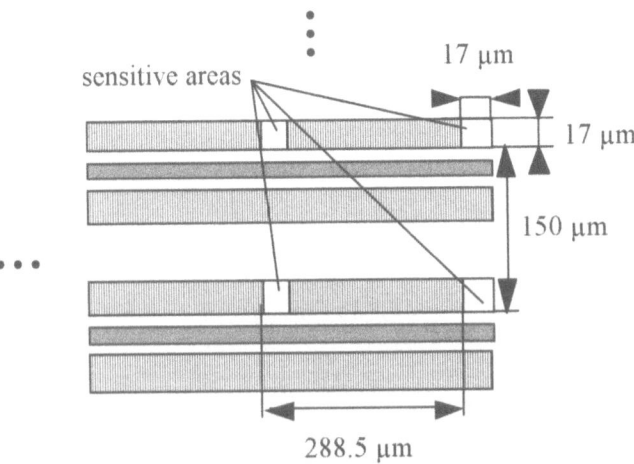

sensitive areas

17 μm

17 μm

150 μm

288.5 μm

Fig. 3. Layout of the two-dimensional lock-in CCD

3   Multiple-Wavelength Source

Multiple-wavelength interferometry can be operated with fixed wavelengths or in a wavelength tuning mode. As already mentioned, the range of non-ambiguity is given by the optical frequency difference. The stability and the calibration of the source will limit the absolute accuracy of the measurement. Moreover, the maximal distance which can be measured by multiple-wavelength interferometry is limited by the coherence length of the source. In addition, distance measurement on rough surfaces may be limited by the source power due to the scattering of the light. The design and the realization of the source are thus of great importance, since the performance of the measuring set-up will be given by its properties (coherence, stability, power).

3.1   Overview of Existing Laser Sources for MWI

Many gas lasers have the advantages of emitting light at different wavelengths. For instance, CO_2 lasers emit a large number of wavelengths between 9 μm and 11 μm, corresponding to transitions for the various vibrational-rotational levels. Bourdet and Orszag measured a length of 50 cm with an accuracy of 0.1 μm [19]. He:Ne lasers, which are traditionally used in most high-performance interferometers, can also be applied to MWI, using different laser lines, e.g., 629.4 nm and 632.8 nm [2,3]. This results in synthetic wavelengths of about 117 μm. In addition, commercially available stabilized He:Ne lasers (e.g. the Renishaw SL10) can produce two stabilized longitudinal modes with a frequency separation of about 1 GHz, corresponding to a synthetic wavelength of 30 cm. Other gas lasers, such as Ar lasers [8], He:Xe lasers [20], and Kr-ion lasers [5] have also been used in MWI experiments.

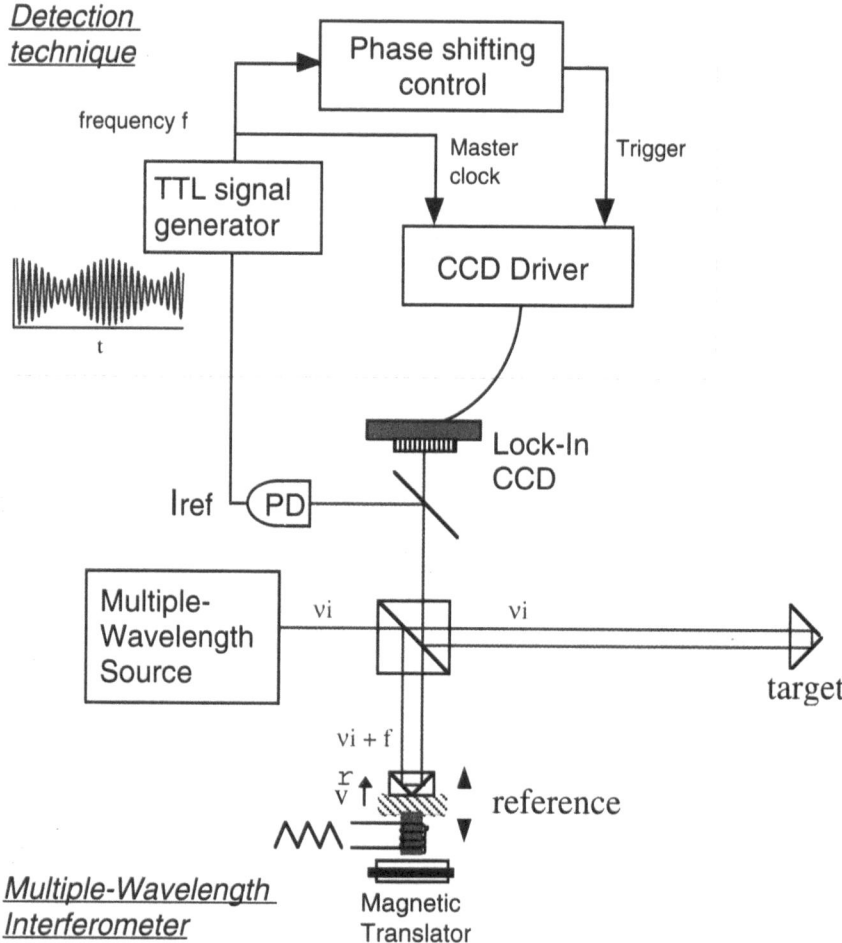

Fig. 4. Multiple-wavelength interferometer using a movable retro-reflector on a magnetic translator to produce simultaneously the heterodyne frequency and the phase shifting at the synthetic wavelength

Semiconductor laser diodes are currently the most energy efficient and the most compact lasers. Moreover, the emitted frequency can be tuned by changing the injection current and the temperature. Tunable lasers are of great interest since the most appropriate synthetic wavelength can be chosen with more flexibility. However, they have to be frequency stabilized on an external reference.

Standard single-mode AlGaAs diode lasers, such as lasers for CD-players, have often been used for MWI experiments [4,9,10,21]. In these lasers, the feedback is obtained by the cleaved-end facets. They are known as Fabry–

Pérot lasers. The linewidth is moderate (typically 10 MHz) and the frequency tunability with temperature is characterized by mode-hops. The temperature tuning behavior can vary from device to device. These discontinuities may therefore limit the choice of synthetic wavelengths.

Distributed Bragg Reflector (DBR) diode lasers are devices where at least one of the cleaved facets is replaced by a Bragg grating. The Bragg grating acts as a frequency-selective mirror. In distributed feedback (DFB) diode lasers the grating is manufactured along the active layer and acts as a distributed selective reflector. The Bragg grating allows an increase of the mode-hop free tuning range. Moreover, the selective mirror leads to high side-mode suppression (> 25 dB). This allows a substantial reduction of the power-independent contribution to the linewidth, which is mainly due to the mode partition noise in standard laser diodes [22]. DBR and DFB laser diodes are thus very promising for multiple-wavelength interferometry.

Tunable external cavity diode lasers may also provide a wide mode-hop free tuning range with a small linewidth. For instance, a tuning range of at least 10 nm with less than three mode-hops can be obtained by using commercially available external cavity diode lasers (NewFocus, velocity tunable diode laser) [23]. In addition, the linewidth may be less than 300 kHz. The main drawback is the complexity of the mechanical cavity.

Tunable Nd:YAG lasers are of great interest for interferometry. The phase fluctuations and the linewidth of such lasers are smaller than for standard diode lasers [24]. The frequency tunability is about 50 GHz. However, Nd:YAG lasers exhibit poor efficiency since they require optical pumping by laser diodes. In summary, Table 1 shows a list of modern tunable laser sources which are suitable to be used in MWI.

Table 1. List of modern tunable lasers with their corresponding performance

Laser types	Standard diode lasers (Sharp)	DBR diode lasers (SDL 5722)	External cavity diode lasers (NewFocus)	Tunable Nd:YAG (Lightwave)
Max. tuning range	1 THz	600 GHz	8 THz	50 GHz
Synthetic wavelength	> 250 μm	> 0.5 mm	> 30 μm	> 6 mm
Mode-hop free tuning range	max. 80 GHz	600 GHz	1 THz	10 GHz
Linewidth	10 MHz	2.5 MHz	300 kHz	5 kHz
Coherence length (calculated)	10 m	40 m	300 m	20 km

3.2 Calibrated Multiple-Wavelength Sources

MWI can be operated with fixed wavelengths or in a wavelength tuning mode [8]. Instead of one phase measurement for a fixed separation $\Delta\lambda = \lambda_1 - \lambda_2$ of the two wavelengths, two phase measurements are performed before and after a change of the wavelength difference $\Delta\lambda$ between the two sources. If the phase ϕ of the variable synthetic wavelength Λ is monitored during the wavelength tuning, the 2π cycles can be counted and the total phase difference is known absolutely. This now allows an absolute determination of the ranging distance L. The evaluation of the ranging distance from ϕ requires the exact knowledge of the wavelength tuning. This may be determined with the help of an additional Michelson interferometer with a precisely known, calibrated optical path difference [8,24]. In the case of fixed wavelengths, the laser sources for the different wavelengths must be stabilized with respect to each other. This can be done with the help of a common reference length in the form of a Fabry–Pérot resonator. Absolute accuracy can be obtained if the Fabry–Pérot is stabilized with respect to a frequency stabilized master laser [2,3]. Another concept for a stabilized multiple-wavelength source, for which the calibration of the different synthetic wavelengths is obtained by the use of opto-electronic beat-frequency measurements, has been reported recently [3,4].

Figure 5 shows the concept of a three-wavelength source with absolute calibration by electronic beat frequency measurement [4]. This source consists of three diode lasers operating at the frequencies ν_1, ν_2 and ν_3. Two of them (ν_1 and ν_2) are stabilized on two consecutive resonances of a common stable Fabry–Pérot resonator. In our experiment, the Fabry–Pérot resonator has a free spectral range of 0.75 GHz, as shown in the bottom part of Fig. 5. The corresponding beat frequency $\nu_{21} = \nu_2 - \nu_1 = 0.75$ GHz is detected and measured by a frequency counter with electronic accuracy. The third laser is tuned (without mode hopping) over N resonances of the Fabry–Pérot. The frequency difference $\nu_{31} = \nu_3 - \nu_1$ is then known with the same relative accuracy as the electronically calibrated beat frequency ν_{21}. For $N = 100$ we obtain $\nu_{31} = N \times \nu_{21} = 75$ GHz ($\Lambda_{31} = 4$ mm). To get an accuracy of $\delta L/L = 10^{-6}$, the free spectral range of the Fabry–Pérot must be calibrated with an accuracy of $\delta\nu_{21}/\nu_{21} = 10^{-6}$, or in our case $\delta\nu_{21} = 0.75$ kHz, which is feasible for a stable Fabry–Pérot by long time averaging. With such a multiple-wavelength source it would be possible to measure distances within 200 mm ($\Lambda_{21} = 400$ mm) without ambiguity and with a resolution of 20 μm ($\Lambda_{31} = 4$ mm with 2/100 interpolation). Experimental investigation was performed with commercial GaAlAs monomode laser diodes (Sharp LT027MD) emitting at 780 nm with a maximum optical power of 10 mW. The result proves that an absolute calibration of the synthetic wavelength with an accuracy of at least 1.25×10^{-5} is achieved by means of electronic beat-frequency measurements at ν_{21} [4]. The synthetic wavelength Λ_{31} can be chosen anywhere within the tuning range of the laser diode LD$_3$, which is about 100 GHz,

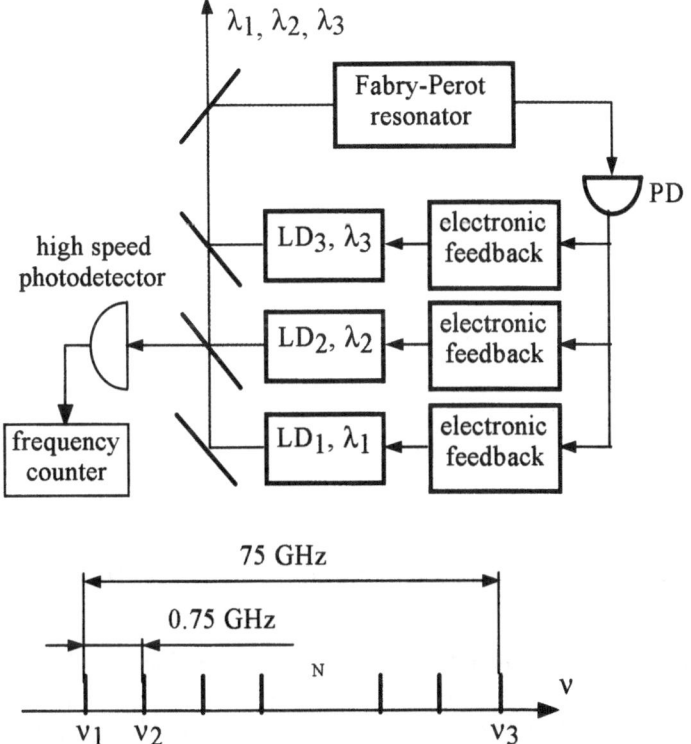

Fig. 5. Stabilized three-wavelength source calibrated by electronic means

by selecting the number N of resonances of the Fabry–Pérot counted while tuning the laser diode LD$_3$ from ν_1 to ν_3.

3.3   High-Power Multiple-Wavelength Source

Absolute distance measurement can be performed on rough surfaces by means of MWI. The problems with non-cooperative targets (rough surfaces) are: statistical properties of returning light (speckles) and low coherent power (diffused light, speckles) [25]. The multiple-wavelength source which is described above is composed of low optical power diode lasers (i.e. about 7 mW). Only a few mW of optical power are available to illuminate the diffusing target. For distance measurements beyond 1 m, the received coherent power on the detector would not be enough for measurements with a reasonable response time. We have studied two possibilities of increasing the source power.

The first one consists of using a semiconductor optical amplifier after the low power source. For this task, we used a tapered amplifier chip for 785 nm (SDL 8630). The tapered design allows a high efficiency and a near-diffraction limited output beam [26]. For a few mW of input power, we obtained an output power of about 100 mW for each wavelength, corresponding to a gain

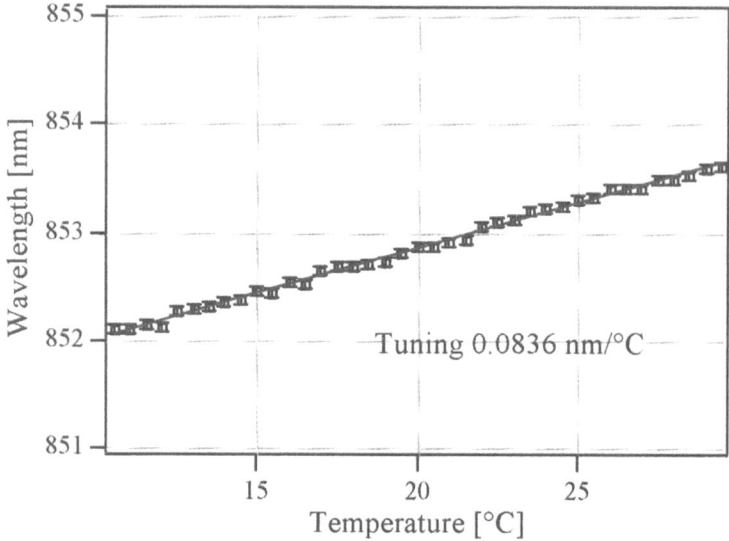

Fig. 6. Wavelength tuning versus temperature for DBR diode lasers SDL 5722

of about 100. In this MOPA (master oscillator power amplifier) configuration, the spectral characteristics are only determined by the low-power diode lasers. In this way, the multiple-wavelength source is composed of only one high-power device, which is an advantage for a low-cost measuring system. The main drawback is the high optical isolation (> 70 dB) which is required between the low-power source and the optical amplifier because of the feedback sensitivity of diode lasers. The second solution consists of using directly distributed high-power Bragg-reflector laser diodes (SDL-5722H). The maximal output power is about 150 mW. A multiple-wavelength source similar to the low-power source described in Sect. 3.2 was mounted with such diode lasers. The frequency tunability with temperature is mode-hop free over at least 1.5 nm (620 GHz), as shown in Fig. 6, which allows one to choose the most appropriate synthetic wavelength with a large flexibility between 0.5 mm and 200 mm, depending on the number of Fabry–Pérot resonances between the two frequencies. Moreover, the measured linewidth is only 2.5 MHz, which corresponds to a coherence length of about 40 m, about four times longer than for the standard low power laser diodes.

3.4 Fiber Optic Fabry–Pérot Resonator

In order to obtain a compact all-fiber system we investigated different solutions to replace the bulky confocal resonator, which had been used for the experiments described in Sect. 3.2, with a fiber optic resonator. An obvious solution would be a Fabry–Pérot with photo-induced Bragg gratings in the optical fiber [27]. However, our experimental and theoretical investigations

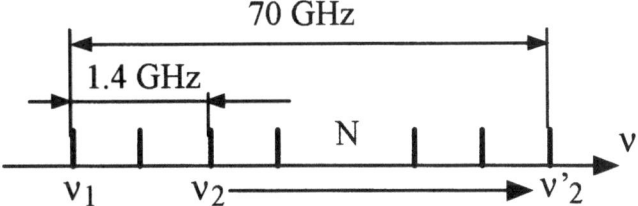

Fig. 7. Two-wavelength source stabilized on a fiber Fabry–Pérot etalon

revealed that due to the dispersion of the Bragg gratings the frequency separation (free spectral range) between the resonances is not sufficiently constant over the required tuning range. Fiber optic Fabry–Pérot resonators are also fabricated by depositing directly two highly reflective multi-layer mirrors onto each end of an optical fiber. Such devices are commercially available (e.g. from Micron Optics, Inc.). Since the fiber is birefringent, a polarization controller should be used to match the input light with one of the two polarization eigenmodes of the resonator. The length of the Fabry–Pérot cavity is about 15 cm, corresponding to a free spectral range (FSR) of 700 MHz. A finesse of 200 was specified by the manufacturer, corresponding to a resonance width of about 3.5 MHz. The fiber resonator was tested by using a DBR diode laser (SDL5722) as a light source. The resonance width is slightly broadened to 6 MHz because of the linewidth of the laser (2.5 MHz).

Two DBR diode lasers (SDL5722) and the fiber Fabry–Pérot were used to realize a two-wavelength source with calibration by beat-frequency measurements, similar to the source shown in Fig. 5. During the calibration procedure, both diode lasers are stabilized with a frequency difference of two resonances (Fig. 7). The beat-frequency $\nu_2 - \nu_1$ was measured by means of a high-speed photodetector (Newport AD-300AC) and a frequency counter (HP 53131A). Results are shown in Fig. 8 for an integration time of 10 s. The frequency difference $\Delta\nu_{21}$ is calibrated with an accuracy of $\delta\nu_{21}/\nu_{21} = 5 \times 10^{-6}$. Note that the beat frequency remained stable over more than 20 minutes, indicating a high stability of the resonator length. The thermal expansion of the silica fiber is only about $6 \times 10^{-6}/°C$. After the calibration procedure, the second diode laser can be tuned from ν_2 to ν'_2 over N resonances, as shown in Fig. 7. A larger frequency difference is generated in this way, corresponding to a smaller synthetic wavelength. The synthetic wavelength can be chosen anywhere within the tuning range of the DBR laser diode, which is about 600 GHz, by selecting the number of resonances, N, of the Fabry–Pérot counted while tuning the laser diode from ν_2 to ν'_2. In our experiment, we tuned the frequency over 98 resonances in order to generate a frequency difference of 70 GHz, corresponding to a synthetic wavelength of 4.3 mm. The same calibration accuracy is expected for $\nu'_2 - \nu_1$, since the cavity length should remain stable during the tuning time (a few seconds). However, the calibration procedure should be repeated every twenty minutes to overcome

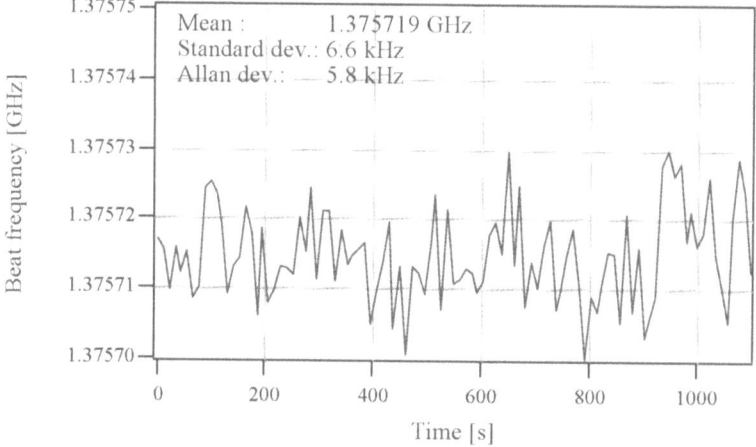

Fig. 8. Beat-frequency measurements for an integration time of 10 s

the problem of ambient temperature variations. This procedure allows the use of two diode lasers instead of the three lasers described in Sect. 3.2; therefore the cost of the system will be reduced.

4   Absolute Distance Measurement

During more than 25 years of commercial application, incremental laser interferometry has advanced to a high technical level. For more than two decades there have been experiments aimed at replacing incremental laser interferometry with absolute distance interferometers of comparable technical specifications [28]. MWI can overcome the drawback of the incremental mode of laser interferometry [3].

4.1   Cooperative Targets

We tested a three-wavelength interferometer for absolute distance measurements, using the calibrated multiple-wavelength source described in Sect. 3.2 and the simplified heterodyne detection technique presented in Sect. 2.2 [9]. Assuming a fringe interpolation of at least $2\pi/200$ for the synthetic wavelength, it would be possible to measure distances within 200 mm ($\Lambda_{21} = 400$ mm) without ambiguity and with a resolution of 10 μm ($\Lambda_{31} = 4$ mm). The technique consists of two successive two-wavelength interferometric measurements, the first one using ν_3 and ν_1 and the second one using ν_3 and ν_2 to simultaneously illuminate the Michelson-type interferometer. The corresponding synthetic wavelengths are $\Lambda_{31} = 4$ mm and $\Lambda_{32} = 4.04$ mm, respectively. The detection of the modulation power is performed by a photodiode followed by a lock-in amplifier. For an interferometric path difference L, the phases ϕ_{31} and ϕ_{32}, obtained by moving the reference mirror in steps of $\Lambda_{31}/8 = 0.5\,\text{mm}$, are given by

$$\phi_{31} = \frac{4\pi}{\Lambda_{31}}L \quad \text{and} \quad \phi_{32} = \frac{4\pi}{\Lambda_{32}}L. \tag{18}$$

The phase difference $\phi_{21} = \phi_{31} - \phi_{32}$ can then be calculated and is related to the path difference L by

$$\phi_{21} = \frac{4\pi}{\Lambda_{21}}L, \tag{19}$$

which is now sensitive to the synthetic wavelength $\Lambda_{21} = 400\,\text{mm}$. Assuming that the resolution of ϕ_{21} is better than $2\pi/200$, this phase measurement can be used to evaluate the fringe order M of the synthetic wavelength Λ_{31} without ambiguity, using the algorithm

$$M = \text{Round}\left[\frac{1}{2\pi}(N\phi_{21} - \phi_{31}\right], \tag{20}$$

where $N = 100$ is the number of resonances of the Fabry–Pérot between ν_1 and ν_3. The path difference L can then be calculated by

$$L = \left(M + \frac{\phi_{31}}{2\pi}\right)\frac{\Lambda_{31}}{2}. \tag{21}$$

The multiple-wavelength source allows the accurate calibration of the frequency difference $\nu_3 - \nu_1$, as explained in Sect. 3.2. The refractive index of air was estimated from the atmospheric conditions [29].

We checked the resolution of the setup by measuring distances over a range of 2 mm, corresponding to a displacement of $\Lambda_{31}/2$. We used the HP-laser interferometer as a reference. The measuring time was 500 ms (5 × 100 ms, using a five-phase stepping algorithm [18]). For each distance, the measurement was repeated ten times, yielding a repeatability in the order of 8 µm, which corresponds to a phase accuracy of approximately $2\pi/250$. We then measured the phase ϕ_{21} as described above over a range of 200 mm. The standard deviation $\delta\phi_{21}$ was about $2\pi/400$, which is better than expected for independent measurements of $\delta\phi_{31}$ and $\delta\phi_{32}$. After fringe order estimation, the distance was determined using (21). The lower trace of Fig. 9 shows the MWI results versus the HP-laser results, whereas the upper graph shows the corresponding deviation of the MWI mean value from the reference value given by the HP-interferometer. The standard deviation is about 9 µm.

Combining time-of-flight distance measurements and the described MWI allows a micrometer accuracy over even larger distances. However, the coherence length of the laser diodes may become a limiting factor for the maximum distance to be measured. With the measured linewidth of 9 MHz for the Sharp LT027MD, which corresponds to a coherence length of $L_c = 10.6\,\text{m}$, absolute distance measurements up to 5.3 m should be possible. However, the phase

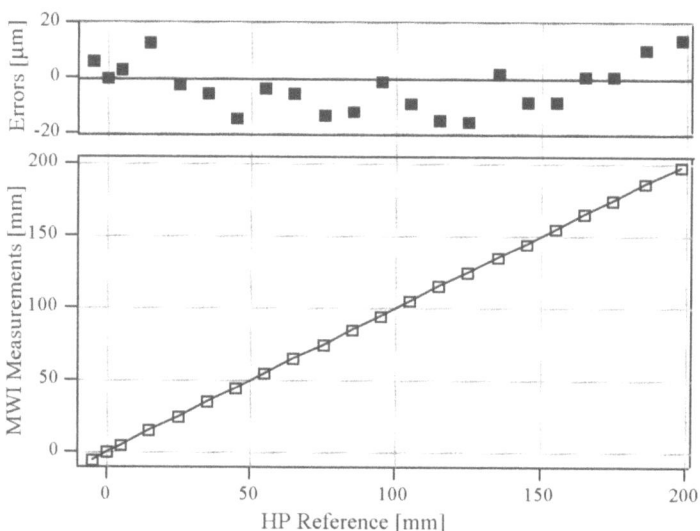

Fig. 9. Absolute distance measurements over a range of 200 mm using three-wavelength interferometry

fluctuations due to the frequency noise of the lasers become more important for larger distances. From the measured frequency noise spectrum we estimate a phase error of about $0.02\,\text{rad} \approx 2\pi/300$ for a path difference equal to the coherence length and an observation time of $T = 100\,\text{ms}$ [30,31].

MWI can also be operated in a wavelength tuning mode, using an optical reference path for calibration [8]. Bechstein and Fuchs [24] reported recently on an absolute distance interferometer using a variable synthetic wavelength and a material length standard as the optical reference path for calibration. As light sources they employed two piezo-electrically tunable diode lasers with external cavity, made by New Focus. The tuning range was 40 GHz with a repetition rate of 20 tunings per second. Measurements made over distances of several meters in a workshop environment with an averaging time of 5 s gave an uncertainty of $\pm 10\,\mu\text{m}$. As expected, the method is insensitive to reflector vibration. Further improvements hinge on the availability of fast-tuning single-frequency lasers.

4.2 Non-cooperative Targets

As stated in Sect. 3.3, the problems with non-cooperative targets (rough surfaces) are: statistical properties of returning light (speckles) and low coherent power (diffused light, speckles). The problem of the speckle effect is partially overcome in MWI, since the diffusely scattering target looks like a reflecting surface at the synthetic wavelength, at least from the point of view of the synthetic phase. The statistical properties of speckle fields, applied to rough surface interferometry, have been rigorously investigated by Fercher

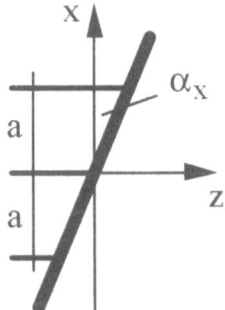
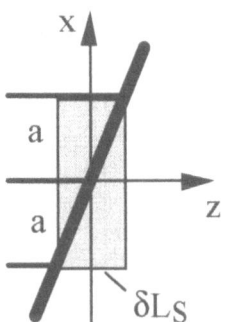

Fig. 10. Limitation by the illumination spot size for distance measurements of diffusing targets using coherent light. The position of a diffusing surface, tilted by the angle α_x, is only determined within the spot size $2a$ and the depth δL_S

and Vry [32]. Assuming two laser beams with wavelengths λ_1 and λ_2 and a Gaussian beam profile with diameter $2a$ at the $(1/e^2)$ intensity value, illuminating exactly the same area of the same rough surface, one obtains for the cross-correlation between the two speckle fields

$$|\mu| = \exp\left[\frac{4\pi^2}{\Lambda^2}\left(2\sigma_h^2 + \frac{s^2a^2}{2}\right)\right]. \tag{22}$$

The rough surface is described by a roughness function $h(x)$ with variance σ_h. The surface may be tilted with respect to the direction of illumination z by the tilt angle $s = \tan\alpha_x$ (Fig. 10), so that the target surface is finally described by $z(x) = z_0 + h(x) + sx$. Equation (22) shows the decrease of the correlation $|\mu|$ by increasing the surface roughness and the tilt. Both factors are independent from each other and depend on the synthetic wavelength Λ. The resulting statistical error for the measured distance then becomes

$$\delta L_S = \sqrt{2\sigma_h^2 + \frac{s^2a^2}{2}}. \tag{23}$$

The position of a diffusing (rough) surface is only determined within the limitation of the surface roughness σ_h and within the limitation of the illuminating spot size, as shown in Fig. 10 for a tilted surface.

Successful application of MWI for distance measurements to non-cooperative targets, using superheterodyne detection and a high power diode laser with acousto-optic modulators for a 500 MHz frequency shift ($\Lambda = 0.6\,\mathrm{m}$), was reported several years ago [10,33]. Nevertheless, the problem of the randomly distributed intensity remains. To improve the probability of detection by averaging, two-dimensional measurement is of great interest [25].

The system presented in Sect. 2.3 was adapted for measurements on non-cooperative targets [9]. The final set-up, shown in Fig. 11, is similar to the one used for ESPI (electronic speckle pattern interferometry) [34], except for the moving reference and the lock-in CCD. The imaging lens has a focal length

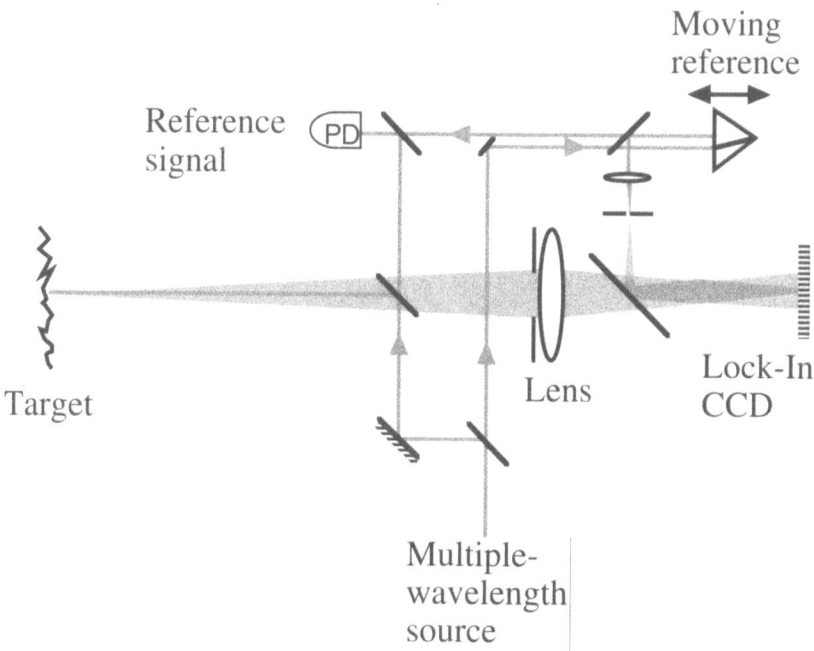

Fig. 11. Multiple-wavelength interferometer set-up with non-cooperative target

of 100 mm. The source consists of the low-power source described in Sect. 3.2 followed by the semiconductor amplifier (see Sect. 3.4). We used a synthetic wavelength of 4 mm in our experiment. The test object was a plane surface of frosted aluminum at 40 cm from the imaging lens. To improve the probability of detecting "bright" speckles, we used the two-dimensional version of the lock-in CCD which is composed of 5×12 pixels (see Fig. 3). The modulation power of the interference signal is detected in ten intervals of 10 ms duration and 6 ms separation, and phase shifted by 45°. Five 90° phase shifted values are obtained by adding two consecutive samples. The phase of the synthetic wavelength is then calculated using the five-frame error compensation algorithm [18]. During the integration phase, the sensitive area of the pixel is active for 5 ms (8×0.625 ms). Therefore, the effective integration time for one measurement is about 50 ms. To improve the statistical signal quality, only 30 pixels (50% of 5×12) with the best modulation power were retained for the phase evaluation in our experiment. The phase measurements for each retained pixel were then fitted to a plane surface. In this way, phase measurements at the center of the illumination area indicate a phase resolution of about $2\pi/200$, as shown in Fig. 12, corresponding to a distance resolution of about 10 μm. In this case, the tilt angle was less than 1° and therefore too small to be measured.

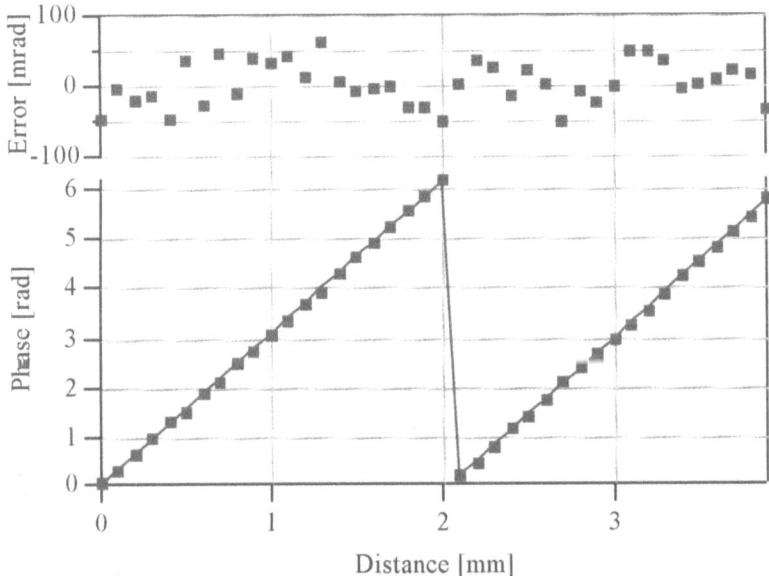

Fig. 12. Phase measurements on non-cooperative targets over a range of 4 mm, after signal processing and thresholding

4.3   Signal-to-Noise Ratio and Power Limitation

The noise of the detected signal will introduce phase fluctuations, which can be explained by considering the phasor representation of a noisy component on a harmonic signal. Therefore, the signal-to-noise ratio (SNR) should be high enough to ensure a good synthetic phase interpolation. The phase noise for a given SNR is [35]

$$\delta\phi = \frac{1}{\sqrt{\text{SNR}}}.\tag{24}$$

From (24), we see that a SNR of about 1000 is required to get a phase interpolation of $2\pi/200$. In the case of heterodyne detection, the optical power of the reference wave can be increased until the shot noise dominates the electronic noise. For a given quantum efficiency η, the required number of photons in the measuring arm then becomes

$$n_{\text{m}} = \frac{\text{SNR}}{\eta},\tag{25}$$

which is consistent with the fact that for shot noise limited detection the SNR is equal to the mean number of photo electrons from the signal [36]. Assuming a quantum efficiency of $\eta = 0.6$, we see that 1670 photons should arrive from the measuring arm onto each pixel to obtain an SNR of 1000. We note that this is true only if the optical power of the reference beam allows a shot noise limited detection. For an integration time of 0.625 ms and a wavelength of 850 nm, we find a minimal optical power of 650 fW per pixel, corresponding to an intensity of 2.2 nW/mm² on the detector (pixel size in Fig. 3 is 17 μm × 17 μm). Therefore, one of the limiting factor for the maximal distance which can be measured with non-cooperative targets is the optical power of the scattered light which returns to the detector. The power which is collected by the lens depends on the solid angle defined by the distance d_0 and the pupil diameter D (Fig. 11). The collected power is

$$P_1 = P_i \rho \left(\frac{D}{d_0} \right)^2 \tag{26}$$

where P_i is the incident power on the target, and ρ is the diffuse reflection coefficient. The intensity in the image plane is given by

$$I = \frac{4P_1}{\pi D_i^2} = \frac{4\rho}{\pi D_i^2} \left(\frac{D}{d_0} \right)^2 P_i , \tag{27}$$

where D_i is the diameter of the image of the illuminated spot on the target. In practice, the illuminated spot is chosen to give a diameter D_i of the image of about 2 mm, corresponding approximately to the size of the lock-in CCD. The pupil diameter D must be chosen so that one pixel includes only a few speckles in order to get a good interference contrast [34]. In our experiment, we used a pupil diameter D of 24 mm, giving a speckle diameter of 7 μm. For $P_i = 90$ mW and a distance of 2.3 m, we measured an intensity of 500 nW/cm² in the pupil plane. With an integration time of 0.625 ms per sample, we obtained a mean voltage of about 0.5 V from the lock-in CCD, which corresponds with the sensitivity of 0.78 μV/photon to about 600 000 photons. From (27), we see that the intensity in the image plane is inversely related to the square of the distance d_0. If the distance is increased by a factor of 5, the intensity is decreased by a factor of 25. At 11 m, one should therefore expect about 25 600 photons, corresponding to an optical power of 9 pW with an integration time of 0.625 ms (typical value). It has been shown above, that at least 1670 photons must be detected to obtain an SNR of 1000, corresponding to a phase noise of $2\pi/200$. For distance measurements at 10 m, the signal quality is therefore high enough to achieve synthetic phase interpolation better than $2\pi/200$.

As already mentioned in Sect. 2.3, the lock-in CCD has a fill-factor of less than 1%. In order to optimize the optical power, only those parts of the target which are seen by the pixels of the CCD should be illuminated. Therefore, the total output power can be substantially decreased by using appropriate beam-shaping optics. For that purpose, a microlens array was used in our experiment to match the illumination pattern with the detector array, as described in Fig. 13. For an optical power of 20 μW per illuminated spot, we obtained a mean voltage of about 0.6 V per pixel at a distance of 2.3 m, corresponding to about 800 000 photons. At a distance of 11 m then,

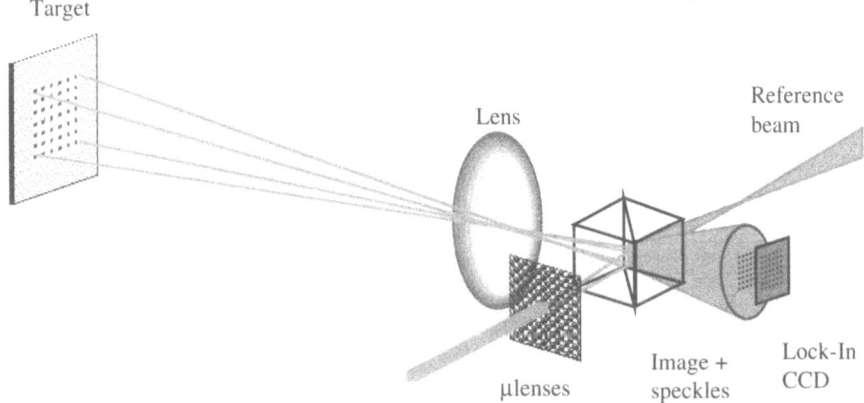

Target

Lens

Reference
beam

μlenses

Image +
speckles

Lock-In
CCD

Fig. 13. Optimization of the optical power by using an appropriate illumination pattern

about 32 000 photons are expected, corresponding to an optical power of 12 pW. Therefore, the corresponding SNR should be high enough to achieve synthetic phase interpolation better than $2\pi/200$. The total optical power which is required by the CCD is $5 \times 12 \times 20\,\mu W = 1.8\,mW$ for measurements at a distance of 10 m.

5 Conclusions

Multiple-wavelength interferometry is the most promising approach for absolute distance measurements with high resolution. A novel concept of a multiple-wavelength source, using tunable diode lasers, with absolute calibration by electronic beat-frequency measurements was developed. Experimental results of the comparison with an HP-interferometer prove that a calibration of the synthetic wavelength in the millimeter range with an accuracy better than 10^{-5} is feasible. With the reported three-wavelength source, absolute distance measurements were performed up to 200 mm (synthetic wavelength of 400 mm), and with a resolution better than 10 μm (synthetic wavelength of 4 mm, phase resolution better than $2\pi/200$).

Low-cost standard laser diodes can be used for measurements up to 5 m on cooperative targets. The main drawback of standard devices is the mode-hop characteristic and the low optical power which may not be sufficient for the measurements on non-cooperative targets. High-power DBR laser diodes, which can be tuned continuously over at least 1.5 nm and have a coherence length of about 40 m, are now commercially available. Using a fiber optic Fabry–Pérot resonator, a compact and portable stabilized multiple-wavelength source was developed. Calibration of the synthetic wavelength better than 10^{-5} was achieved with this source. Distance measurements to non-cooperative targets with a resolution of 10 μm were demonstrated, using a custom designed lock-in CCD and appropriate signal processing.

References

1. Michelson, A.A., Benoit, J.R. (1895) Détermination expérimentale de la valeur du mètre en longueurs d'ondes lumineuses. Trav Mem Bur Int Poids Mes 11: 1
2. Dändliker, R. (1992) Distance Measurements with Multiple Wavelength techniques. In: Linkwitz, K., Hangleiter, U. (Eds.) 2nd Internat. Workshop on High Precision Navigation. Ferd. Dümmlers Verlag, Bonn, pp 159–170
3. Dändliker, R., Hug, K., Politch, J., Zimmermann, E. (1995) High Accuracy Distance Measurement with Multiple-wavelength Interferometry. Opt Eng 34: 2407–2412
4. Zimmermann, E., Salvadé, Y., Dändliker, R. (1996) Stabilized Three-Wavelength Source Calibrated by Electronic Means for High Accuracy Absolute Distance Measurement. Opt Lett 21: 531–533
5. Fercher, A.F., Hu, H.Z., Vry, U. (1985) Rough Surface Interferometry with a Two-wavelength Heterodyne Speckle Interferometer: Appl Opt 24: 2181–2188
6. de Groot, P. (1991) Interferometric Laser Profilometer for Rough Surfaces. Opt Lett 16(6): 357–359
7. Dändliker, R., Thalmann, R., Prongué, D. (1987) Two-wavelength Laser Interferometry Using Superheterodyne Detection. Proc SPIE 813: 9–10
8. Dändliker, R., Thalmann, R., Prongué, D. (1988) Two-wavelength Laser Interferometry Using Superheterodyne Detection. Opt Lett 13: 339–341
9. Dändliker, R., Salvadé, Y., Zimmermann, E. (1998) Distance Measurement by Multiple-wavelength Interferometry. J Opt 29: 105–114
10. Manhart, S., Maurer, R. (1990) Diode Laser and Fiber Optics for Dual-wavelength Heterodyne Interferometry. Proc SPIE 1319: 214–216
11. Sodnik, Z., Fischer, E., Ittner, T., Tiziani, H.J. (1991) Two-wavelength Double Heterodyne Interferometry Using a Matched Grating Technique. Appl Opt 30: 3139–3144
12. Gelmini, E., Minoni, U., Docchio, F. (1994) Tunable, Double-wavelength Heterodyne Detection Interferometer for Absolute-distance Measurement. Opt Lett 19: 213–215
13. Salewski, K.-D., Bechstein, K.-H., Wolfram, A., Fuchs, W. (1996) Absolute Distanzinterferometrie mit variabler synthetischer Wellenlänge. tm-Technisches Messen 63: 5–13
14. Creath, K. (1988) Phase-measurement Interferometry Techniques. In: Wolf, E. (Ed.) Progress in Optics 26. Elsevier, Amsterdam, pp 349–393
15. Schwider, J. (1990) Advanced Evaluation Techniques in Interferometry. In: Wolf, E. (Ed.) Progress in Optics 28. North Holland, Amsterdam, pp 271–359
16. Spirig, T., Seitz, P., Vietze, O., Heitger, F. (1995) The Lock-in CCD – Two-dimensional Synchronous Detection of Light. IEEE J Quantum Electron 31: 1705–1708
17. Spirig, T., Marley, M., Seitz, P. (1997) The Multitap Lock-in CCD with Offset Subtraction. IEEE Trans Electron Devices 44: 1643–1647
18. Schmit, J., Creath, K. (1995) Extended Averaging Technique for Derivation of Error-compensating Algorithms in Phase-shifting Interferometry. Appl Opt 34: 3610–3619

19. Bourdet, G.L., Orszag, A.G. (1979) Absolute Distance Measurements by CO_2 Laser Multiwavelength Interferometry. Appl Opt 18: 225–227

20. Matsumoto, H. (1981) Infrared He-Xe Laser Interferometry for Measuring Length. Appl Opt 20: 231–234

21. de Groot, P., Kishner, S. (1991) Synthetic Wavelength Stabilization for Two-color Laser Diode Interferometry. Appl Opt 30: 4026–4033

22. Petermann, K. (1988) in Laser Diode Modulation and Noise. Kluwer, Dordrecht, chap. 7.6.5

23. Day, T., Brownell, M., Wu, I.-F. (1995) Widely Tunable External Cavity Diode Lasers. Proc SPIE 2378: 35–41

24. Bechstein, K.H., Fuchs, W. (1998) Absolute Interferometric Distance Measurements Applying a Variable Synthetic Wavelength. J Opt 29: 179–182

25. Dändliker, R., Geiser, M., Giunti, C., Zatti, S., Margheri, G. (1995) Improvement of Speckle Statistics in Double-wavelength Superheterodyne Interferometry. Appl Opt 34: 7197–7201

26. Mehuys, D., Welch, D.F., Goldberg, L. (1992) 2.0 W cw, Diffraction-limited Tapered Amplifier with Diode Injection. Electron Lett 28: 1944–1946

27. Morey, W.W., Ball, G.A., Meltz, G. (1994) Photoinduced Bragg Gratings in Optical Fibers. Optic & Photonics News, February 1994, 6–14

28. Gillard, C.W., Buholz, N.E. (1983) Progress in Absolute Distance Interferometry. Opt Eng 22: 348–353

29. Edlén, B. (1966) The Refractive Index of Air. Metrologia 2: 71–79

30. Salvadé, Y., Zimmermann, E., Dändliker, R. (1996) Limitations of Multiple-wavelength Interferometry Due to Frequency Fluctuations of Laser Diodes. Proceedings of the International Workshop on Interferometry. Optical Society of Japan, Waco, pp 9–10

31. Salvadé, Y., Zimmermann, E., Dändliker, R. (1997) Limitations of Interferometry Due to Frequency Fluctuations of Laser Diodes. Proceedings of the Topical Meeting on Optoelectronic/ Displacement Measurements and Applications. EOS, Nantes

32. Vry, U., Fercher, A.F. (1986) Higher-order Statistical Properties of Speckle Fields and Their Application to Rough-surface Interferometry. J Opt Soc Am A 3: 988–998

33. Fischer, E., Dalhoff, E., Heim, S., Tiziani, H.J. (1995) High Precision Absolute Interferometry Up to 100 m. In: Linkwitz, K., Hangleiter, U. (Eds.) International Workshop on High Precision Navigation. Ferd. Dümmlers Verlag, Bonn, pp 531–538

34. Jones, R., Wykes, C. (1983) in Holographic and Speckle Interferometry. Cambridge University Press, London, chap. 4.3

35. Dändliker, R. (1980) Heterodyne Holographic Interferometry. In: Wolf, E. (Ed.) Progress in Optics 17. North-Holland, Amsterdam, pp 1–84

36. Saleh, B.E.A., Teich, M.C. (1991) in Fundamental of Photonics. Wiley, New York, chap. 17.5

Speckle Metrology:
Some Newer Techniques and Applications

R. S. Sirohi

Summary. A speckle pattern arises due to the self-interference of a multitude of waves with random amplitudes and phases which originate from the scattering centers on the surface, illuminated by a coherent wave. A fully developed speckle pattern results when the rms value of the surface roughness is greater than a wavelength. The contrast of the speckle pattern is unity. The speckle pattern undergoes both positional shift and irradiance changes when the generating surface is displaced or deformed. In order to relate the changes in the speckle pattern to those at the surface, imaging geometry is most often used. The measurement methods are divided into two groups, namely, speckle photography and speckle interferometry. Speckle photography measures the positional changes of the speckles while speckle interferometry measures the phase changes, and consequently the irradiance changes. It may be noted that both these changes occur simultaneously with one possibly predominating over the other.

1 Speckle Photography

This technique involves recording a speckle pattern before and after the object deformation on the same photographic plate. The double-exposure specklegram is interrogated to extract information regarding the spatial shift of the speckles. This information is extracted by filtering, which can be performed either at the space or frequency plane. Point-wise filtering yields both the magnitude and direction of the speckle shift on a tiny region of specklegram which is illuminated by a narrow beam. This shift is then related to the local shift at the object. The specklegram is interrogated at a large number of points to generate the deformation pattern. The technique is laborious, but has now been fully automated. On the other hand, frequency plane filtering generates fringes over the whole specklegram and hence over the object: the fringes correspond to the constant displacement of speckles.

A simple analysis shows that in-plane displacement sensitivity is at least an order of magnitude higher than that of the out-of-plane displacement. Speckle photography, therefore, has been used for in-plane displacement measurement. The sensitivity and range of measurement are governed by speckle size and decorrelation effects.

1.1   Speckle Shear Photography

We are often interested in strain rather than displacement. Strain is obtained from displacement data by differentiation, a procedure that is error-prone. However, strain can be obtained by optical differentiation. A double-exposure specklegram when illuminated by two parallely-shifted narrow beams will generate two superposed Young's fringe patterns: the moiré pattern formed due to superposition is used to obtain strain. In a method used to record a double-exposure shear specklegram, the object is imaged by a lens having a two-aperture mask; each aperture carries a wedge and a sheet polarizer. The transmission axes of the sheet polarizers are crossed and the object is illuminated either by a circularly polarized light or a linearly polarized light with its azimuth at 45° to the polarizer transmission axis. Therefore, two sheared images of the object are recorded in the first exposure. The object is loaded, and another record is made. This double-exposure specklegram can be interrogated by a single narrow beam to generate the moiré pattern. Figure 1 shows a photograph of the moiré pattern thus obtained. Strain information over the whole object can be obtained by interrogating the specklegram at large number of points. The sensitivity of the technique is not variable as the shear is fixed during recording. The fringe contrast is also poor as four speckled beams participate in the moiré formation.

Recently photorefractive crystals, particularly barium titanate crystal, have been used for deformation measurement. Figure 2a shows a schematic of a two beam coupling arrangement. The object is illuminated by a narrow beam. The illuminated region of the object is imaged inside the crystal. A pump beam is added. After a short time, a refractive index grating is formed inside the crystal. This grating decays slowly when the object beam is blocked. During this period, the object wave is created by the interaction of the pump beam with the index grating. When the object is deformed,

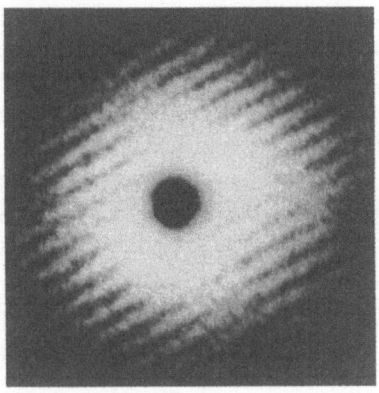

Fig. 1. Moiré fringes depicting strain

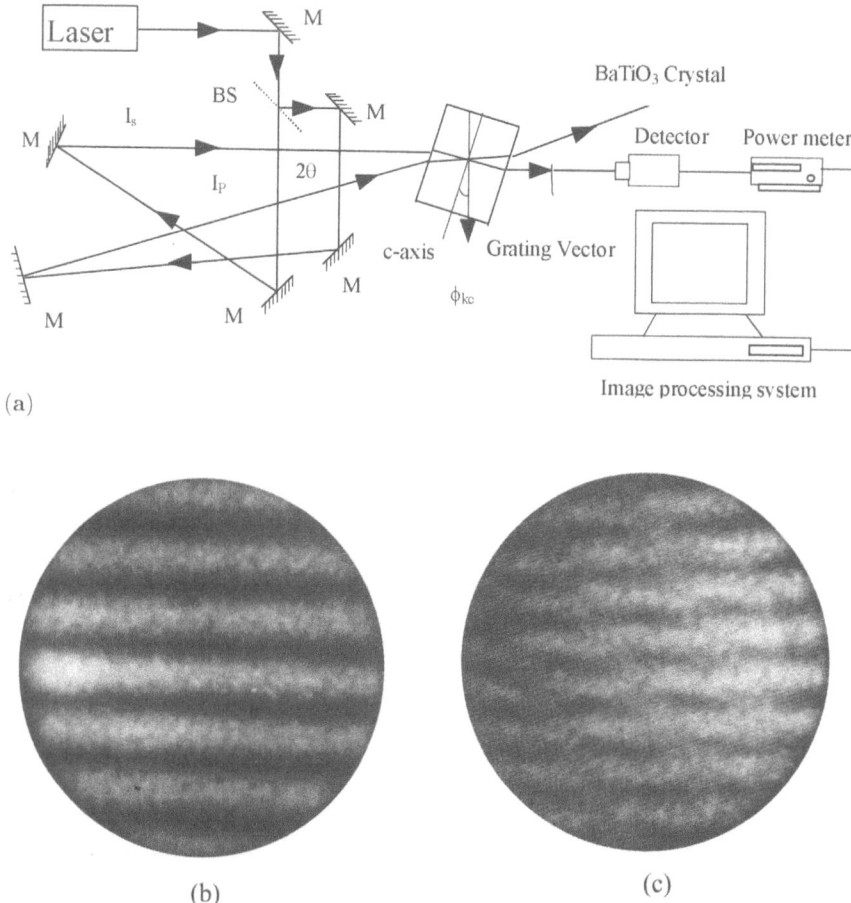

(a)

(b)   (c)

Fig. 2. Two-beam coupling. (**a**) Experimental set up. (**b**) In-plane displacement fringes. (**c**) Moiré fringes

the directly transmitted wave interferes with the diffracted wave producing fringes that represent in-plane displacement. Figure 2b shows a photograph of such a pattern. Instead of illuminating the object with a single beam, two shifted beams are used to obtain strain value. Due to loading, deformation is different at the two illuminated points, and hence two patterns corresponding to two different displacements are produced due to interference. These patterns form a moiré pattern, which is used to obtain strain. Figure 2c shows an interferogram depicting a moiré pattern.

2   Speckle Interferometry

Though the speckle pattern arises due to interference, we need to use a reference wave to code the phases of speckles in speckle interferometry. Initially, speckle interferometry was used to measure in-plane displacement. The object was illuminated by two symmetrical collimated beams with respect to the direction of observation. The arrangement is known as the Leendertz method. Here, one illumination wave acts as a reference for the other. The arrangement is sensitive to the in-plane component that contains the illumination beams, and the sensitivity is governed by the interbeam angle. The orthogonal in-plane component is measured by rotating the experimental set up by 90°. Another arrangement to measure in-plane displacement was proposed by Duffy. Unlike the Leendertz arrangement (dual illumination direction and single observation direction), Duffy's method uses a single illumination beam and two symmetric directions with respect to the optical axis for observation. Therefore, the imaging lens carries a pair of apertures that define the observation directions. The sensitivity is now governed by the lens aperture. The Leendertz method has a very high sensitivity ($\lambda/2 \sin \theta$, 2θ is the interbeam angle) but a small range due to decorrelation, while Duffy's method has a low sensitivity ($\lambda/2 \sin \alpha$, 2α is the inter-aperture angle) and a large range to due large speckle size. Both these methods can be combined into one thereby extending both the sensitivity and range of measurement. Figure 3a shows an experimental set up. Essentially four beams participate in the formation of a speckle pattern. The double-exposure specklegram, when whole field filtered, generates the system of fringe patterns shown in Fig. 3b. The fine fringe pattern due to Leendertz is modulated by a coarse pattern due to Duffy. The object is a cantilever that has been in-plane loaded between exposures.

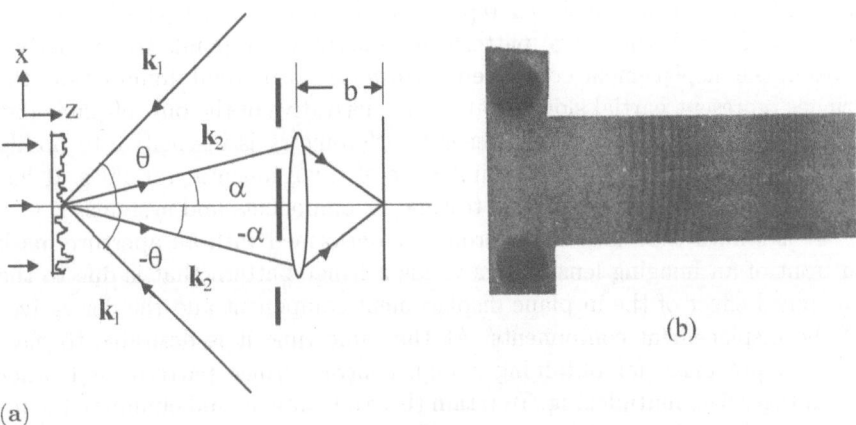

(a)   (b)

Fig. 3. Speckle interferometry. (**a**) Experimental arrangement. (**b**) In-plane fringes on a cantilever

Speckle interferometry is also used to measure the out-of-plane displacement component. Essentially, a Michelson interferometer with one of its mirrors replaced by the object is used for measurement. The object is imaged on the recording plane and the mirror provides a specular reference wave.

Duffy was the first to introduce the mask in front of the imaging lens, which amounts to using a band of frequencies for image formation. This is equivalent to pre-filtering thereby facilitating whole field filtering as one of the first order halos easily used for information retrieval. Duffy's arrangement is sensitive to the in-plane displacement components along the line joining the centers of the two apertures. Therefore, it is natural to use three or four apertures to obtain both the components of in-plane displacement. A modification of the two-aperture arrangement in which a ground glass is placed on an aperture and is illuminated by an unexpanded beam to generate a diffuse reference beam makes this configuration out-of-plane sensitive. Therefore, it becomes possible to retrieve all three components of the deformation vector by a judicious choice of the aperture configuration. Indeed with the use of shear elements on additional apertures, the slope fringes can also be obtained in addition to the displacement fringes corresponding to the three components. On the other hand, several states of the object can be recorded either by shifting apertures longitudinally (frequency multiplexing) or angularly (θ-multiplexing). The total number of states that can be recorded is limited by the dynamic range of the recording medium, and the number of apertures such that their spectra do not overlap.

2.1   Speckle Shear Interferometry

In speckle shear interferometry, a point on the object is imaged as two points, or, a single point on an image receives contributions from two points on the object. Essentially, two displaced images are recorded: one speckle field acts as a reference to its displaced replica or vice versa. When the images are laterally sheared, the fringe pattern, in general, corresponds to the derivatives of the displacement components. However, for normal illumination, the fringes represent partial slope, that is, the derivative of the out- of-plane displacement component. Speckle shear interferometry is insensitive to bodily displacements and tilts, and the influence of environmental variables. It has been used for the nondestructive testing of components and systems.

It is known that shear interferometry performed with an aperture mask in front of an imaging lens always yields a fringe pattern that is due to the combined effect of the in-plane displacement component and the derivatives of the displacement components. At the same time it is desirable to have an aperture mask for obtaining a high contrast fringe pattern, and other advantages like multiplexing. To retain these advantages and eliminate the in-plane displacement sensitivity, the configuration of Fig. 4 is used. The object is illuminated normally and viewed axially. The object beam is divided and arranged so that it passes through the apertures to form two images. The

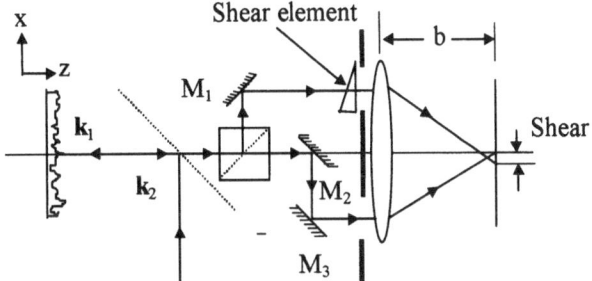

Fig. 4. Experimental set up for speckle shear interferometry, which is insensitive to in-plane displacement

shear is introduced by placing a wedge plate in front of an aperture. Shear could also be introduced by tilting any one of the mirrors M_1, M_2, or M_3. The mirror combination M_2, M_3 compensates for any extra path difference. This arrangement yields pure slope fringe pattern, and, in addition, it is amenable to spatial phase-shifting.

3   Electronic Speckle Pattern Interferometry (ESPI)

In the methods described so far, either photographic material or PR crystals are used for recording. The underlying assumption was that the speckles are resolved by the recording material. In fact, the speckle size in a subjective speckle pattern can be controlled by the F# of the imaging lens. Therefore, speckle methods can be used with wide variety of recording media such as electronic detectors like CCD arrays. Use of electronic detectors eliminates the messy development process and the data can be processed at video rates. Further, the two exposures can be handled independently. The technique of performing speckle interferometry with electronic detectors is known as electronic speckle pattern interferometry (ESPI) or simply video holography. The availability of fast PCs and large density CCD detectors makes ESPI a very attractive technique; perhaps it is an alternative to holographic interferometry in the industrial environment. Further phase-shifting is easily adopted with ESPI. All the techniques known under speckle interferometry and speckle shear interferometry are amenable for electronic detection.

The response of two objects, say a master and a test object, to an external agency can be compared by comparative ESPI. Figure 5a shows a schematic of the experimental set up. The experimental configuration is sensitive to out-of-plane displacement only. The objects A and B are placed equidistant from the beam splitter, and are imaged on the CCD array. The first frame is stored in the frame grabber. The objects are then subjected to loading and the second frame is grabbed. The intensity values are subtracted pixel by pixel and the difference signal, after rectification, is displayed on the monitor.

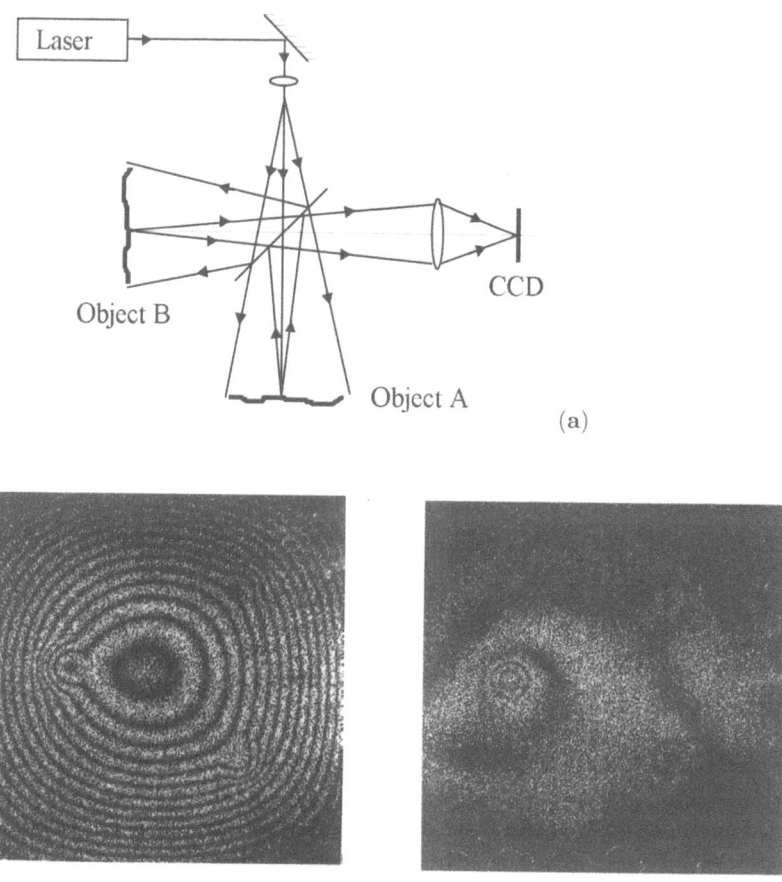

(a)

(b)   (c)

Fig. 5. Experimental set up for comparative ESPI. (**b**) Uncompensated interferogram. (**c**) Compensated interferogram

Since the configuration is sensitive to out-of-plane displacement only, the dark fringes are due to

$$2[w_A(x, y) - w_B(x, y)] = m\lambda$$

where w_A and w_B are the out-of-plane displacement components of the objects A and B respectively. If the responses of the objects are identical, $w_A = w_B$, a fringe-free field appears irrespective of the loading. On the other hand, those areas of the object that respond differently to the loading exhibit fringe patterns. The technique, therefore, has promise as a nondestructive method for comparing the response of a test object with that of the master object at least in the laboratory conditions. It is possible to compensate fairly large deformations. Figures 5b and 5c show interferograms, one uncompensated and the other compensated. The objects were edge-clamped diaphragms with one having programmed defects. They are uniformly loaded by an application of pressure between the frames.

The measurement of residual stress of a component is essential for knowing its performance over an extended period. The stress is released either by (i) increasing the temperature or (ii) hole drilling. Heating usually results in decorrelation and hence is not often attempted. On the other hand, hole drilling, both blind and through, is frequently used. Both the out-of-plane and in-plane displacement components due to stress release have been measured. The following set up produces in-plane displacement fringes after stress release. The experimental set up shown in Fig. 6a is a Leendertz configuration with 2D-sensitivity. A telephoto lens with a sufficiently long working distance is used for imaging. The first frame is captured and stored in the frame grabber. Using a high-speed diamond drill, a tiny blind hole is very carefully drilled. The drill is removed and then the second frame is captured. The two frames are subtracted pixel by pixel to yield in-plane sensitive fringe patterns. Figure 6b,c shows interferograms for a u-family and v-family of fringes. From these interferograms, the u- and v- components of in-plane displacement are calculated. An excellent agreement between theory and experimental results is obtained.

Shear ESPI is best suited for nondestructive testing as it is insensitive to bodily displacements and tilts, and the influence of environmental variables. Like in holographic interferometry, its success depends on the judicious choice of the loading method. In some situations, it may be desirable to obtain the size of the defect. Using plate theory, we present a method to obtain both the diameter and thickness of a defect, which is assumed to be a diaphragm on a rigid boundary. For this purpose, shear ESPI using Michelson interferometer is performed. The object is a thick plate with programmed circular defects. This is edge-clamped and uniformly loaded by an application of pressure. Using phase-shifting, the slope map at the defective region and then the midsection of this slope map is obtained. From this distribution, an approximate diameter of the defect is obtained. The gradient of this curve at the center $w''|_{x=0}$ is measured and can be expressed as

$$w''|_{x=0} = -4\frac{w_0}{a^2}$$

where a is the radius of the diaphragm and w_0 is its central deflection. It is thus seen that the gradient of the slope curve at the center of defect is proportional to w_0. Further it can be shown that

$$\frac{w''|_{x=0}a}{p} = -\frac{3(1-\mu^2)}{4E}\left(\frac{a}{t}\right)^3$$

where p is the applied pressure and t is the thickness of the diaphragm. A calibration curve between $(w''|_{x=0}a)/p$ and $(a/t)^3$ can be generated for a given material. The thickness of the unknown defect, assumed to be a

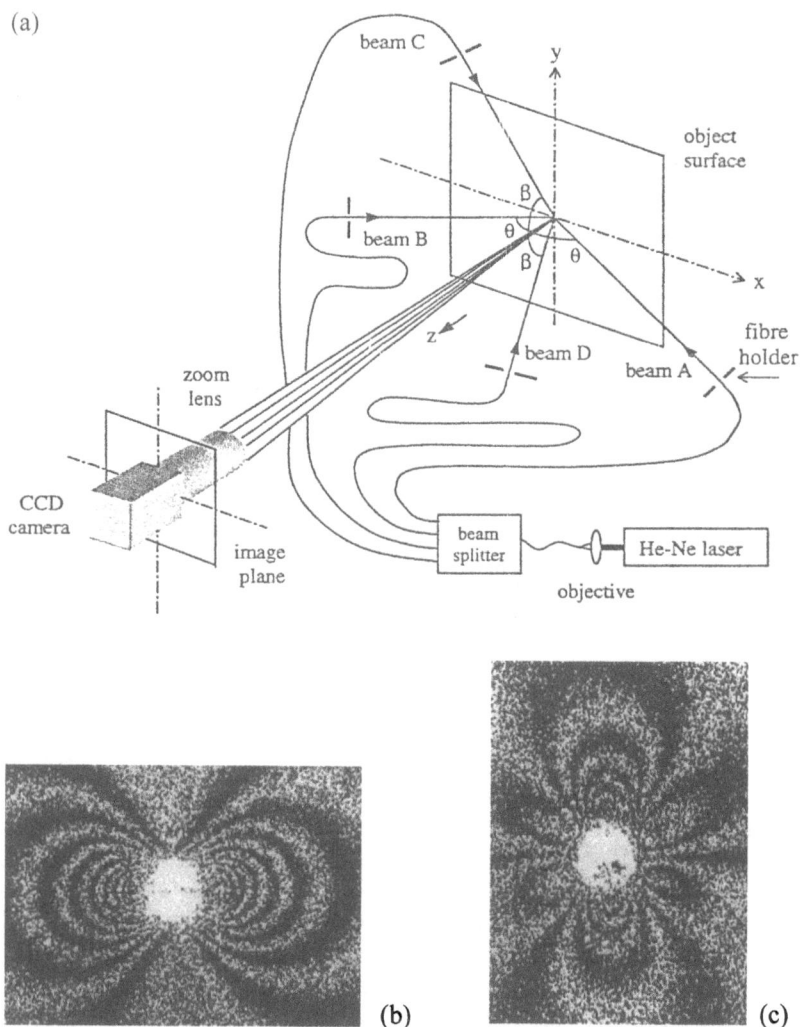

Fig. 6. (a) Experimental set up. (b) u-family and (c) v-family of in-plane displacement fringes (Courtesy Zhang, Data Storage Institute, Singapore)

diaphragm, is then obtained from the measured value of $w''|_{x=0}$, a, and p, using this calibration curve.

The natural surfaces are good candidates for speckle metrology. Treatment of surfaces, however, does help in several cases. For example, coating a dark surface with white paint improves fringe contrast. Further, when retroreflective paint is applied to the surface, we can essentially enhance the sensitivity for in-plane displacement measurement by a factor of two, but the sensitivity remains practically unchanged for out-of-plane displacement measurement. The increase in sensitivity for in-plane measurement is realized at the expense of perspective error, which becomes serious when large angles of illumination are used.

Speckles are very sensitive to environmental variables. For example, they twinkle when the temperature in the path varies. Also, when a speckle phenomenon is used for measurement in a hostile environment, temporal phase-shifting yields highly erroneous results. It is, therefore, useful to employ spatial phase-shifting. Various means are available to introduce a carrier for spatial phase-shifting. For example, when ESPI is used for out-of-plane displacement measurement, a diverging reference wave is made to originate from a point slightly shifted from the center of the exit pupil of the imaging lens. For in-plane measurement and also for shear ESPI, Duffy's two-aperture arrangement with appropriate aperture separation generates the carrier.

It has been pointed out that the aperturing of the imaging lens yields separation of fringe patterns corresponding to displacement components and their derivatives in speckle interferometry. It should be possible to extend this to ESPI provided the detector has sufficient resolution. In addition, using the Fourier transform method with phase-shifting, in-plane and out-of-plane displacement components and slope can be obtained. Therefore, the availability of large density CCD arrays and still faster PCs would soon make it possible to obtain the displacement components and slopes using ESPI. At the same time, PR crystals may compete with electronic detection.

Acknowledgements

A part of the work reported here was carried out during the stay at the National University of Singapore.

References

1. Dainty, J.C. (Ed.) (1975) Laser Speckle and Related Phenomena. Springer-Verlag, Berlin
2. Erf, R.K. (Ed.) (1978) Speckle Metrology. Academic Press, New York
3. Françon, M. (Ed.) (1979) Laser Speckle and Applications in Optics. Academic Press, New York
4. Jones, R., Wykes, C. (1989) Holographic and Speckle Inteferometry. Cambridge University Press, Cambridge
5. (1991) Selected Papers on Speckle Metrology. In: Sirohi, R.S. (Ed.) Society for Photo-Optical Instrumentation Engineers Milestone Series MS 35. SPIE Optical Engineering Press Bellingham, WA
6. Sirohi, R.S. (Ed.) (1993) Speckle Metrology. Marcel Dekker, New York
7. (1991) Selected Papers on Electronic Speckle Pattern Interferometery. In: Meinlschmidt, P., Hinsch, K.D., Sirohi, R.S. (Eds.) Society for Photo-Optical Instrumentation Engineers Milestone Series MS 132. SPIE Optical Engineering Press Bellingham, WA

Limits of Optical Range Sensors and How to Exploit Them

G. Häusler, P. Ettl, M. Schenk, G. Bohn, and I. Laszlo

Summary. It appears that only three different physical principles are required to classify all range sensors: (classical) interferometry on smooth surfaces (type III), broad-band interferometry on rough surfaces (type II), and triangulation (type I). We will show that signal formation in classical interferometry is completely different from signal formation in white-light interferometry on rough surfaces. The latter (type II) principle has some remarkable features: the distance uncertainty does not depend on the aperture of observation, and it displays some kind of "superresolution" by acquiring surface roughness, even if the lateral microtopology of the object is unresolved. We will demonstrate this by experiments and theory.

1 Introduction

We will discuss the signal formation of different types of range sensors, for a couple of reasons. The first reason is curiosity: how many sensor principles are necessary to cover the vast range from atomic distances to stellar distances? The second is of great practical value: what is the ultimate physical limit of different principles? It will turn out that the question about the limits is extremely helpful to get a deep insight into the underlying physical principles, and to build a new sensor with unique properties.

Not to overload the patience of the reader, we will summarize the results now. It appears that all range sensors might be classified into one of only three types:

- Type I: triangulation on rough surfaces, such as laser triangulation, phase-measuring triangulation, focus sensing, non-fluorescent confocal microscopy. The basic source of noise on the measured distance signal is – for these sensors – coherent noise. As a fatal consequence, the ultimate limit of distance uncertainty δz scales with the square of the measuring distance.

- Type II sensors are sensors based on multi-wavelength interferometry applied to optically rough surfaces. This new type of sensor is specifically interesting and will be discussed in detail. One unique property of these sensors is that the basic source of noise on the measured distance is the roughness of the object. This property enables us to measure surface roughness without laterally resolving the microtopology. Furthermore, the ultimate limit of distance uncertainty does not depend on the measuring distance. Hence, these sensors can measure down into deep boreholes, without loss of longitudinal accuracy.

- Type III sensing is classical interferometry on optically smooth surfaces. Sensors of this type average the measured distance over the area of the point spread function. Hence, the standard deviation of the measured signal decreases inversely with the measuring distance.

2   About Smooth and Rough Surface Interferometry

Multiple-wavelength interferometry goes back to Michelson [1], who measured the smooth end surfaces of the standard meter. After the invention of the laser, a very early two-wavelength method for surface profilometry on rough surfaces was proposed by Hildebrandt and Haines [2]. Vry and Fercher [3] investigated the potential of fast two-wavelength distance measurements on rough surfaces. The method was improved by heterodyning (Sodnik et al. [4]).

A generalization of "multiple-wavelength" is "white-light". Although white-light microscope interferometers have been known for more than 50 years, white-light interferometry experience a renaissance for about 10 years, because it is a valuable tool in the semiconductor industry, as pointed out by Davidson et al. [5]. Lee and Strand [6] demonstrated for optically *smooth* surfaces that the longitudinal resolution of a coherence scanning microscope is decoupled from the lateral resolution.

White-light interferometry can measure the topology of surfaces by acquiring the maximum of the so-called correlogram, for each pixel in the image. Our measuring device (we called it coherence radar [7]) is a white-light Michelson interferometer where we replace one mirror by the object, which jointly with the reference mirror is imaged on the CCD camera. The coherence radar was introduced as a method to measure the topology of optically *rough* surfaces (e.g. technical surfaces, human skin [8] for medical applications), while classical white-light interferometry can only measure smooth surfaces. To emphasize the difference to classical white-light interferometry, we mention below that for certain applications we can replace the reference mirror by a ground glass [18].

Since the signal formation from smooth or rough surfaces is completely different, we have clearly to distinguish both cases. What does *smooth* and *rough* surface mode mean?

Let us consider an imaging device which images the object into the image plane, where we have a point spread function (lateral resolution cell) of width δx (see Fig. 1a). If the height variation of the object within the resolution cell is smaller than $\pm \lambda/8$ then we do not have a phase variation greater than $\pm \pi/2$ and the elementary phasors always add up constructively. This means that in the image plane we will have only constructive interference in any

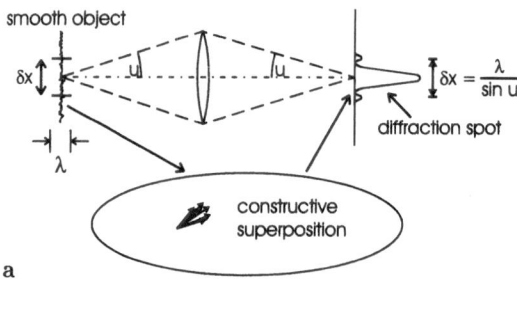

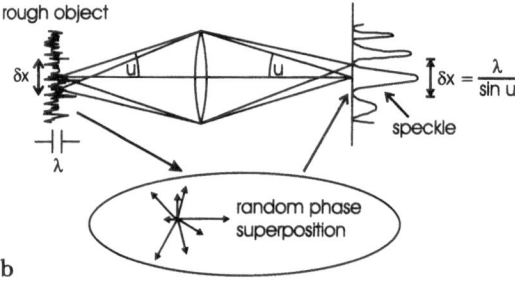

Fig. 1. (**a**) Signal formation in smooth surface mode; (**b**) signal formation in rough surface mode

pixel. No speckles appear. We will call this mode "smooth surface mode" or "resolved mode". For smooth surfaces, the amplitude

$$A = \frac{1}{N} \sum_{k=1}^{N} \exp(\mathrm{i}\varphi_k) \tag{1}$$

for N scatterers in a resolution cell can be approximated by the Taylor expansion of the exponential, because all the individual phases are small. Therefore the amplitude A simplifies to the average:

$$A \approx 1 + \mathrm{i}\langle\varphi_k\rangle. \tag{2}$$

The total amplitude is the superposition of A and the reference wave. It can be seen that $\mathrm{i}\langle\varphi_k\rangle$ introduces a shift in the distance signal which is equal to the mean distance of the individual scatterers inside the resolution cell. This means that white-light interferometry averages the distance over the object area within the resolution cell. Or, in other words, the standard deviation of the measured shape is proportional to the observation aperture $\sin(u)$ (see Fig. 1a). If we close the observation aperture, or increase the observation distance, the object appears more smooth. We call this scaling behavior "type III" [9,10].

   If the height variations are greater than $\pm\lambda/4$ (see Fig. 1b) phase variations greater than $\pm\pi$ appear and we may have destructive interference (we skip the transient range between $\lambda/8$ and $\lambda/4$). Fully developed speckles can appear in the image plane if the height variations are big enough. This is only possible if we take care to achieve spatially coherent observation [7]. We call this mode "rough surface mode" or "unresolved mode", in which we do not average over *the phase* emitted by laterally unresolved details, but over the complex amplitude. In this mode the lateral resolution cell of the sensor optics is larger than the lateral correlation length of the object microtopology. Hence, the elementary waves emitted by the object that are accumulated in the lateral resolution cell of the sensor have large statistical phase differences. The phase of each resolution cell is statistically independent of its neighbors, because the approximation of the amplitude in (2) is no longer valid. We want to emphasize that the property of "smooth" or "rough" surface does not depend only on the object. It depends on the resolution of the imaging system as well. For example, even a ground glass can be smooth for a high-aperture microscope.

3   Smooth Surface Mode

First, let us consider the smooth surface mode (classical interferometry). For the evaluation of the maximum of the correlogram envelope the Fourier transformation method [11,12] is established. We have developed two other evaluation methods [13,14]: the "three-point Gaussian interpolation" and the "enhanced three-point Gaussian interpolation", which are faster. This advantage, however, has sometimes to be paid for with less accuracy for non-cooperative objects [13,14]. To demonstrate the accuracy of our methods we will give some measurement results. The calibration gauge in Fig. 2a has a groove of 254 nm depth and the data have a noise level of about 5 nm. Figure 2b shows a mirror measured with a standard deviation of only 3 nm. We do not know whether this is the object itself, the reference, or if it is caused by the measuring device or by the evaluation.

The following interesting measurement results will clearly confirm that for the smooth surface mode white-light interferometry averages the *distance* over the object area within the resolution cell. Furthermore, we will calculate and compare roughness parameters to the values given by the PTB (Physikalisch Technische Bundesanstalt, i.e. the German office of standards).

Figures 3a,b show the PTB roughness standard SR13. It is not as evident as before whether this is a smooth object or a rough object because the object has a total height variation of about 250 nm ($\approx \lambda/3$). But *in each resolution cell* the height variation is below $\pm \lambda/8$. Hence, we have to consider it as a smooth object, according to our considerations above. Looking at the object with a higher resolution (Fig. 3b), it is obvious that we are in the smooth surface mode. In this case the width of the resolution cell is about 1.6 μm, much smaller than the lateral correlation length of the sample. The measurement uncertainty is a few nanometers only, in both cases. Another roughness standard, PTB RNO893, was measured with different resolutions

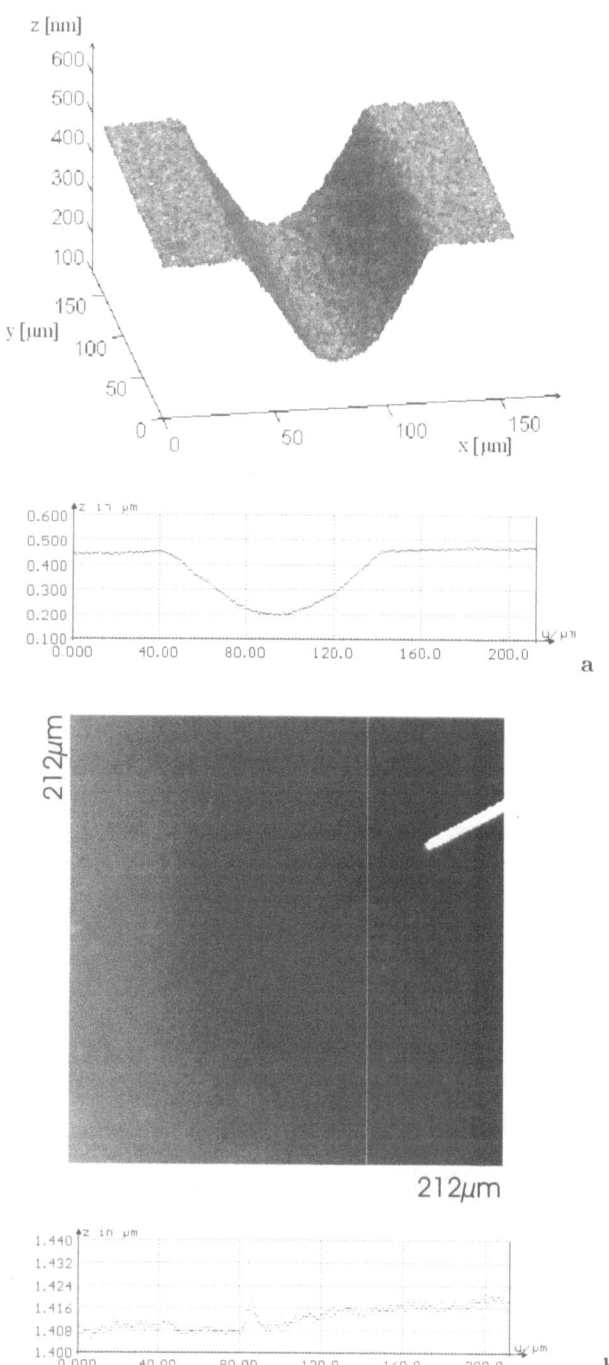

Fig. 2. (a) Measurement of a calibration gauge in smooth surface mode; (b) measurement of a mirror in smooth surface mode

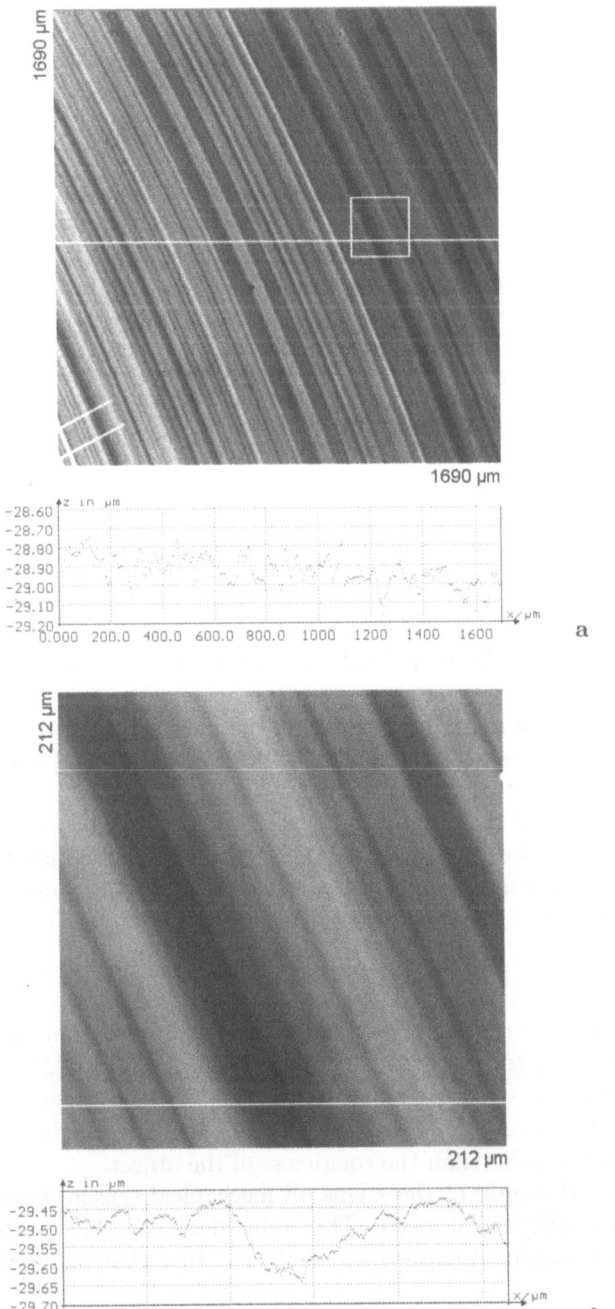

Fig. 3. (a) Measurement of a PTB roughness standard SR13 with low resolution (aperture 0.12); (b) measurement of a PTB roughness standard SR13 with high resolution (aperture 0.60)

(see Figs. 4a,b). With high resolution (Fig. 4b) the fine details are resolved, but as the resolution becomes lower (Fig. 4a) we can no longer laterally resolve the fine structure. The fine details are "averaged" as stated above, i.e. they get lost, but the coarse shape of the object still remains the same, e.g. the large groove on the right-hand side or even the smaller on the left one are clearly visible in both profiles.

For this object we calculated roughness parameters from the data acquired with the coherence radar, which in this case degenerates to a classical white-light interferometer. We compared them to the PTB values measured by a mechanical stylus. For the measurement and evaluation of the acquired data with mechanical stylus sensors, rules are given by the German DIN standards 4762, 4768 and 4777, and the ISO standard 4287. These rules define the roughness parameters R_a (arithmetical mean deviation) and R_q (root mean square deviation) as follows:

$$R_q = \sqrt{\frac{\sum_{i=1}^{N}(z_i - \bar{z})^2}{N}} \tag{3a}$$

$$R_a = \frac{1}{N}\sum_{i=1}^{N}|z_i - \bar{z}| \tag{3b}$$

where z_i denotes the height of the individual scatterer relative to the mean surface height $\bar{z}$ and N is the total number of scatterers.

Hence, R_q measures the standard deviation σ_h of the surface height distribution. For clarity we call these parameters R_a^c and R_q^c when measured with the coherence radar, in order to distinguish them from the standard stylus measurements. We determined the roughness parameter R_a^c from our optical measurements analogous to the rules given in the DIN and ISO standards. The results are displayed in Fig. 5. R_q is not specified by the PTB, but we determined R_q^c from our data. The variation for different resolutions is similar to that of R_a^c.

Deviations can be explained as follows:

- For different lines across the sample, the surface height distribution can vary. Hence, the determined roughness parameters can vary as well.
- Although the "roughness" of the reference mirror can introduce some error, we do not take it into account, because we assume that the roughness of the mirror is much smaller than the roughness of the object.
- Because of the finite diameter of the stylus all mechanical sensors have an inherent low-pass filtering property. The stylus does not "fall" into narrow grooves. The measured surface will look smoother than the real surface.

We conclude two major results, for *smooth* surface measurements:

1. There is a lateral averaging of the surface topology over the area of the resolution cell.

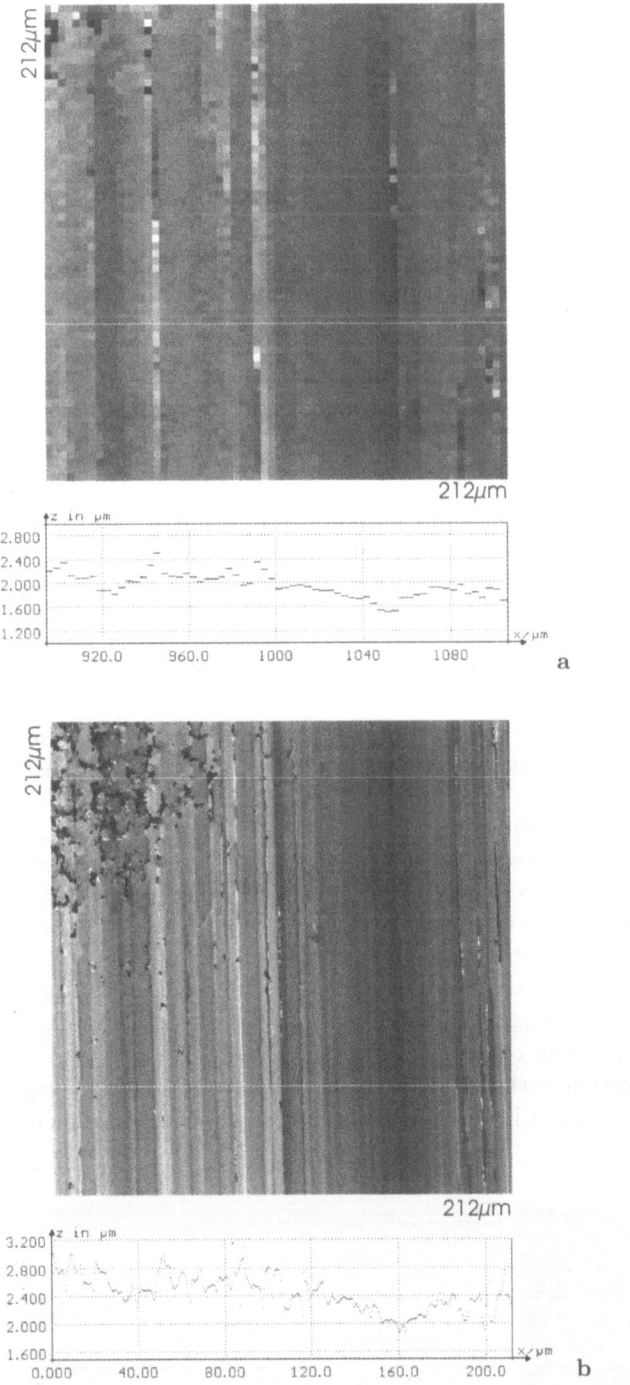

Fig. 4. (a) Measurement of a PTB roughness standard RN0893 (aperture 0.12); (b) measurement of a PTB roughness standard RN0893 (aperture 0.60)

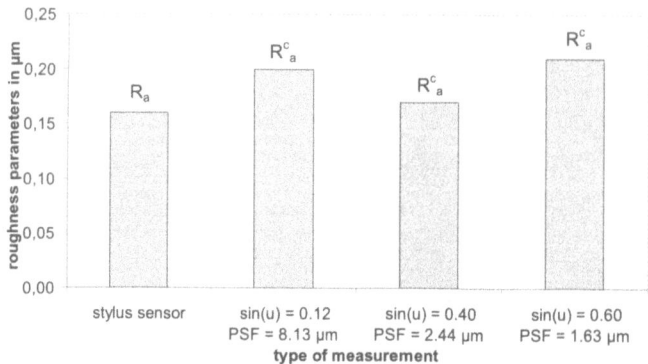

Fig. 5. Roughness parameter R_a and R_a^c measured with a mechanical sensor and the coherence radar, respectively

2. The roughness parameters can be determined optically, as well as with a mechanical stylus, but with the advantage of non-contact and extremely high lateral resolution.

4   Rough Surface Mode

So far we have considered "classical white-light interferometry" with high longitudinal resolution evaluation. Let us now discuss the more challenging case of the rough surface mode. We have to remember now that over the resolution cell the object height variations are greater than $\pm\lambda/4$ (see Fig. 1) and the phase variation introduced by this height variation is greater than $\pm\pi$. This results in speckles in the image plane (fully developed for a sufficiently wide height distribution). For this case, the system does not acquire the microtopology itself, because this is not resolved. What we see is a statistical representation of the microtopology. The unresolved microtopology introduces a "noise" on the measured distance signal. For each speckle, the correlogram is independent of its neighbors (see Fig. 6).

The first observation is that the phases of the correlograms are uncorrelated. That makes classical interferometry impossible, because phase evalu-

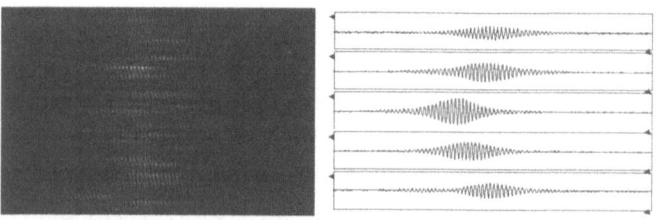

Fig. 6. Typical correlograms for a rough surface

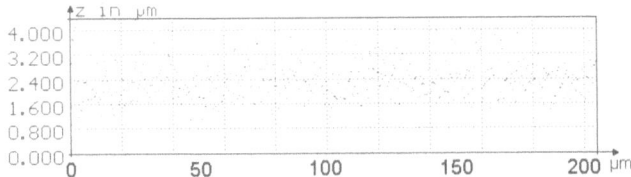

Fig. 7. Profile of an object measured in rough surface mode

ation does not give any information about the object shape as was already pointed out in [7]. Instead, we have to evaluate the contrast of the correlogram. The second observation is that the location of the correlogram is not the same for each speckle. As depicted in Fig. 6, the position of the envelope maximum (our distance signal) may vary by several wavelengths. The interesting question is: how does this *distance signal "noise"* relate to the surface topology? An example of a measurement in the rough surface mode is given in Fig. 7. The distance signal "noise" is obvious.

However, we will demonstrate that although we do not have access to the (optically unresolved) microtopology (because we image with very low aperture), we can nevertheless calculate the standard deviation σ_h of the surface height variation from the measured data.

It was shown by Ettl [15] and Dresel [16] that we can calculate σ_h from the mean absolute deviation $\langle |z| \rangle$ of the measured distance signal z. In fact, it turns out that σ_h is equal to $\langle |z| \rangle$:

$$\sigma_h = \langle |z| \rangle . \tag{4}$$

This result is extremely surprising. Although in the rough surface mode we only detect a distance signal which does not directly represent the real surface height distribution, we can still determine σ_h from the statistical variations, i.e. from the mean distance signal $\langle |z| \rangle$. This means that although we do not optically resolve the lateral correlation length of the object, we nevertheless have access to the standard deviation of the unresolved topology, which we call R_q^c when measured with the coherence radar. In a statistical sense, we have some kind of "superresolution".

For our experimental tests we used the roughness standard RUGOTEST No. 3 (the French roughness standard). We measured eight samples [13] (N6a, N6b, N7a, N7b, N8a, N8b, N9a, N9b) of the roughness standard in the rough surface mode. For each we calculated the roughness parameter R_q^c from our measurement results and compared it to the parameter R_q measured by a mechanical stylus. With the mechanical stylus we measured three lines on each sample. For each measurement we calculated R_q in accordance with the DIN and ISO standards. From the three measurements we calculated the mean value of R_q for each sample. With the coherence radar we measured each sample only once, because the system supplies data for a whole surface area, not just for a line. From these data, for each sample we take five lines.

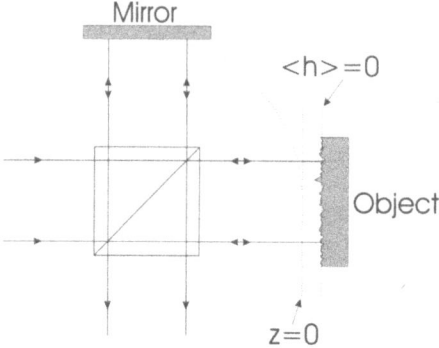

Fig. 8. Setup of the coherence radar

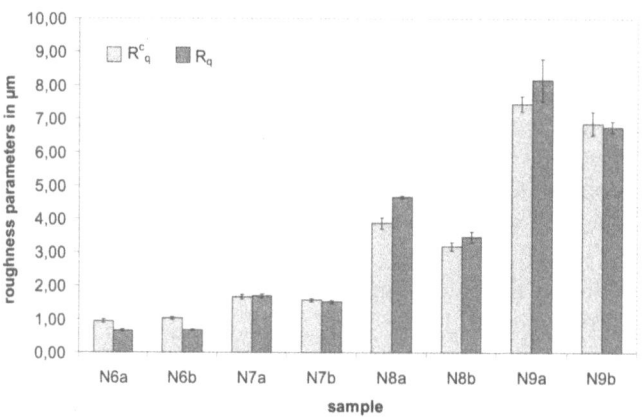

Fig. 9. Root mean square deviation measured by a mechanical stylus (R_q), and by the coherence radar (R_q^c)

First we calculate $\bar{z}$ which is then defined as the zero level (see Fig. 8). Then we calculate $\langle |z| \rangle$ in order to get R_q^c. Again, from the five individual values we calculate the mean value of R_q^c for each sample. The results are shown in Fig. 9.

It can be seen that the measured values correspond to each other. The standard deviations of the results show that the roughness parameter variations are bigger for higher roughness, both for the mechanical sensor as well as for the coherence radar. This can lead to deviations in the measurement results between the coherence radar and the mechanical sensor. In addition, other reasons for deviations are similar to those explained in the smooth surface mode.

We also determined the roughness parameters R_z^c and R_{max}^c analogous to roughness parameters R_z and R_{max} for mechanical stylus sensors as described in the DIN and ISO standards. Although in the rough surface mode we mea-

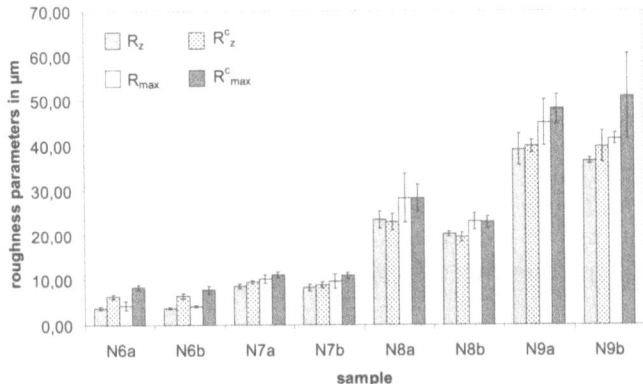

Fig. 10. Roughness parameters R_z, R_{max} and R_z^c, R_{max}^c measured with a mechanical sensor and the coherence radar, respectively. The lateral correlation length of the microtopology is *not* resolved by the sensor

sure a statistical representation of the surface, the maximum deviation from the mean surface level is close to the data measured by a mechanical sensor. The results are displayed in Fig. 10.

R_z and R_{max} relate to the maximum deviation from the mean surface height level. Therefore these roughness parameters are highly dependent on the extrema of the measurement data. For optical sensors it is therefore very important to exclude all erroneous data points (outliers). Specifically, for very rough surfaces it is sometimes not easy to distinguish between correct and erroneous data points. There is no straightforward standard to decide this, which means that the method to exclude the erroneous data points has an influence on the determined roughness parameters R_z^c and R_{max}^c. But no optical method can avoid single measurement errors, due to the intensity variation of the speckle (zero intensity may occur). Therefore a standard is needed to remove these single measurement errors in order to get unambiguous roughness parameters R_z^c and R_{max}^c. Such a standard does not yet exist.

5   Conclusions

The result of the previous section is quite remarkable. Although we have no means to measure the sub-resolution microtopology of the object, we can nevertheless determine its standard deviation from the acquired signal. Thus we achieve some kind of "superresolution". This behavior is apparently different from almost all other measuring systems in optics, mechanics, or electronics, where the measuring device introduces some low-pass filtering on the resolution. The underlying reason why classical interferometry and the coherence radar act so differently is that the first averages over phases, while the second averages over complex amplitudes.

The independence of the size of the lateral resolution cell means that we can measure surface roughness independently of the measuring distance and independently of the observation aperture. We have described this property of the coherence radar in earlier papers [18–20]. We can clearly distinguish this type of sensor behavior from classical interferometry, and from all triangulation methods. That is why we gave the method an individual name: "coherence radar" [7]. It turns out that there are three types of sensor principles, which are different in how the physical measuring uncertainty scales with the observation aperture. We will call them type I, II, and III [9].

5.1   Type I Sensors

These are all triangulation sensors, such as in laser triangulation, phase-measuring triangulation, stereo-photogrammetry, focus sensing, (non-fluorescent-)confocal scanning microscopy. There the measuring uncertainty is given by [21]

$$\delta z = C \frac{\lambda}{2\pi \sin(\theta) \sin(u)} \tag{5}$$

where C is the contrast of the coherent noise, θ is the angle of triangulation, λ is the wavelength and $\sin(u)$ is the observation aperture. The source of this physical lower limit of uncertainty is coherent noise. This can easily be shown by covering the surface with a thin fluorescent layer and observing only the incoherently scattered light [22]. In this case, the measuring uncertainty can be much better than that given by (5).

5.2   Type II Sensors

These are all types of multi-λ-sensors working at rough surfaces, such as the coherence radar, or two-λ-interferometers, as well as methods that measure the time of flight. Here we have the remarkable results:

- δz is independent of the observation aperture or measuring distance;
- the angle of triangulation can be zero, illumination and observation can be coaxial.

5.3   Type III Sensors

These are all sensors based on classical interferometry, working on smooth surfaces, including white-light interferometry. Here we have the interesting fact that the standard deviation of the measured signal will be smaller with larger distance, or with smaller observation aperture. This is caused by local averaging over the resolution cell:

$$\delta z \propto \sin(u) . \tag{6}$$

This is probably the reason why researchers do not image the interferometer mirror. Instead the Fourier plane is used to average over the mirror surface. Hence, high accuracy is possible, e.g. gravitational wave detectors.

It appears that all distance sensors might be placed in one of only three types, in spite of the fact that there exists a wide spectrum of sensors capable of measuring either the distance of stars or the distance of atomic layers, thus covering some 24 orders of magnitude. The categorization may help to bring some order to the vast number of sensors described in the literature, or available on the market. Knowing the physical limits, furthermore, enables us to see if a given sensor can be technically improved or still work at the limit allowed by nature.

References

1. Michelson, A.A. (1895) Determination Experimentale de la Valeur du Metre en Longueurs d'Ondes Lumineuses. Trav Mem Bur Int Poids Mes 11: 1–42
2. Hildebrandt, B.P., Haines, K.A. (1967) Multiple-Wavelength and Multiple-Source Holography Applied to Contour Generation. J Opt Soc Am 57: 155–162
3. Vry, U., Fercher, A. (1986) Higher-order Statistical Properties of Speckle Fields and Their Application to Rough Surface Interferometry. J Opt Soc Am A 3: 998
4. Sodnik, Z., Fischer, E., Ittner, T., Tiziani, H.J. (1991) Two-wavelength double Heterodyne Interferometry Using a Matched Grating Technique. Appl Opt 30: 3139
5. Davidson, M., Kaufman, K., Mazor, I. (1988) First Results of a Product Using Coherence Probe Imaging for Wafer Inspection. Proc SPIE 921: 100–114 (1988)
6. Lee, B.S., Strand, T. (1990) Profilometry with a Coherence Scanning Microscope. Appl Opt 29: 3784
7. Dresel, T., Häusler, G., Venzke, H. (1992) Three-dimensional Sensing of Rough Surfaces by Coherence Radar. Appl Opt 31: 919–925
8. Lindner, M.W., Häusler, G. (1998) Coherence Radar and Spectral Radar – New Tools for Dermatological Diagnosis. J Biomed Optics 3: 21–31
9. Häusler, G. (1994) Range Sensing of the First, the Second, and the Third Kind. Proceedings of the EOS Topical Meeting on Optical Metrology and Nanotechnology, Engelberg
10. Ammon, G., Andretzky, P., Blossey, S., Bohn, G., Ettl, P., Habermeier, H.P., Harand, B., Häusler, G., Laszlo, I., Schmidt, B. (1997) New Modifications of the Coherence Radar. Proceedings of the 3rd International Workshop on Automatic Processing of Fringe Patterns, Bremen
11. Takeda, M., Mutoh, K. (1983) Fourier Transform Profilometry for the Automatic Measurement of 3-D Object Shapes. Appl Opt 22: 3977–3982
12. Chim, S.S.C., Kino, G.S. (1991) Phase Measurements Using the Mireau Correlation Microscope. Appl Opt 30: 2197–2200
13. Ettl, P., Häusler, G., Schmidt, B., Schenk, M., Laszlo, I. (1998) Roughness Parameters and Surface Deformation Measured by Coherence Radar. Proc SPIE 3407: 133–140
14. Schmidt, B. (1998) Kohärenzradar für mikroskopische Anwendungen. Diploma Thesis, University of Erlangen

15. Ettl, P. (1995) Studien zur hochgenauen Objektvermessung mit dem Kohärenzradar. Diploma Thesis, University of Erlangen
16. Dresel, T. (1991) Grundlagen und Grenzen der 3D-Bildverarbeitung mit dem Kohärenzradar. Diploma Thesis, University of Erlangen
17. Jähne, B. (1989) Digitale Bildverarbeitung. Springer, Berlin
18. Ammon, G., Andretzky, P., Blossey, S., Bohn, G., Ettl, P., Habermeier, H.P., Harand, B., Häusler, G. (1997) Coherence Radar – New Modifications of White Light Interferometry for Large Object Shape Acquisition. Proceedings of the EOS Topical Meeting on Optoelectronics Distance Measurements and Applications, Nantes
19. Häusler, G., Leuchs, G. (1997) Physikalische Grenzen der optischen Formerfassung mit Licht. Physikalische Blätter 53: 417–421
20. Häusler, G. (1997) About the Scaling Behavior of Optical Range Sensors. Proceedings of the 3rd International Workshop on Automatic Processing of Fringe Patterns, Bremen
21. Dorsch, R., Häusler, G., Herrmann, J. (1994) Laser Triangulation: Fundamental Uncertainty in Distance Measurement. Appl Opt 33: 1306–1314
22. Häusler, G., Kreipl, S., Lampalzer, R., Schielzeth, A., Spellenberg, B. (1997) New Range Sensors at the Physical Limit of Measuring Uncertainty. Proceedings of the EOS Topical Meeting on Optoelectronics Distance Measurements and Applications, Nantes

Imaging Spectroscopy for the Non-invasive Investigation of Paintings

A. Casini, F. Lotti, and M. Picollo

Summary. This chapter describes a portable apparatus for the non-invasive examination of paintings based on a lead sulphide vidicon camera that works in the 400–2000 nm range. Imaging spectroscopy is achieved by collecting a multi-wavelength sequence of near-monochromatic images in the investigated area. After amplitude normalization and geometrical registration, the reflectance spectra can be reconstructed. Pigment identification may be made by comparing these spectra with those of a database of reference pigments and paintings. More detailed investigations on the pigment layers are afforded by appropriate algorithms, which supply distribution maps that can easily be correlated with the visual aspect of the painting. Examples of applications on test objects and on a XVI century Italian painting are reported. A brief review is also given of the modern infrared solid-state cameras that are most suited to replace the noisy and instable lead sulphide vidicon cameras in these applications.

1 Introduction

The application of physical and chemical examination methods is a current practice in identifying materials that constitute works of art and in monitoring their state of conservation. These methods provide curators and conservators with useful information for establishing the best conservation conditions, revealing the artists' techniques, and comprehending the history of the objects.

Many analytical techniques provide accurate results on the composition of matter by operating on very small samples (a few micrograms). Moreover, by working with microscopes, single pigment granules and homogeneous pieces of binding materials or varnish can be isolated and analysed. On the other hand, the results obtained with these techniques may not be representative of larger areas, due to the size of the specimens investigated.

However, starting from the fact that the work of art is a unique piece, the removal of samples from it for external examination must be carried out with much care and only when sampling is necessary for providing more detailed information in its regard. Therefore, non-invasive techniques play an important role in the analysis of artworks as they can be applied extensively and the measurements can be repeated, even on the same area.

Among these non-invasive methods, spectroscopic techniques are widely applied to the examination of works of art. These techniques provide information on pigment characterisation, material stratification, painting techniques and, more generally, the state of preservation of paintings, by taking advantage of the interaction between radiation and matter. This interaction depends essentially on the radiation wavelength and on the chemical and physical structure of the irradiated matter. Consequently, the materials are revealed through their spectral features. By working in the visible (Vis, 0.38–0.75 µm), in the near infrared (NIR, 0.75–1.0 µm) and in the short wavelength infrared (SWIR, 1.0–2.5 µm) regions, a variety of electronic (crystal–field, charge–transfer, valence–conduction band transitions) and vibrational processes (overtones and combination transitions) may produce typical spectral features, which appear in the absorption or reflectance spectra [1].

Both one- and two-dimensional spectroscopic techniques can operate in the abovementioned spectral regions. The former supply spectroscopic information on small portions of the object (a few square millimetres) and are usually realised by means of fibre optics spectrophotometers [2,3]. The latter show their data as a sequence of images recorded at different wavelengths and make it possible to extract the reflectance spectrum at any point within the image (imaging spectroscopy, IS). In this case, the spectral features of the investigated areas can be directly placed in correspondence with the visual aspect of the painting and can help conservators in forming an overall understanding of the artwork. In fact, IS can provide information not only on the composition and distribution of the materials, but also on the techniques used by the artists to lay them.

2   Historical Background

Since the 1930's, the application of black-and-white IR-photographic techniques has had a great importance in the study of painting [4–7]. IR luminescence was also applied to the studies of paintings and other artistic objects, providing excellent results [8]. In the 1970's, pigments and other artistic materials were examined by means of false-colour IR photography [9].

This technique was based on the use of IR films (e.g. Kodak 2236 IR film) and continued to be improved up until the early years of the 1990's [10–12]. However, despite its easy use, false-colour IR photography has a weak reproducibility.

It was also in the early 1970's that the first works appeared dealing with the application of an IR camera based on a lead sulphide (PbS) vidicon tube for the reflectographic study of paintings [13,14].

After van Asperen de Boer's first applications, wide-band reflectography in the NIR and SWIR ranges, performed by means of video cameras, has become one the most used imaging techniques for curators and conservators. In fact, it has provided an extensive knowledge of underdrawings, retouches, and *pentimenti*. By using these vidicon-based cameras, it has been possible to extend the upper wavelength limit of the reflectography from 0.9 µm to about

2 μm. In this way, additional information on underlying conditions has been obtained, because layers that appear opaque in IR photographs can become transparent at longer wavelengths [14–18]. With these cameras, which are simultaneously sensitive to Vis, NIR, and SWIR radiation, it is simple to record the same view in the visible and IR regions. This fact facilitates the correlation between underdrawings or *pentimenti* and visual details.

Imagery of static objects such as paintings may even be performed using a single detector and a scanning mechanism. An example of such a device for obtaining a high-resolution wide-band infrared reflectogram is given in [19]. However, even if these devices can provide unequalled results, their slowness limits their usage for high-quality documentation.

The progress made during the last decade in the technology of solid-state image sensors for the Vis, NIR and SWIR regions has provided operators with new cameras for investigating artworks.

Silicon CCD's have greatly improved the performance of imaging systems [18,20,21]. However, as their operative spectral range is limited to the Vis–NIR range, these cameras are not sufficient to cover the whole field of interest for the reflectography of artworks.

Technology is now available for the production of large focal plane arrays (FPA's) of SWIR solid-state sensors whose quality is not far from that of silicon CCD's [22–25]. The most interesting SWIR FPA's are made with the following solid-state detector materials:

- platinum silicide (PtSi; 1.2–5.0 μm)
- indium gallium arsenide (InGaAs; 1.0–2.5 μm)
- mercury cadmium telluride (HgCdTe; 1.0–25 μm)
- indium antimonide (InSb; 0.9–6.0 μm)

PtSi cameras provide good visualisation of underdrawings because the spatial uniformity of their FPA's is excellent and their spatial resolution results close to the optical limits of the lens [22,23]. Even if they are designed for the MWIR range (Mid Wavelength Infrared; 2.5–5.0 μm) to be used as thermal cameras, they can also be applied in the SWIR range by replacing germanium windows with silicon windows. However, when tested, some PtSi cameras exhibited a strong reduction in sensitivity near 1.9 μm. This fact was probably due to a reflecting cavity built into the sensor array in order to maximise its response in the thermal region (3–5 μm) [22].

SWIR InGaAs cameras are very promising, as they can work well without cooling. It follows that their cost may be reduced and their lifetime may be longer. However, the sizes of InGaAs area arrays are too small for applications to works of art, where enlarged views are advantageous in order to alleviate the mosaic assembly procedure. Only very recently, an InGaAs camera has been produced with a 340 × 240 pixel array (Sensor Unlimited, US).

Fully assembled HgCdTe cameras are not commercially available in SWIR configurations. SWIR HgCdTe sensors have been put to use only within expensive custom-made equipments (e.g. in astrophysics: NICMOS, NIR camera and multi-object spectrometer, onboard the Hubble Space Telescope and ARNICA, Arcetri Near Infrared Camera, at the Gornergrat Infrared Observatory) [24].

A NIR-SWIR camera developed by Amber Raytheon for the Clementine Deep Space Program (US Department of Defence) is based on an InSb FPA, likely derived from the FPA assembled in their commercial thermal camera Radiance-1 (3–5 μm). Until the end of 1996, however, SWIR versions of InSb cameras could not be purchased, even though the situation may since have changed.

To summarise, it can be said that the costs of these cameras, which are based on solid-state FPA's, are still too high for their extensive use in the field of cultural heritage conservation. In fact, only very large institutions (e.g. the Conservation Department of the National Gallery of Art in Washington, DC, USA) can afford to purchase these devices. Therefore, vidicon-tube cameras are still widely in use among art conservators for wide-band reflectography.

3   Imaging Spectroscopy Equipment

A natural evolution of wide-band reflectography is multi-wavelength reflectance imaging in the Vis–NIR–SWIR range, which may be equivalent to IS if the spectral sampling is sufficiently dense.

Since 1990, the authors have been working to set up and improve an IS system based on a Hamamatsu C-2400 camera [15,26]. This camera was equipped with a vidicon tube (PbO-PbS N2606-6) operating in the 0.4–2.0 μm range. A detailed description of the entire experimental set-up of the system is given elsewhere [26]; therefore, only a brief account is included here.

This system is able to generate a calibrated and registered sequence of near-monochromatic images taken at different wavelengths, from which the reflectance spectrum can be evaluated at any point.

A pair of light beams generated by quartz tungsten halogen lamps provides radiation that is projected symmetrically onto the painting at 45°, to avoid specular reflectance. To avoid useless illumination of the painting, the sources are shuttered instead of switching the lamp on and off with consequent variations in the temperature.

The spectral sampling in the Vis, NIR, SWIR regions sufficient for detecting the main spectral features of the pigments used in the ancient painting is about 20 nm. This can be achieved by means of a limited number of narrow-band filters placed in front of the camera lens. To date, the IS system implemented at IROE works with 32 interferential filters with 10 nm of bandwidth at half-maximum transmittance between 400 and 1700 nm.

To keep the actual operating centre wavelength of the filters as close as possible to the nominal one, the full-view cone angle is limited to about 15°. The camera uses a 50 mm focal length, f/2.8 photographic Pentax lens. Its transmittance in the NIR-SWIR range is 50% of its maximum and is almost constant in the 1.0–2.0 μm range [20]. For each wavelength (filter), 100 video frames are averaged by the frame grabber in order to reduce the influence of random noise.

Due to the many drawbacks of the vidicon technology with respect to solid state sensors, many corrections have been necessary in order to obtain reliable measurements. The main problems faced have been: temperature instability, high dark current, unavailable lens corrections for the SWIR region, high image persistence (lag). Most of these have been controlled by hardware adjustments and by software corrections and calibrations. Nevertheless, the lag means that each acquisition must wait about 90 seconds before the next one.

Because of the vidicon instabilities, camera calibration is ineffectual, therefore images have been normalised by using reference targets with lambertian diffuse-reflectance (Spectralon™ calibration standards).

Refocusing is necessary for the wavelengths in the NIR-SWIR range. This in turn affects the size of the acquired images, which must be geometrically corrected using a bilinear interpolation based on four reference points, in order to register the image sequence as a coherent three-dimensional array. For this purpose, a cardboard frame which surrounds the recorded area is used that contains both high-contrast patterns, which are necessary for autofocus, and the set of calibration standards (Fig. 1).

The images obtained with this simple band-sequential approach are sharper than those obtained using sophisticated wavelength-selection schemes.

4 Applications

The main purpose of the IS system is to extract reliable spectral information concerning an extended surface of a painting in order to identify pigments and to monitor the state of preservation by comparing measurements at different times.

Moreover, once a sequence of near-monochromatic images is available, corrected and made geometrically coherent, a number of further investigations can be performed. For instance, simple image processing makes it possible to map the pigment distribution and particular features can be enhanced by subtracting or by dividing images at selected wavelengths, thus improving the specificity and efficiency of the traditional wide-band reflectography. Information regarding the layer distribution can also be extracted in some cases by exploiting the deeper penetration shown by the radiation at higher wavelengths in the NIR-SWIR range.

More complex statistical image analysis can help in building maps of the areas showing similar spectral behaviour or in evidencing local non-homogeneity or discrepancies with the visual aspect of the painting. In the following, some laboratory tests are presented which have proved the ability

Fig. 1. Test tablet containing five wooden strips painted with (from top to bottom): *lapis lazuli*, azurite, smalt, Prussian blue and indigo. Pigment thickness increases from left to right. The surrounding cardboard frame contains the patterns for automatic focus control and geometrical registration. The disks are Spectralon™ reflectance calibration standards

of the IS system to carry out these functions, while others are described as case studies on Italian Renaissance paintings.

4.1   Laboratory Tests

Reference panels were prepared in order to acquire sets of images on standard materials with variable thickness. Wooden tablets were used as test objects and were prepared with pigments over a substrate made of gypsum and skin glue using egg tempera.

Figure 1 shows the test tablet for five blue pigments: *lapis lazuli*, azurite, smalt, Prussian blue and indigo (horizontal strips, from top to bottom); the tablet is surrounded by the cardboard frame containing some of the calibration standards (discs) used during the measurement. Several strokes of black carbon chalk were drawn on the preparatory layer in order to investigate the pigment transparency at different wavelengths and thicknesses. This is of particular interest for checking the ability of detecting underdrawings.

For each strip, from 1 to 5 paint layers were overlapped in such a way as to cover 5 different areas with increasing thickness (from left to right in Fig. 1). The light column at the left end of the tablet shows the preparatory layer with the carbon strokes and without pigment.

The reconstructed reflectance spectra of 5 points belonging to different pigments in the areas with medium thickness are presented in Fig. 2; these spectra were obtained after acquisition of 32 multi-wavelength images in the 400–1700 nm range using the described IS system. These spectra were in good agreement with the spectra of the same points collected with a 1-D high-resolution spectrum analyser [27]. In addition to the intrinsic shapes of the pigment spectra, some characteristics of the underlying preparatory layer also appear in the spectra for the medium- and low-thickness areas, as can be seen in Fig. 2. Indeed, gypsum has a typical absorption band (split into three sub-bands) in the 1450–1550 nm range, due to the overtones and combinations of the vibrational modes of the OH$^-$ ion. As the three sub-bands are very narrow and close to each other, they appear as a unique peak in Fig. 2,

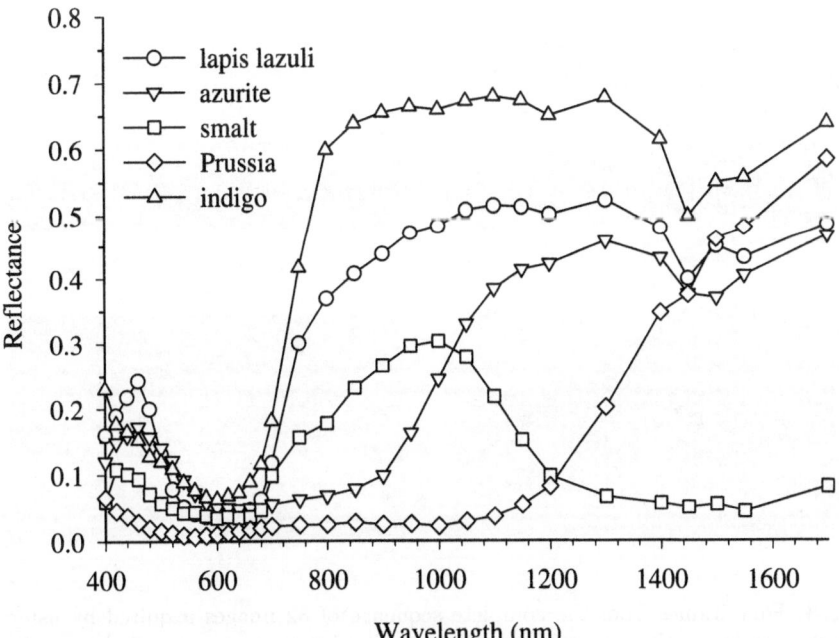

Fig. 2. Reflectance spectra of the five pigments in the test tablet of Fig. 1 (medium-thickness area), computed from a 32 multi-wavelength image sequence acquired in the 400–1700 nm range

due to the limited number of filters in that region (only three). However, their appearance depends on the spectral characteristics and thickness of the overlaid materials.

Figure 3 shows in detail four of the 32 images acquired by the IS system in the 400–1700 nm range. For each wavelength, the different hiding power of the various pigments can be checked while varying their thickness by looking at the black strokes below the painted layers. The black carbon chalk constituting the strokes has a strong absorption throughout the investigated spectral regions. Its reflectance spectrum is therefore very low and flat, evidencing its presence when the radiation penetrates the paint layers. When the thickness is sufficiently high, the penetrating radiation – even in the NIR–SWIR region – no longer interacts with the preparatory layer or with the black strokes. In this case, the recorded spectral features are totally due to the pigment–medium system, and no more patterns can be observed. As shown in the figure, each pigment has a different hiding power at different wavelengths.

These considerations can help in finding the best wavelengths for the observation of underdrawings, thus improving the performance of the wide-band reflectography traditionally used for this purpose.

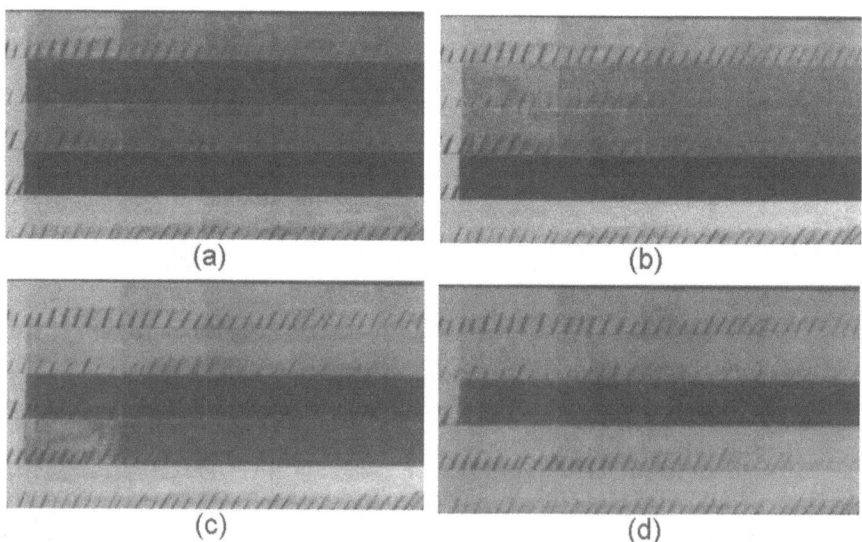

Fig. 3. Four frames from the complete sequence of 32 images acquired by using the IS system on the test tablet in Fig. 1. The black strokes, painted over the preparatory layer, show the different transparency of the 5 blue pigments at different wavelengths in the NIR–SWIR range. (**a**) 800 nm, (**b**) 1050 nm, (**c**) 1300 nm, (**d**) 1550 nm. Pigment thickness increases from left to right

4.2 Study of a Work of Art

Identification of paint materials in actual works of art by means of Vis–NIR–SWIR reflectance spectrum analysis is not an easy task. This is because, in most cases, the artists combined more than one pigment to enhance lights and shades in the scene represented. An extended reference database containing reflectance spectra of both laboratory test samples and actual paintings, including those taken during the different phases of previous restorations, has been set up by the authors to help in the identification problems.

For example, some results are reported from a study made on the *Adorazione dei Magi*, a panel by Jacopo Carucci, better known as Pontormo (c. 1519–1520, permanent collection of the Galleria Palatina, Florence). The measurements were made during restoration in 1998, after the old varnish had been removed (Fig. 4).

Some spectra, extracted from a sequence of 32 multi-wavelength images acquired in the 400–1700 nm range with the IS system described above, are presented in Fig. 5. Two of these belong to blue areas.

The curve marked with circles comes from the sleeve of the man on the right side of Fig. 4 (point **a**). As previously remarked, it is not simple to

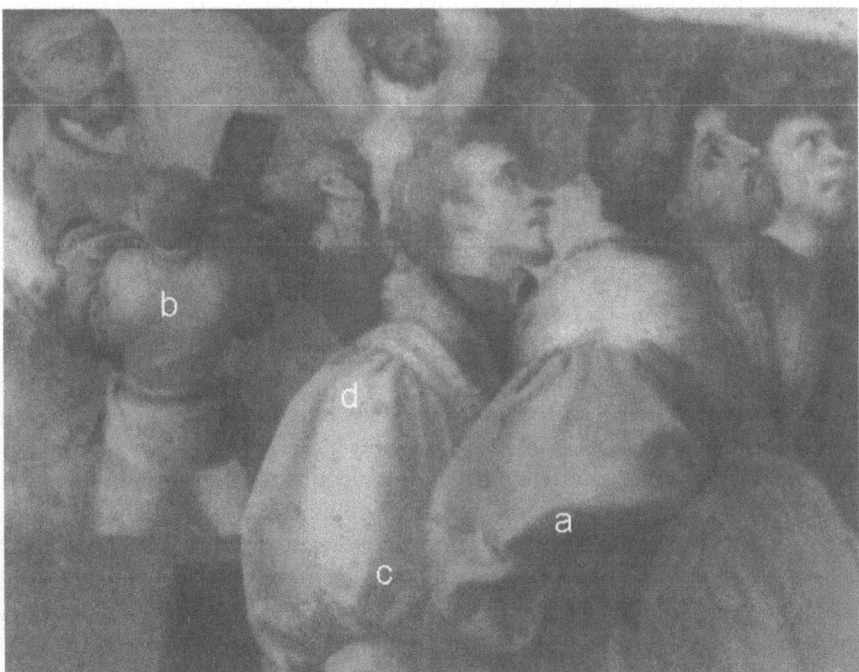

Fig. 4. Detail of the *Adorazione dei Magi* by Jacopo Carucci, known as Pontormo (c. 1519–1520, permanent collection of the Galleria Palatina, Florence). (**a**) and (**b**): tested blue points; (**c**) and (**d**): tested yellow points

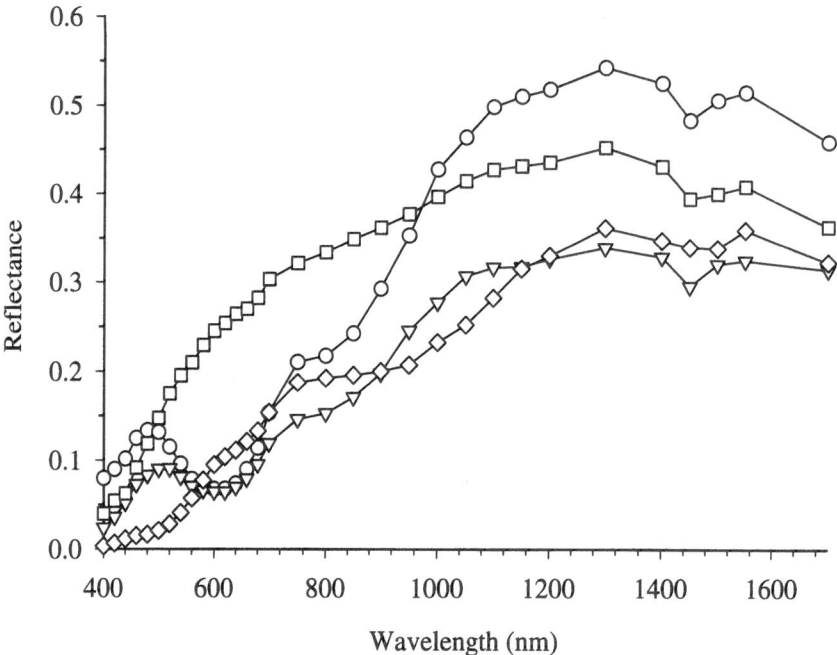

Fig. 5. Reflectance spectra extracted from a 32 multi-wavelength image sequence of the detail shown in Fig. 4. Blue pigments: *circles* from point (**a**) and *triangles* from point (**b**).Yellow pigments: *diamonds* from point (**c**) and *squares* from point (**d**)

identify which kind of blue pigment this spectrum refers to. A probable identification can be attempted by looking at the spectra resulting from the test tablet of blue pigments (Fig. 2). The spectrum examined might be due to a combination of *lapis lazuli* and azurite. In fact, the shape of the spectrum in the Vis–NIR up to 800 nm could be due to the presence of *lapis lazuli*, while, starting from 800 nm to 1000 nm, a relevant increase in the reflectance value points to the presence of azurite. This pigment, in fact, has a rising slope towards higher reflectance values starting from 850 nm, as is shown in Fig. 2. It follows that the artist probably used both blue pigments; in this specific case, *lapis lazuli* was put over the azurite layer.

The second curve related to a blue pigment is marked with triangles in Fig. 5; it comes from a greenish blue area (point **b** in Fig. 4). This spectrum shows a behaviour similar to that of the previous curve (circles), apart from a modest shift of the main reflectance band in the visible spectrum towards the green region, probably due to the presence of small quantities of a yellow pigment.

The other two curves in Fig. 5 come from the sleeve of the young man at the centre of Fig. 4 which shows *chiaroscuro* yellow hues. It was painted using two different pigments, probably lead-tin yellow (*giallorino*) and yellow ochre, as found in other works of the artist.

The spectrum of point (**c**) in Fig. 4 is marked with diamonds in Fig. 5, while that of point (**d**) is marked with squares. The identification of the two pigments has been confirmed by comparing these spectra with those of the authors' reference database. In fact, the curve of point (**c**) shows the spectral behaviour of the mineral goethite (α-FeOOH), which is the main constituent of yellow ochre (or natural Siena). Typical absorption bands can be noted around 400–500 nm, 650 nm and 900–930 nm, due to the transitions of the ferric ion. The other curve, relative to point (**d**), can be attributed to lead-tin yellow, which shows an approximately S-shaped reflectance spectrum, with a high positive gradient in the range 400–550 nm.

In an attempt to map the prevalence of one yellow pigment over the other, wide-band reflectography did not give useful results, due to the similarity of the NIR–SWIR reflectograms. A more fruitful result was achieved by exploiting the 2-D narrow-band information obtained by IS. In fact, if we examine the reference spectra of lead-tin yellow and yellow ochre, two spectral regions, around 440–600 nm and 800–1000 nm, can be noted in which the reflectance curves show different degrees of concavity.

To quantify this characteristic, the spectrum curvature in the area examined was evaluated by means of a robust, second-order polynomial regression algorithm applied to the spectrum of each pixel and limited to the selected spectral regions. The normalised values of the curvature were then converted as an image into grey levels, as shown in the example of Fig. 6, in which the first band (440–600 nm) is considered.

The prevalence of the two pigments can be associated with the curvature values. Positive curvature values (yellow ochre) are represented as lighter tones, and negative curvature values (lead-tin yellow) as darker ones.

This map can help in evaluating the distribution of the two pigments. In fact, the artist enhanced the *chiaroscuro* by using *giallorino* for the bright yellow areas and yellow ochre for the darker ones. Further qualitative information about the relative depth of the two layers can be inferred by comparing the distribution maps computed from different bands, taking into account the higher penetration at longer wavelengths; prevalence of one pigment at longer wavelengths means that the pigment is deeper. Following this line, interesting results on the yellow pigments used by Pontormo are described in [26].

Other mapping methods resort to multivariate analysis of variance, as, for example, principal component analysis (PCA), which has the characteristic of concentrating most of the variability contained in a sequence of many multi-wavelength images in a few PC's (usually less than 5) [28]. PC images can reveal underlying features which cannot be seen using wide-band reflectography. Moreover, cluster analysis can help in grouping spectra with particular statistical features; clusters are used, therefore, to map areas with similar spectral behaviour [29].

Fig. 6. Map of the curvature of the spectra from Fig. 5 in the 440–600 nm region. Values are reported on a normalised grey scale, where medium grey (50%) is zero curvature, lighter is positive, and darker is negative. The map indicates the distribution of yellow ochre (light) and lead-tin yellow (dark) on the yellow sleeve of the man in the middle

5   Conclusions

Among the non-invasive techniques for investigating paintings, imaging reflectance spectroscopy in the Vis–SWIR range realises a powerful extension of current techniques using video cameras for reflectography, going far beyond the mere observation of underdrawings and *pentimenti.*

The results obtained in laboratory experiments and on art paintings by the IS system described prove the high potential of this approach, even though the system was built using old and cheap sensor technology.

With this technique, it is possible to characterise pigments from reliable and geometrically coherent spectral information concerning extensive areas of the painting. By taking advantage of the numerical three-dimensional organisation of the multi-wavelength image sequences acquired, and by using suitable image processing techniques, it is possible to build maps of pigment distribution, enhance areas with artefacts, and develop problem-oriented procedures, for example, to understand how the pigments are layered.

The drawbacks of the equipment used (slowness and limited spatial resolution) can easily be overcome by using focal plane arrays of a sufficient number of sensors (512 × 512 at least), the cost of which will probably become reasonable in the near future.

Acknowledgements

The authors are grateful to Dr. Marco Chiarini (Soprintendenza Beni Artistici e Storici, Galleria Palatina, Florence), Dr. Magnolia Scudieri and Ms. Diane Kunzelmann (Restoration Laboratory of Soprintendenza Beni Artistici e Storici, Florence) for allowing the recording of spectra from the *Adorazione dei Magi*. They also wish to thank Prof. Pier Andrea Mando' (Physics Dept, University of Florence) for providing the painted test tablets. This work was partially supported by the Special Project *Safeguard of Cultural Heritage* of the Italian National Research Council (CNR).

References

1. Hunt, G.R. (1977) Spectral Signatures of Particulate Minerals in the Visible and Near Infrared. Geophysics **42**, 501–513
2. Bacci, M., Picollo, M., Radicati, B., Bellucci, R. (1994) Spectroscopic Imaging and Non-Destructive Reflectance Investigations Using Fiber Optics. In: Non-Destructive Testing of Works of Art, 4th International Conference at Berlin, Germany, 162–174
3. Bacci, M., Picollo, M. (1996) Non-Destructive Spectroscopic Detection of Cobalt (II) in Paintings and Glasses. Stud. in Conserv. **41**, 136–144
4. Lyon, R.A. (1934) Infra-Red Radiations Aid Examination of Paintings. Tech. Stud. Field Fine Arts **2**, 203–212
5. Rawlins, F.I.G. (1938) A Novel Infra-Red Camera for Art Gallery Work. Museums J. **38**, 186–187
6. Farnsworth, M. (1938) Infra-Red Absorption of Paint Materials. Tech. Stud. Field Fine Arts **7**, 88–98
7. Keck, S. (1941) A Use of Infra-Red Photography in the Study of Technique. Tech. Stud. Field Fine Arts **9**, 145–152
8. Bridgman, C.F., Gibson, H.L. (1963) Infrared Luminescence in the Photographic Examination of Paintings and Other Art Objects. Stud. Conserv. **8**, 77–83
9. Olin, C.H., Carter, T.G. (1970) Infrared Color Photography of Paintings Materials. In: Technical Papers from 1968 Through 1970, IIC–America Group, New York, 83–88
10. Matteini, M., Moles, A., Tiano, P. (1978) Infrared Colour Films as an Auxiliary Tool for the Investigation of Paintings. In: ICOM Committee for Conservation, 5th Triennial Meeting at Zagreb, Yugoslavia, 1–19
11. Eastman Kodak Company (1986) Applied Infrared Photography. Technical publication No. M–28
12. Moon, T., Schilling, M.R., Thirkettle, S. (1992) A Note on the Use of False-Color Infrared Photography in Conservation. Stud. Conserv. **37**, 42–52
13. van Asperen de Boer, J.R.J. (1969) Reflectography of Paintings Using an Infra-Red Vidicon Television System. Stud. Conserv. **14**, 96–118
14. van Asperen de Boer, J.R.J. (1974) A Note on the Use of an Improved Infrared Vidicon for Reflectography of Paintings. Stud. Conserv. **19**, 97–99
15. Bacci, M., Baronti, S., Casini, A., Lotti, F., Picollo, M., Casazza, O. (1992) Non-Destructive Spectroscopic Investigations on Paintings Using Optical Fibers. Proc. Mat. Res. Soc. Symp. **267**, 265–283

16. Kossolapov, A. (1993) An Improved Vidicon TV Camera for IR–Reflectrography. In: ICOM Committee for Conservation, 10th Triennial Meeting at Washington, DC, USA, 25–31

17. Burmester, A., Bayerer, F. (1993) Towards Improved Infrared Reflectograms. Stud. Conserv. **38**, 145–154

18. Saunders, D., Cupitt, J. (1995) Elucidating Reflectograms by Superimposing Infra-Red and Colour Images. Nat. Gallery Tech. Bull. **16**, 61–65

19. Bertani, D., Cetica, M., Poggi, P., Puccioni, G., Buzzegoli, E., Kunzelman, D., Cecchi, S. (1990) A Scanning Device for Infrared Reflectography. Stud. Conserv. **35**, 113–117

20. Walmsley, E., Fletcher, C., Delaney, J.K. (1992) Evaluation of System Performance of Near-Infrared Imaging Devices. Stud. Conserv. **37**, 120–131

21. Aldrovandi, A., Bertani, D., Cetica, M., Matteini, M., Moles, A., Poggi, P., Tiano, P. (1988) Multispectral Image Processing of Paintings. Stud. Conserv. **33**, 154–159

22. Norton, P.R. (1991) Infrared Image Sensors. Optic. Eng. **30**, 1649–1663

23. Walmsley, E., Metzger, C., Fletcher, C., Delaney, J.K. (1993) Evaluation of Platinum Silicide Cameras for Use in Infrared Reflectography. In: ICOM Committe for Conservation, 10th Triennial Meeting at Washington, DC, USA, 57–62

24. Baffa, C., Gennari, S., Hunt, L.K., Lis,i F., Tofani, G., Vanzi, L. (1995) TIRGO and Its Instrumentation. Optic. Eng. **34**, 2731–2735

25. Dereniak, E.L., Boreman, G.D. (1996) Infrared Detectors and Systems. John Wiley, New York

26. Casini, A., Lotti, F., Picollo, M., Stefani, L., Buzzegoli, E. (1999) Image Spectroscopy Mapping Technique for Non-Invasive Analysis of Paintings. Stud. Conserv. **44**, 39–48

27. Bacci, M., Picollo, M., Radicati, B. (1996) Fiber Optics Reflectance Spectroscopy: A Non-destructive and Non-invasive Technique for the Identification of Blue Pigments. In: Non-Destructive Testing of Works of Art, 5th International Conference at Budapest, Hungary, 89–100

28. Geladi, P., Grahn, H. (1996) Multivariate Image Analysis. Wiley, New York

29. Baronti, S., Casini, A., Lotti, F., Porcinai, S. (1998) Multispectral Imaging System for the Mapping of Pigments in Works of Art by Use of Principal-Component Analysis. Appl. Optics **37**, 1299–1309

Part VII

Biomedical Optics

Optical Coherence Tomography in Medicine

A. F. Fercher and C. K. Hitzenberger

Summary. This contribution presents a concise overview of optical coherence tomography (OCT) and partial coherence interferometry (PCI). Basic principles and important techniques are explained. Recent papers describing applications of OCT and PCI in the medical field are reviewed.

1 Introduction

Optical tomography techniques have the ability to peer inside the body non-invasively. Tissue absorption and scattering coefficients are derived from the characteristics of scattered light transmitted and/or re-emitted by the body. Three basic optical tomography approaches have been developed: The most straightforward optical technique would be diffraction tomography [1]. Due to problems encountered with the detection and inversion of the diffraction field data, two other techniques have been put forward: diffusing photon tomography, deriving spatial maps of absorption and scattering coefficients from the characteristics of multiply scattered light transmitted through the body [2], and optical coherence tomography (OCT) [3], where these data are derived from singly scattered light detected by partial coherence interferometry (PCI) [4–6]. Both techniques can yield morphological and functional images. Several instruments based on diffuse optical tomography and on OCT are commercially available.

"Tomographic" imaging techniques derive two-dimensional data from a three-dimensional object to obtain a slice image of the internal structure. Depending on the physical probe used to interact with the object, approaches like transmission, reflection, diffraction, emission, or back-scattering tomography are used. If x-rays or γ-rays are used as probes, of course, straight ray propagation can be assumed. If the wavelength of the radiation is not appreciably smaller than the structure elements of the object, diffraction plays an important role. At present, however, OCT is a back-scattering straight-ray approach. In OCT the depth position of light-re-emitting sites is detected by a PCI depth scan while the probe beam laterally scans across the object.

Partial coherence interferometry has found important medical applications, in particular in the ophthalmologic field of intraocular distance measurement [7–9]. For example, it has recently been shown that the refractive outcome of cataract surgery can be improved substantially if PCI biometry data are used [10]. The first application of PCI in ophthalmology was described by Fercher and co-workers in 1986 [11] (based on a German patent application filed in 1982 [12]). They used the dual-beam technique in a Fabry–Perot interferometer configuration and a multi-mode laser diode. By operating the laser at low injection current levels, it was possible to obtain a coherence length in the 10 µm range. No heterodyne technique was used in these early experiments. The Fabry–Perot configuration is particularly simple and easy to adjust. One problem of this interferometer is that zero path difference between the two interferometer plates can hardly be obtained. Hence, this technique is limited to the measurement of distances larger than a few tens of micrometers. Later, Hitzenberger described a dual-beam Michelson interferometer using the Doppler heterodyne technique [13]. He too used a multi-mode laser diode at $\lambda = 780$ nm and a coherence length of $l_C \approx 100$ µm. Later on he used SLD's at $\lambda = 830$ nm with a coherence length of $l_C \approx 20$ µm as light sources [14].

It was recognized very early on that interferometry can be performed in direct space (time domain) as well as in Fourier space (frequency domain). The physical basis is the Fourier transform relation between the coherence function and the power spectrum of light [15]. Hence, there are two basic possibilities to measure the position of light-re-emitting sites within an object by using partially coherent light [16]: first, as in classical interferometry, the mutual coherence function of the re-emitted object beam and the reference beam can be used; second, the mutual power spectrum of the re-emitted object beam and the reference beam can be used (spectral interferometry). In the past, spectral interferometry was used (channelled spectrum technique) to measure the thickness of thin films [17] and for the absolute measurement of small displacements [18]. In OCT either a PCI depth scan is performed in direct space to directly obtain the mutual coherence function of the re-emitted object beam and the reference beam or the mutual power spectrum of the reference beam and the light beam re-emitted by the object is Fourier-transformed.

In a first step towards tomography, topograms of the fundus of the eye have been generated. Figure 1 shows the first topogram of the human fundus obtained with dual-beam PCI, as presented at the ICO-15 SAT Conference 1990 [13,19]. Reflectometry OCT was pioneered by Fujimoto and co-workers. In 1991, the first (in vitro) OCT images, obtained by using reflectometer PCI, were presented by this group [3,6]. In 1993, the first in vivo tomogram (of the human optic disc), based on dual-beam PCI, was published by Fercher et al. [4,11,13]. After initial studies [20] the great potential of OCT in ophthalmic diagnostics [21,22] was demonstrated. Later, successful OCT applications were reported from several fields of medicine [23–28].

As mentioned above, OCT is based on scanning. The object is scanned in azimuth by an illuminating light beam derived from a PCI device, and back-scattered light is detected. Transverse resolution is obtained from the

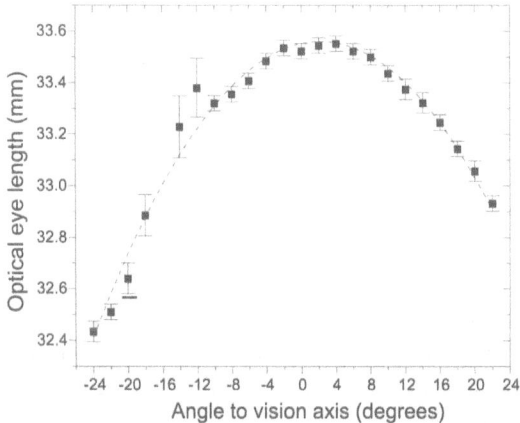

Fig. 1. First optical coherence topogram. Multimode laser diode: $\lambda = 780\,\text{nm}$. Profile of Bruch's membrane of a human eye obtained in vivo by PCI [13,19]. Note the external fovea at $0°$

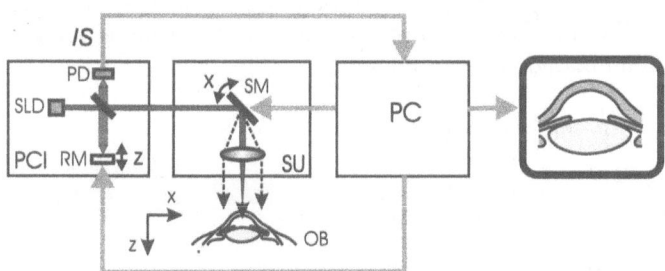

Fig. 2. General OCT scheme. OB = object; PC = personal computer; PD = photodetector; RM = reference mirror; SLD = superluminescent diode; SM = scanning mirror. A partial coherence interferometry (PCI) depth scan performed by the reference mirror is used to detect the range or depth position (z) of light remitting sites in an object; a conventional scanning unit (SU) detects the azimuthal position (x). A PC controls the position of the scanning mirror SM and the reference mirror RM. The interferometer signal IS modulates (after some electronic processing) the pixel brightness at the monitor

azimuthal scanning procedure; depth or longitudinal resolution is obtained by PCI (or spectral interferometry). A computer synthesizes the OCT image using azimuth data from the scanning device, depth data from the reference mirror position, and modulating pixel brightness from the photoelectric PCI depth scan signal (indicated by IS in Fig. 2).

Since the introduction of OCT several new techniques have been developed. The next sections give a survey on basic OCT principles and techniques and describe applications in medicine.

2   Back-Scattering

Let an object be illuminated by part of the light wave emitted by a source; see Fig. 3. Let this wave be a Gaussian beam and the object be positioned at the beam waist. If the object depth D is of the order of the Rayleigh length the illuminating light (light source field E_{LS}) is an approximately plane wavefront with wave vector $k^{(i)}$:

$$E_{LS}(r, t) = A_{LS} \exp\left(ik^{(i)}r - i\omega t\right),\tag{1}$$

where A_{LS} is the light wave amplitude, $|k^{(i)}| = k = 2\pi/\lambda = \omega/c$ is the wave number, λ is the wavelength, ω is the frequency, c is the velocity of light, and r is the position vector. We ignore any field quantization and represent the electric field E as a complex scalar, i.e. we also ignore polarization effects. Let $E_O(r, t)$ be the wave scattered by the object. The sum of the two waves $E_{LS}(r, t) + E_O(r, t)$ satisfies the scalar Helmholtz equation. Basically, the Helmholtz equation is obtained from the wave equation by separating off a sinusoidal time dependence. Therefore, a sinusoidal time dependence is added below in $E_{LS}(r, t)$. In the case of weakly scattering objects the scattered field can be obtained by the first Born approximation [29] as a volume integral extended over the illuminated object volume:

$$E_O(r, t) = -\frac{1}{4\pi} \int_{V(r')} F_O(r') E_{LS}(r', t) G\left(|r - r'|\right) d^3 r',\tag{2}$$

with the Green's function of the Helmholtz equation

$$G\left(|r - r'|\right) = \frac{\exp\left(ik^{(S)}|r - r'|\right)}{|r - r'|}.\tag{3}$$

$k^{(S)} = |k^{(S)}| = k$, $k^{(S)}$ is the wave vector of the (coherently) scattered wave. Equation (2) can be considered as a quantification of Huygens' principle. The Green's function represents the secondary wavelets which combine to form

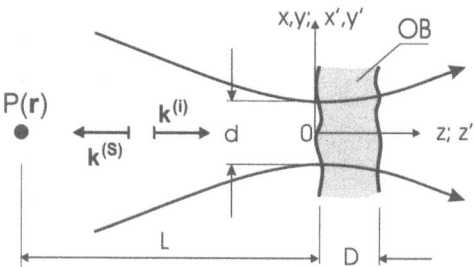

Fig. 3. Back-scattering. The object OB is illuminated by a focused (Gaussian) light beam with wave vector $k^{(i)}$; shown are the $1/e^2$ intensity contours. Back-scattered light (wave vector $k^{(s)}$) is detected at P

the scattered light. $F_O(\boldsymbol{r})$ is the scattering potential of the object; we shall also call it the "object structure". It determines the relative amplitudes and phases of the scattered wavelets.

$$F_O(\boldsymbol{r}) = -k^2 \left[m^2(\boldsymbol{r}) - 1 \right] , \tag{4}$$

where $m(\boldsymbol{r})$ is the complex refractive index of the object. In general, of course, m will depend on the wavelength λ; here, m is assumed to be isotropic and independent of λ. Furthermore, because in OCT the object is illuminated by a rather narrow light beam ($d \ll L$ in Fig. 3), far-field scattering is a reasonable approximation. In this case the exponent in G can be written [30] as

$$\mathrm{i}k^{(S)} |\boldsymbol{r} - \boldsymbol{r}'| = \mathrm{i}k^{(S)} (\boldsymbol{r} - \boldsymbol{r}') . \tag{5}$$

From (2) we obtain

$$E_O(\boldsymbol{r}, \boldsymbol{K}, t) = -\frac{A_{LS}}{4\pi L} \exp\left(\mathrm{i}k^{(S)}\boldsymbol{r} - \mathrm{i}\omega t\right) \int\limits_{V(\boldsymbol{r}')} F_O(\boldsymbol{r}') \exp\left(-\mathrm{i}\boldsymbol{K}\boldsymbol{r}'\right) \mathrm{d}^3\boldsymbol{r}', \tag{6}$$

i.e. the wave scattered by the object is modified by the Fourier transform of the scattering potential of the object [31]. $\boldsymbol{K} = \boldsymbol{k}^{(S)} - \boldsymbol{k}^{(i)}$ is the scattering vector. L is the distance of the detection point P at position $\boldsymbol{r}$ from the origin near the object. Hence, the three-dimensional scattering potential $F_O(\boldsymbol{r})$ can be obtained from the scattered field $E_O(\boldsymbol{r}, \boldsymbol{K}, t)$ by an inverse Fourier transform.

There are some limitations in this imaging procedure. These can be seen from the data geometry in $\boldsymbol{K}$ space (Fig. 4). For any direction of the scattered light the scattering vector $\boldsymbol{K}$ points to the surface of a sphere. This is the Ewald sphere, well known from crystallography [32,33]. This description sheds some light on the imaging properties of recent OCT techniques which are mainly determined by the properties of back-scattering.

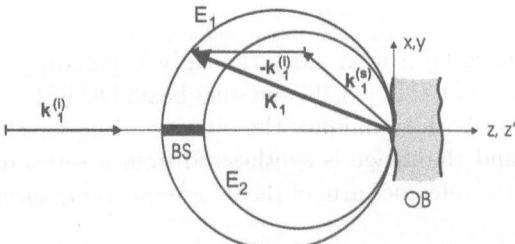

Fig. 4. $\boldsymbol{K}$-space geometry of back-scattering. $\boldsymbol{k}_1^{(i)}$ = wave vector of illuminating wave at wavelength λ_1; $\boldsymbol{k}_1^{(S)}$ = wave vector of scattered wave at λ_1; $\boldsymbol{K}_1$ = scattering vector corresponding λ_1; E_1, E_2 = Ewald spheres corresponding to wavelengths λ_1 and λ_2; OB = object; BS = Fourier data accessible by back-scattering

For example, to obtain a true three-dimensional reconstruction of the scattering potential $F_O(r)$ a certain range of data $E_O(r, t)$ in the three-dimensional Fourier space has to be measured. It can be seen from Fig. 4, however, that back-scattering with monochromatic light gives access to only one high spatial frequency, i.e. discontinuities of the scattering potential. A wavelength range from λ_1 to λ_2 provides Fourier components of the scattering potential in the spatial frequency range $[K_{1Z}, K_{2Z}]$ and, therefore, access to a finite range of Fourier components. Nevertheless, again only relatively abrupt changes in the scattering potential, occuring within a few wavelengths, can be seen by back-scattering tomography. Note, however, that according to the basic theorem of diffraction tomography, near-field data do also give access to some Fourier components in the interior of the corresponding Ewald sphere [1].

Finally, let us consider the direction of the scattering vector K. From (6) the back-scattered light is composed of components generated by the three-dimensional scattering potential $F_O(r)$ of the form $\exp(-iKr)$. For illustration, let us split $\exp(-iKr)$ in the cosine and sine parts. The cosine component $\cos(Kr)$ represents a cosinusoidal distribution (of the scattering potential) in space with K as its normal vector, i.e. as a vector normal to planes Kr = constant. The additional sinusoidal components allow the spatial phases to be adjusted properly [34]. Hence the scattering experiment yields a decomposition of the scattering potential into a three-dimensional spectrum of harmonics that is constant in planes with the plane normal K. Back-scattered light gives access to only those harmonics in the spatial Fourier spectrum of the scattering potential that have their plane normals oriented more or less parallel to the direction of illumination. So, back-scattering with narrow-frequency-band light gives access to high-frequency Fourier data. Hence, OCT is a high-pass imaging technique.

3   Time-Domain OCT

3.1   Reflectometry OCT

Two time-domain techniques have been used since the early beginnings of OCT: reflectometry OCT [3] based on PCI [35,36] and dual-beam OCT [4,37] based on dual-beam PCI [11]. In both techniques the object is transversely scanned by the probing light and the image is synthesized from a series of PCI depth scans performed by the reference arm of the interferometer at each transverse position.

In the reflectometry technique (Fig. 5) the probing light is one beam of a Michelson interferometer. The depth position of light remitting sites within the object is detected by the PCI depth scan: The reference mirror scans along the reference beam axis. Interference occurs at the interferometer exit if the optical distance of the reference mirror to the beam splitter of

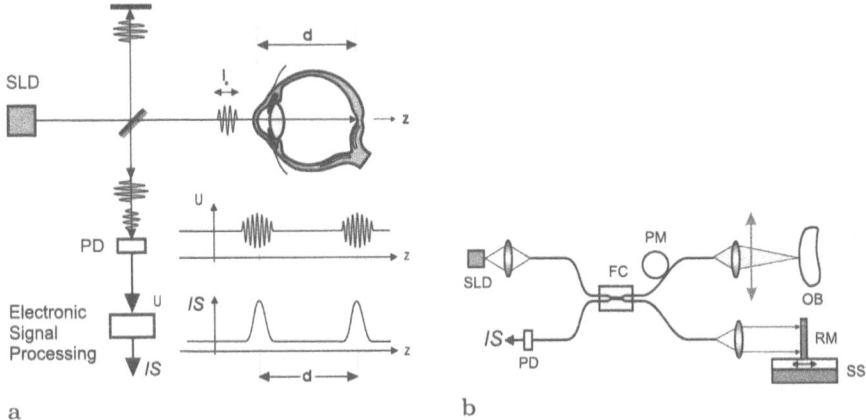

Fig. 5. Time-domain reflectometry OCT [3]. SLD = superluminescent diode; PD = photodetector. (**a**) Reflectometry PCI depth scan signal generation. IS = electronic interferometer signal. (**b**) Fiber optics technique; scheme of the first OCT scanner. FC = fiber coupler; OB = object; PM = phase modulator; RM = reference mirror; SM = scanning mirror; SS = scanning stage

the interferometer matches the distance of a light-re-emitting site within the coherence length l_C of the light in the object arm to the beam splitter.

An AC signal (U) is detected by the photodetector with the Doppler frequency corresponding to the speed of the reference mirror. Therefore, noise-reducing electronic AC techniques can be used to obtain high signal quality. Usually, amplification, followed by band-pass frequency filtering, rectification, and low-pass frequency filtering, is used to obtain the electronic interferometer signal (IS). This signal can be used either to determine distances between object interfaces or as the basic PCI depth scan signal for OCT. In the reflectometry technique, however, the inherent mechanical instability of in vivo objects can lead to position errors in the interferometer signal. This problem can be overcome if the reference mirror is moved at high speed [38] (or by the dual-beam technique).

As shown below in (17), the electric interferogram signal equals the convolution of the object scattering potential along the PCI depth scan with the complex degree of the time coherence. Usually, this signal is used to synthesize tomograms as shown, for example, in Fig. 6. However, like in classical microscopy, many modifications of the original interferometer scheme are possible, either to enhance image contrast or to introduce new image parameters.

Figure 7 shows an example of birefringence OCT imaging obtained by de Boer et al. [39]. Using polarization-sensitive detection of the light re-emitted by the object the magnitude of the birefringence as a function of depth is imaged. Birefringence of collagen, a constituent of many biological tissues, can be used as the optical marker for thermal tissue damage.

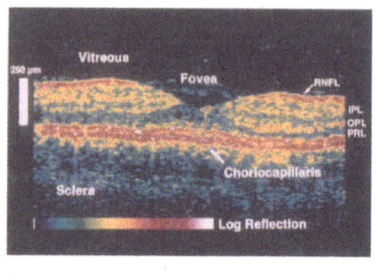
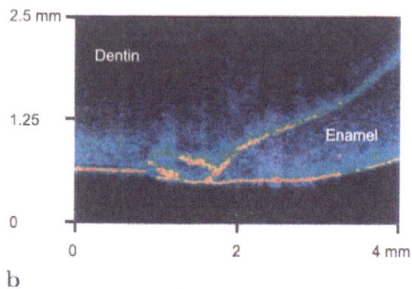

a   b

Fig. 6. (a) Cross-sectional OCT scattering potential image of a human eye in the macular region. IPL = inner plexiform layer; OPL = outer plexiform layer; PRL = photoreceptor layer. SLD: $\lambda = 843$ nm; $\Delta\lambda = 30$ nm. Courtesy of J. G. Fujimoto, M.I.T. Reprinted from [6] by permission of the Optical Society of America. (b) OCT scattering potential image of a human periodonitum at the cemento-enamel junction. SLD: $\lambda = 1280$ nm; $\Delta\lambda = 34$ nm

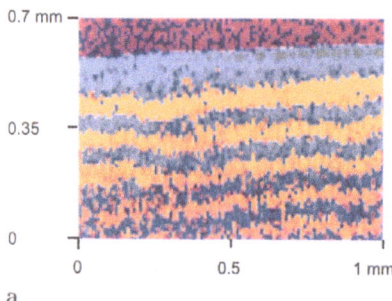
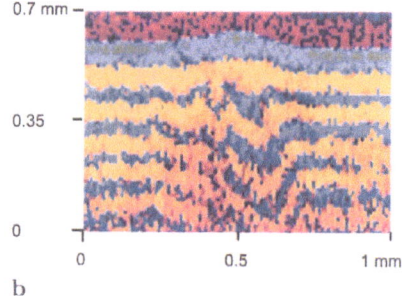

a   b

Fig. 7. Birefringence OCT images of bovine tendon 1 mm wide and 0.7 mm deep. SLD: $\lambda = 862$ nm; $\Delta\lambda = 21$ nm. The banded false-color structure indicates birefringence. (a) Birefringence image of fresh bovine tendon; (b) Birefringence image of bovine tendon following exposure to laser pulses incident from the upper left causing an initial surface temperature of 77 °C. Courtesy of J. F. de Boer, Beckman Laser Institute, Irvine. Reprinted from [39] by permission of the Optical Society of America

Another important image parameter is absorption. Using Fourier transformation of the interference signals obtained with a pair of light-emitting diodes with center wavelengths at 1.3 and 1.46 mm, Schmitt et al. [40] demonstrated the ability of this technique to distinguish lipid and water inclusions in a scattering material.

3.2   Dual-Beam OCT

In contrast to the reflectometer technique, here the object is not part of the interferometer. It is illuminated by the two beams ("dual beam") exiting from a Michelson interferometer (or any other two-beam interferometer). The dual beam (DB in Fig. 8) is used to transversely scan the object with the help of the

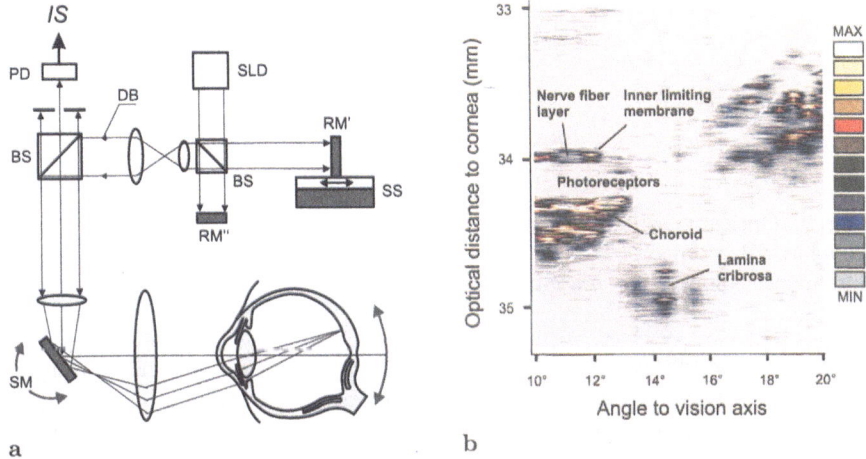

Fig. 8. Time-domain dual-beam OCT [37]. (**a**) Optical scheme. IS = interferometer signal. BS = beam splitter; DB = dual beam; PD = photodetector; RM′, RM″ = reference mirrors; SLD = superluminescent diode; SM = scanning mirror; SS = scanning stage. (**b**). First in vivo OCT tomogram [4]. SLD: $\lambda = 830$ nm; $\Delta\lambda = 25$ nm. A logarithmic color scale has been used to encode the scattering potential

scanning mirror (SM), whereas the depth scan is performed by shifting one of the reference mirrors (RM′) with a constant speed v. The corresponding Doppler shift $f_D = 2v/\lambda$ modulates the intensity of the interference pattern at the photodetector (PD). The photodetector signal is then amplified and filtered by a bandpass filter that transmits only signals with f_D. The envelope of this signal is recorded as a function of the difference in the interferometer's arm length with a PC. Obviously, this path difference does not depend on the distance between object and interferometer. Hence, this technique is not dependent on this distance and, therefore, may be used for high-precision depth and thickness measurement.

The unique stability of the dual-beam technique in the depth direction facilitates high-precision depth measurements. In fact, this feature of the dual-beam technique has already lead to important medical applications of the dual-beam PCI technique itself. In particular, in the ophthalmologic field of intraocular distance measurement, a wide range of applications in physiologic studies and in therapeutic techniques has been established [7–14].

3.3 En-Face OCT

Usually, OCT generates depth images, i.e. images in planes normal to the object surface. En-face images can be obtained by fixing the reference mirror and transversely scanning either the object or the sample arm of the interferometer [41]. A corresponding arrangement has also been described by Podoleanu et al. [42]. In their experimental arrangement a single-mode

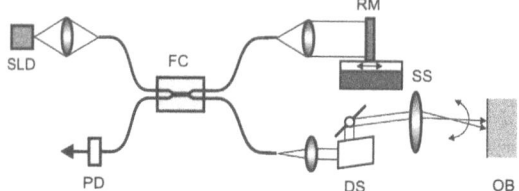

Fig. 9. Optical scheme of en-face OCT. DS = dual galvo scanner; FC = fiber coupler; OB = object; PD = photodetector; RM = reference mirror; SLD = super-luminescent diode; SS = scanning stage

directional coupler was used, where the output fibers form a Michelson inter-ferometer. The reference arm consists of a fixed microscope objective and a mirror mounted on a micrometer scanning stage. The sensing arm contains a microscope objective collimating the beam exiting from the fiber. A two-directional galvo scanner is used to transversely scan the target. Scanning the light beam across the (plane) target surface leads to a modulation of the corresponding path length and thus to a frequency modulation of the photoelectric signal similar to that generated by the PCI depth scan.

A disadvantage of this technique is that, due to the nonlinear path length dependence on the scanning angle, a frequency spread of the photoelectric signal occurs leading to a reduced signal-to-noise ratio. This disadvantage can be overcome by either frequency-shifting the reference beam or laterally displacing the target beam from the axis of rotation of one of the galvo mirrors leading to an additional modulation, which is essentially linearly dependent on the deflection angle.

4   Resolution in OCT

OCT uses PCI. PCI is based on spatially coherent but temporally low- or non-coherent light, which is typically broad-bandwidth light emitted by su-perluminescent diodes. In a first step of analysis we shall again assume the object is illuminated by a monochromatic wave parallel to the z axis and use the Helmholtz equation solution, (6), to describe back-scattering along the negative z axis. In the second step we shall then adapt the results to low-time-coherence light. Figure 3 depicts the scattering configuration of the coherence scan in OCT. We replace the integrations over x' and y' in (6) by a constant factor W, chosen to be proportional to the cross section of the beam waist of the illuminating beam. From (2) and (6) the field scattered by the object at P ($\boldsymbol{k}^{(S)} = -\boldsymbol{k}^{(i)}$; $|\boldsymbol{k}^{(S)}| = |\boldsymbol{k}^{(i)}| = k$) is, apart from a constant factor,

$$E_{\mathrm{O}}(z,k,t) = -\frac{A_{\mathrm{LS}}W}{4\pi L}\exp(-\mathrm{i}kz - \mathrm{i}\omega t)\int_{0}^{D} F_{\mathrm{O}}(z')\exp(2\mathrm{i}kz')\mathrm{d}z'$$

$$= -\frac{A_{\mathrm{LS}}W}{8\pi L} \int\limits_{0}^{2D} F_{\mathrm{O}}\left(\frac{z'}{2}\right) \exp[\mathrm{i}k(z' - z) - \mathrm{i}\omega t]\mathrm{d}z' . \tag{7}$$

This equals a spatial convolution of the two-fold dilated object structure with a back-travelling light wave, $E_{\mathrm{B}} = A_{\mathrm{LS}}\exp(-\mathrm{i}kz - \mathrm{i}\omega t)$, which, apart from of a constant factor, equals the source wave:

$$E_{\mathrm{O}}(z,t) \propto F_{\mathrm{O}(1/2)}(z) \otimes E_{\mathrm{B}}(z,t) . \tag{8}$$

The following convolution notation has been used:

$$f_{(\alpha)}(t;p) \otimes g_{(\beta)}(t;p) = \int\limits_{-\infty}^{+\infty} f_{(1)}(\alpha x; p)g_{(1)}(t - \beta x; p)\mathrm{d}x$$

$$= \int\limits_{-\infty}^{+\infty} f(\alpha x; p)g(t - \beta x; p)\mathrm{d}x .$$

Hence, the back-scattered object wave equals the reflected light source wave convolved with the scattering potential of the object. The scattering potential has to be extended by a factor of two in space because of the double transit of the light. From (3) and (6) it is clear that z is the additional path length the light experiences within the object.

Again, we have used a solution of the Helmholtz equation. However, as already mentioned, the waves used in OCT are not true sinusoidal waves but quasi-monochromatic waves with statistical properties. (A more basic treatment would have to solve the two wave equations for the mutual coherence function of the light scattered by the object [15].) The interferometer is illuminated by a light source wave with intensity

$$I_{\mathrm{LS}} = \frac{1}{2}\varepsilon_0 c\frac{1}{T}\int\limits_{T} E^*(t)E(t + \tau)\mathrm{d}t = \frac{1}{2}\varepsilon_0 c\left\langle E_{\mathrm{LS}}^*(t)E_{\mathrm{LS}}(t)\right\rangle . \tag{9}$$

The complex degree of time coherence is

$$\gamma_{\mathrm{LS}}(\tau) = \frac{\left\langle E_{\mathrm{LS}}^*(t)E_{\mathrm{LS}}(t + \tau)\right\rangle}{\left[\left\langle E_{\mathrm{LS}}^*(t)E_{\mathrm{LS}}(t)\right\rangle\left\langle E_{\mathrm{LS}}^*(t + \tau)E_{\mathrm{LS}}(t + \tau)\right\rangle\right]^{1/2}}$$

$$= \frac{(1/2)\varepsilon_0 c\left\langle E_{\mathrm{LS}}^*(t)E_{\mathrm{LS}}(t + \tau)\right\rangle}{I_{\mathrm{LS}}}, \tag{10}$$

where ε_0 is the electric permittivity [43]. The observable quantities, like the intensity of light, are time averages. If the object wave (E_{O}) and the reference wave (E_{R}) are colinear at the interferometer exit, the resulting intensity depends on only the time delay t between the two waves:

$$I_{\mathrm{E}}(\tau) = \frac{1}{2}\varepsilon_0 c\left\langle|E_{\mathrm{R}}(t) + E_{\mathrm{O}}(t + \tau)|^2\right\rangle . \tag{11}$$

We still have to incorporate the proper time delays encountered by the light back-scattered by the object.

Time-domain OCT is based on the transit time of wave groups. Wave groups show specific propagation properties. Their phase velocity equals the (phase) velocity of the corresponding carrier wave, whereas the amplitudes propagate with the group velocity [44]. Therefore, the time delay τ between the object and reference waves according to the group delay time caused by the additional path length of the light within the object is

$$\tau = \frac{\Delta z\, n_{\mathrm{G}}}{c} . \tag{12}$$

n_{G} is the group index; Δz is the geometrical path difference between the object and reference waves introduced by the additional path length the light experiences through the object. Therefore, the spatial convolution of the back-travelling wave with the scattering potential of the object introduces a time delay τ (replacing the z dependence) to the object wave:

$$F_{\mathrm{O}(1/2)}(z) \otimes E_{\mathrm{B}}(z,t) \propto E_{\mathrm{O}}(t+\tau) . \tag{13}$$

The reference wave is derived directly from the light source wave by a beam splitter and reflection at the reference mirror, $E_{\mathrm{R}}(t) = \alpha E_{\mathrm{B}}(t)$; α is a constant depending on the properties of the beam splitter and the reference mirror. Hence, the resulting intensity I_{E} at the interferometer exit is

$$\begin{aligned} I_{\mathrm{E}}(\tau) &= \frac{1}{2}\varepsilon_0 c \langle |E_{\mathrm{R}} + E_{\mathrm{O}}|^2 \rangle \\ &= I_{\mathrm{R}} + I_{\mathrm{O}} + \alpha\varepsilon_0 c \Re\left\{\langle E_{\mathrm{B}}^*(t) E_{\mathrm{O}}(t+\tau)\rangle\right\} . \end{aligned} \tag{14}$$

The desired depth information is contained in the mutual coherence function of the re-emitted object wave and the reference wave in the interferogram term. We assume the path lengths of the reference beam and the object beam are matched. Then τ is caused by the time delay introduced by the object structure; see (13). Usually, the reference wave is Doppler-shifted and consequently the photodetector signal at the interferometer exit is modulated at the same frequency. Band-pass frequency filtering is used to separate the photoelectric interferogram signal $U(z)$ from noise and unimportant DC terms. Using the commutative law of convolution we obtain

$$U(\tau) \propto \Re\left\{\langle E_{\mathrm{B}}^*(t) E_{\mathrm{B}}(t+\tau) \otimes F_{\mathrm{O}(1/2)}(\tau)\rangle\right\} . \tag{15}$$

The time average $\langle ... \rangle$ depends only on the time coherence function $\Gamma_{\mathrm{LS}}(\tau)$ or the complex degree $\gamma_{\mathrm{LS}}(\tau)$ of coherence of the source wave

$$\frac{1}{2}\varepsilon_0 c \langle E_{\mathrm{B}}^*(t) E_{\mathrm{B}}(t+\tau)\rangle = I_{\mathrm{B}}\gamma_{\mathrm{LS}}(\tau) . \tag{16}$$

Furthermore, apart from a constant factor, the interferogram term is the real part of the convolution of the object structure with the complex degree of the time coherence:

$$U(\tau) \propto \Re\left\{\gamma_{\mathrm{LS}}(\tau) \otimes F_{\mathrm{O}(1/2)}(\tau)\right\} . \tag{17}$$

Therefore, in OCT the complex degree $\gamma_{LS}(\tau)$ of the time coherence plays the role of a longitudinal or depth point spread function.

For example, most of the light sources used in OCT at present have an approximately Gaussian frequency spectrum. Therefore, they also have an approximately Gaussian degree of coherence [43]:

$$\gamma_{LS}(\tau) = \exp(-i\omega_0\tau - (1/2)\delta^2\tau^2) \,, \tag{18}$$

with the full width at half maximum (FWHM) value or coherence time $\tau_C = 1.177/\delta$, and the longitudinal coherence length $l_C = c\tau_C$:

$$l_C = \frac{4\ln 2}{\pi} \frac{\lambda^2}{\Delta\lambda} \,, \tag{19}$$

where $\Delta\lambda$ is the wavelength FWHM. In OCT, however, where light travels along the same path twice – there and back – through the object, depth resolution equals $l_C/2$. (Therefore, a so-called round-trip coherence length $l_C/2$ has also been used as a definition of coherence length [38,45]; in fact, $l_C/2$ defines depth resolution in OCT.) At present, for example, superluminescent diodes at $\lambda \approx 830$ nm, have spectral widths of about $\Delta\lambda = 25$ nm. Thus a corresponding coherence length of $l_C = 24\,\mu m$ and a depth resolution of approximately $12\,\mu m$ are obtained.

In imaging optics, the transverse resolution is usually defined by the Rayleigh criterion: two object points are resolved if the center of the transverse point-spread function (Airy disk) of one object point falls on the first zero of the point-spread function generated by the second [46]. Most beams used in OCT are approximately Gaussian beams in the fundamental transverse mode. In this case, however, there is no zero intensity. If the transverse sampling rate is high enough, i.e. if it satisfies the sampling theorem, the transverse resolution in OCT is determined by the transverse width of the beam performing the coherence scan. High transverse resolution demands focussing of the beam to a small beam waist radius w_0. w_0 is the radius at which the beam intensity falls to $1/e^2$ of its central value [47]. The minimal transverse separation of two resolved image points is of the order of this value. Hence, w_0 can be used to define the minimal transverse separation δ resolved in OCT:

$$\delta = w_0 = \frac{\lambda}{\pi\vartheta} \,. \tag{20}$$

Of course, a small beam waist radius demands a large asymptotic angle ϑ of beam divergence (or "aperture"). But high divergence results in a reduced confocal beam parameter z_0. Twice the confocal beam parameter can be used to define the depth of focus DOF:

$$\text{DOF} = 2z_0 = 2\frac{\pi w_0^2}{\lambda} \,. \tag{21}$$

Therefore, a compromise has to be found between the desired depth of focus DOF and the desired transverse resolution δ. For example, a transverse resolution of $\delta = 20\,\mu\text{m}$ at a wavelength of $\lambda = 830$ nm leads to a depth of focus of DOF $= 3$ mm.

For the sake of completeness we mention that the above resolution arguments apply for all OCT techniques described in this chapter.

5   Frequency-Domain OCT

5.1   Complex Spectral OCT

In back-scattering with monochromatic waves a direct implementation of the Fourier transform relation between the scattering potential of the object structure and the complex scattered field amplitude (6) can be used. Such a monochromatic implementation is achieved, for example, with the help of a spectrometer displaying the monochromatic components of the re-emitted light. In this case we can again use the monochromatic Helmholtz equation solution (7):

$$E_\mathrm{O}(z, K, t) = -\frac{A_\mathrm{LS}W}{4\pi L} \exp(-\mathrm{i}kz - \mathrm{i}\omega t) \int\limits_0^D F_\mathrm{O}(z') \exp\left[\mathrm{i}Kz'\right] \mathrm{d}z' , \qquad (22)$$

with $K = 2k$. Hence the object wave equals a multiplication of the back-scattered wave with the inverse Fourier transform of the object structure in K space:

$$E_\mathrm{O}(z, K, t) = -\frac{A_\mathrm{LS}W}{4\pi L} \exp\left(-\mathrm{i}kz - \mathrm{i}\omega t\right) \hat{F}_\mathrm{O}(-K) , \qquad (23)$$

$$\hat{F}_\mathrm{O}(K) = \mathrm{FT}\left\{F_\mathrm{O}(z)\right\} . \qquad (24)$$

Therefore, the object structure can be obtained by a Fourier transform of spectrally resolved amplitude and phase data of the back-scattered object wave:

$$F_\mathrm{O}(z) = \mathrm{FT}\left\{\hat{F}_\mathrm{O}(-K)\right\} , \qquad (25)$$

where $\hat{F}_\mathrm{O}(-K)$ is obtained with the help of an interferometric spectrometer. This instrument is a spectrometer with phase-measuring capability.

In addition to the usual spectral intensity data, spectral phase data can also be measured with this instrument. The optical scheme of this instrument is shown in Fig. 10. Basically, it corresponds to a Michelson interferometer with a spectrometer at the interferometer exit. The spectrometer displays the monochromatic field components. An additional piezoelectric translator is used at the reference arm to perform the phase measurement. A superluminescent diode is used as a light source.

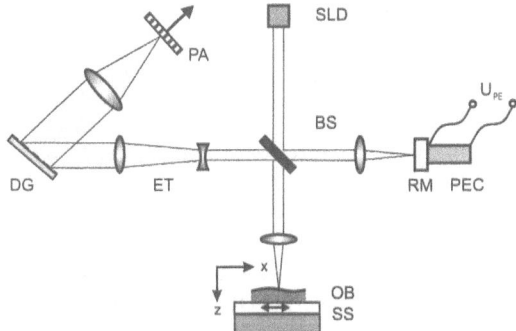

Fig. 10. Complex spectral OCT [48]. U_{PE} is the piezo-electric driving voltage. BS = beam splitter; DG = diffraction grating; ET = beam expanding telescope; OB = object; PA = photodetector-array; PEC = piezo-electric crystal; RM = reference mirror; SLD = superluminescent diode; SS = scanning stage

This technique gives direct access to both amplitude and phase data of the complex scattering potential of the object. This is an important issue. In the macroscopic theory of electrodynamics the absorption coefficient is [43]

$$\alpha = \frac{2\omega\kappa}{c} \, , \qquad (26)$$

where $\kappa(\boldsymbol{r})$ is the extinction coefficient; it is part of the complex refractive index m:

$$m(\boldsymbol{r}) = \eta(\boldsymbol{r}) + \mathrm{i}\kappa(\boldsymbol{r}) \, . \qquad (27)$$

$\eta(\boldsymbol{r})$ is the phase refractive index. Hence, α can be obtained from the complex scattering potential F_O of the object:

$$F_O(\boldsymbol{r}) = -k_0^2 \left[\eta^2(\boldsymbol{r}) - \kappa^2(\boldsymbol{r}) - 1 + 2\mathrm{i}\eta(\boldsymbol{r})\kappa(\boldsymbol{r}) \right] \, . \qquad (28)$$

By splitting the complex scattering potential into real and imaginary parts, for example, the extinction coefficient and the phase refractive index can be obtained by a straightforward computation. However, the high pass imaging property of OCT has still to be taken into account.

As such, all recent spectral techniques, like spectral interferometry [31] and spectral radar [49], add a reference beam with a fixed path length or phase, respectively. This reference wave encodes the object phase in the resulting interferogram term. It is not used, however, to determine the phase of the scattered field explicitly. It serves to introduce a carrier frequency in the spectrum in order to separate the auto-correlation terms in the reconstruction (see next subsection).

To obtain the complex scattering potential of the object, both the spectral amplitudes and phases of the object wave have to be measured. This can be done by standard phase-measurement interferometry techniques [50]. A

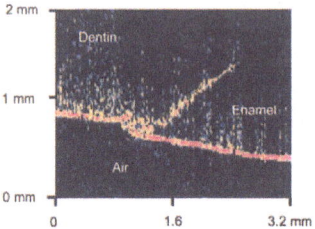

Fig. 11. Tomographic image of the air/tooth interface and the dentin and enamel junction of a human tooth obtained by complex spectral OCT. SLD: $\lambda = 830\,\mathrm{nm}$, $\Delta\lambda = 20\,\mathrm{nm}$. The magnitude of the scattering potential is logarithmically encoded in false colours. Depth dimensions are in terms of optical path length. Refractive indices are approximately 1.6 for enamel and 1.5 for dentin [80]. The lower imaging depth of only approximately 1 mm in comparison to Fig. 6b is due to the shorter wavelength used here

standard detector array is used in a spectrometer at the interferometer exit; see Fig. 10. Known phase changes are induced between the object and reference beam by shifting the reference mirror at fixed amounts. The spectral object wave phase and its intensity can be calculated directly from the spectral intensity data at the interferometer exit obtained at discrete reference mirror positions [48].

Several algorithms have been developed that calculate the phase of a light wave by using this technique and compensate phase-shift miscalibration and detector nonlinearity [51]. Even in the case of a phase-shift miscalibration of 20% these algorithms can compensate the phase error to less than 1%. This tolerance is of particular interest in complex spectral OCT, first, because it enables an achromatic phase measurement with a single phase-shifting reference mirror and, second, because it provides some tolerance to unstable objects. Figure 11 presents an example of a complex spectral OCT image of a human tooth at the gingival margin.

5.2   Spectral Interferometry OCT and Chirp OCT

As shown by (6), the scattering potential along the depth scan can be obtained by a Fourier transform of the scattered wave at a range of wavenumbers k. Two corresponding techniques have been used in the past. The first is a "white"-light source technique based on the modulation of the spectrum of the re-emitted light [16,18]; this technique uses an interferometer together with a spectrometer, as indicated in Fig. 12a. The second is a tunable laser technique simply using a tunable laser together with an interferometer, as indicated in Fig. 12b.

It was recognized quite early on that the need for a mechanically moving reference mirror in OCT can be avoided by using spectral interferometry or channeled spectra, and that the inverse Fourier transform of the spectral

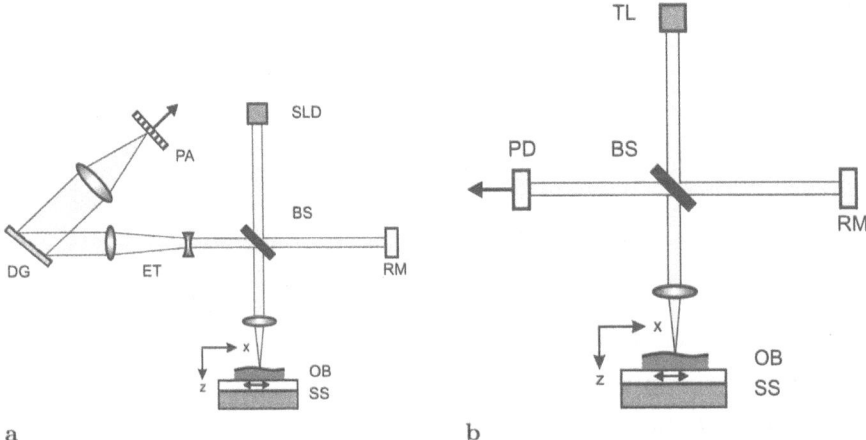

Fig. 12. (a) Spectral interferometry or white-light source OCT, and (b) tunable laser or chirp OCT. BS = beam splitter; DG = diffraction grating; ET = expansion telescope; OB = object; PA = photodetector array; PD = photodetector; RM = reference mirror; SLD = superluminescent diode; SS = scanning stage; TL = tunable laser

intensity yields the coherence function of the light re-emitted from the object and thus the same signal as obtained in the PCI technique used in time-domain OCT [52].

In the "white"-light source technique (also called "spectral radar" [49]), a light source emitting spatially coherent light with large spectral bandwidth, like a superluminescent diode, is used in an interferometer. The spectral intensity of the light exiting from the interferometer is detected by a photodetector array at the exit plane of a spectrometer.

In the tunable laser technique (also called "chirp" OCT [53]; the name comes from the early days of the electronic radar technique when a sound produced by a pulse resembled the chirps of a bird in song) the wavelength is tuned over a range of wavenumbers and the spectral light intensity re-emitted from the object is detected by a photodetector.

Apart from a constant C the wavenumber-dependent intensity spectrum $I_O(K)$ of the back-scattered light is equal to the square of the Fourier transform of the scattering potential, (23), of the object:

$$I_O(K) = |E_O(K)|^2 = C|\hat{F}_O(-K)|^2 . \tag{29}$$

Taking the inverse Fourier transform of $I_O(K)$ would yield the auto-correlation function (ACF) of the scattering potential:

$$\mathrm{FT}^{-1}\{I_O(K)\} = C\langle F_O^*(z)F_O(z+Z)\rangle = C\,\mathrm{ACF}_\mathrm{F}(Z) . \tag{30}$$

However, autocorrelation is not reversible. There are two possibilities to obtain the scattering potential of the object. First, the scattering potential of the object can be obtained if an additional singular light remitting interface (reference mirror RM, for example, in Fig. 12a,b) is positioned at an optical path difference z_R from the object. Second, the true scattering potential can also be obtained if the object itself contains one interface with a relatively large reflectivity R acting as a reference mirror [31]. Here we shall describe only the situation with an additional reference mirror. In this case the scattering potential can be described as a sum of the object structure $F_O(z)$ plus a delta-like potential at z_R (with amplitude reflectivity R):

$$F(z) = F_O(z) + R\delta(z - z_R) .\tag{31}$$

Then the autocorrelation yields four terms:

$$\begin{aligned}
&\langle F^*(z)F(z+Z)\rangle \\
&= \langle F_O^*(z)F_O(z+Z)\rangle + \langle F_O^*(z)R\delta(z+Z-z_R)\rangle \\
&\quad + \langle R\delta^*(z-z_R)F_O(z+Z)\rangle + \langle R^2\delta^*(z-z_R)\delta(z-z_R+Z)\rangle \\
&= \mathrm{ACF_F}(Z) + RF_O^*(z_R-Z) + RF_O(z_R+Z) + R^2\delta(Z) .
\end{aligned}\tag{32}$$

The third term yields a reconstruction of the complex object structure, centered at $Z = -z_R$. Encoded in this reconstruction is, of course, also the object phase. Any overlap between the four terms of the ACF is avoided by choosing the distance between the reference mirror and the object larger than the object depth D: $|z_R| > D$. However, from the similarity theorem it follows that expansion in the space domain corresponds to a compression in the Fourier domain [34]. Hence, increasing the path difference between object and reference beam increases the spatial frequency in the spectrometer plane and therefore reduces the accessible depth of the object field.

Spectral OCT techniques may play a role in dermatologic OCT for acquisition of skin morphology. An example is shown in Fig. 13. These techniques do not need interferometer reference arm scanning. The PCI depth scan is replaced by a Fourier transform of the spectral intensity of the re-emitted light.

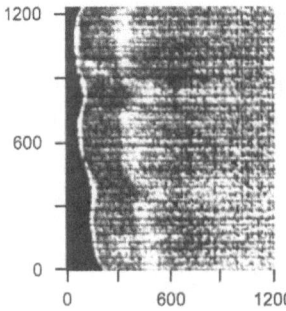

Fig. 13. 1.2 mm × 1.2 mm spectral radar tomogram of normal skin at the finger tip. From left: Stratum corneum, Stratum papillare with dermal papillae, and Stratum germinativum. SLD: $\lambda = 849$ nm, $\Delta\lambda = 50$ nm. Courtesy of G. Häusler, University of Erlangen-Nürnberg, Germany

With spectral radar superficial spreading melanoma have been distinguished from normal skin [49].

6   Doppler OCT

Doppler OCT (DOCT) was pioneered by Wang et al. [54]. In this technique Doppler velocimetry is combined with optical coherence tomography. Doppler velocimetry detects the frequency shift ω_D in the light scattered back from moving particles. This frequency shift is

$$\omega_D = \left(\boldsymbol{k}^{(S)} - \boldsymbol{k}^{(i)}\right)\boldsymbol{v} = \boldsymbol{K}\boldsymbol{v} \ . \tag{33}$$

$\boldsymbol{v}$ is the velocity of the moving particle. A frequency shift in the object beam can be accounted for by a corresponding phase term $f(t) = \exp(i\omega_D t)$, normalized to unity for convenience:

$$E_O(t) = f(t)E_B(t) \ . \tag{34}$$

The intensity at the interferometer exit is (with the reference field strength $E_R(t)$):

$$I(\tau) \propto \left\langle \left| E_R(t) + E_O(t+\tau) \right|^2 \right\rangle \propto I_R + I_O + 2\Re\left\{ \left\langle E_R^*(t)E_O(t+\tau) \right\rangle \right\} \ . \tag{35}$$

The frequency-dependent intensity or power spectrum is obtained from the Fourier transform of the correlation function of the electric fields:

$$\begin{aligned} I(\omega) &\propto \mathrm{FT}\left\{ \left\langle E_R^*(t)E_O(t+\tau) \right\rangle \right\} = \mathrm{FT}\left\{ f(t) \left\langle E_R^*(t)E_R(t+\tau) \right\rangle \right\} \\ &= \hat{f}(\omega) \otimes S(\omega) \ . \end{aligned} \tag{36}$$

$\hat{f}(\omega)$ is the spectrum of the light scattered by the moving particles, $S(\omega)$ is the power spectrum of the source light.

   In contrast to conventional OCT techniques the photodetector signal is not frequency filtered to obtain the envelope of the interferometer signal. In the implementation by Chen et al. [55] data acquisition is performed by a lateral scan across the sample surface (x in Fig. 14), whereas the depth scan (z in Fig. 14) is obtained by a longitudinal incremental movement of the sample along the surface normal.

   By stretching the fiber wrapped around a piezoelectric cylinder an optical phase modulation at frequency ω_P is generated. Depending on the direction of flow the Doppler effect adds an additional positive or negative frequency shift. At each x position Doppler OCT pixel data are collected at a series of depth positions z_i defined by the position of the reference mirror. At each depth position z_i the real part of the correlation function of the electric fields is recorded and the power spectrum of the back-scattered light at this pixel is obtained by a short-time fast Fourier transform.

   A tomographic structural image is obtained from the power spectrum intensity at $\omega = \omega_P$ (Fig. 15a). Because the magnitude of the back-scattered

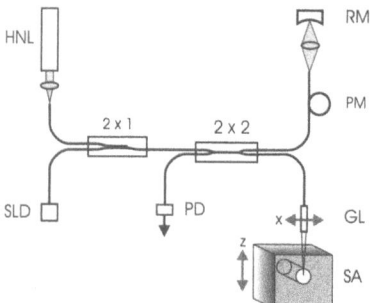

Fig. 14. Doppler OCT instrumentation (adapted from [55]). $2 \times 1 = 2 \times 1$ fiber optic coupler; $2 \times 2 = 2 \times 2$ coupler; GL = gradient lens; HNL = He-Ne-laser; PD = photodetector; PM = phase modulator; RM = reference mirror; SA = sample; SLD = superluminescent diode

intensity decreases exponentially as the depth in the sample increases, a logarithmic scale is used to display that image. A fluid velocity image is obtained from the frequency of the centroid of the measured power spectrum (Fig. 15b). As in standard OCT, the structural transverse resolution is limited by the aperture of the illuminating beam and the depth resolution is limited by the source coherence length. From the Fourier uncertainty relation the frequency resolution is inversely proportional to the temporal width of the short-time fast Fourier transform. Furthermore, the velocity resolution depends on the angle between the scattering vector and the direction of motion of scatterers in the sample [56].

Izatt et al. [57] use a direct modification of the conventional OCT technique including an axially scanning reference mirror to perform the PCI depth scan. In their implementation the interferometric fringe frequency arises from the net sum of Doppler shifts generated by the moving reference mirror and moving scatterers. Coherent detection is used to obtain the envelope of the interferometric photodetector signal. These authors also showed that the accuracy of the velocity measurement can be substantially improved by ensemble averaging the power spectrum [58].

DOCT can be used for noninvasive imaging of in vivo blood flow dynamics and tissue structures. From the power spectrum calculated at each pixel both a flow-velocity image and a tomographic structural image can be obtained by the same scan. This has been demonstrated on the biological model of the chick chorioallantoic membrane; see Fig. 15. The use of DOCT to monitor in vivo blood flow dynamics and vessel structure changes in response to vasoactive drug and photodynamic therapy has been demonstrated by Chen et al. [59].

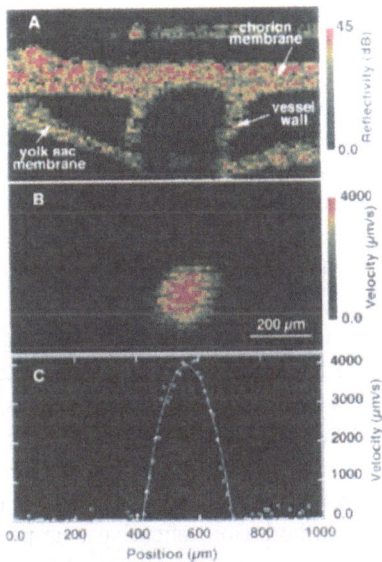

Fig. 15. (**A**) Color-coded structural image and (**B**) color-coded velocity image in a chorioallantoic vein. (**C**) velocity profile across the center of the vein: solid line, fit to a parabolic function. SLD: $\lambda = 850$ nm, $\Delta\lambda = 25$ nm. Courtesy of Z. Chen, Beckman Laser Institute, Irvine. Reprinted from [60] by permission of the Optical Society of America

7   OCT Macroscopy and Morphometry

OCT resolution is at the threshold of microscopy. Hence, the distinction between macroscopy and micoscopy is somewhat arbitrary. In this section, however, we concentrate on OCT techniques that mainly use the high depth resolution provided by PCI, i.e. techniques relying more or less on the identification of layers.

The first field of application and still the main medical field of OCT is ophthalmology. The first description of the diagnostic potential of OCT in this field appeared in 1993 [20]. The clinical potential of OCT in the diagnosis of macular diseases was described by Puliafito et al. [22]: OCT can clearly visualize macular holes, epiretinal membranes, macular edema, intraretinal exsudates, idiopathic central serous chorioretinopathy, and detachments of pigment epithelium and neurosensory retina. OCT is noninvasive and uses infrared light and thus it is more comfortable for and better tolerated by patients than, for example, fluorescein angiography or fundus photography. Although there might be fields in ophthalmology where OCT cannot replace the corresponding conventional diagnostic techniques or further studies might be necessary [61], OCT can usually add new information [62]. OCT typically has the advantage of high depth resolution. Thus it has a particular potential to discover layered structures [63], find changes in layered structures [64], and yield quantitative depth and layer thickness data [37].

In the past, no in vivo imaging technique was capable of resolving the small structures that comprise the retinal layers. Therefore, no direct correlation of OCT images with, for example, corresponding in vivo light microscopy images is possible. Hence, for clinicopathologic correlation a direct comparison of OCT and histologic images of the Macaca mulatta macula was made [65]. In another approach, tomograms of pathologic retinae were compared with those of healthy, normal subjects to elucidate the origin of the PCI depth scan signal peaks [66].

OCT paves the way towards quantitative medical morphology. Here, too, ophthalmology is the most advanced medical field. Dimensions, profiles, and quantitative interrelationships between ocular structures play an increasingly important role in modern ophthalmology. The first step was PCI. PCI can be used to measure the corneal tickness of the human eye, anterior chamber depth and lens thickness with submicrometer precision [8,14,67]. The axial eye length and its changes during accommodation, effective lens position and lens–capsule distance after cataract surgery, and the thickness of retinal layers have been determined with micrometer precision [37,68–70]. High-resolution imaging is important in ophthalmology to identifiy intraocular abnormalities and monitor ocular therapies. Quantitative assessment of the retinal thickness and its layers is of particular importance in glaucoma diagnosis. OCT measurements may become a sensitive diagnostic test for the early detection of macular thickening in patients with diabetic retinopathy [71]. Retinal thickness and nerve fiber layer thickness have been measured by OCT with standard deviations of $17\,\mu m$ [37,72,73] and less. Schuman et al. [74] demonstrated a high correlation of OCT nerve fiber layer thickness measurements with the functional status of the optic nerve. OCT layer thickness measurements turned out to be of rather high reproducibility [75] as well as a sensitive and an early indicator of nerve fiber layer thinning in eyes with optic nerve head drusen [76]. OCT has also been shown to be successful in monitoring choroidal neovascularizations [77].

Another field of OCT in ophthalmology is the in vivo assessment of tissue lesions generated by laser radiation. For example, the in vivo assessment of argon laser retinal lesions with OCT gives a new insight in the pathologic response [78]. In laser thermokeratoplasty thermally induced lesions are used to change the radius of curvature of the cornea. OCT could provide a non-invasive method for immediate and follow-up control. Figure 16 shows an example from a corresponding study.

Another promising field of OCT application is in vivo and in vitro imaging of hard and soft tissue structure in the oral cavity. Extensive imaging of the various types of oral mucosa and the structure of healthy and decayed teeth has been performed by F. I. Feldchtein et al. [80]. A dual-wavelength OCT scanner (at $\lambda = 830\,nm$ and $\lambda = 1230\,nm$) has been used for imaging oral

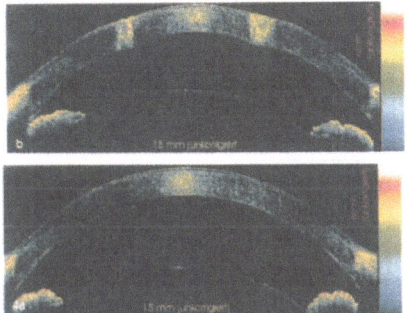

Fig. 16. OCT images of the cornea of enucleated porcine eyes before (*above*) and after thermokeratoplastic laser irradiation (*below*) with a diode laser beam ($\lambda = 1.86$ nm). SLD: $\lambda = 830$ nm; $\Delta\lambda = 22$ nm. These images depict the scattering potential. Here scattering is an approximately monotonic function of the extent of the thermal lesion. Courtesy of N. Koop, LMZ Lübeck. Reprinted from [79] by permission of Springer

mucosa. Hard tissue images were acquired with a single-wavelength OCT device operating at 1280 nm. These techniques could accurately depict and detect anomalities in dental tissue structures.

A novel dental OCT system has been developed by Colston et al. [81] which incorporates the interferometer sample arm and transverse scanning optics into a handpiece that can be used intraorally. With a wavelength of 1310 nm, the average imaging depth varied from 3 mm in hard tissues to 1.5 mm in soft tissues. Cross-sectional images of dental microstructure demonstrated the potential of this system for the diagnosis of periodontal disease, detection of caries, and evaluation of dental restorations.

Dermatological diagnosis is another medical field closely related to layer thickness problems. After a study [82] of the application of OCT in dermatology, several authors presented additional results. Welzel et al. [83] correlated the structures seen in OCT to light microscopic morphology. Experimentally induced bullous changes were used to distinguish different skin layers seen in OCT images. With 830 nm superluminescent diodes, a penetration depth of 1.5 mm has been achieved. Sweet gland ducts and melanoma nests could be localized. Spectral radar has also been used to distinguish a superficially spreading melanoma from normal skin [49].

OCT imaging of skin at two wavelengths (830 nm and 1285 nm) simultaneously has been shown to increase the potential of OCT for in vivo imaging of skin [84]. The longer wavelength can reduce the influence of multiple scattering on image contrast and resolution and thus increase the effective penetration depth in skin to 2 mm. This is of particular importance in the field of melanoma diagnosis; here it has long been known that it is the growth in the depth direction that best correlates with clinical outcomes. Sweet glands and deep dermal vessels could be imaged by using that wavelength. Three-dimensional OCT images of human skin have also been constructed from stacks of two-dimensional slices and allow full reconstruction of the skin and of features present immediately below the skin surface [85].

8 OCT Microscopy and Optical Biopsy

At present OCT does not have transverse and depth resolution in the $1\,\mu m$ range like standard optical microscopy does. However, there have already been some attempts towards this goal. The main obstacles are the limited aperture of the scanning beam and the rather large coherence length of the spatially coherent light sources available. With the currently used superluminescent diodes, depth resolutions on the order of 10 to $15\,\mu m$ have been achieved. The combination of two LEDs by using a fiber optic coupler yielded a depth resolution of $7.2\,\mu m$ [86]. With a broadband $Ti:Al_2O_3$ source, a resolution of $3.7\,\mu m$ has been demonstrated in biological tissue with a total thickness of a few hundred microns [87]. To obtain high depth resolution in thick dispersive media, like the human eye, the object dispersion has to be compensated. Hitzenberger et al. [88] have shown that, depending on the remaining uncertainty of the object dispersion, an optimum spectral bandwidth can be found, leading to minimal coherence length and hence optimal depth resolution. Using a synthesized light source generated by superimposing two superluminescent diodes with different center wavelengths a depth resolution of 6 to $7\,\mu m$ has recently been achieved at the fundus of a human eye in vivo [73].

To obtain high transverse resolution a short confocal parameter must be used. This concept limits the depth of field. Alternatively, the focusing lens can be shifted synchronously with the reference mirror during the depth scan [86]. This, however, involves the (rapid) movement of at least two optical components. Hence, a dynamic coherent focus technique has been proposed, in which only one mirror has to be moved. This technique renders the depth of field independent of the confocal beam parameter. A transverse resolution of better than $10\,\mu m$ has been demonstrated over a depth of $430\,\mu m$ [89].

Conventional biopsy and histopathology relies on microscopic inspection of excised tissue specimen. However, excision of tissue specimen is often contraindicated or impossible. Since OCT provides a resolution comparable to conventional histology, but in real time, it has a high potential to be used as a type of real time "optical biopsy" [90]. OCT does not require excision of tissue with its associated complications, cost, and delay in obtaining a diagnosis. Moreover, large regions may be surveyed for pathology and conventional biopsy can be reserved only for those regions in which abnormalities are detected by OCT imaging. In fact, OCT imaging has already been used as a biopsy technique in a wide range of biological systems to detect diseases. These include the tomographic imaging of the internal microstructure of in vitro atherosclerotic plaques with a level of resolution not previously

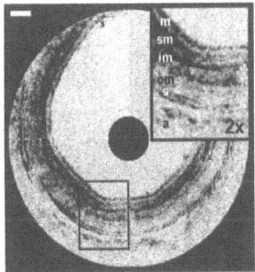

Fig. 17. Endoscopic OCT image of the rabbit esophagus in vivo. Kerr-lens mode-locked Cr^{4+}:forsterite laser: $\lambda = 1280\,nm$; $\Delta\lambda = 75\,nm$. Bar, $500\,\mu m$. m = mucosa; sm = submucosa; im = inner muscular layer; om = outer muscular layer; s = serosa; a = adipose and vascular supportive tissue. Courtesy of J. G. Fujimoto, M.I.T. Reprinted from [93] with permission. Copyright 1999 American Association for the Advancement of Science

achieved by other imaging modalities as well as the tomographic real-time diagnostics for intraoperative monitoring [24] and in microsurgical intervention [91]. Optical biopsy based on OCT also provides diagnostic information by differentiating the architectural morphology of urologic tissue [23], gastrointestinal tissue [27], and respiratory tissue [92].

Imaging of internal organ systems can be performed with the help of an endoscope. To avoid parasitic speckle noise a single-mode fiber optic beam delivery is necessary. A prototype endoscopic OCT system has been described by Tearney et al. [93,94]. A device with a diameter as small as 1.1 mm has been developed; Fig. 17 shows an OCT image obtained with that instrument. Feldchtein et al. [95] demonstrated that OCT data give more information about organs with epithelial tissue separated from their underlying stroma by a smooth basal membrane as, for example, in the larynx, the bladder, and the uterine cervix.

9   Conclusion

OCT is a quickly developing new imaging modality. OCT provides in vivo images at near-cellular-level resolution in real time. OCT can image scattering potential, birefringence, spectral absorption, and other object parameters. The Doppler version of this technique can be used to obtain structural images as well as velocity images. The great potential of OCT has lead to extensive clinical applications. Some of these might disappear; new ones will be introduced. There is no doubt about a bright future for OCT in the medical field as well as in related technological fields.

Acknowledgements

The authors acknowledge the permission from several authors and editors to reproduce figures out of their work. Thanks are also due to our co-workers at the Institute of Medical Physics. Our own work is based on projects financed by the Austrian Science Foundation (FWF projects P7300-MED, P9781-MED, and P10316-MED).

References

1. Wolf, E. (1996) Principles and development of diffraction tomography. In: Consortini, A. (Ed.) Trends in Optics, Vol. 3. Academic Press, San Diego, London, pp 83–110
2. Tromberg, B., Yodh, A., Sevick, E., Pine, D. (1997) Diffusing photons in turbid media: introduction to the feature. Appl Opt 36: 9
3. Huang, D., Swanson, E.A., Lin, C.P., Schuman, J.S., Stinson, W.G., Chang, W., Hee, M.R., Flotte, T., Gregory, K., Puliafito, C.A., Fujimoto, J.G. (1991) Optical coherence tomography. Science 254: 1178–1181
4. Fercher, A.F., Hitzenberger, C.K., Drexler, W., Kamp, G., Sattmann, H. (1993) In Vivo Optical Coherence Tomography. Am J Ophthalmol 116: 113–114
5. Fercher, A.F. (1996) Optical Coherence Tomography. J Biomed Opt 1: 157–173
6. Swanson, E.A., Izatt, J.A., Hee, M.R., Huang, D., Lin, C.P., Schuman, J.S., Puliafito, C.A., Fujimoto, J.G. (1993) In vivo retinal imaging using optical coherence tomography. Opt Lett 18: 1864–1866
7. Hitzenberger, C.K., Drexler, W., Dolezal, C., Skorpik, F., Juchem, M., Fercher, A. F., Gnad, H.D. (1993) Measurement of the Axial Length of Cataract Eyes by Laser Doppler Interferometry. Invest Ophthalmol Vis Sci 34: 1886–1893
8. Drexler, W., Baumgartner, A., Findl, O., Hitzenberger, C.K., Sattmann, H., Fercher, A.F. (1997) Submicrometer precision biometry of the anterior segment of the human eye. Invest Ophthal Vis Sci 38: 1304–1313
9. Hitzenberger, C.K., Drexler, W., Baumgartner, A., Lexer, F., Sattmann, H., Esslinger, M., Kulhavy, M., Fercher, A.F. (1997) Optical measurement of intraocular distances: a comparison of methods. Lasers Light Ophthalmol 8: 85–95
10. Drexler, W., Findl, O., Menapace, R., Rainer, G., Vass, C., Hitzenberger, C.K., and Fercher, A.F. (1998) Partial coherence interferometry: A novel approach to biometry in cataract surgery. Am J Ophthalmol 126: 524–534
11. Fercher, A.F., Roth, E. (1986) Ophthalmic laser interferometry. Proc SPIE 658: 48–51
12. Fercher, A.F. (1983) Verfahren und Anordnung zur Messung der Teilstrecken des lebenden Auges. Offenlegungsschrift DE 3201801A1
13. Hitzenberger, C.K. (1991) Optical measurement of the axial eye length by laser Doppler interferometry. Invest Ophthalmol Vis Sci 32: 616–624
14. Hitzenberger, C.K. (1992) Measurement of corneal thickness by low-coherence interferometry. Appl Opt 31: 6637–6642
15. Mandel, L., Wolf, E. (1995) Optical coherence and quantum optics, Cambridge University Press, Cambridge

16. Fercher, A.F., Hitzenberger, C., Juchem, M. (1991) Measurement of intraocular optical distances using partially coherent laser light. J Mod Opt 38: 1327–1333

17. Francon, M. (1966) Optical Interferometry, Academic Press, New York

18. Montgomery Smith, L., Dobson, C.C. (1989) Absolute displacement measurements using modulation of the spectrum of white light in a Michelson interferometer. Appl Opt 28: 3339–3342

19. Fercher, A.F. (1993) Ophthalmic interferometry. In: von Bally, G., Khanna, S. (ed) Optics in Medicine, Biology and Environmental Research. Elsevier, Amsterdam (Selected contributions to the OWLS-I-conference 1990, Garmisch-Partenkirchen, Germany, pp 221–235)

20. Izatt, J.A., Hee, M.R., Huang, D., Fujimoto, J.G., Swanson, E.A., Lin, C.P., Shuman, J.S., Puliafito, C.A. (1993) Ophthalmic diagnostics using optical coherence tomography. SPIE Proc 1877. 136–144

21. Izatt, J.A., Hee, M.R., Swanson, E.A., Lin, C.P., Huang, D., Schuman, J.S., Puliafito, C.A., Fujimoto, J.G. (1994) Micrometer-scale resolution imaging of the anterior eye in vivo with optical coherence tomography. Arch Ophthalmol 112: 1584–1589

22. Puliafito, C.A., Hee, M.R., Lin, C.P., Reichel, E., Schuman, J.S., Duker, J.S., Izatt, J.A., Swanson, E.A., Fujimoto, J.G. (1995) Imaging of Macular Diseases with Optical Coherence Tomography. Ophthalmology 102: 217–229

23. Tearney, G.J., Brezinski, M.E., Southern, J.F., Bouma, B.E., Boppart, S.A., Fujimoto, J.G. (1997) Optical biopsy in human urologic tissue using optical coherence tomography. J Urol 157: 1915–1919

24. Boppart, S.A., Bouma, B.E., Pitris, C., Tearney, G.J., Southern, J.F., Brezinski, M.E., Fujimoto, J.G. (1998) Intraoperative assessment of microsurgery with three-dimensional optical coherence tomography. Radiology 208: 81–86

25. Brezinski, M.E., Tearney, G.J., Weissman, N.J., Boppart, S.A., Bouma, B.E., Hee, M.R., Weyman, A.E., Swanson, E.A., Southern, J.F., Fujimoto, J.G. (1997) Assessing atherosclerotic plaque morphology: comparison of optical coherence tomography and high frequency intravascular ultrasound. Heart 77: 397–403

26. Schmitt, J.M., Yadlowsky, M.J., Bonner, R.F. (1995) Subsurface imaging of living skin with optical coherence microscopy. Dermatology 191: 93–98

27. Tearney, G.J., Brezinski, M.E., Southern, J.F., Bouma, B.E., Boppart, S.A., Fujimoto, J.G. (1997) Optical biopsy in human gastrointestinal tissue using optical coherence tomography. Am J Gastroent 92: 1800–1804

28. Herrmann, J.M., Brezinski, M.E., Bouma, B.E., Boppart, S.A., Pitris, C., Southern, J.F., Fujimoto, J.G. (1998) Two- and three-dimensional high-resolution imaging of the human oviduct with optical coherence tomography. Fertil-Steril 70: 155–158

29. Wolf, E. (1969) Three-dimensional structure determination of semi-transparent objects from holographic data. Opt Commun 1: 153–156

30. Fercher, A.F., Bartelt, H., Becker, H., Wiltschko, E. (1979) Image formation by inversion of scattered field data: experiments and computational simulation. Appl Opt 18: 2427–2439

31. Fercher, A.F., Hitzenberger, C.K., Kamp, G., El-Zaiat, S.Y. (1995) Measurement of Intraocular Distances by Back-scattering Spectral Interferometry. Opt Commun 117: 43–48

32. Dändliker, R., Weiss, K. (1970) Reconstruction of the three-dimensional refractive index from scattered waves. Opt Commun 1: 323–328

33. Cowley, J.M. (1975) Diffraction Physics. North-Holland, Amsterdam
34. Bracewell, R. (1965) The Fourier Transform and Its Applications. Mc Graw-Hill, New York
35. Youngquist, R.C., Carr, S., Davies, D.E.N. (1987) Optical coherence-domain reflectometry: a new optical evaluation technique. Opt Lett 12: 158–160
36. Tanaka, K., Yokohama, I., Chida, K., Noda, J. (1987) New measurement system for fault location in optical waveguide devices based on an interferometric technique. Appl Opt 26: 1603–1606
37. Drexler, W., Hitzenberger, C.K., Sattmann, H., Fercher, A.F. (1995) Measurement of the thickness of fundus layers by partial coherence tomography. Opt Engng 34: 701–710
38. Swanson, E.A., Huang, D., Hee, M.R., Fujimoto, J.G., Lin, C.P., Puliafito, C.A. (1992) High-speed optical coherence domain reflectometry. Opt Lett 17: 151–153
39. de Boer, J.F., Srinivas, S.M., Malekafzali, A., Chen, Z., Nelson, J.S. (1998) Imaging thermally damaged tissue by polarization sensitive optical coherence tomography. Opt Express 3: 212–218
40. Schmitt, J.M., Xiang, S.H., Yung, K.M. (1998) Differential absorption imaging with optical coherence tomography. J Opt Soc Am A 15: 2288–2296
41. Izatt, J.A., Hee, M.R., Owen, G.M., Swanson, E.A., Fujimoto, J.G. (1994) Optical coherence microscopy in scattering media. Opt Lett 19: 590–592
42. Podoleanu, A.G. et al. (1996) Coherence imaging by use of a Newton rings sampling function. Opt Lett: 1789–1791
43. Loudon, R. (1985) The Quantum Theory of Light. Clarendon Press, Oxford
44. Hariharan, P. (1985) Optical Interferometry. Academic Press, San Diego
45. Bouma, B., Tearney, G.J., Boppart, S.S., Hee, M.R., Brezinski, M.E., Fujimoto, J.G. (1995) High-resolution optical coherence tomographic imaging using a mode-locked Ti: Al_2O_3 laser source. Opt Lett 20: 1486–1488
46. Goodman, J.W. (1969) Introduction to Fourier Optics. McGraw-Hill, New York
47. Gerrard, A., Burch, J.M. (1975) Introduction to Matrix Optics Methods in Optics. Wiley, London
48. Fercher, A.F., Leitgeb, R., Hitzenberger, C.K., Sattmann, H., Wojtkowski, M. (1999) Complex spectral interferometry OCT. Proc SPIE 3564: 173–178
49. Häusler, G., Lindner, M.W. (1998) "Coherence RADAR" and "spectral RADAR" – New tools for dermatological diagnosis. J Biomed Opt 3: 21–31
50. Robinson, D.W., Reid, G.T (1993) Interferogram analysis. Institute of Physics Publishing, Bristol and Philadelphia
51. Schmit, J., Creath, K. (1995) Extended averaging technique for derivation of error-compensating algorithms in phase-shifting interferometry. Appl Opt 34: 3610–3619
52. Fercher, A.F., Hitzenberger, C.K., Drexler, W., Kamp, G., Strasser, I., Li, H. C. (1993) In Vivo Optical Coherence Tomography in Ophthalmology. In: Müller, G., et al. (Eds.) Medical Optical Tomography: Functional Imaging and Monitoring. SPIE Press, Bellingham (SPIE Institutes of Advanced Optical Technologies, Vol. IS11, pp 355–370)
53. Haberland, U.H.P., Blazek, V., Schmitt, H.J. (1998) Chirp optical coherence tomography of layered scattering media. J Biomed Opt 3: 259–266
54. Wang, X.J., Milner, T.E., Nelson, J.S. (1995) Characterization of fluid flow velocity by optical Doppler tomography. Opt Lett 20: 1337–1339

55. Chen, Z., Milner, T.E., Dave, D., Nelson, J.S. (1997) Optical Doppler tomography imaging of fluid flow velocity in highly scattering media. Opt Lett 22: 64–66

56. Yazdanfar, S., Kulkarni, M.D., Izatt, J.A. (1997) High resolution imaging of in vivo cardiac dynamics using color Doppler optical coherence tomography. Opt Express 1: 424–431

57. Izatt, J.A., Kulkarni, M.D., Yazdanfar, S., Barton, J.K., Welch, A.J. (1997) In vivo bidirectional color Doppler flow imaging of picoliter blood volumes using optical coherence tomography. Opt Lett 22: 1439–1441

58. Kulkarni, M.D., van Leeuwen, T.G., Yazdanfar, S., Izatt, J.A. (1998): Velocity-estimation accuracy and frame-rate limitations in color Doppler optical coherence tomography. Opt Lett 23: 1057–1059

59. Chen, Z., Milner, T.E., Wang, X., Srinivas, S., Nelson, J.S. (1998) Optical Doppler tomography: Imaging in vivo blood flow dynamics following pharmacological intervention and photodynamic therapy. Photochem Photobiol 67: 1–7

60. Chen, Z., Milner, T.E., Srinivas, S., Wang, X., Malekafzali, A., van Gemert, M.J.C., Nelson, J.S. (1997) Noninvasive imaging of in vivo blood flow velocity using optical Doppler tomography. Opt Lett 22: 1119–1121

61. Spraul, C.W., Lang, G.E., Lang, G.K. (1998) Die Bedeutung der optischen Kohärenz-Tomographie in der Diagnostik der altersbezogenen Makula-Degeneration. Korrelation von fluoreszenzenzangiographischen mit OCT-Befunden. [Value of OCT in diagnosis of age-related macular degeneration. Correlation of fluorescein angiography and OCT findings]. Klin Monatsbl Augenheilk 212: 141–148

62. Ho, A.C., Guyer, D.R., Fine, S.L. (1998) Macular hole. Surv Ophthalmol 42: 393–416

63. Lincoff, H., Kreissig, I. (1998) Optical coherence tomography of pneumatic displacement of optic disc pit maculopathy. Br J Ophthalmol 82: 367–372

64. Trabucchi, G., Sannace, C., Introini, U., Brancato, R. (1998) Partial lipodystrophy with associated fundus abnormalities: an optical coherence tomography study. Br J Ophthalmol 82: 326

65. Toth, C., Narayan, D.G., Boppart, S.A., Hee, M.R., Fujimoto, J.G., Birngruber, R., Cain, C.P., DiCarlo, C.D., Roach, W.P. (1997) A comparison of retinal morphology viewed by optical coherence tomography and by light microscopy. Arch Ophthalmol 115: 1425–1428

66. Drexler, W., Findl, O., Menapace, R., Kruger, A., Wedrich, A., Rainer, G., Baumgartner, A., Hitzenberger, C.K., Fercher, A.F. (1998) Dual beam optical coherence tomography: Signal identification for ophthalmologic diagnosis. J Biomed Opt 3: 55–65

67. Wälti, R., Böhnke, M., Gianotti, R., Bonvin, P., Ballif, J., Salathe, R.P. (1998) Rapid and precise in vivo measurement of human corneal thickness with optical low-coherence reflectometry in normal human eyes. J Biomed Opt 3: 253–258

68. Baumgartner, A., Möller, B.A., Hitzenberger, C.K., Drexler, W., Fercher, A.F. (1997) Measurements of the posterior structures of the human eye in vivo by partial-coherence interferometry using diffractive optics. Proc SPIE 2981: 85–93

69. Drexler, W., Baumgartner, A., Findl, O., Hitzenberger, C.K., Fercher, A.F. (1997) Biometric investigation of changes in the anterior eye segment during accommodation. Vision Res 37: 2789–2800

70. Findl, O., Drexler, W., Menapace, R., Bobr, B., Bittermann, S., Vass, C., Rainer, G., Hitzenberger, C.K., Fercher, A.F. (1998) Accurate determination of effective lens position and lens capsule-distance with 4 intraocular lenses. J Cataract Refract Surg 24: 1094–1098

71. Hee, M., Puliafito, C.A., Duker, J.S., Reichel, E., Coker, J.G., Wilkins, J.R., Schuman, J.S., Swanson, E.A., Fujimoto, J.G. (1998) Topography of diabetic macular edema with optical coherence tomography. Ophthalmology 105: 360–370

72. Hee, M.R., Izatt, J.A., Swanson, E.A., Huang, D., Schulman, J.S., Lin, C.P., Puliafito, C.A., Fujimoto, J.G. (1995) Optical Coherence Tomography of the Human Retina. Arch Ophthalmol 113: 325–332

73. Baumgartner, A., Hitzenberger, C.K., Sattmann, H., Drexler, W., Fercher, A.F. (1998) Signal and Resolution Enhancements in Dual Beam Optical Coherence Tomography of the Human Eye. J Biomed Opt 3: 45–54

74. Schuman, J.S., Hee, M.R., Puliafito, C.A., Wong, C., Pedut-Kloizman, T., Lin, C.P., Hertzmark, E., Izatt, J.A., Swanson, E.A., Fujimoto, J.G. (1995) Quantification of nerve fiber layer thickness in normal and glaucomatous eyes using optical coherence tomography. Arch Ophthalmol 113: 586–596

75. Schuman, J.S., Pedut-Kloizman, T., Hertzmark, E., Hee, M.R., Wilkins, J.R., Coker, J.G., Puliafito, C.A., Swanson, E.A. (1996) Reproducibility of nerve fiber layer thickness measurements using optical coherence tomography. Ophthalmol 103: 1889–1898

76. Roh, S., Noecker, R.J., Schuman, J.S., Hedges, III, T.R., Weiter, J.J., Mattox, C. (1998): Effect of optic nerve head drusen on nerve fiber layer thickness. Ophthalmol 105: 878–885

77. Hee, M.R., Baumal, C.R., Puliafito, C.A., Duker, J.S., Reichel, E., Wilkins, J.R., Coker, J.G., Schuman, J.S., Swanson, E.A., Fujimoto, J.G. (1996) Optical coherence tomography of age-related macular degeneration and choroidal neovascularization.. Ophthalmol 103: 1260–1270

78. Toth, C.A., Birngruber, R., Boppart, S.A., Hee, M.R., Fujimoto, J.G., DiCarlo, C.D., Swanson, E.A., Cain, C.P., Narayan, D.G., Noojin, D.G., Roach, W.P. (1997) Argon laser retinal lesions evaluated in vivo by optical coherence tomography. Am J Ophthalmol 123: 188–198

79. Koop, N., Brinkmann, R., Lankenau, E., Flache, S., Engelhardt, R., Birngruber, R. (1997) Optische Kohärenztomographie der Kornea und des vorderen Augenabschnitts. Ophthalmologe 94: 481–486

80. Feldchtein, F.I., Gelikonov, G.V., Gelikonov, V.M., Iksanov, R.R., Kuranov, R.V., Sergeev, A.M., Gladkova, M.D., Ourutina, M.N., Warren, J.A., Reitze, D.H. (1998) In vivo OCT imaging of hard and soft tissue of the oral cavity. Opt Express 3: 239–250

81. Colston, B.W., Sathyam, U.S., DaSilva, L.B., Everett, M.J., Stroeve, P., Otis, L.L. (1998) Dental OCT. Opt Express 3: 230–238

82. Schmitt, J.M., Yadlowsky, M.J., Bonner, R.F. (1995) Subsurface imaging of living skin with optical coherence microscopy. Dermatology 191: 93–98

83. Welzel, J., Lankenau, E., Birngruber, R., Engelhardt, R. (1997) Optical coherence tomography of the human skin. J Am Acad Dermatol 37: 958–963

84. Pan, Y., Farkas, D.L. (1998) Noninvasive imaging of living human skin with dual-wavelength optical coherence tomography in two and three dimensions. J Biomed Opt 3: 446–455

85. Lendenrink, E. (1998) Optical coherence tomography for three-dimensional imaging of skin features. In: Proc SPIE Vol. 3564: Medical Applications of Lasers in Dermatology, Cardiology, Ophthalmology, and Dentistry

86. Schmitt, J.M., Lee, S.L., Yung, K.M. (1997) An optical coherence microscope with enhanced resolving power in thick tissue. Opt Commun 142: 203–207

87. Bouma, B., et al. (1995) High Resolution Optical Coherence Tomographic Imaging Usind a Mode-locked Ti: Al_2O_3 Laser Source. Opt Lett 20: 1486–1488

88. Hitzenberger, C.K., Baumgartner, A., Drexler, W., Fercher, A.F. (1999) Dispersion effects in partial coherence interferometry: Implications for intraocular ranging. J Biomed Opt 4: 144–151

89. Lexer, F., Fercher, A.F., Sattmann, H., Drexler, W., Molebny, S. (1998) Dynamic coherent focus for transversal resolution enhancement of OCT. Proc SPIE 3251: 85–90

90. Fujimoto, J.G., Brezinski, M.E., Tearney, G.J., Boppart, S.A., Bouma, B., Hee, M.R., Southern, J.F., Swanson, E.A. (1995) Optical biopsy and imaging using optical coherence tomography. Nature Medicine 1: 970–972

91. Brezinski, M.E., Tearney, G.J., Boppart, S.A., Swanson, E.A., Southern, J.F., Fujimoto, J.G. (1997) Optical biopsy with optical coherence tomography: Feasibility for surgical diagnostics. J Surg Res 71: 32–40

92. Pitris, C., Brezinski, M.E., Bouma, B.E., Tearney, G.J., Southern, J.F., Fujimoto, J.G. (1998) High resolution imaging of the upper respiratory tract with optical coherence tomography: a feasibility study. Am J Respir Crit Care Med 157: 1640–1644

93. Tearney, G.J., Brezinski, M.E., Bouma, B.E., Boppart, S.A., Pitris, C., Southern, J.F., Fujimoto, J.G. (1997) In vivo endoscopic optical biopsy with optical coherence tomography. Science 276: 2037–2039

94. Tearney, G.J., Boppart, S.A., Bouma, B.E., Brezinski, M.E., Weissman, N.J., Southern, J.F., Fujimoto, J.G. (1996) Scanning single-mode fiber optic catheter-endoscope for optical coherence tomography. Opt Lett 21: 543–545

95. Feldchtein, F.I., Gelikonov, G.V., Gelikonov, V.M., Kuranov, R.V. Sergeev, A.M., Gladkova, N.D., Shakhov, A.V., Shakhova, N.M., Snopova, L.B., Terent'eva, A.B., Zagainova, E.V., Chumakov, Yu.P., Kuznetzova, I.A. (1998) Endoscopic applications of optical coherence tomography. Opt Express 3: 257–270

The Spectral Optimization of Human Vision: Some Paradoxes, Errors and Resolutions

B. H. Soffer and D. K. Lynch

Summary. The peak brightness of the solar spectrum is in the green when plotted in wavelength units. It peaks in the near-infrared when plotted in frequency units. Therefore, the oft-quoted notion that evolution led to an optimized eye whose sensitivity peaks where there is the most available sunlight is misleading and erroneous. The confusion arises when spectral density distribution functions like the spectral radiance are compared with ordinary functions like the sensitivity of the eye. Spectral radiance functions, excepting very narrow ones, can change peak positions greatly when transformed from wavelength to frequency units, but sensitivity functions do not. Expressing the spectral radiance in terms of photons per second, rather than power, also causes a change in the shape and peak of the distribution even while keeping the choice of bandwidth units fixed. The confusion arising from comparing simple functions to distribution functions occurs in many parts of the scientific and engineering literature aside from vision. Some examples are given. The eye does not appear to be optimized for detection of the available sunlight, including the surprisingly large amount of infrared radiation in the environment. The color sensitivity of the eye is discussed in terms of the spectral properties and the photo and chemical stability of available biological materials. It is likely that we are viewing the world with a souvenir of the human evolutionary voyage.

1 Introduction

Many people believe that evolution has produced a human eye whose color sensitivity roughly matches the sunlight spectrum [1–9]. Some authors only hint but others state the case even more strongly, i.e., that both the solar spectrum and the color sensitivity of the eye peak very nearly together at around 560 nm in the green. Such agreement could hardly be accidental, so the implication and reasoning goes, and therefore the human eye must have evolved to possess a near-optimum color sensitivity. The framed text insert shows a sampling of quotes from the vision literature illustrating how pervasive this idea has become. Many more authors cause this idea to spread by further quoting and paraphrasing these ideas, without sufficient reflection, in fresh publications of their own. For example Sekuler and Blake paraphrase Mollon [2] in their textbook "Perception" [10].

We will show that the apparent wavelength coincidence between the solar spectral radiance and the eye's spectral sensitivity is artificial and often

CAVEAT LECTOR

"The peak of the [solar spectral irradiance] curve is located at the visible
    wavelengths we see with our eyes" [1].

"For the main business of vision ... most mammals depend on a single class
    of cone, which has its peak sensitivity near the peak of the solar
    spectrum, in the range 510–570 nm" [2].

"Figure 1.3 compares the spectral content of light ... with the spectral
    sensitivities of the rod and cone systems of human vision" [3].

"Sunlight comprises wavelengths ranging from .... For humans visible light
    ranges from approximately 400 to 700 nm" [4].

"The spectral response of the human eye is closely matched to the peak of
    the sun's radiation (5500 Å) in daylight" [5].

"This shows that the eye is sensitive to a region of the spectrum where the
    radiation reaching the earth from the sun is most plentiful" [6].

"Note that the maximum available energy from sunlight peaks in the same
    region of the spectrum where the eye is most sensitive. This coinci-
    dence is probably not accidental, but is more likely the product of
    biological evolution" [7].

"It is no accident that this [human cone] sensitivity is centered on the peak of
    the energy distribution of light from the sun. Evolution of the eye has
    obviously taken advantage of the spectral character of daylight" [8].

misleading. It results from the choice of units in which the solar spectrum
is plotted. Comparing radiance [11] to sensitivity is like "comparing apples
and oranges": they are fundamentally different quantities and their shapes
and peaks should not be compared to one another (even though they can
legitimately be *multiplied* together for some purposes, as we will show). In
particular we will show how the wavelength of peak emission depends on the
units used in computing and displaying a spectral distribution. Furthermore,
we will demonstrate that, on the contrary, the spectral sensitivity of the eye
does not depend on the units used, and suggest that the eye is poorly optimized to take full advantage of all the visible and the enormous amount
of infrared light that is available in the environment. We will then discuss
evolution as it relates to color vision. Examples of similar confusions from
fields other than vision will also be given.

2   Spectral Radiant Density Distributions Contrasted with Sensitivity

Figure 1 shows the spectrum of the sun [12] at sea level for a daytime mid latitude summer with the sun at the zenith and with nominal values for Rayleigh scattering, water vapor absorption, boundary layer, stratospheric aerosols, etc. Also shown in Fig. 1 is a 5800 K Planck function scaled to approximately match the sunlight. We concentrate on sunlight rather than daylight because that is what all the authors referred to above have done. Daylight, of course, is highly variable, and has been much studied [13]. It depends on many factors, including, the direction and degree of sky exposure, weather conditions, time of day, and polarization.

Several aspects of the solar spectrum are noteworthy. First, sunlight shows significant departures from a Planck function. The features are due to absorption by the earth's atmosphere and to absorption in the solar photosphere. Figure 1 also shows that the brightest part of the spectrum seems to occur in the green near 0.5 µm (500 nm). But rather than being a pronounced peak

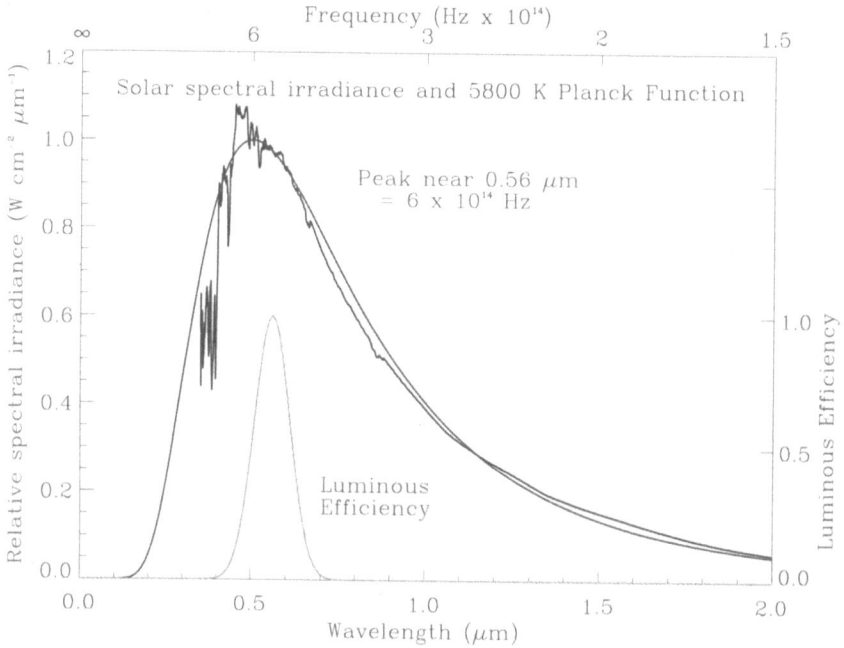

Fig. 1. The solar spectrum (*black*) plotted in wavelength units peaks near 500 nm. Also shown is an approximate fit of a 5800 K Planck function (*purple*) which has been scaled to match the solar spectrum. This shows that the solar spectrum is roughly Planckian in the optical part of the spectrum. The luminous efficiency of the eye (*green*) peaks at 560 nm. All three curves appear to peak near 500–560 nm, a wavelength region generally perceived as being green

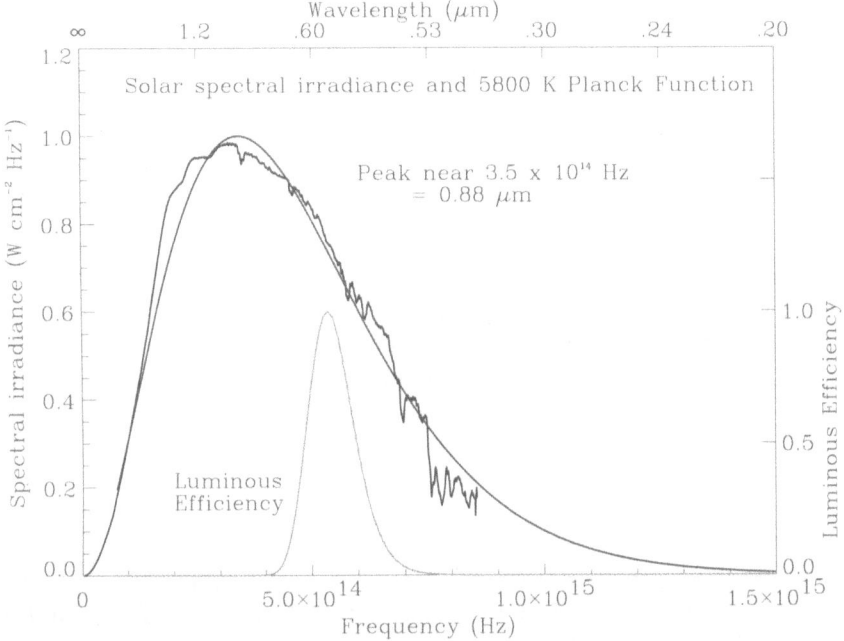

Fig. 2. The same data shown as in Fig. 1 except plotted in frequency units. Here the sun and Planck functions peak near 880 nm in the near infrared while the luminous efficiency curve still peaks at 560 nm. The solar irradiance and Planck function transform differently than the luminous efficiency

here, there is a broad plateau that is nearly flat between 450 nm and 610 nm. This plateau is due to the combined influence of thousands of solar and atmospheric absorption lines, which, though not resolved here, serve to alter the shape from a pure black body.

Let us examine the position of the peaks. Figure 1 shows the sun's spectral irradiance plotted in units that are most commonly used for visible spectra, i.e., $W\,cm^{-2}\,\mu m^{-1}$ vs. μm (wavelength units). In these units the peak of the solar spectrum is unquestionably at a wavelength that by itself would appear green. Also shown in Fig. 1 is the luminous efficiency of the eye adapted from Judd and Wyszecki [14]. The peak of the luminous efficiency is also in the green at 560 nm.

The spectrum can also be plotted in frequency units, i.e., $W\,cm^{-2}\,Hz^{-1}$ as a function of frequency (Fig. 2). In this case, the spectrum no longer peaks in the green but rather in the near infrared equivalent to $0.88\,\mu m$ (880 nm). Yet the peak of the luminous efficiency of the eye remains in the green. As frequency and wavelength distributions are both equally valid representations of the very same physical phenomenon, we *seem* to be left with the question: "Where does the solar spectrum 'really' peak, in the green or in the near-infrared?"

The answer is that it depends on the choice of the independent variable for the bandwidth. To see this, let us first approximate the solar spectrum by a Planck function because it is analytic and closely matches the solar spectrum. The arguments will hold for the solar spectrum as well. In wavelength units the Planck function spectral irradiance $B_\lambda(T)$ is

$$B_\lambda(T) = 2hc^2\lambda^{-5}/(e^{hc/\lambda kT} - 1) \tag{1}$$

in units of power per unit area per unit wavelength interval. As Wien's displacement law says, the wavelength of peak emission is $0.2897/T = 0.2897/5800 = 4.99 \times 10^{-5}\,\mathrm{cm} \approx 500\,\mathrm{nm}$, corresponding to a frequency ν of $\nu = c/\lambda = 6 \times 10^{14}\,\mathrm{Hz}$. This is in the green part of the spectrum and agrees with most peoples' idea of the shape and peak of the solar spectrum. Wien's law with the conventional constant, however, only works when the spectrum is plotted *per unit wavelength* interval. When the same spectrum is plotted as irradiance *per unit frequency* interval $\mathrm{W\,cm^{-2}\,Hz^{-1}}$,

$$B_\nu(T) = 2h\nu^3 c^{-2}/(e^{h\nu/kT} - 1), \tag{2}$$

the distribution peaks at $3.4 \times 10^{14}\,\mathrm{Hz}$, corresponding to $8.8 \times 10^{-5}\,\mathrm{cm}$ ($0.88\,\mu\mathrm{m} = 880\,\mathrm{nm}$). The peak wavelength of the Planck distribution in frequency is easily shown to be 1.76 longer than the peak of the wavelength distribution for any temperature. See Figs. 1 and 2.

Although (1) and (2) are equivalent representations, converting one to the other is not simply a matter of making the substitution $\nu = c/\lambda$. This is because the Planck function is a density distribution function and is defined differentially

$$B_\lambda\,\mathrm{d}\lambda = B_\nu\,\mathrm{d}\nu. \tag{3}$$

This is simply conservation of energy. Since $\mathrm{d}\nu/\mathrm{d}\lambda = -c/\lambda^2$ (and ignoring the minus sign because it is merely an artifact of the directions of integration of (3), then

$$B_\lambda\,\mathrm{d}\lambda = B_\nu c/\lambda^2\,\mathrm{d}\lambda, \tag{4}$$

and thus $B_\lambda = B_\nu c/\lambda^2$ or conversely $B_\nu = B_\lambda \lambda^2/c$. The apparent "shift" in peak wavelength between B_λ and B_ν is not simply due to a substitution of variables, $\nu = c/\lambda$, but to the $1/\lambda^2$ Jacobian weighting factor as well. This is a necessary result of the differential nature of the Planck distribution function.

The relation between B_λ and B_ν is illustrated in Figs. 3 and 4. Figure 3 shows a 5800 K Planck function as a wavelength distribution and Fig. 4 shows the same function as a frequency distribution. Figure 3 is divided into equal intervals of wavelength in the amount of $0.1\,\mu\mathrm{m}$. The same intervals are marked in frequency units in Fig. 4. In Fig. 4 they clearly are unequally spaced because there are more wavelengths per unit frequency at longer wavelengths than at shorter ones. Conversely, there are more frequencies per unit

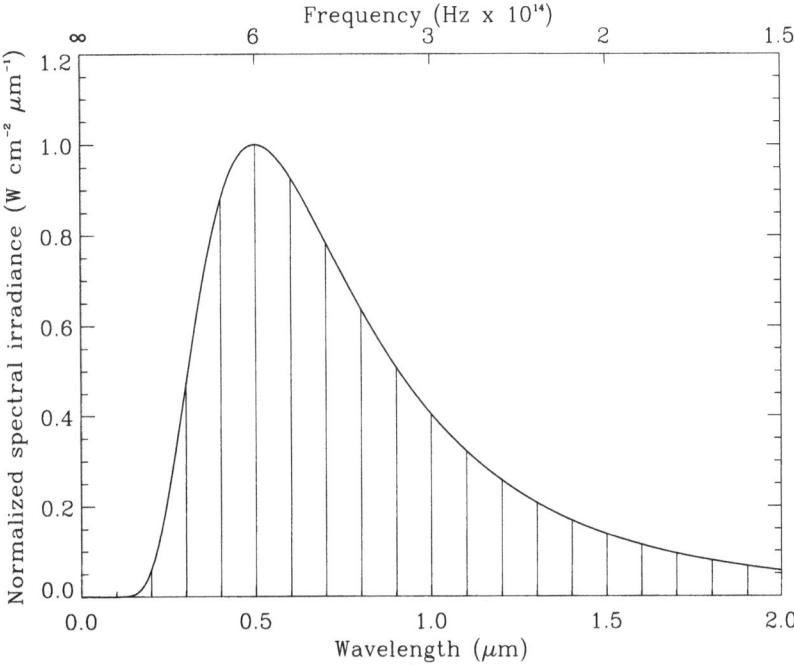

Fig. 3. A 5800 K Planck distribution function divided into equal 100 nm wavelength intervals

wavelength at shorter wavelengths. Clearly then, a plot of irradiance per unit frequency would skew the curve to longer wavelengths, which is exactly what we have just seen happen (Figs. 1 and 2).

We have used Planck's function only as a convenient example. Most distribution functions would suffer some change in shape, depending, not only on the transformation itself, but also on the original shape and width of the distribution to be redistributed as well. For example, a very narrow distribution with little to redistribute, such as a spectral line, would shift much less than 1.76 times its wavelength. A function broader than the Planck distribution could shift more.

Another commonly used representation of irradiance functions describes the spectral irradiance in terms of the number N of photons per second, rather than in watts. No matter what choice is made for the units used to describe the spectral bandwidth, e.g., wavelength or frequency, the transformation to photons per second by itself engenders a change in the shape and a shift in the distribution's peak position. Even though there is no Jacobian involved, the new function itself now has a different form and a different dependence on the independent variable. For example, again for Planck's function B_λ, in

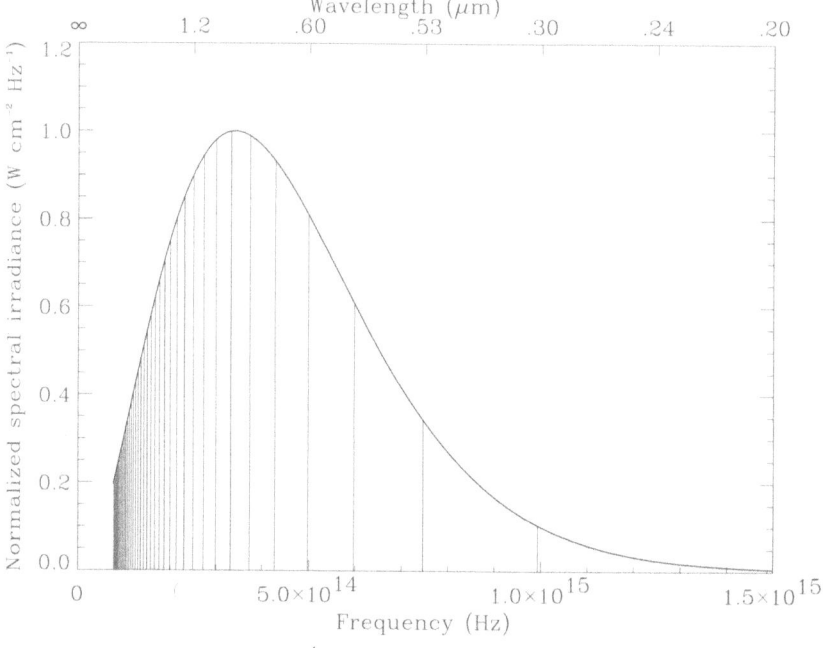

Fig. 4. The same Planck function and wavelength intervals as Fig. 3 transformed into frequency intervals. Note that the frequency intervals are not equal

terms of the number of photons per second and noting that $N = \text{power}/h\nu$ and $B_N = B_\lambda/h\nu$, then we have

$$B_N(T) = 2hc^2\lambda^{-4}/(e^{hc/\lambda kT} - 1) \tag{5}$$

and the photon number distribution B_N stretches vertically and *nonlinearly* by the factor of λ in comparison to the power distribution $B_\lambda(T)$. Wien's displacement law for this distribution has a different constant: $\lambda T = 0.3670$. The peak of the 5800 K Planck spectrum shifts to 633 nm in this representation.

There are a myriad of equally-valid ways of representing density distribution functions. For example, plotting the Planck distribution semi-logarithmically as in Fig. 5, for either log wavelength or log frequency, reveals yet a different shape and a different peak, this time near 720 nm. The Jacobians for these logarithmic transformations are $1/\lambda$ and $1/\nu$ respectively. The wavelength–frequency pair of these two semi logarithmic representations of Planck's function or any other distribution using these variables including sunlight, have exactly the same left-right mirrored function form. This can easily be seen by noting that $d(\log \lambda) = -d(\log \nu)$ and again ignoring the minus sign, so that $B_{\log \lambda} = B_{\log \nu}$. No special physical significance should be attached to this curious symmetry nor should the logarithmic form be singled out as a preferred physical or physiological representation [15].

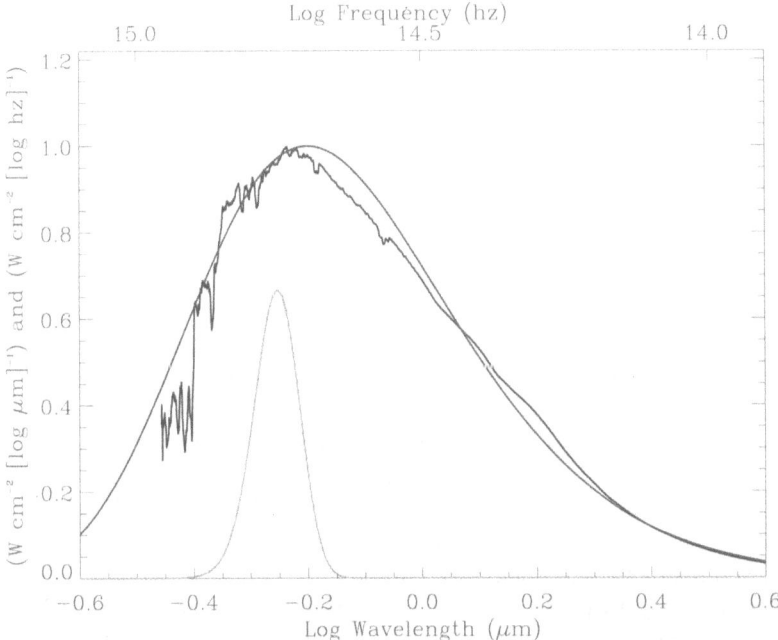

Fig. 5. Relative spectral irradiance. A semi-logarithmic plot of the Planck function and solar spectrum compared with the luminous efficiency of the eye

The spectral behavior of optical filters and detectors are not described by density distributions and so they transform in a much simpler way. Spectral sensitivity of any detector including the eye is expressed in units of amps/watt, volt/watt, or in the case of the eye, lumens/watt, each at a given *wavelength*. Filter transmission, being a unitless ratio between zero and one *at each wavelength*, behaves identically. This is fundamentally different than spectral irradiance which by virtue of being a density distribution function, is expressed per unit bandwidth, for example, as a *value per unit wavelength interval*. Consequently, sensitivities possess no Jacobian differential weighting factor as when transforming the representation of the eye's sensitivity from wavelength intervals to frequency intervals. One need only use the substitution $\nu = c/\lambda$. This is why the peaks in the sensitivity curve remain at the same frequency (and wavelength) when plotted in either frequency or wavelength units (Figs. 1 and 2).

The fact that all measurements are necessarily made with instruments that must have finite bandwidth resolution or, expressed in conjugate space, finite convolutional spread functions, may cause some confusion about the distinction we are making. Measuring the spectral transmission of a filter, for example, will result in apparent values measured in finite bandwidth intervals, but this is merely a sampling issue. The measured values in the intervals are averages for the finite intervals and represent the transmission at each point in the interval. This does not make the measured transmission in any sense a density distribution.

The eye's spectral response can be likened to a filter. The spectral absorption of the eye is nearly linear. It is nearly intensity independent over many orders of magnitude, and each absorbed photon is equally effective, although only about 10% of the incident photons are absorbed [5]. So we can treat the eye's spectral sensitivity just like the transmission of a colored linear filter. Multiplying the sensitivity by the Planck function will result in the spectral radiance which gets through the filter (i.e., the radiance actually detected by the eye and appropriately weighted). This is an example of when it is perfectly legitimate to multiply, point by point, an ordinary function and a distribution function in order to get the desired resultant distribution function. Yet, as we have seen, it can be very misleading to compare shapes and peaks and draw inferences from them.

This explanation may leave some people feeling a little uneasy. If we are designing a detector of broad band light and want to know at what wavelength to position a filter of a fixed bandwidth in order to transmit the most power, intuition might say to put it at the peak of the source's spectrum. Yet we seem to be saying that this mental procedure of sliding the filter back and forth to maximize power near the peak would not work as the peak's position is ambiguous and somewhat arbitrary. What's going on here?

The answer is that for a fixed filter bandwidth in wavelength, maximizing in wavelength space is not the same as maximizing in frequency space. What may not be apparent is that as the wavelength filter is sliding back and forth, its width in frequency space is changing. Consider a filter (like the eye) whose bandwidth is 100 nm centered at 520 nm which just happens to be near the maximum in the wavelength representation of sunlight. The same filter is not near the peak in frequency space. If we were to take the filter in frequency space with the same fixed wavelength bandwidth that was used to optimize in wavelength space, and move it instead as a fixed frequency bandwidth filter toward smaller frequencies to attempt to further maximize the signal, we would find that its width in wavelength space had increased. We would also find a maximum where the filter and source spectrum align, but it would be a *different* maximum than found before because the optimization constraint was different. This is all a result of the relation $d\lambda = -c\,d\nu/\nu^2$. An optimization somewhat similar to the one described above was done analytically for the Planck distribution by Benford [16].

In summary thus far, we have shown that the peak wavelength of the solar spectrum depends on how the spectral distribution is plotted. The fact that in wavelength units the spectrum roughly agrees with the peak sensitivity of the eye is an accidental and meaningless quirk involving the units in which the spectrum is plotted. Computing it in frequency units, for example, is just as valid and results in a peak equivalent to 880 nm, well away from the peak

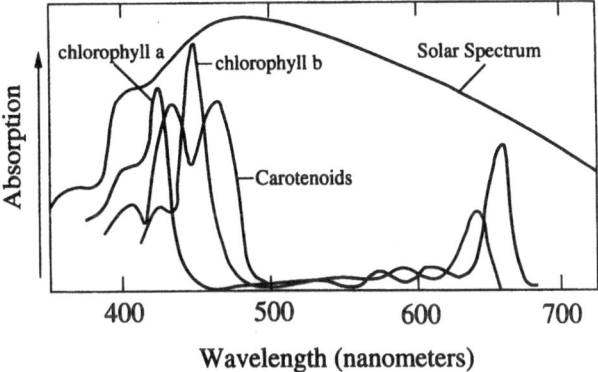

Fig. 6. The absorption spectra of various chromophores involved in photosynthesis compared with the solar spectrum as a function of wavelength. After Szalai and Brudvig [18]

sensitivity of the eye. There is, however, no paradox or inconsistency in this. While we firmly believe in the modern theory of evolution, we wish to warn others, including one of the authors of this article [8], of the dangers of glibly assigning Darwinian significance to what is merely an accidental wavelength coincidence.

To end this section we will give two examples of the error that we have been describing from different scientific disciplines. The first comes from the study of photosynthesis. The error in this example [17] is rather close to the immediate subject of our paper. Here the authors plot the absorption spectra of various chromophores involved in photosynthesis together with the solar spectrum in wavelength units (Fig. 6). In the wavelength representation the curves overlap strongly. The authors incorrectly state: "... almost the entire spectrum of light coming from the sun can be absorbed for use in photosynthesis."

The second example concerns the so called microwave window. Radio astronomers are interested in determining the most transparent spectral region where they might best hope to receive weak signals. The radio sky has many noise components that would interfere with detection. Their density distribution spectra are plotted separately and added together in Fig. 7. Variants of this figure are so ubiquitous as to defy making a proper original attribution. One fascinating place to find it is in the Project Cyclops [18] study for detecting extraterrestrial intelligent life. The spectral noise power densities are described by their "noise temperatures" which relate to the Planck distribution function. These density distributions are here further scaled by the square root of frequency. The three noise sources: galactic non-thermal, the 2.7 K cosmic background, and the quantum noise of coherent detection, define a broad minimum called the free space microwave window. The location and shape of this minimum in the density distribution will depend on the

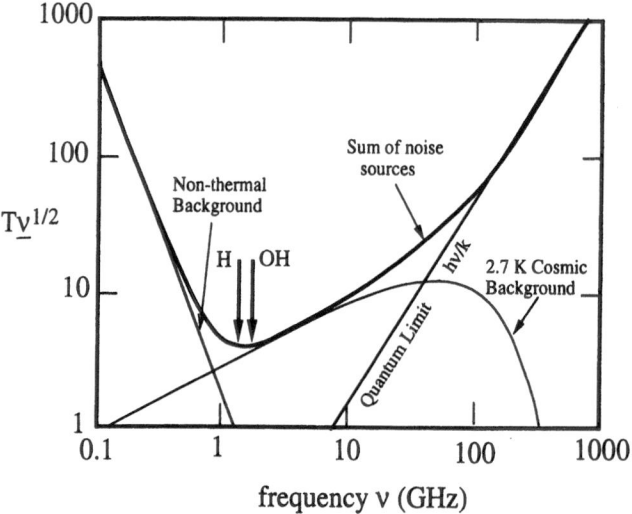

Fig. 7. The sky noise power density distribution at galactic latitude 10°, scaled by the square root of frequency, plotted as noise temperature vs. frequency, showing the minimium in the free space microwave window

choice of representation in the same ways as we described at length above. The common error is to locate the emission frequency of some hoped to be detected narrow line, such as the 21 cm H hyperfine line or the OH line, in that window as we do with the arrows in the figure. Being very narrow lines, their position would *not* change with a change in representation. Granted that the errors in this case are trivial and do not have much practical consequence, they are nevertheless errors in principle.

3   Regarding Evolution

The many opinions noted earlier about the evolution of color vision in the literature were based in part on a simple misunderstanding about density distribution functions. But there is another underlying bias that we would like to mention. The consensual, canonical belief in the power of evolution to optimize absolutely, globally, and without constraint is so strongly held that it is sometimes invoked as sufficient causal explanation of whatever the facts at hand may be. "Visible light ... has just the right wavelengths to reflect light from objects in useful ways, and to permit the evolution of... the eye" [6]. This is a Panglossian view of evolution, a modern Darwinian reading of Leibnitz's world view: the eye is the best of all possible optimizations and every thing about it has sufficient cause. In fact evolution has historically traced many irreversible pathways to reach its present state. Any potentially more favorable global optimum might be too energetically difficult to achieve and would thus be very unlikely to ever occur. A quantitative measure of the optimization of spectral utilization that was achieved by the eye can be obtained by integrating under the curves shown in Fig. 1 to get the fraction of the available light between 320 nm (the atmospheric transmission cut-off) and 1400 nm (the thermodynamic noise limit) detected by the eye [5]. It is only 19%, hardly optimal in a spectral sense. There is also the question of the availability of suitable biological materials which would posses the necessary photochemical and thermochemical stability to achieve a more optimum state. All of these factors have mediated and constrained the process of the evolution of the eye and its optimization to the sun's light.

Sight is clearly a survival advantage and so it seems reasonable to suppose that the better one sees, the more able one is to survive. It might be expected that a very efficiently optimized eye should be able to see both longer and shorter wavelengths with greater sensitivity. There is considerable solar radiation in the near-infrared which the eye does not detect. The full width at half maximum of the solar spectrum in wavelength units (Fig. 1) is approximately 500 nm while that of the eye is only 100 nm. Should not an optimized eye be very sensitive to radiation longward of 700 nm? After all, during the day there is plenty of sunlight at these wavelengths and at night the OH airglow [19] between 0.6 and 1.2 μm would light up the landscape tremendously. Thermal excitations in the retina only begin to compete with any incoming photons at wavelengths longer than 1.4 μm at the absolute threshold of vision [5]. Sensitivity in the infrared 600–1200 nm range would seem like an obvious advantage with significant evolutionary potential. Boynton [6] quotes Pirenne [20] in suggesting that strong infrared sensitivity would be undesirable because the infrared in the eyeball, the body's own heat radiation, would cause the external world to be obscured by a luminous fog of this radiation. This is precisely why we do not expect to see beyond 1.4 μm. That is where the fog would begin to manifest itself. But this effect does not rule out the possibility of near infrared sensitivity up to 1.4 μm. Similarly, there would also seem to be some advantage to having a greater blue and ultraviolet sensitivity. Yet neither of these things happened in humans. Why? If our vision is not well matched to the light of our present environment including the large infrared component bathing us now, could we be mired in some evolutionary backwater, where once we were indeed better adapted? This may put a damper on the enthusiasm of those who prefer a story of continual linear progress, but those familiar with the evolution of the vertebrate eye and how its structure and function mirror its evolutionary history will not be surprised at the following compelling suggestion by Duke–Elder [21]:

"So far as the evidence goes, the eyes of all vertebrates including man are stimulated by approximately the same range of the spectrum (760–390 μm) with the greatest sensitivity at a band with a wavelength varying between 500 and 550 μm; it is no coincidence that this corresponds roughly with the transmission spectrum of water. The visual mechanism of Vertebrates was first evolved in water

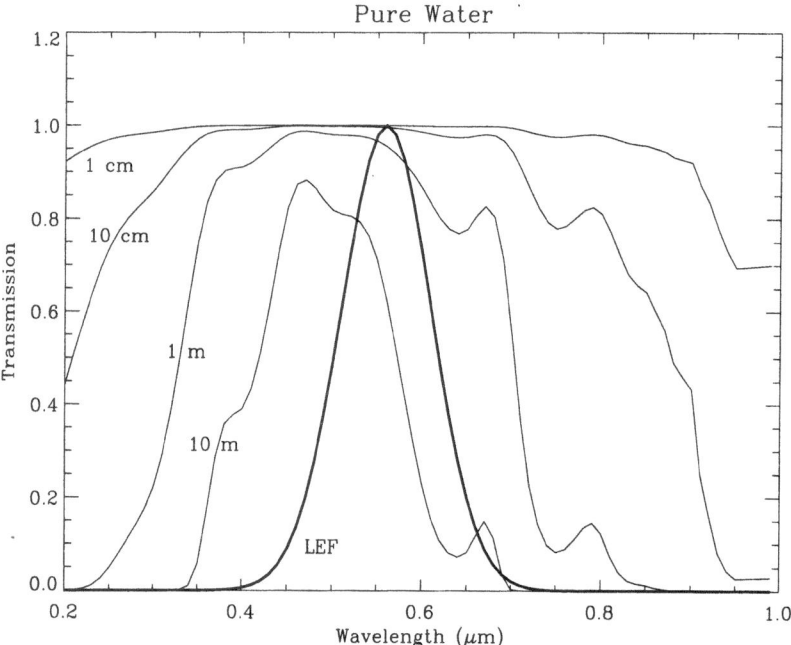

Fig. 8. The transmission of pure water compared with the luminous efficiency function (LEF) of the eye. This calculation was done for absorption and scattering was ignored

and their photo-pigments were presumably developed as sensitizers to allow their possessors to leave the brightly-lit surface and penetrate more deeply into the darker depths of the sea."

This opinion was echoed very recently [3], however mistakenly illustrating the point by wrongly superposing spectral densities and cone color sensitivities, a comparison that we have explained is not appropriate! It is, however, perfectly legitimate and appropriate to compare transmission and sensitivity as we do in Fig. 8, as neither one is a density distribution. Figure 6 shows the transmission of water through different path lengths, plotted together with the luminous efficiency of the eye. Note how much narrower the water's transmission curves are in comparison to the solar spectrum. As a result, water imposed a tighter constraint on the eye's spectral evolution than did sunlight.

Other constraints on the optimization of the visual spectral response come from the properties of available biological materials. To get sensitivities further in the infrared via photoisomerization, larger molecules with longer conjugated chains which would have their first electronic excited states at lower energies would be needed. But such large molecules are unstable and subject to dissociation and bleaching by thermal processes at body temperature where the thermal energy (kT) is comparable to the excited state energies

[22–24]. This is why it is extremely difficult to produce stable photographic sensitizing dyes or laser giant pulse Q-spoiling dyes at wavelengths longer than 1 μm in aqueous solutions. Although the tails of animal photopigment sensitivity become uselessly small beyond about 850 nm [25], except under artificially extreme brightness conditions, organic dye molecules have been synthesized with longer peak wavelength sensitivities. They, however, are unstable in aqueous media, all the more so when traces of oxygen species and free radicals are present.

At the short wavelength side, all organic molecules are susceptible to direct UV damage or indirect damage from UV-generated free radicals in their proximity. Those are not promising prospects for evolutionary candidates. People who have had their corneas or lenses removed can see a bit further in the UV, but they often develop UV-related ocular damage and dysfunction [26]. Interestingly, many insects, animals, and birds do see a bit further in the UV and IR than we do [27]. For instance, compared to other frogs, *Rana Tempoaria* has high sensitivity down to 330 nm [28], which coincidentally corresponds very nearly to the atmospheric short wavelength transmission sharp cutoff at 320 nm.

4   Summary

We have shown that, contrary to the belief expressed by many authors, the eye is only weakly optimized to take full advantage of the available solar spectrum. The erroneous belief often arises from a blind faith in the power of evolution to optimize absolutely, coupled with a misunderstanding of the nature of density distribution functions. That misunderstanding appears in a diverse range of scientific literature. Other constraints upon the eye's evolutionary optimization besides the sun's radiance were also important, such as the historically significant influence of the transmission of water, the susceptibility of potentially available biological materials such as photopigments to UV damage, and the instability of possible infrared sensitive photopigments. Contemplating why we did not evolve to use other mechanisms to produce a broader band visual sensitivity, such as electron-hole pair generation, for an unlikely example, would seem to be a futile exercise in a counterfactual history of evolution. But the question of how and why our vision evolved to employ its equally unlikely photoisomerization scheme for vision remains an interesting and open issue.

Acknowledgments

We would like to thank Drs. James Hecht, Mark German, Kent Ashcraft, and Jay Neitz for stimulating discussions on the eye, based on an earlier version [29] of this paper. This work was supported in part by the Aerospace Independent Research and Development Program.

References and Notes

1. Lang, K. (1995) Sun Earth and Sky. Springer, Berlin
2. Mollon, J.D. (1989) The Uses and Origins of Primate Colour Vision. J exp Biol 146: 21–38
3. McIlwain, J.T. (1996) An Introduction to the Biology of Vision. Cambridge University Press, Cambridge
4. Thompson, E. (1995) Colour Vision. Routledge, London, New York
5. Rose, A. (1973) Vision Human and Electronic. Plenum, New York and London
6. Boynton, R. (1979) Human Color Vision. Holt, Rienhart and Winston, New York
7. Boynton, R.M. (1990) Human Color Perception. In: Leiboveic, K.N. (Ed.) Science of Vision. Springer, New York
8. Lynch, D.K., Livingston, W.C. (1995) Color and Light in Nature. Cambridge University Press, Cambridge
9. White, L. (1996) Infrared Photography Handbook. Amherst Media, Amherst New York
10. Sekuler, M., Blake, R. (1994) Perception, 3rd edn. McGraw-Hill, New York
11. Several similar but distinct spectral radiometric quantities: radiant power, radiant emittance, irradiance, radiant intensity, and radiance are employed to describe spectral density distributions. As they all are spectral densities, i.e., quantities per unit bandwidth interval, e.g., per unit wavelength interval or per unit frequency interval, they illustrate the argument of this paper equally well.
12. Anderson, G.P., et al. (1994) MODTRAN2: Evolution and Applications. Proc SPIE 222: 790–799
13. Henderson, S.T. (1970) Daylight and Its Spectrum. Amer. Elsevier, New York
14. Judd, D.B., Wyszecki, G. (1975) Color in Business. Science and Industry, 3rd edn. Wiley, New York
15. Incidentally, human hearing, on the other hand, does have a physiologically preferred spectral representation. We have an intrinsic logarithmic spectral response to acoustic radiance. This can be appreciated by noticing that we associate essentially the same musical pitch and note name to all the n octaves (2^n multiples) of a given note's frequency. If we pick, for example, middle C = 256 Hz, the frequency of the note one octave higher, by definition, is 512 Hz. We hear all those n higher octave notes, ignoring intensity dependent effects, still as a kind of replica of C, although higher in pitch. In representing musical spectral power or noise density distributions, the very same representational issues that we have been discussing for optical power naturally arise. But here, if the intention is to represent human perceptual hearing, then the logarithmic representation is clearly to be preferred and thus, by settling on it, many of the representational pitfalls and paradoxes can be avoided. Audio engineers have introduced several spectral logarithmic measures and quantities. For example the logarithmic equivalent of equally distributed, or so called "white," noise is appropriately called "pink" noise. It preferentially weights the lower frequencies logarithmically, thereby putting equal noise power into each octave. To the ear, pink noise sounds uniformly distributed.
16. Benford, F. (1939) Laws and Corrolaries of the Black Body. J Opt Soc Am 29: 92–96. He solved for the maximal power, relative to the total radiated power, in the neighborhood of a given wavelength of interest, λ_M, at which the absorption

The Spectral Optimization of Human Vision   405

(or sensitivity) is highest. In the wavelength representation this occurs when $\lambda_M T_M = .3666$. Note that the optimum wavelength, for a given temperature, does not coincide with the peak of the Planck distribution.

17. Szalai, V.A., Brudviig, G.W. (1998) How Plants Produce Dioxygen. Amer Scientist 86: 542–551
18. Project Cyclops (1973) CR 114445, NASA/Ames Research Center, Moffett Field, California, 41
19. Krasovsky, V.I., Shefov, N.N., Yarin, V.I. (1962) Atlas of the Airglow Spectrum 3000–12400 Å. Planet Space Sci 9: 883–915
20. Pirenne, M.H. (1948) Vision and the Eye. Chapman and Hall, London. This quote does not appear in the second edition (1967)
21. Duke-Elder, S. (1958) The Eye in Evolution. Henry Kimpton, London
22. Pastor, R.C., Kimura, H., Soffer, B.H. (1971) Thermal Stability of Polymethine Q-Switch Solutions. J Appl Phys 42: 3844–3847
23. Pastor, R.C., Soffer, B.H., Kimura, H. (1972) Photostability of Polymethine Saturably Absorbing Dye Solutions. J Appl Phys 43: 3530–3533
24. Ali, M.A. (1975) Temperature and Vision. Revue Canadienne de Biologie V34: 131–186
25. Knowles, A., Dartnall, H.J.A. (1977) In: Davson, H. (Ed.) The Eye 2B. Academic Press, London
26. Van Kuijk, F.J.G.M. (1991) Effects of Ultraviolet Light on the Eye: Role of Protective Glasses. Environmental Health Perspectives 96: 177–184
27. Bowmaker, J.K. (1991) The Evolution of Vertebrate Visual Pigments and Photoreceptors. In: Cronly-Dillon, J.R., Gregory, R.L (Eds.) Vision and Visual Dysfunction 2. CRC, Boca Raton
28. Govardovskii, V.I., Zueva, L.V. (1974) Spectral Sensitivity of the Frog in the Ultraviolet and Visible Region. Vision Res 14: 1317–1321
29. Soffer, B.H., Lynch, D.K. (1997) Has Evolution Optimized Vision For Sunlight. Annual Meeting of the OSA, Long Beach CA, Oct 12–17, SUE4 p 70;
    Lynch, D.K., Soffer, B.H. (1999) On the Solar Spectrum and the Color Sensitivity of the Eye. Optics and Photonics News, March

Part VIII

Others

Optical Methods for Reproducing Sounds from Old Phonograph Records

J. Uozumi and T. Asakura

Summary. During the long history of the development of sound recording technology, a large number of recordings have been made for academic purposes in various fields, as well as for amusement purposes. In utilizing those sound materials, however, there is a serious problem that early recording media such as wax cylinders and analog disks are less robust and have been, or can be, damaged easily by poor preservation conditions, careless handling and repeated reproductions. As a solution to such a problem, we have developed optical reproduction methods for old recording media: a laser-beam reflection method for wax cylinders, a laser diffraction method for old disc records, and an extension of the laser-beam reflection method for negative cylinders. Principles and properties of these methods are described.

1 Introduction

Since the advent of the first phonograph in 1877, the technology of recording and reproducing sounds has been developed extensively in response to strong public demand. During this history, various kinds of records appeared, from wax phonograph cylinders, a primitive and analog type of record, to optical disks based on modern digital technologies. At each stage of the development of records, a large number of recordings have been made for academic purposes in various fields such as musicology, linguistics and social sciences, as well as for amusement purposes, and they now provide us with a huge number of valuable sound materials.

However, there is a serious problem when we want to use those sound materials. Early recording media such as wax cylinders and analog disks are less robust and tend to be damaged more easily by poor preservation conditions, careless handling and repeated reproductions. Therefore, it is inadequate and sometimes almost impossible to reproduce sounds from records with severe damage by using traditional phonographs or record players. Moreover, even for disks in good condition, the use of a traditional phonograph gives rise to further wearing. Consequently not only is the quality of recorded sounds deteriorating but also the records themselves are being spoiled. This is a major problem since the records are sometimes regarded as a part of a valuable cultural inheritance.

Fig. 1. Piłsudski's wax cylinder (*right*) and its case (*left*)

To overcome such difficulties, we have developed optical reproduction methods for old recording media, for which traditional reproduction instruments are inadequate. In this chapter, we present the history and principles of such optical methods developed by our research group.

2   Wax Cylinder: Laser Beam Reflection Method

2.1   Prologue

As is well known, Thomas Edison, a very famous scientist and engineer in the United States, invented a recording machine which was called the "Phonograph." After 10 years' improvement, the phonograph, based on wax cylinders, became very popular. These wax cylinder phonographs were distributed all over the world for about 40 years from 1887 to 1932. In the United States, wax cylinder phonographs were used mainly for the purpose of amusement. In Europe, on the other hand, these phonographs were used for recording not only the voices of famous people but also well-known music and songs. In addition, the phonographs were used for the academic purpose of recording the various languages of (especially) minority races.

Using the phonograph over the years from 1902 to 1905, B. Piłsudski (1866–1918), a polish anthropologist, recorded on wax cylinders the speech and songs of the Ainu people in Sakhalin and Hokkaido in order to study their culture. In 1977, Piłsudski's 65 wax cylinders were discovered in Poland and brought to the Research Institute for Electronic Science (RIES), Hokkaido University, in 1983 for the purpose of reproduction and investigation of the sounds recorded on them [1].

Somewhat later, using the phonograph over the years from 1920 to 1935, Takashi Kitazato (1870–1960), a language professor at Osaka University, recorded the speech and songs of many people in Japan, Taiwan, the Philippines, Malaya, Singapore and Indonesia to investigate the origin of the Japanese language. In 1985, Kitazato's 240 wax cylinders were discovered in Kyoto and also brought to the RIES for reproduction of their sounds.

2.2   Wax Cylinder

Phonograph cylinders of the Edison type used by Piłsudski were 55 mm in diameter and around 105 mm long (Fig. 1). On their surface, nearly 400 turns of grooves were cut with a pitch of 254 µm. Since the rotational velocity of a phonograph in his time was some 140–160 rpm, the sound for 2–3 min was recorded on one wax cylinder. Since, in the recording for the wax cylinders, the grooves were cut only by the sound energy, the depth of the grooves was usually very shallow. For example, a groove of the wax cylinder recorded by a professional engineer during his time had a maximum depth of 50 µm. In the case of a recording by an amateur, the maximum groove depth could be much shallower; for example, it was 10–30 µm for Piłsudski's cylinders.

The stylus of the Edison-type phonograph puts a heavy pressure of approximately 20 g on the groove of the wax cylinder in the reproduction process and, therefore, there is a great risk of damaging the wax cylinder. Consequently, a reproduction system using a very light-pressure stylus was developed in our laboratory, and some cylinders in good condition were reproduced successfully by this system [2]. However, some cylinders remained unprocessed because they were cracked or in pieces. These broken wax cylinders were repaired as shown in Fig. 2. Even after the repair, however, the stylus method was inapplicable because the remaining gaps or missing parts could damage the stylus. To reproduce the sounds from the repaired cylinders, we developed a noncontact and nondestructive method based on the reflection property of a laser beam, and called the laser beam reflection method [1,3].

Fig. 2. Wax cylinder (*left*) in pieces and (*right*) after repair

2.3   Laser Beam Reflection Method

Figure 3 shows the principle of the laser beam method. The laser beam is incident onto the grooves cut on the surface of the wax cylinder and reflected at an angle obeying the reflection law. The reflected beam reaches the detecting plane, placed perpendicularly to the optical axis. The intersection point of the reflected beam is separated from the origin by a distance proportional

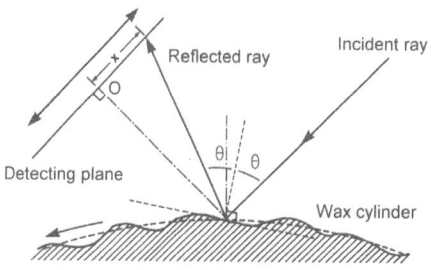

Fig. 3. Schematic 1-D diagram of the laser beam reflection method

to the reflection angle. When the wax cylinder is rotated, the intersection point moves temporally on the detecting plane. This temporal variation of the intersection position is detected by a position-sensitive device (PSD) as a sound signal. The detected signal corresponds to the time-differentiated sound signal since the angle of the reflected ray depends on the inclination of the groove at the illuminated spot position. This differential property can be compensated in the frequency domain electronically, e.g. by using an audiographic equalizer.

A Gaussian beam emerging from the single-mode He–Ne laser with a wavelength of 0.633 μm is focused by an objective lens with a waist diameter of 30 μm. The wax cylinder, which is translated with rotation, is illuminated by a diverging Gaussian beam at a distance z from the beam waist. The illuminated spot diameter can be easily adjusted by moving the objective lens along the optical axis and, hence, by changing the distance z. This corresponds to changing the diameter of a stylus tip, in the stylus method, which cannot be changed easily due to its structure.

The driving part of the reproduction system is shown in Fig. 4 while the illuminating and detecting parts are partly shown in Fig. 5. The wax cylinder is set onto a rotary shaft 150 mm long, with a 50 mm maximum diameter on one side, and a 40 mm minimum diameter on the other. On the surface of the shaft, the screw thread is cut to prevent the wax cylinder from slipping off during the reproduction. The shaft is connected by a belt and reels to the AC motor. The rotating wax cylinder is placed on a movable stage traversed by the stepping motor in a direction perpendicular to the optical axis. This

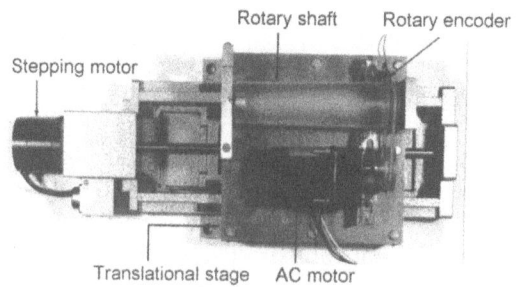

Fig. 4. Top view of the driving part of the reproduction system for wax cylinders

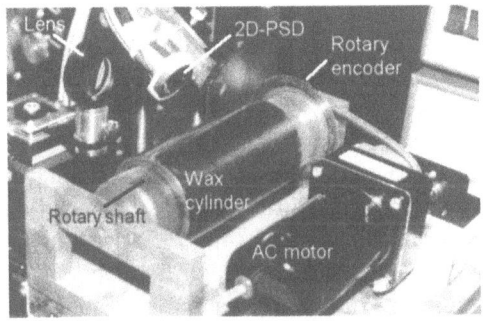

Fig. 5. Constructed reproduction system. A 2-D PSD is used for the detector, though a 1-D PSD was used in the first system. The lens driver for the tracking error compensation mentioned in Sect. 2.5 was removed in this configuration

stage can be moved with an accuracy of $2\,\mu m$ per pulse, and it is set to be translated by one pitch of grooves with $254\,\mu m$ for every rotation of the wax cylinder. The translation and rotation of the wax cylinder by means of the stepping motor and the AC motor are synchronized by using a rotary encoder.

2.4 Properties and Problems of Reproduced Sounds

The laser beam method is based on geometrical optics. However, the actual reflection phenomenon from the grooves does not exactly obey the law of geometrical optics because of the finite diameter of the illuminating laser beam. We have found several problems in the development of this method:

1. Fidelity of the reproduced sound
2. Noise characteristics
3. Existence of echo
4. Occurrence of tracking error

We solved these problems by quantitative investigation of the reproduced sound signals.

To study the problems of fidelity and noise characteristics, we investigated the long-time frequency spectra of the sound signal produced by using the stylus method and the laser beam method. In the stylus method, we used the Edison-type phonograph; in the laser beam method, the illuminating spot diameter is $80\,\mu m$. As a result, we found that, in the case of the stylus method, the sound is in the frequency range from $250\,Hz$ to $6\,kHz$. Especially, the resonant frequencies are at $400\,Hz$, $2\,kHz$ and $4\,kHz$. In the laser beam method, the sound intensity with the low resonant frequency at $400\,Hz$ is strong but the sound intensity with the high frequency is weak. The lack of high-frequency components makes the consonant indistinct.

The existence of noise inherent in the laser beam method was also examined. The low-frequency noise below $300\,Hz$ is very strong and gives rise to a great deal of degradation on the articulation. On the other hand, the high-frequency noise above $1\,kHz$ masks the reproduced sounds and becomes more obstructive to hearing the reproduced sounds.

To investigate the cause of the noise in the laser beam method, we studied the reflected spot at the detecting plane and found a random granular intensity pattern together with the reflected beam spot. This granular pattern may be produced from interference of the laser light reflected by the micro-structure distributed over the surface of the wax cylinder.

We also investigated the fidelity of the reproduced sound by changing the diameter of the illuminating laser beam. Figure 6 shows the variation of the long-time frequency spectra of the reproduced sounds with the spot diameters. With an increase in the beam diameter, the high-frequency components are greatly reduced. This may be due to the smoothing effect for the time-varying directions of the beams reflected from the groove within the illuminating beam spot. As mentioned above, the lack of high-frequency components makes the consonant indistinct. From this investigation, we conclude that the most suitable spot diameter is 30–130 µm from the viewpoint of the fidelity of the reproduced sound.

Figure 6 also shows that the high-frequency noise suddenly decreases with an increase in the illuminating beam spot diameter. On the other hand, the low-frequency noise exists independently of any variation in the spot diameter. Therefore, the high-frequency noise can be effectively suppressed by using an illuminating beam with a large diameter. However, the noise signals in the low-frequency region have a constant intensity independent of the beam diameter. This low-frequency noise below 300 Hz may be suppressed by using a high-pass filter since the sound information in this region was not recorded originally. From the viewpoint of noise reduction, the noise is suppressed by using the laser beam with a spot diameter of over 80 µm.

Another problem is an echo. The echo is overlapped on the reproduced sounds by an increase in the illuminating beam diameter. Figure 7 shows the autocorrelation functions of time-varying sound signals as a function of the distance z. The maximum peaks at $\tau = 0$ s result from the sound signals reproduced from the grooves illuminated by the laser beam. The second peaks at $\tau = 0.4$ s come from the echo and their magnitudes correspond to the intensity. It can be seen that the intensity of the echo increases with an increase in the beam diameter. The laser beam with a spot diameter larger

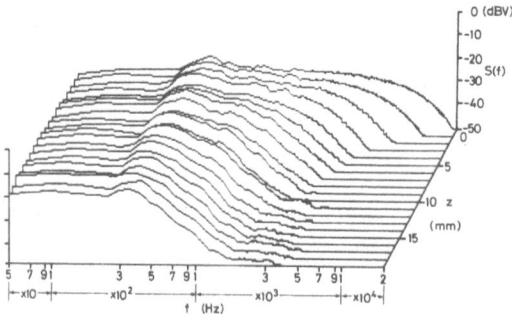

Fig. 6. Long-time frequency spectra obtained from sound signals reproduced by the laser beam reflection method as a function of the distance z from the beam waist

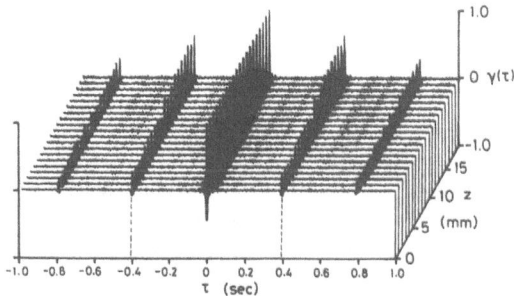

Fig. 7. Autocorrelation functions of reproduced sound signals as a function of the distance z from the beam waist

than the groove width illuminates the adjacent grooves and, therefore, causes the echo superposed on the main signal. We found that the intensity of the echo is to be under 30% of the main sound intensity for a spot diameter of less than 100 μm. Under this condition, the existence of the echo does not disturb hearing the reproduced sounds.

2.5 Tracking Error

There is a possibility that the illuminating beam leaves the grooves, producing a tracking error, because it does not directly trace the grooves of the wax cylinder as in the case of the stylus method. The tracking error results mainly from the position error of the driving part in the reproduction system. As shown in Fig. 8, if the tracking error occurs, the incident beam is reflected in the y-direction. In this case, the intensity of the reproduced sounds decreases suddenly because the intersection point of the reflected beam gets out of the one-dimensional (1-D) PSD.

To avoid the tracking error, we used the two-dimensional (2-D) PSD and the lens driver of a compact disc player. The 2-D PSD allows independent detection of the x- and y-coordinates of the beam spot position. Using the electrical conversion system shown in Fig. 9, the time-varying values of the x- and y-coordinates of the reflected beam become the sound and the tracking error signals, respectively. The tracking error signal is fed to the lens driver

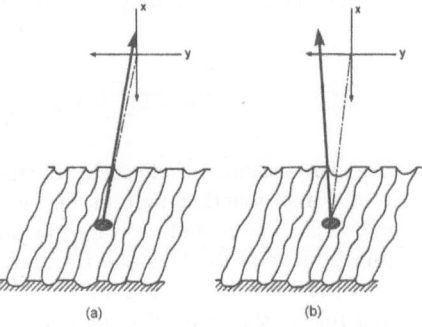

Fig. 8. Optical rays reflected from the grooves (**a**) without and (**b**) with tracking error

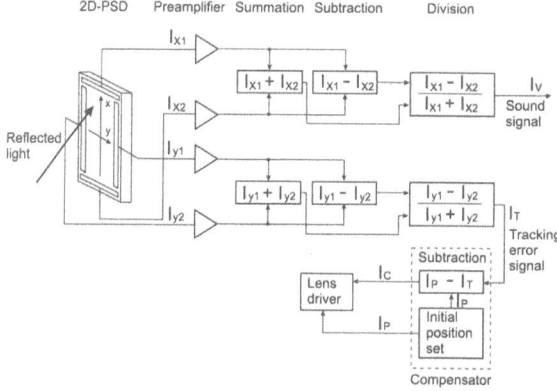

Fig. 9. Schematic diagram of the electrical conversion system of the position signals detected by the 2-D PSD into sound and tracking error signal currents

to move the lens. By moving the lens, the illuminating spot moves to the center of the groove to keep the illumination normal.

2.6   Further Development of the Method

The laser beam reflection method introduced above has been developed further to improve its performance and to be applied to old records of various types.

In the latest system developed in our laboratory, a laser diode (LD) with a wavelength of $0.78\,\mu m$ and a maximum power of $20\,mW$ is used as the light source to make the optical system compact as shown in Fig. 10. In addition, an imaging lens and a pinhole are inserted between the wax cylinder and the 2-D PSD. In this configuration, the lens produces an image of the cylinder surface on the pinhole plane. The pinhole transmits only the image of the illuminating spot and rejects the stray light coming from surrounding parts of the cylinder surface, thus reducing the noise due to the ambient light, e.g. from a fluorescent lamp. A photograph of the latest system is shown in Fig. 11. This compact system was used in 1996 to reproduce the sound from a wax cylinder on which a piano tune, "Hungarian Dance" played by Johannes Brahms, was recorded on 1889. In this new system, the tracking is realized

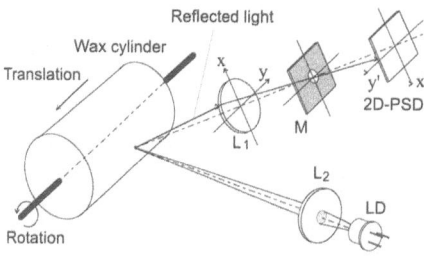

Fig. 10. Schematic diagram of the recent reproduction system for wax cylinders. The pinhole is placed in the image plane of the lens and transmits only the image of the illuminating spot and rejects the ambient stray light

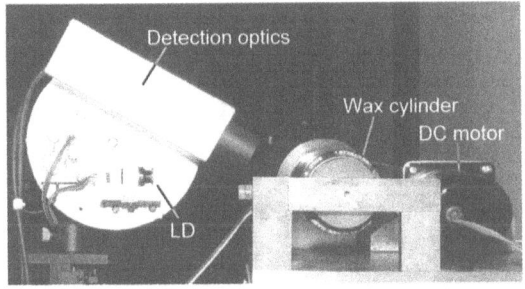

Fig. 11. Photograph of the recent reproduction system for wax cylinders. Detection optics includes the imaging lens, pinhole and 2-D PSD as shown in Fig. 10. A lens is mounted on the front end of the holder of the LD

by adjusting the translation speed of the stage on which the wax cylinder is mounted, rather than using the lens drive.

Some interesting attempts to improve the method have been made by Nakamura et al. [4,5]. As an alternative way of improving the tracking, a contact method based on a lightweight optical fiber was proposed. To solve the noise problem, an incoherent light source was utilized, by which the granular pattern in the beam spot at the detection plane is suppressed.

2.7 Replication of Cylinders

There is a quite different way of reproducing sounds from a wax cylinder without any risk of damaging the cylinder by the heavy stylus of a traditional phonograph, i.e. making a replica of the cylinder and playing it directly on a phonograph. The replication also has the advantage of preserving the current conditions of the cylinder without further deterioration. For the purpose of preserving the Piłsudski cylinders, replicas of them were made using the epoxy resin and molding technique developed in the dental field [2].

First, the wax cylinder was mounted on the axis of an outer cylindrical mold with a 10 cm i.d. Then, dental impression material (Exaflex injection type F, GC LTD) was poured into the space between the cylinder and the wall of the mold, as shown in Fig. 12 (left). The impression material, which was made of vinyl-silicone rubber, has many advantages, such as high accuracy, fluidity and flexibility. After the silicon rubber hardened, the wax cylinder

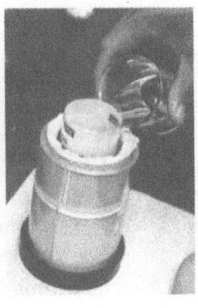

Fig. 12. Replication procedures for the wax cylinder: (*left*) the dental impression material is poured into the space between the cylinder and the wall of the mold; and (*right*) the epoxy resin is poured into the silicone rubber mold

was carefully removed. Next, the epoxy resin (Spurr Resin, TAAB LTD), which was thoroughly mixed and degassed in a vacuum chamber, was poured into the silicon rubber mold, as shown in Fig. 12 (right). The resin was completely cured in a constant temperature oven for 36 hours. It was shown to be possible to reproduce usable sounds from the replicas using a working model of the original Edison-type machine.

3   Disk: Laser Diffraction Method

3.1   Disk

After the project of sound reproduction from the Piłsudski wax cylinders, many old disk records were found in Japan on which old speech and songs of native Formosan tribes had been recorded. Some of these disks are made of ebonite, some of aluminum, and the others of aluminum alloy. All the disk records were monaural, being recorded with a rotational frequency of 78 rpm. The width of the sound grooves and the separation between two adjacent grooves were 50 and 200 μm, respectively, though they vary depending on the type of records and the recorded sound intensity. Since their contents are important from linguistic and folkloric viewpoints, reproduction of the sound from these disks is strongly desirable. Apparently due to improper storage of these disks, however, almost all are covered with mold, and most of the aluminum and aluminum alloy disks are rusted, some lightly and some seriously, and have many scratches. These undesirable changes to the disk surfaces prevent ordinary reproduction of the sound using conventional means such as phonographs and record players. In particular, the rust on the aluminum and aluminum alloy disks, whether slight or serious, disturbs the tracking of the sound grooves by the stylus and makes reproduction of the sound almost impossible.

Therefore, a noncontact optical reproduction method was again considered. However, since the recording principles of disks and wax cylinders are different, the laser beam reflection method developed for the cylinders is inapplicable to the disk. That is, the sound signals of disk records are encoded as a horizontal variation of the sound grooves as shown in Fig. 13. Therefore,

Fig. 13. Sound grooves of a disk. Grooves wind laterally corresponding to the sound signals encoded. The pitch of grooves is not constant and varies according to the recording level

we have developed another reproduction method that exploits the diffraction phenomenon of the laser light by the sound grooves [6].

3.2 Laser Diffraction Method

The principle of detection of sound from the records is illustrated in Fig. 14. In the case of disks of this type, sound signals are encoded as lateral variations of the groove around its mean curve that corresponds to a groove involving no sound signal. A cross section of the sound grooves is approximately V-shaped, though its details depend on the cutting stylus used for the recording. Consider a Gaussian beam from a laser illuminating a sound groove with an illumination angle ϕ so that the plane of illumination includes the axis of the disk. Then, the laser light on reflection suffer from phase modulations proportional to the depth variations across the illuminated area and produces a specific diffraction pattern that consists of a central specular spot and two linear wings extending toward opposite directions from the central spot as shown in Fig. 15. The angle θ of the linear wings from the illumination plane is equal to the angle between the tangents of the mean curve and the groove at the illuminated spot. Therefore, by setting a 1-D PSD perpendicular to the plane of illumination at a moderate distance from the specular spot, the variation of θ is detected as a coordinate at which the linear wing intersects

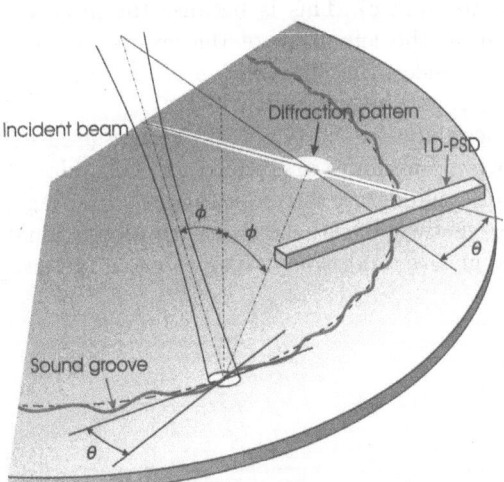

Fig. 14. Principle of sound reproduction from a disk. The mean curve of a groove is nearly a circle (*broken curve*) concentric with the disk. If the illuminating spot is large enough to cover the groove but not too large to illuminate adjacent grooves, the direction θ of the groove at the illuminated spot is converted to the direction of the linear part of the diffraction pattern. In the actual system, a normal illumination ($\phi = 0$) was employed

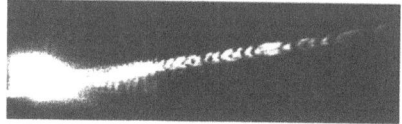

Fig. 15. Diffraction pattern from the sound groove of a disk. The central specular spot and one of two linear wings are shown

the 1-D PSD. Here, it is noted that the PSD detects $\tan \theta$, which is the first derivative of the sound signal. In the system actually constructed, we used a 2-D PSD instead of the 1-D PSD in order to have the sufficient light intensity and to reduce the speckle noise as shown in Fig. 17.

3.3   Tracking

In the case of disks, it is much more difficult to track the sound groove with the laser spot than in the case of wax cylinders. This difficulty comes from the property that, due to sound signals, the sound groove winds, in the same plane in which the tracking direction lies. Therefore, we needed to develop a different principle for tracking the disks.

The tracking principle we employed is as follows. If the illuminated spot is just on the center of the groove as shown in Fig. 16b, the two linear wings of the diffraction pattern have almost the same intensity. When the groove deviates slightly toward one side from the center of the illuminated spot, however, the intensity in the wing on that side increases compared with the wing on the opposite side (Figs. 16a and c). This is because the intensity in the right wing arises mainly from the left slope of the groove and the deviation of the groove to the right makes the illumination of the left slope stronger than that of the right slope, and vice versa.

This change in intensity balance between the two wings can be used for compensating tracking errors. Two photodiodes are placed on both sides of the central spot as shown in Fig. 16d, and detect the intensities of both wings. The difference of the outputs from the two PDs is approximately proportional to the amount of deviation of the groove, and is employed to drive the disk

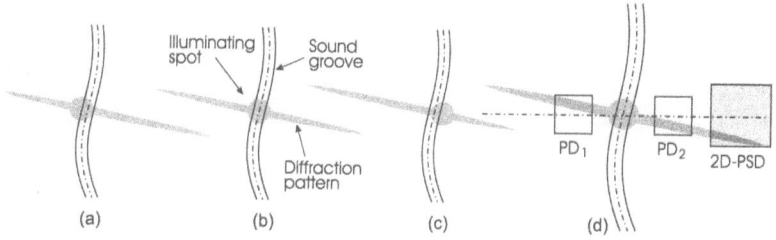

Fig. 16. Principle of detecting tracking errors. As the groove (**a**) deviates to the right, (**b**) meets the spot center, and (**c**) deviates to the left, the intensity balance in the two wings of the diffraction pattern changes as shown. (**d**) The intensity balance is detected by two phododiodes PD_1 and PD_2

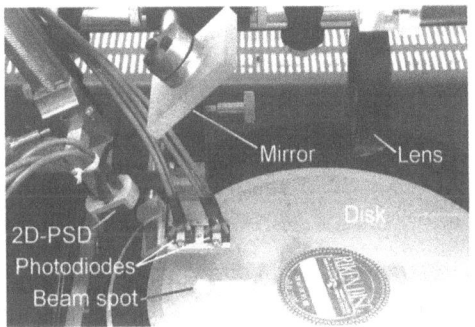

Fig. 17. Illumination optics of the reproduction system for disks. The laser beam coming from the right passes through the lens, is turned downward by the mirror, passes through the hole between the two photodiodes and illuminates the disk

laterally to compensate for the deviation of the groove from the illuminated spot. In an actual system, the differential electric signal was processed with a low-pass filter to give an appropriate response rate and to reduce the effect of noise before being fed into the driving circuit of the translation stage on which a turntable is mounted with the disk.

3.4 Reproduction System

On the basis of the principle given above, an actual system was constructed as shown in Fig. 17. The system consists of three parts: the optical system, sound detection system, and driving and tracking system.

In the optical system, a Gaussian beam from an He–Ne laser at $0.633\,\mu m$ impinges on a disk record at a distance z from its beam waist after passing through a lens system forming a suitable beam-waist width. The illuminated spot size at the disk surface can be adjusted by changing the distance z, which is controlled by varying the position of the last lens of the lens system. The illuminating beam is set normal to the disk surface by a mirror ($\phi = 0$ in Fig. 14). In the sound detecting system, an output signal from the 2-D PSD is fed into the reconstruction circuit and converted into an electric signal to produce a sound signal. This conversion process and the electronic circuit are the same as those used in the laser beam reflection method. In the driving and tracking system, outputs from the two photodiodes are fed into the driving circuit of the translation stage through a differential amplifier, while the disk is rotated at a constant rate of 78 rpm by a turntable mounted on the translation stage.

4 Negative Cylinder: Modification of the Laser Beam Reflection Method

4.1 Negative Cylinder

Recently, we discoverded that many phonograph cylinders are preserved in Germany in the form of metallic negative cylinders on which the folk music of

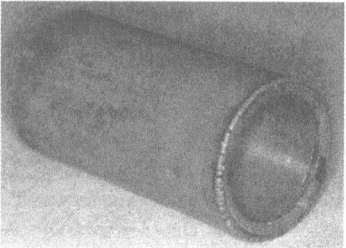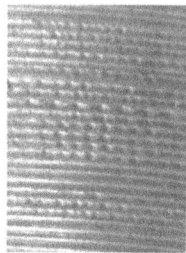

Fig. 18. (*Left*) Negative cylinder. (*Right*) Inside surface of the negative cylinder showing the height variation of the sound bank which is a replica of the sound groove of the original wax cylinder

various countries in the world are recorded. The negative cylinder was made by plating a wax cylinder with copper and then by melting down the original wax cylinder. In this process, the sound is transferred into convex portions on the inside surface. Consequently, any stylus method is ineffective and these negatives have been left for a long time without any investigation of the valuable sounds recorded on them. It was expected, however, that the laser beam reflection method developed for wax cylinders could also be effective for the negatives. To apply the method to the negatives, some modifications were made and a new system was constructed [4,7].

The negative cylinders we employed for the development of the instrument seem to be produced from typical wax cylinders and have nearly the same dimensions of 56, 54 and 110 mm in the outer and inner diameters and length, respectively (Fig. 18 (left)). The typical rotation rates were 144 and 160 rpm, and a sound lasting 2–3 min and the sound signal were encoded on the inside surface, not as a groove but as a spiral *bank* with a pitch of 1/100 inch (Fig. 18 (right)).

4.2 Principle and Reproduction System

The principle of sound reproduction from negative cylinders is basically the same as that from wax cylinders. With reference to Fig. 19, a laser beam is incident on the center of the sound bank and the sound signal recorded as height variations of the sound bank are converted into a position x of the

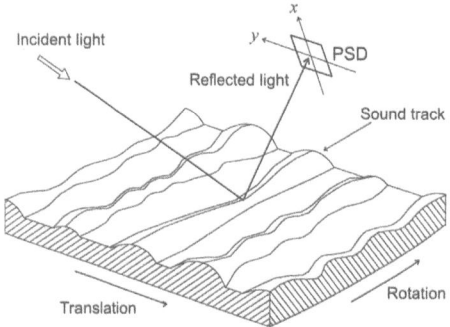

Fig. 19. Principle of the laser beam reflection method. As the cylinder rotates, the reflection direction varies in the x-direction due to the height variation of the sound bank (sound track). The deviation of the illuminated spot from the center of the sound bank gives the variation in the y-direction

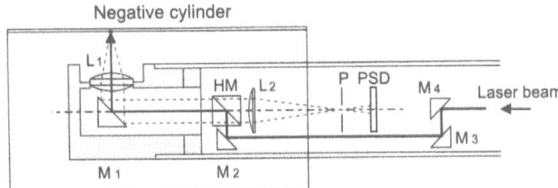

Negative cylinder

Fig. 20. Schematic diagram of the optical head. The lens L_1, mirrors M_1 and M_2, and beam splitter HM are common to the illumination and detection optics

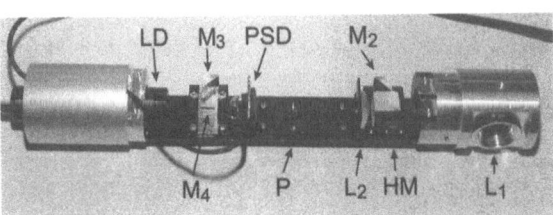

Fig. 21. Photograph of the optical head with the cover removed for display. The mirror M_1 is not shown in this photograph. P indicates the position of the pinhole which is unmounted in this image

reflected beam impinging on the PSD. On the other hand, if the illuminated spot deviates from the center of the bank, the reflected beam varies in the y-direction, which produces a signal for the tracking error and can be used to compensate for the deviation of the illuminated spot by adjusting the translation speed of the cylinder. These two signals in the x- and y-directions can be detected independently by setting a 2-D PSD.

In the case of the negative cylinder, a compact optical system that can be inserted into the cylinder is required. A schematic diagram and a photograph of a reconstructed optical head are shown in Figs. 20 and 21, respectively. A laser beam from an LD with a wavelength of 670 nm is guided by mirrors M_4, M_3, M_2 and M_1, and a beam splitter HM, to a lens L_1, which focuses the beam onto the inside surface of the cylinder. The beam reflected from the cylinder surface goes back through the same part of the system up to HM, and then is led by the lens L_2 to the PSD. A pinhole P is placed in the focal plane of L_2, and its role is the same as in the system shown in Fig. 10.

A schematic diagram of the developed system is shown in Fig. 22. The system consists of the drive and control units, the optical head being mounted on the former. The signal converter processes the output of the PSD and yields the sound and tracking error signals, V_x and V_y. V_y is sent to the pulse motor driver via the low-pass filter and the V–F converter. The rotation rate is adjustable in the range of 140–160 rpm. A photograph of the drive unit is shown in Fig. 23.

Using this system, we successfully reproduced the sounds from some negatives, including performances of Japanese musical instruments, *shamisen* and *koto*, which were recorded in Berlin in 1901. The reproduced sounds have a much better quality than existing wax cylinders. This is partly because wax cylinders currently available have been played many times and are considerably worn out while the negatives preserve the initial quality of their original

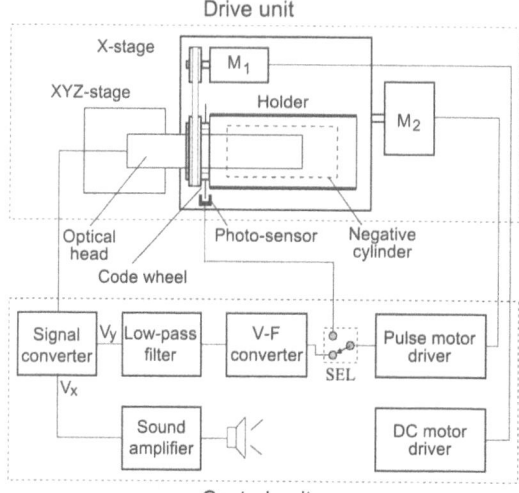

Fig. 22. Diagram of the optical player for negatives. The DC motor M_1 rotates the holder while the pulse motor M_2 drives the x-stage. The output of the PSD is processed to yield the sound and tracking error signals, V_x and V_y. V_x is amplified and drives the speaker, while V_y is sent to the pulse motor driver via the low-pass filter and the V-F converter (auto-tracking mode). The pulse motor can also be controlled by the signal from the photosensor (constant translation mode)

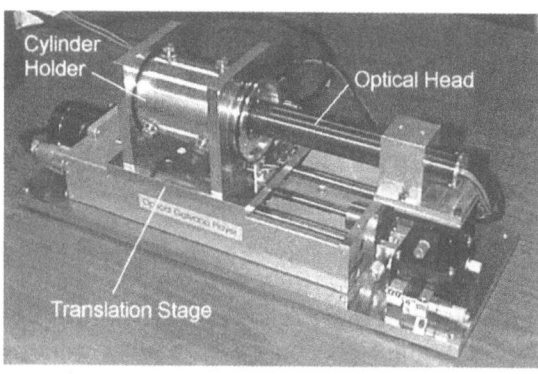

Fig. 23. Drive unit of the optical negative player. The optical head is inserted into the negative held inside the cylinder holder which is mounted on the translation stage (x-stage)

wax cylinders. The present instrument is expected to be used to reveal valuable sound information that has been left undetected for a long time.

5   Conclusion

Several optical methods have been developed for reproducing sounds from old recording media of various types. The quality of the reproduced sounds varies depending on the type of record. For wax cylinders, the quality is satisfactory unless the original sound quality itself is poor, while it is very good for negative cylinders. On the other hand, disks do not always give satisfactory sound quality due to the difficulty of tracking and need further improvement. A further reduction of noise is another problem in all the cases. In spite of some remaining problems, these noncontact and nondestructive methods are powerful tools for playing damaged records, for which traditional phonographs are inadequate. They may well be used to reproduce valuable historical sounds.

References

1. Iwai, T., Asakura, T., Ifukube, T., Kawashima, T (1986) Reproduction of sound from old wax phonograph cylinders using the laser-beam reflection method. Appl. Opt. **25**, 597–604
2. Ifukube, T., Kawashima, T., Asakura, T. (1989) New methods of sound reproduction from old wax phonograph cylinders. J. Acoust. Soc. Am. **85**, 1759–1766
3. Asakura, T., Uozumi, J., Iwai, T., Nakamura, T. (1999) Study on reproduction of sound from old wax phonograph cylinders using the laser. Proc. OWLS V (Springer, Berlin) (to be published)
4. Nakamura, T., Ushizaka, T., Uozumi, J., Asakura, T. (1997) Optical reproduction of sounds from old phonographic wax cylinders. Proc. SPIE **3190**, 304–313
5. Nakamura, T., Asakura, T. (1999) Reproduction of sounds from an old Russian phonographic wax cylinder by various optical methods. Proc. OWLS V (Springer, Berlin) (to be published)
6. Uozumi, J., Asakura, T. (1988) Reproduction of sound from old disks by the laser diffraction method. Appl. Opt. **27**, 2671–2676
7. Uozumi, J., Ushizaka, T., Asakura, T. (1999) Optical reproduction of sounds from negative phonograph cylinders. Proc. OWLS V (Springer, Berlin) (to be published)

Springer Series in
OPTICAL SCIENCES

Springer Series in
OPTICAL SCIENCES